BIORESOURCE UTILIZATION AND MANAGEMENT

Applications in Therapeutics, Biofuels, Agriculture, and Environmental Science

Edited by
Hrudayanath Thatoi, PhD
Swagat Kumar Das, PhD
Sonali Mohapatra

AAP | APPLE
ACADEMIC
PRESS

First edition published 2022

Apple Academic Press Inc.
1265 Goldenrod Circle, NE,
Palm Bay, FL 32905 USA

4164 Lakeshore Road, Burlington,
ON, L7L 1A4 Canada

CRC Press
6000 Broken Sound Parkway NW,
Suite 300, Boca Raton, FL 33487-2742 USA

2 Park Square, Milton Park,
Abingdon, Oxon, OX14 4RN UK

Library and Archives Canada Cataloguing in Publication

Title: Bioresource utilization and management : Applications in Therapeutics, Biofuels, Agriculture, and Environmental Science / edited by Hrudayanath Thatoi, PhD, Swagat Kumar Das, PhD, Sonali Mohapatra.

Names: Thatoi, Hrudayanath, editor. | Das, Swagat Kumar, editor. | Mohapatra, Sonali, editor.

Description: First edition. | Includes bibliographical references and index.

Identifiers: Canadiana (print) 20210090111 | Canadiana (ebook) 20210090227 | ISBN 9781771889339 (hardcover) | ISBN 9781003057826 (ebook)

Subjects: LCSH: Biotechnology industries.

Classification: LCC HD9999.B442 B56 2021 | DDC 338.4/76606—dc23

Library of Congress Cataloging-in-Publication Data

Names: Thatoi, Hrudayanath, editor. | Das, Swagat Kumar, editor. | Mohapatra, Sonali, editor.

Title: Bioresource utilization and management : Applications in Therapeutics, Biofuels, Agriculture, and Environmental Science / edited by Hrudayanath Thatoi, Swagat Kumar Das, Sonali Mohapatra.

Description: First edition. | Palm Bay, FL : Apple Acdemic Press, [2021] | Includes bibliographical references and index. | Summary: "The need for exploration, conservation, and sustainable utilization of bioresources is undeniable for the survival and growth of mankind. This new book throws light on new and recent research on and development of effective strategies for sustainable utilization of bioresources using modern tools and techniques to help meet this challenge. Bioresource Utilization and Management: Applications in Therapeutics, Biofuels, Agriculture, and Environmental Science is divided into four sections that cover utilization of bioresources in therapeutics, in biofuel, in agriculture, and in environmental protection. Beginning with the diverse potential applications of bioresources in food, medicine, and cosmetics, the volume goes on to address the various different underutilized bioresources and their sustainable uses. It discusses important advances in biofuel and patents that highlight recent developments that address the energy crises and the continuously fluctuating cost of petroleum. It explores new energy renewable sources from bioresources and their sustainable utilization in the bioenergy and biofuel industry. Several chapters focus on the sustainable utilization of bioresources in the agricultural sector. The volume considers that developing countries have huge agricultural resources that could be employed for production of value-added byproducts for the sustainable development of a bio-based economy. The book discusses efficient use of underexploited natural bioresources, new chemical approaches for the generation of novel biochemicals, and the applications of genetics approaches for bioresource conservation and production of value-added products. Further, strategies for the production of biopesticides utilizing bioresources are also discussed. Offering comprehensive information, the volume also looks at strategies for using bioresources for environmental protection, such as the development of effective treatment processes of industrial effluents, treatment of xenobiotic compounds, and more. This book is a valuable and up-to-date reference resource for university and industry scientists in the area of therapeutics, biofuel, agriculture, and environmental management and research"-- Provided by publisher.

Identifiers: LCCN 2020056544 (print) | LCCN 2020056545 (ebook) | ISBN 9781771889339 (hardcover) | ISBN 9781003057826 (ebook)

Subjects: LCSH: Biotechnology.

Classification: LCC TP248.2 .B374975 2021 (print) | LCC TP248.2 (ebook) | DDC 660.6--dc23

LC record available at https://lccn.loc.gov/2020056544

LC ebook record available at https://lccn.loc.gov/2020056545

ISBN: 978-1-77188-933-9 (hbk)
ISBN: 978-1-77463-813-2 (pbk)
ISBN: 978-1-00305-782-6 (ebk)

BIORESOURCE UTILIZATION AND MANAGEMENT

Applications in Therapeutics, Biofuels, Agriculture, and Environmental Science

About the Editors

Hrudayanath Thatoi, PhD, is a Professor in the Department of Biotechnology of North Orissa University, Baripada, India, where he is also working as the Director of the Centre for Similipal Studies. Dr. Thatoi has also served as Reader and Head of Department of Biotechnology, College of Engineering and Technology, Biju Patnaik University of Technology, Odisha and Senior Scientist, M.S. Swaminathan Research Foundation, Chennai. The teaching and research interest of Dr. Thatoi include microbiology, molecular biology, industrial and pharmaceutical biotechnology. He obtained his M.Sc, M.Phil, and Ph.D degrees in Botany from Utkal University, India. Dr. Thatoi has implemented several research projects funded by UGC-DAE Govt. of India, DST Govt. of Odisha and Forest Department, Govt. of Odisha. To date, Dr. Thatoi is credited with publication of 18 books, including two textbooks, and more than 250 research papers published in national and international journals, proceedings of conferences, and book chapters. He serves as member of several scientific societies and editorial boards of the national and international journals. He is a Fellow of the Society of Applied Biotechnology. He has guided 18 Ph.D scholars and a number of M.Tech, B.Tech, and M.Phil students for their thesis/dissertation works. He has visited Malaysia, Bangkok, and Nepal to attend international conferences on biotechnology and biodiversity. Dr. Thatoi is a Functional Area Expert in Ecology and Biodiversity, being accredited by Quality Council of India (QCT), Government of India.

Swagat Kumar Das, PhD, is an Assistant Professor in the Department of Biotechnology at College of Engineering and Technology, Biju Patnaik University of Technology (BPUT), Odisha, India. He obtained his BPharm degree from BPUT, Rourkela, and MTech degree from Rajiv Gandhi Proudyogiki Vishwavidyalaya, Bhopal, India, and his PhD from Ravenshaw University, Cuttack, India. He has more than eleven years of teaching and research experience. He has guided a number of BTech and MTech students for their dissertation works. His research activities involved phytochemical analysis and drug development from mangrove plants for diabetes and oxidative stress. His research area also focused on green synthesis of nanoparticles and evaluation of their pharmacological potentials. To his credit, he has published 24 research and review articles in national and international journals of repute and 8 book chapters. He has co-authored three text books and one edited book. He is a Fellow (FSIESRP) of Innovative Scientific Research Professional Malaysia, India chapter.

Mrs. Sonali Mohapatra, is an Assistant Professor in the Department of Biotechnology, CET, Bhubaneswar, India, and submitted her PhD thesis to Biju Patnaik University, Odisha. Mrs. Mohapatra pursued her BTech and MTech degrees in industrial biotechnology from Dr. M.G.R University, Chennai, India. She has significantly contributed in the field of biofuel technology and environmental microbiology. Mrs. Mohapatra has been working on metabolomic approach for improving the ethanol tolerance capacity of ethalogenic microorganisms. She has also received young scientist award for best paper presentation in 2018. She has more than eight years of research and teaching experience. She has guided BTech and MTech students for their dissertation works. To her credit, she has published 18 research and review articles in reputed international journals and 10 book chapters and has filed one patent. She has co-authored two edited books. She is a Fellow (FSIESRP) of Innovative Scientific Research Professional Malaysia, India chapter.

Contents

Contributors

Chinmayee Acharya
Department of Biotechnology, North Odisha University, Takatpur, Baripada, Odisha 757003, India
Environment and Sustainability Department, CSIR-Institute of Minerals and Materials Technology, Bhubaneswar, Odisha 751013, India

Sam Sunday Adefila
Department of Chemical Engineering, Covenant University, Ota, Ogun State, Nigeria

Basant Kumar Agarwala
Department of Zoology, Tripura University, Suryamani Nagar, West Tripura, Tripura 799022, India
Tripura State Pollution Control Board, Parivesh Bhawan, Pandit Nehru Complex, Agartala, West Tripura, Tripura 799006, India

Oluranti Agboola
Department of Chemical Engineering, Covenant University, Ota, Ogun State, Nigeria

Peter Alaba
Department of Chemical Engineering, Covenant University, Ota, Ogun State, Nigeria

Purabi Baidya
Department of Microbiology, Tripura University, Suryamaninagar, West Tripura, Tripura 799022, India

P. Balasubramanian
Agricultural and Environmental Biotechnology Group, Department of Biotechnology and Medical Engineering, National Institute of Technology Rourkela, Odisha 769 008, India

Sinchini Barman
Department of Microbiology, Tripura University, Suryamaninagar, West Tripura, Tripura 799022, India

Bunushree Behera
Agricultural and Environmental Biotechnology Group, Department of Biotechnology and Medical Engineering, National Institute of Technology Rourkela, Odisha 769 008, India

Laxmipreeya Behera
Department of Agricultural Biotechnology, College of Agriculture, Odisha University of Agriculture and Technology, Bhubaneswar, Odisha 751003, India

Swatismita Behera
Department of Biotechnology, College of Engineering and Technology, Biju Patnaik University of Technology, Techno Campus, Ghatikia, Bhubaneswar, Odisha 751003, India

H. G. Behuria
Department of Biotechnology, North Orissa University, Baripada 757003, Odisha, India

Gourav Bhattacharjee
Department of Microbiology, Tripura University, Suryamaninagar, West Tripura, Tripura 799022, India

Birendra Kumar Bindhani
School of Biotechnology, KIIT Deemed to be University, Bhubaneswar, Odisha 751024, India

Karabi Biswas
Environmental Microbiology Research Laboratory, Department of Botany, University of Kalyani, Kalyani, West Bengal 741235, India

Tethi Biswas
Department of Microbiology, Tripura University, Suryamani Nagar, Tripura West 799022, India
Centre of Excellence in Environmental Technology and Management, Maulana Abul Kalam Azad University of Technology, Haringhata, Nadia, West Bengal 741249, India

Abhispa Bora
Department of Microbiology, Tripura University, Suryamaninagar, West Tripura, Tripura 799022, India

Gautam Bose
Division of Mechanical Processing, ICAR-National Institute of Natural Fibre Engineering and Technology, Kolkata, West Bengal 700040, India

Amrita Chakraborty
Department of Microbiology, Tripura University, Suryamaninagar, West Tripura, Tripura 799022, India

Chaitali Chanda
Centre of Excellence in Environmental Technology and Management, Maulana Abul Kalam Azad University of Technology, Haringhata, Nadia, West Bengal 741249, India

Priyasankar Chaudhuri
Department of Zoology, Tripura University, Suryamani Nagar, West Tripura, Tripura 799022, India

Shaon Ray Chaudhuri
Department of Microbiology, Tripura University, Suryamani Nagar, Tripura West 799022, India

Sukanya Chowdhury
Department of Microbiology, Tripura University, Suryamaninagar, West Tripura, Tripura 799022, India

Gautam Das
Department of Microbiology, Tripura University, Suryamaninagar, West Tripura, Tripura 799022, India

Gitishree Das
Research Institute of Biotechnology and Medical Converged Science, Dongguk University, Seoul, Gyeonggi-do, Republic of Korea

Swagat Kumar Das
Department of Biotechnology, College of Engineering and Technology, Biju Patnaik University of Technology, Techno Campus, Ghatikia, Bhubaneswar, Odisha 751003, India

Kingsuk Das
Post Graduate Department of Botany, Serampore College, Serampore, Hooghly, West Bengal 712201, India

Sumona Deb
Department of Microbiology, Tripura University, Suryamaninagar, West Tripura, Tripura 799022, India

Purnasree Devi
Department of Microbiology, Tripura University, Suryamaninagar, West Tripura, Tripura 799022, India

S. Dikshit
Department of Biotechnology, IIT, Bombay, India

Swapan Kumar Ghosh
Post Graduate Department of Botany, Molecular Mycopathology Lab. Biocontrol Unit, Ramakrishna Mission Vivekananda Centenary College (Autonomous), Rahara, 24 Parganas (North), West Bengal 700118, India

Mandakini Gogoi
Department of Microbiology, Tripura University, Suryamaninagar, West Tripura, Tripura 799022, India

Ronald Jamatia
Department of Microbiology, Tripura University, Suryamaninagar, West Tripura, Tripura 799022, India

R. Jayabalan
Department of Life Science, National Institute of Technology Rourkela, Odisha 769001, India

Srikanta Jena
Department of Zoology, School of Life Sciences, Ravenshaw University, Cuttack 753003, Odisha, India

Saurabh N. Joglekar
Department of Chemical Engineering, Visvesvaraya National Institute of Technology, Nagpur, Maharashtra 440010, India

Meghavi Kathpalia
Department of Microbial Technology, Amity University, Noida, Uttar Pradesh 201301, India

Megha Kaviraj
Microbiology Laboratory, Crop Production Division, ICAR-National Rice Research Institute, Cuttack, Odisha 753006, India

I. Kohli
Department of Microbial Technology, Amity University, Noida, Uttar Pradesh 201301, India

Bhaskar D. Kulkarni
CSIR-National Chemical Laboratory, Pune, Maharashtra 411008, India

Ashutosh Kumar
Department of Microbiology, Tripura University, Suryamaninagar, West Tripura, Tripura 799022, India

R. Kumar
Department of Environmental Sciences, H.N.B. Garhwal University (A Central University), Srinagar Garhwal, Uttarakhand 246174, India

Sanjeet Kumar
Ambika Prasad Research Foundation, Bhubaneswar 751006, Odisha, India

Upendra Kumar
Microbiology Laboratory, Crop Production Division, ICAR-National Rice Research Institute, Cuttack, Odisha 753006, India

Swastika Kundu
Microbiology Laboratory, Crop Production Division, ICAR-National Rice Research Institute, Cuttack, Odisha 753006, India

S. Mahajan
Department of Microbial Technology, Amity University, Noida, Uttar Pradesh 201301, India

Chanchal Majumder
Department of Civil Engineering, Indian Institute of Engineering Science and Technology, Shibpur, Botanic Garden, Howrah, West Bengal 711103, India

Sachin A. Mandavgane
Department of Chemical Engineering, Visvesvaraya National Institute of Technology, Nagpur, Maharashtra 440010, India

Neelam Meher
Academy of Management and Information Technology, Khordha, Odisha 752057, India

Leena Mishra
Division of Mechanical Processing, ICAR-National Institute of Natural Fibre Engineering and Technology, Kolkata, West Bengal 700040, India

Suruchee Samparna Mishra
Department of Biological Sciences, Indian Institute of Science Education and Research, Berhampur, Odisha 760010, India

Debashrita Mittra
Post Graduate Department of Biotechnology, Utkal University, Vani Vihar, Bhubaneswar, Odisha 751004, India

Ajoy Modak
Department of Microbiology, Tripura University, Suryamaninagar, West Tripura, Tripura 799022, India

D. Mohanty
Department of Microbial Technology, Amity University, Noida, Uttar Pradesh 201301, India

S. Mohapatra
Department of Microbial Technology, Amity University, Noida, Uttar Pradesh 201301, India

Sonali Mohapatra
Department of Biotechnology, College of Engineering and Technology, Biju Patnaik University of Technology, Bhubaneswar, Odisha 751003, India

Ranjan Kumar Mohapatra
School of Biotechnology, KIIT Deemed to be University, Bhubaneswar, Odisha 751024, India

Emetere Moses
Department of Physics, Covenant University, Ota, Ogun State, Nigeria
Department of Petroleum Engineering, Covenant University, Ota, Ogun State, Nigeria

Amitava Mukherjee
Centre for Nanobiotechnology, Vellore Institute of Technology, Vellore, Tamil Nadu 632914, India

Indranil Mukherjee
Centre of Excellence in Environmental Technology and Management, Maulana Abul Kalam Azad University of Technology, Haringhata, Nadia, West Bengal 741249, India

P. K. Nanda
Indian Veterinary Research Institute, Eastern Regional Station, Belgachia Road, Kolkata 700037, India

Ranjan Kumar Naik
ICAR-Central Research Institute for Jute & Allied Fibres, Barrackpore, Kolkata, West Bengal 700120, India

Hari Narayan
Microbiology Laboratory, Crop Production Division, ICAR-National Rice Research Institute, Cuttack, Odisha 753006, India

Rajib Nath
Department of Agronomy, Bidhan Chandra Krishi Viswavidyalaya, Mohanpur, West Bengal 741252, India

A.K. Nayak
Microbiology Laboratory, Crop Production Division, ICAR-National Rice Research Institute, Cuttack, Odisha 753006, India

Suman Nayak
Agricultural and Environmental Biotechnology Group, Department of Biotechnology and Medical Engineering, National Institute of Technology Rourkela, Odisha 769 008, India

S. K. Nayak
Department of Biotechnology, North Orissa University, Takatpur, Mayurbhanj, Odisha 757003, India

Emeka Okoro
Department of Mechanical Engineering, University of Johannesburg, Auckland Park, South Africa

Chittaranjan Panda
Environment and Sustainability Department, CSIR-Institute of Minerals and Materials Technology, Bhubaneswar, Odisha 751013, India

Pankaj Kumar Parhi
School of Biotechnology, KIIT Deemed to be University, Bhubaneswar, Odisha 751024, India
Department of Chemistry, Fakir Mohan University, Balasore, Odisha 756089, India

Pranav D. Pathak
Department of Chemical Engineering, Visvesvaraya National Institute of Technology, Nagpur, Maharashtra 440010, India
MIT School of Bioengineering Sciences & Research, Pune, Maharashtra 440010, India

Jayanta Kumar Patra
Research Institute of Biotechnology and Medical Converged Science, Dongguk University, Seoul, Gyeonggi-do, Republic of Korea

Ritesh Pattanaik
School of Biotechnology, KIIT Deemed to be University, Bhubaneswar, Odisha 751024, India

Piyali Paul
Department of Microbiology, Tripura University, Suryamaninagar, West Tripura, Tripura 799022, India

Sushmita Paul
Mycology and Plant Pathology Laboratory, Department of Botany, Guwahati University, Guwahati 781014, Assam, India

Krishna Pramanik
Department of Biotechnology and Medical Engineering, National Institute of Technology, Rourkela, Odisha 769008, India

Himani Priya
Microbiology Laboratory, Crop Production Division, ICAR-National Rice Research Institute, Cuttack, Odisha 753006, India

Jayashree Prusty
Department of Zoology, School of Life Sciences, Ravenshaw University, Cuttack 753003, Odisha, India

S. Rangabhashiyam
Department of Biotechnology, School of Chemical and Biotechnology, SASTRA University, Thanjavur, Tamil Nadu 613401, India

Sangeeta Raut
Centre for Biotechnology, School of Pharmaceutical Sciences, Siksha 'O' Anusandhan (Deemed to be University), Bhubaneswar, Odisha 751003, India

Snehasini Rout
Microbiology Laboratory, Crop Production Division, ICAR-National Rice Research Institute, Cuttack, Odisha 753006, India

Yasaswinee Rout
Ambika Prasad Research Foundation, Bhubaneswar 751006, Odisha, India

Rotimi Sadiku
Department of Chemical, Metallurgical and Materials Engineering, Tshwane University of Technology, Pretoria, South Africa

Amrita Saha
Department of Environmental Science, Amity Institute of Environmental Sciences, Amity University, Kolkata, West Bengal 700135, India

Saurav Saha
Department of Microbiology, Tripura University, Suryamaninagar, West Tripura, Tripura 799022, India

Sabuj Sahoo
Post Graduate Department of Biotechnology, Utkal University, Vani Vihar, Bhubaneswar, Odisha 751004, India

S. K. Sahu
Department of Biotechnology, North Orissa University, Baripada 757003, Odisha, India

Shitarashmi Sahu
Department of Biotechnology and Medical Engineering, National Institute of Technology, Rourkela, Odisha 769008, India

K.C. Samal
Department of Agricultural Biotechnology, College of Agriculture, Odisha University of Agriculture and Technology, Bhubaneswar, Odisha 751003, India

Aneeya K. Samantara
National Institute of Science Education and Research, Jatni, Khordha 752050, India

D. P. Samantaray
Department of Biotechnology, IIT Bombay, Mumbai, Maharashtra 400076 India

Dibyajyoti Samantaray
Department of Biotechnology, College of Engineering and Technology, Biju Patnaik University of Technology, Techno Campus, Ghatikia, Bhubaneswar, Odisha 751003, India

Samuel Eshorame Sanni
Department of Chemical Engineering, Covenant University, Ota, Ogun State, Nigeria

Priya Sarkar
Department of Microbiology, Tripura University, Suryamaninagar, West Tripura, Tripura 799022, India

Abhisek Sasmal
College of basic Science and Humanities, OUAT, Bhubaneswar-751003

Sunil K. Sett
Department of Jute and Fibre Technology, Calcutta University, Kolkata, West Bengal 700019, India

S. Sharma
Department of Economics, OUAT, Bhubaneswar 751003, Odisha, India

Sankar Narayan Sinha
Environmental Microbiology Research Laboratory, Department of Botany, University of Kalyani, Kalyani, West Bengal 741235, India

Amarpreet Singh
Ramie Research Station, Sorbhog, Barpeta, Assam 781317, India

Dhananjay Soren
Department of Zoology, Ravenshaw University, Cuttack, Odisha 753003, India

Mathumal Sudarshan
Trace Element Laboratory, Inter University Consortium, Kolkata, West Bengal 700098, India

Amrita Swain
Department of Zoology, Ravenshaw University, Cuttack, Odisha 753003, India

Luna Samanta
Department of Zoology, Ravenshaw University, Cuttack, Odisha 753003, India

Rajreepa Talukdar
Mycology and Plant Pathology Laboratory, Department of Botany, Guwahati University, Guwahati 781014, Assam, India

Kumananda Tayung
Mycology and Plant Pathology Laboratory, Department of Botany, Guwahati University, Guwahati 781014, Assam, India

Ashoke Ranjan Thakur
School of Science, Sister Nivedita University, New Town, West Bengal 700156, India

Hrudayanath Thatoi
Department of Biotechnology, North Orissa University, Baripada, Odisha 757003, India

Abbreviations

3HB	3-Hydroxybutyrate
4-HBA	4-Hydoxybenzoic acid
5-FdUMP	5-Fluorodeoxyuridylic acid monophosphate
AC	*Azolla caroliniana*
AChE	acetylcholinesterase
AD	Alzheimer's disease
ADF	acid detergent fiber
AF	*Azolla filiculoides*
AGE	advanced glycation end products
Ag-NP	silver nanoparticle
AME	*Azolla mexicana*
AMI	*Azolla microphylla*
AN	*Azolla nilotica*
AP	apple peel
AP	*Azolla pinnata*
ApoE	apolipoprotein E
APP	Aβ protein precursor
AR	*Azolla rubra*
Arg	arginine
Asp	asparagine
ATCC	American Type Culture Collection
ATP	adenosine triphosphate
AWM	agricultural waste management
Aβ	amyloid-β
BACE1	β-amyloid precursor protein cleaving enzyme1
BLESS	direct in situ breaks labeling, enrichment on streptavidin, and next-generation sequencing
BOD	biological oxygen demand
BP	banana peel
BTX	benzene, toluene, and isomers of xylene
BuChE	butyrylcholinesterase
C/N	carbon-to-nitrogen ratio
CA	coal ash
CAP	cellulose acetate phthalate

CBP	consolidated bioprocessing
CCD	central composite design
CCS	CO_2 capture and storage
CD	cylodextrin
CDK5	cyclin-dependent kinase 5
CDP	Clean Development Program
CEC	cation-exchange capacity
CFA	coal fly ash
CGTase	cyclodextrin glycosyltransferase
CNS	central nervous system
CNSA	coconut shell ash
CO_2	carbon dioxide
COD	chemical oxygen demand
COX-2	cyclooxygenase-2
CREB	cAMP response element binding
CRISPR	Clustered Regularly Interspaced Short Palindromic Repeat Sequence
crRNA	CRISPR RNA
CSF	cerebrospinal fluid
DHA	docosahexaenoic acid
DM	diabetes mellitus
DSB	double-stranded break
DSC	differential scanning calorimetry
EAXm	epicarp of *X. moluccensis*
EC	epicatechin
ECG	epicatechin-3-gallate
EDX	energy dispersive X-ray analysis
EDXRF	energy dispersive X-ray fluorescence
EGC	epigallocatechin
EGCG	epigallocatechin-gallate
EIA	Energy Information Administration
EPA	eicosapentaenoic acid
EPS	extracellular polymeric substance
ETC	electron transport chain
FA	fly ash
FAAE	fatty acid alkyl esters
FAME	fatty acid methyl esters
FDA	Food and Drug Administration
FFA	free fatty acid
FPW	fruit peel waste

FT	Fourier transform
FTIR	Fourier transform infrared spectroscopy
GABA	gamma aminobutyric acid
GBSS	granule-bound starch synthase
GC–MS	gas chromatography and mass spectrometry
GDP	gross domestic products
GHG	greenhouse gas
GMO	genetically modified organisms
GP	guava peel
GPAT	glycerol-3-phosphate transferases
GPx	GSH peroxidase
GRE	ginger root extract
gRNA	guide RNA
GSH	glutathione
GST	GSH-S transferase
Hb	hemoglobin
HDL	high-density lipoprotein
HDR	homology directed repair
HPLC	high-performance liquid chromatography
HR	homologous recombination
HRAP	high-rate algal ponds
HSDE	high-speed diesel engine
HTL	hydrothermal liquefication
IL	ionic liquid
iNOS	inducible nitric oxide synthase
IPR	intellectual property rights
LAM-HTGTS	linear amplification-mediated high-throughput genome-wide translocation sequencing
LCA	life cycle assessment
LCIA	life cycle impact assessment
LCSA	life cycle sustainability assessment
Lys	lysine
MALDI	matrix-assisted laser desorption/ionization
MIC	minimum inhibitory concentration
MP	mango peel
mtDNA	mitochondrial DNA
nanoDSF	nano differential scanning fluorimetry
NBS	National Bureau of Standard
NDF	neutral detergent fiber
NER	net energy ratio

NET	negative emission technologies
NFT	neurofibrillary tangle
NF-κB	nuclear factor kappa-B
NHEJ	nonhomologous end joining
NIST	National Institute for Standard and Technology
NMDA	n-methyl-D-aspartate
NMR	nuclear magnetic resonance
NO	nitric oxide
NP	nanoparticle
NSAIDs	nonsteroidal anti-inflammatory drugs
NUC	nuclease
OECD	Organisation for Economic Co-operation and Development
OP	orange peel
OS	oxidative stress
P(3HB)	poly(3-hydroxybutyrate)
P(HB-*co*-HV-*co*-HHx)	poly(hydroxybutyrate-*co*-hydroxyvalerate-*co*-hydroxyhexanoate)
P4HB	poly(4-hydroxybutyrate)
PAM	protospacer adjacent motif
PAP	pineapple peel
PASS	Prediction of Activity Spectra for Substances
PBR	photobioreactors
PEG	polyethylene glycol
PGA	poly(glycolic acid)
PHA	polyhydroxyalkanoate
PHB	polyhydroxybuytyrate
PHC	petroleum hydrocarbon
Phe	phenylalanine
PKC	phosphokinase-C
PLA	poly(lactic acid)
PP	pomegranate peel
PS	phase solubility
PSH	polycyclic aromatic hydrocarbons
PUFA	polyunsaturated fatty acids
QBTU	quadrillion Btu
RBITC	rhodamine B isothiocyanate
REC	recognition
ROS	reactive oxygen species
RSM	response surface methodology

SAC	*S*-allyl-L-cysteine
SBR	Similipal Biosphere Reserve
SCG	spent coffee ground
SDA	saw dust
SEM	scanning electron microscopy
sgRNA	single-guide RNA
SHF	separate hydrolysis and fermentation
S-LCA	social LCA
SLM	sucrose-loaded model
SMD	Sauter mean diameter
SMF	submerged fermentation
SNP	single-nucleotide polymorphism
SOM	soil organic matter
SSCF	simultaneous saccharification and co-fermentation
SSF	simultaneous saccharification and fermentation
SSF	solid-state fermentation
STZ	streptozotocin
SWBR	softwood biomass residue
TALEN	transcription activator-like effector nuclease
TAN	total ammonia nitrogen
TEA	techno-economic analysis
TGA	thermogravimetric analysis
TLC	thin-layer chromatography
trRNA	trans-activating crRNA
TSA	three-step activation
Tyr	tyrosine
UV	ultraviolet
UV–Vis	ultraviolet–visible
VOC	volatile organic compounds
VOO	virgin olive oil
WC	wild Cucurbits
WHO	World Health Organization
WWT	wastewater treatment
XRPD	X-ray powder diffraction
ZFN	zinc-finger nuclease
λ_{max}	absorption maxima

Preface

The exploration, conservation, and sustainable utilization of bioresources are undeniable. The survival and growth of mankind is hugely dependent on the biological resources and capability to sustainably utilize these resources for the benefit of mankind. This book will throw some light on various aspects of research and development in the recent time on effective strategies for sustainable utilization of bioresources using modern tools and techniques. The book has been divided into four parts: (1) Utilization of Bioresources in Therapeutics, (2) Utilization of Bioresources in Biofuel, (3) Utilization of Bioresources in Agriculture, and (4) Utilization of Bioresources in the Environment.

The first section of the book is dedicated to the diverse potential applications of bioresources in food, medicine, and cosmetics. The under-utilized bioresources are being addressed, and the book provides a much-needed boost to the ongoing effort to focus attention on different bioresources and their sustainable uses.

The second section of the book discusses the various aspects of biofuel along with different patents highlighting recent development in biofuel study. Energy crises and the continuously fluctuating cost of petroleum have moved researchers' attention toward new sustainable and renewable energy sources and materials. Biomass, or bioresources, is one of the most abundant and cheapest renewable energy sources and materials and is environmentally friendly. Therefore, finding new energy sources from bioresources and their sustainable utilization will be a major strategy for the growth of the bioenergy and biofuel industry. Several possible routes to provide energy as well as potential value-added products from bioresources have also been discussed in this section. The various pretreatments and characteristics of potential bioresources are presented. Besides direct combustion of biomass, the products from other thermochemical conversion processes could be converted to renewable biofuels, materials, and chemicals.

The third section of the book focuses on different aspects of sustainable utilization of bioresources in the agricultural sector. Developing countries have huge agricultural resources that could be utilized for production of value-added byproducts for the sustainable development of bio-based economy. The book discusses various aspects of efficient utilization of under-exploited natural bioresources, new chemical approaches for the generation of novel biochemicals, and the applications of genetics approaches for bioresource conservation and production of value-added products. Further, strategies to production of biopesticides utilizing bioresources are also discussed.

The fourth section of the book offers comprehensive information about various strategies for environmental protection, development of treatment processes of industrial effluent technologies for climate change, treatment of xenobiotic compounds, etc. It also incorporates recent advances in utilization of bioresources generally associated with environmental issues.

This book should serve as an up-to-date reference resource for university and industry scientists in the area of therapeutics, biofuel, agriculture, and environment management and research.

PART I

Utilization of Bioresources in Therapeutics

CHAPTER 1

Antifungal Drugs from Endophytic Microbes: Present and Future Prospects

RAJREEPA TALUKDAR, SUSHMITA PAUL, and KUMANANDA TAYUNG[*]

Mycology and Plant Pathology Laboratory, Department of Botany, Guwahati University, Guwahati 781014, Assam, India

Corresponding author. E-mail. kumanand@gauhati.ac.in

ABSTRACT

Endophytes are microbes colonizing inner plant tissues without causing any disease symptoms. They have been recognized as repository of novel bioactive metabolites. The bioactive secondary metabolites isolated from endophytic fungi include compounds with antimicrobial, anti-inflammatory, antiproliferative, or cytotoxic activity toward human cancer cell lines, and activity against plant pathogens or plant insect pests. The prevalence of invasive fungal infections has increased significantly among populations and also during medical procedures such as organ transplantation, cancer chemotherapy, and bone marrow transplantation. Moreover, the incidence of fungal infection has also increased as phytopathogens. Plants have also started developing newer fungal diseases due to the developing resistant strains. However, only a limited number of antifungal agents are currently available for the treatment of life-threatening fungal infections as well as phytopathogens. In the recent years, endophytic microbes isolated from medicinal plants have been reported to produce several interesting and effective antifungal metabolites. Thus, the exploration of endophytic microbes from medicinally important plants could be good candidate for obtaining antifungal metabolites to solve growing invasive fungal infection. This assumption becomes more significant considering the myriads of medicinal plants with antifungal activities, and few of these plants have been investigated from endophytic microbes so far.

1.1 INTRODUCTION

Endophytes are described as nonpathogenic plant microorganisms dwelling inside the tissues of a host plant. They are extremely common in almost all plant species and are highly diverse microorganisms that live within healthy tissues of the host without causing visible symptoms of plant diseases (Borges et al., 2009). Both fungi and bacteria are the most common microbes existing as endophytes, but the most frequently isolated are the endophytic fungi (Staniek et al., 2008). Fungal endophytes are a diverse and versatile group of microorganisms that colonize plants in the Arctic, Antarctic, geothermal soils, deserts, oceans, rainforests, mangrove swamps, and coastal forests. They have been isolated from the root, stem, and leaves of a wide range of hosts including algae, bryophytes, pteridophytes, gymnosperms, and angiosperms. There are over 300,000 higher plant species, and it can be assumed that each of these species hosts a complex community of endophytic microbes (Saikkonen et al., 1998). Although the relationship between endophytes and their hosts varies from organism to organism, fungal endophytes are an important component of microbial biodiversity. Some of these endophytes have produced important bioactive metabolites for wide therapeutic applications. Currently, there is growing interest to study microbial endophytes harbored in medicinal plants, as many of these endophytic microbes have been reported to produce bioactive molecules similar to their respective hosts. In recent times, several bioactive metabolites have been reported from endophytic microbes isolated from medicinal plants.

There is an urgent need for new and effective compounds to provide assistance to ever-growing human health problems of the world such as drug resistance in bacteria, the appearance of life-threatening viruses, increased incidence of fungal infections, and new effective anticancer and antiplasmodial agents. Amongst important health problems, invasive fungal infections have been reported to be a serious public hazard, especially for the increasing number of immune-compromised populations including patients suffering from AIDS or cancer, as well as those under organ transplantation and in intensive care units (Rabkin et al., 2000). However, only a limited number of antifungal agents are currently available for the treatment of life-threatening fungal infections. Although new antifungal agents have been introduced in the market, the development of resistance to antifungal drugs has become increasingly apparent, especially in patients with long-term treatment. Thus, there is a general call for new antibiotics and chemotherapeutic agents that are highly effective, possess low toxicity, and have a minor environmental impact. Endophytic microbes are recognized as repository of bioactive metabolites. The bioactive metabolites produced by endophytic organisms originate from different biosynthetic pathways and belong to diverse structural groups such as terpenoids, steroids, quinones, phenols, and coumarins. The functional diverse metabolites produced by endophytes had been well reviewed by Tan and Zou (2001). In the recent years, several effective antifungal metabolites have been reported from endophytes. Considering the myriads of medicinal plants and their associated endophytes, there is a greater possibility of finding endophytic microbes that produce metabolites with promising antifungal activity for treating the invasive fungal infections. Therefore, this review highlights the current antifungal metabolites isolated from endophytic microbes and their possible candidature for future antifungal drugs.

1.2 THE NEED FOR ANTIFUNGAL DRUGS

Medically important fungal infections can be broadly classified into two groups (Richardson, 2005). The first group is superficial mycoses, examples of which include oropharyngeal candidiasis and dermatophyte infections of various regions of the body such as skin and mucosa. Although immuno-compromised people may have increased rates or severity of disease, superficial mycoses are common among those with intact immune function. The second major group is opportunistic infection by fungal species that cause invasive infection. These fungal infections involve sterile body sites such as the bloodstream, central nervous system, or organs including lung, liver, and kidneys. Many of the fungi that cause invasive disease either infect, or colonize, most human beings but the vast majority of the clinically significant diseases occur in people with compromised immune function. As such, the progress of new diseases such as AIDS and severe acute respiratory syndrome in the human population requires discovery and development of new drugs to combat them. Not only do diseases such as AIDS require drugs that target them specifically, but also new therapies are recently needed for treating ancillary infections which are a consequence of a weakened immune system. Furthermore, others who are immuno-compromised (e.g., cancer and organ transplant patients) are at risk for opportunistic pathogens, such as *Aspergillus* spp., *Cryptococcus* spp., and *Candida* spp., which normally are not major problems in the human population.

Serious invasive fungal infections caused by yeasts, such as *Candida* spp., and molds, such as *Aspergillus* spp., have become an increasing threat to human health during the last 20 years with the increasing number of immunosuppressed patients in hematology and oncology wards. The greatest factor for fungal infection is the depth and duration of neutropenia (i.e., decreased neutrophils leading to susceptibility to infection). Other risk factors include the type and status of the primary illness and the use of broad-spectrum antibiotics and antifungal medications. Environmental quality can also affect the likelihood of infection, as defective air-handling systems and building work in facilities housing patients have been implicated in the development of fungal infection. The advent of HIV has also created a pool of patients who are susceptible to both serious invasive and superficial fungal infections. The routine use of cyclosporine A, as an immunosuppressant, aggressive chemo- and radiotherapy regimens, although effective at tumor reduction, also cause massive and prolonged myelosuppression as well as lead to deep and invasive fungal infections in patients. Victims of burns develop damaged dermis and mucous membranes that allow the entry of fungal spores, resulting in invasive fungal infection. The worsening situation of fungal infection is illustrated by the fact that yeasts and molds now rank among the 10 most frequently isolated pathogens, and overall some 7% of all febrile episodes are caused by fungi. The relatively sudden increase in the number of patients with invasive fungal infection means that the supply of novel, potent antifungal agents has failed to keep pace with the therapeutic need. Only a limited number of antifungal agents are currently available for the treatment of life-threatening fungal infections. Although new antifungal agents have been introduced in the market, the development of resistance to antifungal drugs has become increasingly apparent, especially in patients with long-term treatment. Thus, there is a general call for new antifungal drugs that are highly effective, possess low toxicity, and have minor environmental impact.

1.2.1   *MECHANISM OF ACTION OF ANTIFUNGAL AGENTS*

Fungal and human cells, because of phylogenetic similarities, have homologous metabolic pathways to generate energy, to synthesize proteins, and for cell division. Currently available antifungal agents may be categorized according to their molecular targets. Primary molecular targets for antifungal agents are enzymes and other molecules involved in cell wall synthesis, plasma membrane synthesis, fungal DNA synthesis, and mitosis (Figure 1.1) (Ma and Armstrong, 2017). Unlike human cells, fungal cells are surrounded by a cell wall. Its major components are chitin, β-(1,3)-d-glucan, β-(1,6)-d-glucan, and several glycoproteins. Chitins are linear polysaccharides that are bundled into microfibrils and serve as scaffolding for the fungal cell wall. β-(1,3)-d-glucan and β-(1,6)-d-glucan are glucan polymers that are covalently linked to the chitin scaffold. The echinocandin antifungal agents are inhibitors of β-(1,3)-d-glucan synthase (Ma and Armstrong, 2017). Another important biochemical difference between human and fungal cells is related to the sterol used to maintain plasma membrane structure and function. Human cells use cholesterol, whereas fungal cells use the structurally distinct ergosterol. The first step in the biosynthesis of ergosterol involves the conversion of squalene to lanosterol by squalene epoxidase. The *allylamine* and *benzylamine* antifungal agents are inhibitors of squalene synthase.

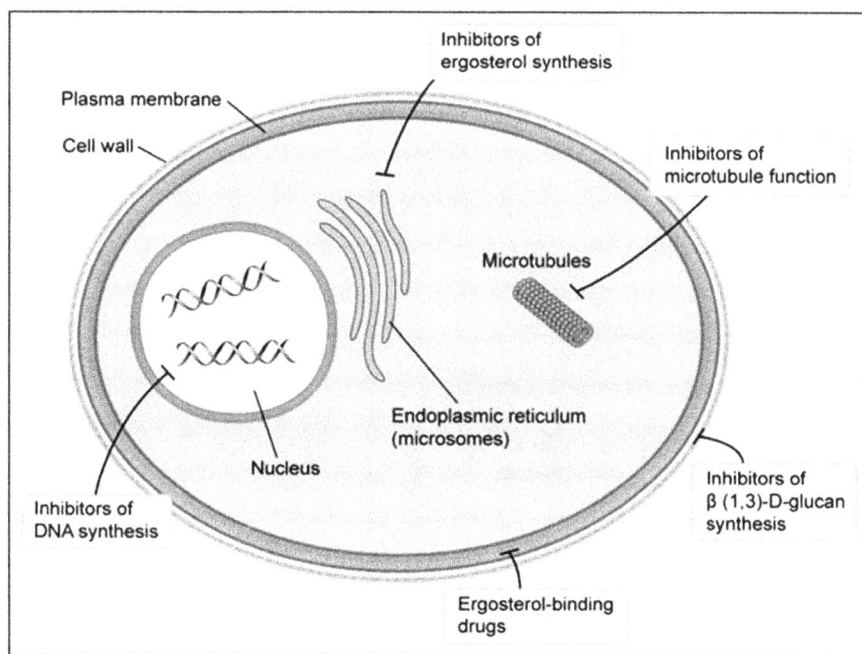

FIGURE 1.1 Molecular targets for antifungal agents.

The second step in plasma membrane synthesis is the conversion of lanosterol to ergosterol. This step is mediated by 14α-sterol demethylase, a fungus-specific cytochrome P450 enzyme. The azole antifungal agents are inhibitors of 14α-sterol demethylase (Ma and Armstrong, 2017). Predictably, the allylamine, benzylamine, and azole antifungal agents sequentially inhibit ergosterol synthesis. The polyene antifungal agents bind to ergosterol and disrupt fungal plasma membrane stability. The antifungal agent *flucytosine* is absorbed via cytosine-specific permeases lacking in human cells. Flucytosine is metabolized to 5-fluorouracil, which is converted to 5-fluorodeoxyuridylic acid monophosphate (5-FdUMP). 5-FdUMP inhibits thymidylate synthase; consequently, it inhibits DNA synthesis and cell division (Table 1.1). The antifungal agent griseofulvin inhibits microtubule function (Table 1.1); consequently, it disrupts the mitotic spindle and inhibits mitosis (Ma and Armstrong, 2017).

TABLE 1.1   Indication for Antifungal Chemotherapy (Pappas et al., 2017)

Mechanism of Action	Drugs	Indications
Inhibitors of β-(1,3)-D-glucan synthase	Echinocandins, Anidulafungin, Caspofungin, Micafungin	Candidiasis, Aspergillosis
Inhibitors of squaline synthase	Allylamines and benzylamines Butenafine Naftifine Terbinafine	*Tinea corporis* or tinea cruris (all) *Tinea capitis* (terbinafine) *Tinea pedis* (butenafine)
Inhibitors of 14α-sterol demethylase	Imidazoles Butoconazole Clotrimazole Econazole Ketoconazole Triazoles Fluconazole Itraconazole Posaconazole Terconazole Voriconazole Isavuconazole	Candidiasis Cryptococcosis Coccidioidomycosis Histoplasmosis Blastomycosis Candidiasis Aspergillosis Blastomycosis Histoplasmosis
Inhibitors of fungal plasma membrane stability	Polyenes, Amphotericin B, Nystatin	Most fungal infections (Amphotericin B) Mucocutaneous candidiasis
Inhibitor of fungal nucleic acid synthesis	Flucytosine	Candidiasis, Cryptococcosis
Inhibitor of fungal mitosis	Griseofulvin	Fungal infections of the skin, hair, or nail caused by *Trichophton*, *Microsporum*, or *Epidermophyton*

1.3   ANTIFUNGAL DRUGS FROM ENDOPHYTIC MICROBES

Endophytic fungi are considered as the resource for new therapeutics and compounds. Fungal secondary metabolites are the compounds with low molecular weight that are not required for growth, but they are produced as an adaptation for specific functions in nature. Endophytes are rich source of natural products displaying a broad spectrum of biological activities. They produce diverse groups of metabolites such as steroids, xanthones, phenols, isocoumarines, perylene derivatives, quinones, furandiones, terpenoids, depsipeptides, cytochalasine, polyketides, alkaloids, peptides, proteins, lipids, shikimates, glycosides, and isoprenoids (Mayer et al., 2011). Microbial endophytes that have been reported to produce bioactive secondary metabolites include bacteria, actinomycetes, and fungi. Endophytic fungi are a poorly investigated group of microorganisms that represent an abundant and dependable source of bioactive and chemically novel compounds. Several attempts have been made to isolate and identify various bioactive metabolites from endophytic fungi. Endophytic fungi can grow in small to large fermenters to provide sufficient supply of bioactive compounds and thus, can be exploited commercially. Presently there is a huge potential for endophytes biologically active natural products that are useful not only in medical but also to agricultural and industrial applications.

There is a whole area of endophytic fungi that has not really been explored or exploited in any great detail. Endophytes living asymptomatically within plant tissues have been found in almost all plant studied to date. They play a major role in physiological activities of host plants influencing enhancement of stress, insect, nematode, and disease resistance (Caroll, 1988). Endophytes can also accelerate plant growth and nitrogen fixing capabilities of host plants (Verma et al., 2001). Endophytes constitute a valuable source of bioactive secondary metabolites (Strobel, 2002) and would be a source of new drugs of biotechnological importance and plant disease management programs (Murray, 1992). Each of the nearly 300,000 species of land plant on earth is likely to host one or more endophyte species. Strobel and Daisy (2003) commented that endophytes could be a goldmine of secondary metabolites. According to them, endophytes are the reservoirs of a variety of chemical structures such as taxol, torreyanic acid, ambuic acid, cryptocandin, subglutinols A and B, and others. New organisms and many novel natural products derived from endophytic fungi have been found to inhibit or kill a wide variety of harmful microorganisms such as bacteria, fungi, viruses, and protozoans that affect humans and animals. Here we are covering the metabolites obtained from endophytic fungi, belonging to various classes of fungi along with unidentified fungi and their potential as an antifungal agent. Many of these compounds are shown in Table 1.2. This search is driven by the development of resistance in infectious microorganisms (e.g., species of *Staphylococcus*, *Mycobacterium*, and *Streptococcus*) to existing compounds and by the menacing presence of naturally resistant organisms.

TABLE 1.2 Some Endophytic Bacteria with Biocontrol Potentials Against Some Fungal Pathogens

Endophytic Bacteria	Host Plants	Target Fungal Pathogens	References
Paenibacillus polymyxa GS01	Panax ginseng	*Rhizoctonia solani*	Cho et al., 2007
Bacillus sp. GS07	Panax ginseng	*Rhizoctonia solani*	Cho et al., 2007
*Pseudomonas poae*JA01	Panax ginseng	*Rhizoctonia solani*	Cho et al., 2007
Bacillus amyloliquefaciens subsp. *plantarum*	Panax notoginseng	*Fusarium oxysporum, Ralstonia sp., Meloidogyne hapla*	Ma et al., 2013
Bacillus methylotrophicus	*Panax notoginseng*	*Fusarium oxysporum, Ralstonia sp., Meloidogyne hapla*	Ma et al., 2013
B. subtilis	*Triticum aestivum L.*	*Gaeumannomyces graminis var. tritici*	Liu et al., 2009
Bacillus licheniformis	*Platycodon grandiflorum*	*Phytophthora capsici, Fusarium oxysporum, Rhizoctonia solani, Pythium ultimum*	Islam et al., 2010
Bacillus pumilus	*Platycodon grandiflorum, Codonopsis lanceolata*	*Phytophthora capsici, Fusarium oxysporum, Rhizoctonia solani, Pythium ultimum*	Islam et al., 2010
Pseudomonas aeruginosa	*Piper nigrum L.*	*Phytophthora capsici*	Aravind et al., 2009
Pseudomonas putida	*Piper nigrum L.*	*Phytophthora capsici*	Aravind et al., 2009
Bacillus megatarium	*Piper nigrum L*	*Phytophthora capsici*	Aravind et al., 2009
Paenibacillus sp.	*Manihot esculenta*	*Rhizoctonia solani*	Canova et al., 2010
Rhodococcus sp. KB6	*Arabidopsis thaliana*	*Ceratocystis fimbriata, P. syringae pv.*tomato DC3000	Hong et al., 2016a
Paenibacillus polymyxa	*Arabidopsis thaliana*	*Phytophthora capsici, Ceratocystis fimbriata, P. syringae pv.* tomato DC3000	Hong et al., 2016b

1.3.1   ANTIFUNGAL METABOLITES FROM ENDOPHYTIC FUNGI

Various works have been reported by scientists proving that endophytes provide benefits to their hosts, including improved drought tolerance, protection against pathogens, enhanced growth, and defense against herbivorous feeding. These features combined with their immense diversity have led drug discovery scientists to consider endophytic fungi as sources of potentially interesting metabolites. According to recent reviews, the characterization of 138 secondary metabolites from endophytic fungi was reported before 2000 (Tan and Zou, 2001) along with the additional 184 metabolites reported by 2006 (Gunatilaka, 2006). These secondary metabolites have been found to include a diverse range of compounds like alkaloids, terpenoids, quinones, peptides, xanthones, and phenols. The secondary metabolites isolated from endophytes has been found to show bioactivity against cancer cell lines (Wang et al. 2011), pathogenic bacteria, and fungi. Various antifungal metabolites have been isolated from different classes of endophytic fungi. Some of them have been described below.

1.3.1.1   COMPOUNDS PRODUCED BY COELOMYCETES

The genus *Pestalotiopsis* exist as an endophyte in most of the world's rainforests and are extremely biochemically diverse. As mentioned elsewhere, *Pestalotiopsis microspora* (*P. microspora*) is a common rainforest endophyte (Strobel et al., 1996). It turns out that enormous biochemical diversity does exist in this endophytic fungus, and as such there seem to be many secondary metabolites produced by a myriad of strains of this widely dispersed fungus. One such secondary metabolite is ambuic acid, an antifungal agent that has been recently described from several isolates of *P. microspora*, found as representative isolates in many of the world's rainforests (Li et al., 2001). In fact, this compound and another endophyte product, terrein, have been used as models to develop new solid-state nuclear magnetic resonance tensor methods to assist in the characterization of molecular stereochemistry of organic molecules (Harper et al., 2003). A strain of *P. microspora* was also isolated from the endangered tree *Torreya taxifolia* that produces several compounds that have antifungal activity, including pestaloside, an aromatic glucoside, and the following two pyrones: pestalopyrone and hydroxypesta-lopyrone. These products also possess phytotoxic properties. A newly described species of *Pestalotiopsis*, namely, *Pestalotiopsis jesteri* (*P. jesteri*), from the Sepik River area of Papua New Guinea, produces jesterone and hydroxy-jesterone (Figure 1.2), which exhibit antifungal activity against a variety of plant-pathogenic fungi. Phomopsichalasin, a metabolite from an endophytic *Phomopsis* sp., represents the first cytochalasin-type compound with a three-ring system, replacing the cytochalasin macrolide ring. This metabolite mainly exhibits antibacterial activity in disk diffusion assays against *Bacillus subtilis* (12-mm zone of inhibition), *Salmonella enteric* serovar Gallinarum (11-mm zone of inhibition), and *Staphylococcus aureus* (8-mm zone of inhibition). It also displays a moderate activity against the yeast *Candida tropicalis* (8-mm zone of inhibition). An endophytic *Fusarium* sp. from the plant *Selaginella pallescens*, collected from the Guanacaste Conservation Area of Costa Rica, was screened for antifungal activity. A new pentaketide antifungal agent, CR377, was isolated from the culture broth of the fungus and showed potent activity against *Candida albicans* (*C. albicans*) in agar diffusion assays performed on fungal lawns (Brady and Clardy, 2000). Colletotric acid, a metabolite of *Colletotrichum gloeosporioides*, an endophytic fungus in *Artemisia mongolica*, displays antimicrobial activity against bacteria as well as against the fungus *Helminthsporium sativum*. Another *Colletotrichum* sp., isolated from *Artemisia annua* (*A. annua*), produces bioactive metabolites that showed varied antimicrobial activity as well. *A. annua* is a traditional Chinese herb that is well recognized for its synthesis of artemisinin (an antimalarial drug) and its ability to inhabit many geographically different areas. The *Colletotrichum* sp. found in *A. annua* produced not only metabolites with activity against human-pathogenic fungi and bacteria but also metabolites that were fungistatic to plant-pathogenic fungi. Some examples of products coming from this group of endophytes are described herein. Ambuic acid is a highly functionalized cyclohexenone and was isolated from endophytic fungi *P. microspora*, *Pestalotiopsis guepinii*, and *Monochaetia* sp. from rainforest plants. It was active against *Pythium ultimum* with a minimum inhibitory concentration (MIC) of 7.5 mcg/mL and also active against several fungi such as *Fusarium* sp., *Diplodia natelensis*, and *Cephalosporium graminineum* but was found inactive against phytopathogenic fungi, namely, *Pyricularia oryzae*, *Rhizoctonia solani* (*R. solani*), *Botrytis cinerea* (*B. cinerea*), and human-pathogenic fungi, suggesting that the role played by this compound in the fungus-plant relationship is that of providing protection to the plant by virtue of its antimycotic activity. Ambuic acid also targets the quorum-sensing-mediated virulence expression of gram-positive bacteria. Pestacin and isopestacin were obtained from culture fluids of *P. microspora*, an endophyte isolated from the Combretaceae family, *Terminalia morobensis*. Pestacin and isopestacin display antimycotic as well as antioxidant activities. Another species of *Pestalotiopsis*, namely, *P. jesteri*, an endophytic fungal species isolated from the inner bark of small limbs of a *Fragraea bodenii* located at the sing–sing grounds of the Aluakambe village from the Septik River area of Papua New Guinea, produces the highly functionalized cyclohexenone epoxides, jesterone, and hydroxyjesterone. Jesterone was found active against *Pythium ultimum, Aphanomyces* sp., *Phytophthora citrophthora, Phytophthora cinnamomi, Sclerotinia sclerotiorum, R. solani, Geotrichum candidum (G. candidum)*, and *Pyricularia oryzae* with MICs of 25, 6.5, 25, 6.5, 100, 25, >100, and 25 mcg/mL, respectively. Hydroxyjesterone was also found active against *Aphanomyces* sp. and *Phytophthora cinnamomi* with MICs of 125 and 62.5 mcg/mL, respectively (Li and Strobel, 2001). Pestalachlorides A-C, three new chlorinated benzophenone derivatives, were isolated from an endophytic fungus. The fungus was isolated from an unidentified tree in Xinglong near Dongzai Hainan Province, Republic of China. These compounds were evaluated for antifungal activity against the plant-pathogenic fungi, namely, *Fusarium culmorum (F. culmorum), Gibberella zeae (G. zeae)*, and *Verticillium aibo-atrum (V. aibo-atrum)*. Pestalachloride A displayed potent antifungal activity against *F. culmorum*, with an IC50 value of 0.89 micromolar, whereas pestalachloride B exhibited remarkable activity against *G. zeae*, with an IC50 value of 1.1 micromolar, whereas pestalachloride C did not show noticeable *in-vitro* antifungal activities against *F. culmorum, G. zeae*, and *V. aibo-atrum* (IC50 > 100 micromolar) (Li et al., 2008). Pestafolide A—a new reduced spiroazaphilone derivative—and pestaphthalides A and B—two new isobenzofuranones— were isolated from an endophytic fungus *Pestalotiopsis foedan*. The fungus was isolated from the branch of an unidentified tree near Dongzai, Hainan Province, Republic of China. The isolated compounds were evaluated for antifungal activity against *C. albicans* (American Type Culture Collection or ATCC 10231), *G. candidum* (AS2.498), and *Aspergillus fumigatus (A. fumigatus)* (ATCC 10894) in agar diffusion assays. Pestafolide A displayed antifungal activity against *A. fumigatus* (ATCC 10894), with a zone of inhibition of 10 mm at 100 mcg/disk. Pestaphthalide A showed activity against *C. albicans* (ATCC 10231), with a 13-mm zone of inhibition and pestaphthalide B showed activity against *G. candidum* (AS2.498) with an 11-mm zone of inhibition when tested at the same level (Fluconazole: 18–28-mm zones of inhibition for *C. albicans, A. fumigatus*, and *G. candidum* at 100 mcg/disk) (Ding et al., 2008). Pestalofones A–E, five new cyclohexanone derivatives, were isolated from *Pestalotiopsis fici*, isolated from the branches of an unidentified tree in the suburb of Hangzhou, Zhejiang Province, Republic of China. Pestalofones A–E were evaluated for activities against *C. albicans* (ATCC 10231), *G. candidum* (AS2.498), and *A. fumigatus* (ATCC 10894).

Pestalofones C and E showed significant antifungal activity against *A. fumigatus*, with the IC50/MIC values of 1.10/35.3 and 0.90/31.2 micromolar, respectively (Liu et al., 2009). *Phomopsis* is another genus that exists as an endophyte, is associated with most of the plants, and is extremely biochemically diverse. Here are some examples of bioactive metabolites products of this endophytic genus. Cytosporones B and C were isolated from an endophytic fungus *Phomopsis* sp. ZSU-H76 obtained from the stem of the mangrove tree *Excoecaria agallocha* Linn from Dongzai, Hainan, China. Both the compounds were found active against *C. albicans* and *Fusarium oxysporum* with an MIC ranging from 32 to 64 mcg/mL (Huang et al., 2008). Cytosporone B was also reported from an endophytic fungus, *Dothiorella* sp., strain HTF3 isolated from the mangrove plant *Avicennia marina* at the estuary of Jiulong River, Fujian Province. It showed activities against fungi, namely, *Aspergillus niger* (*A. niger*), *Trichoderma* sp., and *Fusarium* sp. with MICs of 0.125, 62.5, and 62.5 mcg/mL, respectively. Phomoxanthone A, a dimeric xanthone, was isolated from an endophytic fungus *Phomopsis* sp., isolated from the stem of *Costus* sp. (Costaceae), growing in the rainforest of Costa Rica. It showed the moderate inhibition of *Ustilago violacea* at a concentration of 10 mg/mL. Phomoxanthone A showed strong activity against phytopathogenic fungi, namely, *Phytophthora infestans*, *B. cinerea*, *Pyricularia oryzae*, and *Ustilago violacea*. The metabolites, namely, phomosines A–G, 6-hydroxy-6-isopropylcyclohex-1-enecarboxylic acid, and alternariol were isolated from the endophytic fungus *Phomopsis* sp. isolated from *Adenocarpus foliolosus* from Gomera, Spain. The fungicidal properties of these compounds were evaluated against *Microbotryum violaceum*. Five 10-membered lactones were isolated from *Phomopsis* sp. YM 311483, obtained from the stem of *Azadirachta indica* growing in Yuanjiang Country, a tropical region in Yunnan province, People's Republic of China. These lactones were evaluated for their antifungal activity against seven plant pathogens, namely, *A. niger*, *B. cinerea*, *Fusarium avenaceum*, *Fusarium moniliforme* (*F. moniliforme*), *Helminthosporium maydis*, *Penicillium islandicum*, and *Ophiostoma minus*, using the dose-dependent paper-disk diffusion method. Cerulenin was isolated from an endophytic fungus *Phomopsis* sp., isolated from an asymptomatic leaf of a tropical rainforest tree from French Guyana. It is an inhibitor of fatty acid and polyketide synthases with known anti-*Candida* activity (Hoberg et al., 1986). Two new metabolites, ethyl 2,4-dihydroxy-5,6-dimethylbenzoate and phomopsilactone, were isolated from *Phomopsis cassiae*, an endophytic fungus in *Cassia spectabilis*. Both the compounds displayed strong antifungal activity against the phytopathogenic fungi *Cladosporium cladosporioides* (*C. cladosporioides*) and

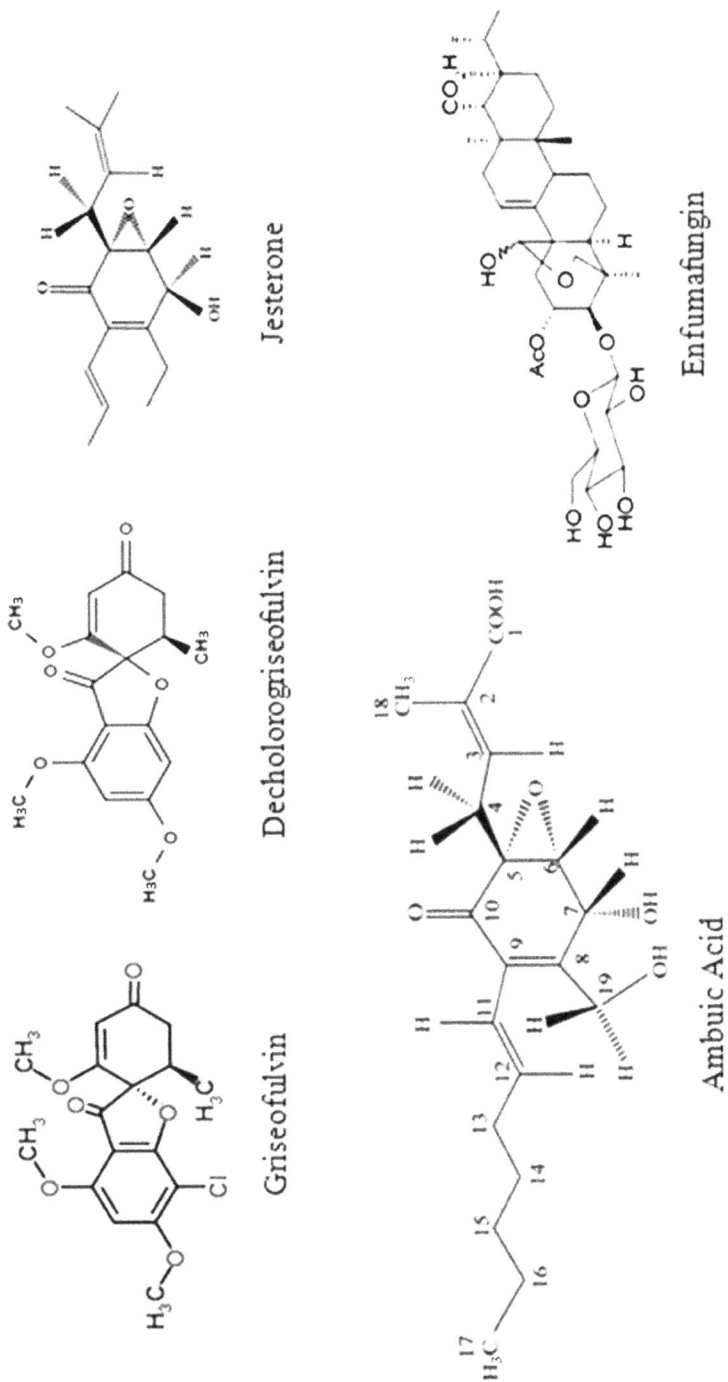

FIGURE 1.2 Structures of some antifungal metabolites obtained from endophytic fungi.

Cladosporium sphaerospermum (*C. sphaerospermum*), and the detection limit for both the compounds was 1 mcg—the same as for the positive control Nystatin. Cycloepoxylactone was isolated from an endophytic fungus *Phomopsis* sp., isolated from the leaves of *Laurus azorica* that grows in Gomera, Spain. It exhibited good antifungal activity against *Microbotryum violaceum* with a radius of 10 mm at the concentration of 50 mcg/disk.

Cadinane sesquiterpenes derivatives, namely, two diastereoisomers of 3,9,12-trihydroxycalamenenes, 3,12-dihydroxycalamenene, 3,12-dihydroxyca-dalene, and 3,11,12-trihydroxycadalene, were isolated from *Phomopis cassiae*, isolated from *Cassia spectabilis* collected from Brazil. The antifungal activity of these compounds was evaluated against the phytopathogenic fungi *C. clado-sporioides* and *C. sphaerospermum*. The compound 3,11,12-trihydroxycadalene was the most active compound, and the detection limit for the compound was found to be 1.0 mcg, comparable with the same amount of the standard Nystatin. The compound 3,12-dihydroxycadalene also exhibited potent activity against two fungi tested (Silva et al., 2006). Phomodione was isolated from *Phoma* sp., an endophyte on a Guinea plant (*Saurauia scaberrinae*). Phomodione exhibited antifungal activity against *Pythium ultimum*, *Sclerotinia sclerotiorum*, and *R. solani*, with MIC between 3 and 8 mcg/mL. Pyrenophorol, a known macrolide, and (–)dihydropyrenophorin-1,4-lactone were isolated from *Phoma* sp., an endophytic fungus isolated from *Lycium intricatum* from Gomera, Spain. All of these compounds exhibited antifungal activity against *Microbotryum viola-ceum*. Moriniafungin, a novel sordarin analog with potent antifungal activity, was isolated from *Morinia longiappendiculata* isolated from stems of four plant species namely *Santolina rosmarinifolia*, *Helichrysum stoechas*, *Thymus mastichina*, and *Calluna vulgarisin* collected from central Spain. Moriniafungin exhibited an MIC of 6 mcg/mL against *C. albicans* and IC50 ranging from 0.9 to 70 mcg/mL against a panel of clinically relevant strains namely *C. albicans* (MY1055), *Cryptococcus neoformans* (*C. neoformans*) (MY 2062), *C. neofor-mans* (ATCC 66031), *Candida glabrata* (MY1381), *C. parapsilosis* (ATCC 22019), *Candida krusei* (ATCC 6258), and *Candida lusitaniae* (MY1396) (Cruz et al., 2005, Collado et al., 2006). Colletotric acid was isolated from *Colletotrichum gloeosporioides*, an endophytic fungus colonized inside the stem of *Artemisia mongolica* and inhibited the growth of the pathogenic fungus *Helminthosporium sativum* with an MIC value of 50 mcg/mL. 3-Beta-hydroxy-5-alpha,8-alpha-epidioxy-ergosta-6,22-diene and 3-beta-hydroxyergosta-5-ene were characterized from the culture of *Colletotrichum* sp., an endophyte isolated from inside the stem of *A. annua*. These compounds exhibited anti-fungal activities against *C. albicans* and *A. niger* (MICs: 50–100 mcg/mL). *Cis*-4-hydroxy-6-deoxyscytalone and (4R)-4,8-dihydroxy-alpha-tetralone were isolated from endophytic fungus *Colletotrichum gloeosporioides* residing in healthy leaves of *Cryptocarya mandioccana*. The antifungal activity of these two compounds was evaluated, and the detection limit of these compounds required to inhibit growth of the phytopathogenic fungi *C. cladosporioides* and *C. sphaerospermum* was 5 mg, comparable with the positive control Nystatin. Exochromone, a highly substituted chromone dimer, was isolated from *Exophiala* sp. isolated from *Adenocarpus foliolosus*, growing on the hills of Baranco de Los Jargus, Gomera. Exochromone showed a radial zone of inhibition of 9 mm at the concentration of 1 mg/mL against *Microbotryum violaceum* (Hussain et al., 2007). The endophytic fungi, namely, *Aspergillus flavus*, *A. niger*, *Colletotrichum gloeosporioides*, and *Alternaria alternata* were isolated from *Lannea coromandelica*. Kojic acid, octadecanoic acid, *n*-hexadecanoic acid, diethyl phylate, and 3-phenyl propionic acid are the known compounds for their antimicrobial activity, isolated from an endophyte *Aspergillus flavus* in this study. Hence, the synergetic action of these compounds might be the possible reason for its activity against the fungal pathogens like *C. albicans* and *Malassezia pachydermis*. Moreover, this is the first report that Kojic acid has been isolated from the endophytic fungi grown in a very common Indian tree *Lannea coromandelica*. Observations in this research have important implications for the production of kojic acid at a low cost, which may turn to be of wide significance in health, pharmaceutical, food and cosmetic industries. Increase in the yields of these products may also be enhanced by studying the biochemical pathways, new developments in fermentation technology, membrane technologies and genetic manipulation.

1.3.1.2 COMPOUNDS PRODUCED BY ASCOMYCETES

Cryptosporiopsis quercina is the imperfect stage of *Pezicula cinnamomea*, a fungus commonly associated with hardwood species in Europe. It was isolated as an endophyte from *Tripterigeum wilfordii*, a medicinal plant native to Eurasia (Zou et al., 2000). On petri plates, *C. quercina* demonstrated excellent antifungal activity against some important human fungal pathogens, for example, *C. albicans* and *Trichophyton* spp. A unique peptide antimycotic, termed cryptocandin, was isolated and characterized from *C. quercina*. This compound contains a number of peculiar hydroxylated amino acids and a novel amino acid: 3-hydroxy-4-hydroxy methyl proline. The bioactive compound is related to the known antimycotics, the echinocandins and the pneumocandins. As is generally true not one but several bioactive and related compounds are produced by a microbe. Thus, other antifungal agents related to cryptocandin are also produced by *C. quercina.* Crypto-candin is also active against a number of plant-pathogenic fungi, including *Sclerotinia sclerotiorum* and *B. cinerea.* Cryptocandin and its related compounds are currently being considered for use against a number of fungi causing diseases of skin and nails. Cryptocin, a unique tetramic acid, is also produced by *C. quercina.* This unusual compound possesses potent activity against *Pyricularia oryzae* as well as a number of other plant-pathogenic fungi (Guo et al., 2008). The compound was generally ineffective against a general array of human-pathogenic fungi. Nevertheless, with MICs of this compound for *P. oryzae* being 0.39 g/mL, this compound is being examined as a natural chemical control agent for rice blast and is being used as a base model to synthesize other antifungal compounds. The ecomycins are produced by *Pseudomonas viridiflava* (*P. viridiflava*), which is a member of a group of plant-associated fluorescent bacteria. It is generally associated with the leaves of many grass species and is located on and within the tissues (Li and Strobel, 2001). The ecomycins B and C represent a family of novel lipopeptides and have molecular weights of 1153 and 1181. Besides common amino acids such as alanine, serine, threonine, and glycine, some unusual amino acids are also involved in the structure of the ecomycins, including homoserine and hydroxyaspartic acid. The ecomycins are active against human-pathogenic fungi such as *Cryptococcus neoformans* and *C. albicans.* Cryptocandin A, an antifungal lipopeptide, was isolated and characterized from the endophytic fungus *Cryptosporiopsis quercina*, isolated from the stem of *Tripterigeum wilfordii* (Strobel et al., 1999). This compound contains a number of unusual hydroxylated amino acids and a novel amino acid, 3-hydroxy-4-hydroxymethylproline, and a β-(1,3)-glucan synthesis inhibitor. Cryptocandin A is active against some important human fungal pathogens, including *C. albicans* and *Trichophyton* sp., and also against a number of plant-pathogenic fungi, including *Sclerotinia sclerotiorum* and *B. cinerea.* Similarly, a group of peptides, Echinocandins A, B, D, and H, also known as β-(1,3)-glucan synthesis inhibitors, were isolated from endophytic *Crypto-sporiopsis* sp. and *Pezicula* sp. in *Pinus sylvestris* and *Fagus sylvatica* and showed to be antifungal. Cryptocin, a tetramic acid, is an antifungal compound also obtained from *Cryptosporiopsis quercina* isolated from the inner bark of the stem of *Tripterigeum wilfordii.* This unusual compound possesses potent activity against *Pyricularia oryzae*, with MIC of 0.39 mcg/mL, a causal agent of rice blast disease, as well as a number of other plant-pathogenic fungi (Li et al., 2000). Enfumafungin, a triterpenoid glucoside, was produced by *Hormonema* sp. (ATCC 74360) an endophytic fungus isolated from *Juni-perus communis.* Enfumafungin shows an interesting antifungal spectrum, and its effect on the morphology of *A. fumigatus* was shown to be comparable to that of the glucan synthase inhibitor pneumocandin B. Epichlicin, a novel cyclic peptide, was isolated from *Epichloe typhina*, an endophytic fungus from *Phleum pratense*. Epichlicin showed inhibitory activity toward the spore germination of *Cladosporium phlei*, a pathogenic fungus of *Epichloe typhina* plant at an IC50 value of 22 nanomoles. New naphthoquinone spiro-ketals, namely, preussomerins EG1–EG3, were isolated from the mycelium of Edenia gomezpompae, isolated from the leaves of *Callicarpa acuminata* (Verbenaceae) collected from the ecological reserve El Eden, Quintana Roo, Mexico. The new spiroketals displayed significant growth inhibition against all the phytopathogens, namely, *Phytophthora capsici*, *P. parasitica*, *Fusarium oxysporum*, and *F. solani* with IC50 values in the range of 20–170 mcg/mL. Weber et al. (2007) reported 5-(1,3-butadien-1l)-3-(propen-1-yl)-2 (5H)-furanone from an endophytic strain E99297, isolated from the twig of *Cistus salvifolius*, which as previously shown by Kopcke et al. (2002) belongs to the Sarcosomataceae (order Pezizales). The compound was isolated on the basis of its anti-*Candida* activity and was identical to 5-(E)-buta-1,3-dienyl-3 (E)-propenyl-5H-furan-2-one isolated by Kopcke et al. (2002). Depsidones, botryorhodines A–D, were produced from the stems of the medicinal plant *Bidens pilosa* (Asteraceae). Botryorhodine A exhibited the MICs of 26.03 and 191.60 micromolar against *Aspergillus terreus* and *Fusarium oxysporum*, respectively, whereas the compound botryorhodine B exhibited the MICs of 49.70 and 238.80 micromolar against *Aspergillus terreus* and *Fusarium oxysporum*, respectively. Dinemasones A–C were isolated from *Dinemasporium strigosum*, isolated from the roots of *Calystegia sepium* from the shores of the Baltic Sea, Wustrow, Germany. Dinemasones A and B exhibited considerable activity against *Microbotryum violaceu*. Chaetoglobosins A and C were characterized from an endophytic fungi *Chaetomium globosum* isolated from the leaves of *Ginkgo biloba*. Both the compounds displayed marked inhibitory activity against *Mucor miehei* by the agar diffusion method with the 25- and 15-mm zones of inhibition of diameter 10 mcg/disk, respectively. Sordaricin, a known antifungal metabolite, was isolated from an endophytic fungus *Xylaria* sp. PSU-D14, isolated from the leaves of *Garcinia dulcis*, collected from Songkhla Province, Thailand. It exhibited moderate antifungal activity against *C. albicans* (ATCC 90028) with an MIC value of 32 mcg/mL. Griseofulvin and 7-dechlorogriseofulvin were isolated from *Xylaria* sp. F0010 isolated from *Abies holophylla*. Compared to 7-dechlorogriseofulvin, griseofulvin showed high in vivo and in vitro antifungal activities and effectively controlled the development of plant-pathogenic fungi, namely, *Magnaporthe grisea*, *Corticium*

sasakii, *Puccinia recondite*, and *Blumeria graminis* f. sp. hordei at doses of
50–150 mcg/mL, depending on the disease. Griseofulvin was also isolated
from an endophytic fungus PSU-N24, isolated from *Garcinia nigrolineata*.
It displayed strong antifungal activity against *Microsporum gypseum*
SH-MU-4 with an MIC value of 2 mcg/mL. 2-Hexyl-3-methyl-butanodioic
acid and Cytochalasin D were isolated from an endophytic fungus *Xylaria*
sp. isolated from *Palicourea marcgravii*. These compounds were found
active against phytopathogenic fungi *C. cladosporioides* and *C. sphaero-*
spermum. 7-Amino-4-methylcoumarin was isolated from an endophytic
Xylaria sp., obtained from *Ginkgo biloba* L. The compound showed strong
in-vitro antifungal activities with MIC of 15, 40, and 25 mcg/mL against *C.*
albicans, *Penicillium expansum*, and *A. niger*, respectively (Liu et al., 2008).
Colletotric acid, a metabolite of *Colletotrichum gloeosporioides*, an endophytic fungus from the plant *Artemisia mongolica*, displayed antimicrobial
activity against bacteria as well as against fungus, *Helminthosporium*
sativum. In another study *Colletotrichum* sp. isolated from *A. annua*,
produces bioactive metabolites that showed varied antibacterial and antifungal activity. Similar to our study, kojic acid (KA), a natural pyrone, is
reported as a potent chemo sensitizing agent of complex III inhibitors
disrupting the mitochondrial respiratory chain in fungi thereby acting as an
antifungal agent.

1.3.1.3   COMPOUNDS PRODUCED BY HYPHOMYCETES

A new pyrone derivative, Penicillone, together with Pyrenocine A and B (PSU-
A71) were isolated from the leaves of *Garcinia atroviridis*, collected in Songkhla Province, Thailand. All the compounds were tested for antifungal activity
against *Microsporum gypseum* SH-MU-4. Arthrichitin, a cyclic depsipeptide,
was isolated from *Arthrinium phaeospermum* obtained from unidentified grass
(Vijayakumar et al., 1996). It was also isolated as LL15G256γ from the marine
fungus *Hypoxylon oceanicum*. Arthrichitin has a broad spectrum of activity
against *Candida* sp., *Trichophyton* sp. and several phytopathogens. *In-vitro*
testing has shown that arthrichitin inhibits membrane preparations of fungal
chitin and glucan synthases, with a greater potency against chitin synthase.
Fungal cells exposed to arthrichitin undergo morphological changes similar to
the effects of chitin synthase inhibitor, polyoxin B. The morphological effects
of arthrichitin occur at concentrations that are 10-fold below those required to
inhibit chitin synthase. This suggests the existence of another target for
arthrichitin, possibly an isozyme of chitin synthase not present in the membrane

preparation. It is also possible that the compound alters regulatory processes in the cell cycle that affect the cell wall. The in vitro potency of arthrichitin is too low for its use in the clinic. However, it has been suggested that the development of analogs, such as those based on the much larger cyclic peptides, the echinocandins, might yield congeners with improved activity. The solanapyrone analogs, solanapyrone C, N, and O, nigrosporalactone, and phomalactone were isolated from *Nigrospora* sp. YB-141, an endophytic fungus isolated from *Azadirachta indica*. All the compounds were tested for their antifungal properties against seven phytopathogenic fungal strains, namely, *Aspergillus niger*, *B. cinerea*, *Fusarium avenaceum*, *F. moniliforme*, *Helminthosporium maydis*, *Ophiostoma minus*, and *Penicillium islandicum*. All five compounds showed antifungal activities against *B. cinerea* with the MIC values of 31.25–250 mcg/ mL. Trichodermin was characterized from *Trichoderma harzianum*, an endophytic fungus from *Llex cornuta*. Plant experimental results showed that trichodermin exhibited significant protective effect to early blight on tomato and damping-off on cucumber. The EC50 values of trichodermin inhibiting the mycelia growth of *Alternaria solani* (*A. solani*) and *R. solani*, respectively. In addition, protective and therapeutic effects of trichodermin at 100 mg/L against *A. solani* and *R. solani* were 97.8% and 98.1%, and 96.7% and 97.3%, respectively. Nodulisporins D–F (3S,4S,5R)-2,4,6-trimethyloct-6-ene-3,5-diol, 5-hydroxy-2-hydroxymethyl-4H-chromen-4-one, 3-(2,3-dihydroxyphenoxy)-butanoic acid, and benzene-1,2,3-triol were isolated from *Nodulisporium* sp. an endophytic fungus isolated from the plant *Erica arborea* from Gomera. These compounds were found active against *Microbotryum violaceum*. Wang et al. (2002) isolated Brefeldin A, with a wide range of biological activities, namely, antifungal, antiviral, antimitotic, anti-inflammatory and antitumor from *Aspergillus clavatus* and *Paecilomyces* sp. endophytic in Chinese *Taxus mairei* and *Torreya grandis*. Brefeldin A was also isolated from an endophytic fungal strain of *Eupenicillium brefeldianum*, isolated from a Chinese traditional medicinal plant *Arisaema erubescens* and from an endophytic fungal strain of *Cladosporium* sp., isolated from a Chinese traditional medicinal plant *Quercus variabilis*. CR377, a new pentaketide, was isolated from fungus *Fusarium* sp., isolated from the interior of a surface-sterilized piece of the *Selaginella pallescens* stem tissue, collected in the Guanacaste Conservation Area of Costa Rica. CR377 was tested against *C. albicans* by the agar diffusion method. The compound exhibited inhibition zones of 20 against *C. albicans* strains. Nystatin, a positive control produced inhibition zones of 19 against the Wisconsin. Four known naphtho-gamma-pyrones, rubrofusarin B, fonsecinone A, asperpyrone B, and aurasperone A were isolated from *A. niger* IFB-E003, an endophyte isolated from the leaves of *Cynodon dactylon* collected from the Yancheng Biosphere

Reserve, People's Republic of China. These compounds exhibited growth inhibition against *Trichophyton rubrum* and *C. albicans* with MICs ranging between 1.9 and 15.6 mcg/mL. Four natural nitro metabolites, 1-hydroxy-5-methoxy-2-nitro-naphthalene, 1,5-dimethoxy-4-nitronaphthalene, 1-hydroxy-5-methoxy-2,4-dinitronaphthalene, and 1,5-di-methoxy-4,8-dinitronaphthalene, known from chemical synthesis but new as natural products, were isolated together with 1-hydroxy-5-methoxynaphthalene from an endophytic fungus *Coniothyrium* sp., isolated from the shrub *Sideritis chamaedryfolia* from an arid habitat near Alicante, Spain. The Compounds extracted showed excellent activity against *Microbotryum violaceum*. Fusicoccane diterpenes, namely, periconicins A and B with antibacterial activities were isolated from an endophytic fungus *Periconia* sp., OBW-15, collected from small branches of *Taxus cuspidate*. Periconicin A showed potent inhibitory activity against the agents of human mycoses, including *C. albicans*, *Trichophyton mentagrophytes*, and *T. rubrum*, with MICs in the range of 3.12–6.25 mcg/mL. Monomethylsulochrin, Rhizoctonic acid, Asperfumoid, Physcion, 7,8-dimethyliso-alloxazine, and 3,5-dichloro-*p*-anisic acid were identified from *Penicillium* sp., isolated from the leaves of *Hopea hainanensis*. All of the six isolated compounds were subjected to antifungal activity against three human-pathogenic fungi namely *C. albicans*, *Trichophyton rubrum*, and *A. niger*. Compounds inhibited the growth of *C. albicans* and *A. niger*. The antifungal metabolites, namely, Asperfumoid, Physcion, Fumigaclavine C, Fumitremorgin C, and Helvolic acid were obtained from an endophytic fungus *A. fumigatus* CY018, isolated from the leaves of *Cynodon dactylon*. All of the five compounds inhibited the growth of *C. albicans* with MICs of 75.0, 125.0, 31.5, 62.5, and 31.5 mcg/mL, respectively. *Aspergillus clavatonanicus*, an endophytic fungal strain from *Taxus mairei*, yielded Clavatol and Patulin. Both compounds exhibited in vitro inhibitory activity against several plant-pathogenic fungi, namely, *B. cinerea*, *Didymella bryoniae*, *Fusarium oxysporum* f. sp. *cucumerinum*, *R. solani*, and *Pythium ultimum*. Asperamide A and B, a sphingolipid, and their corresponding glycosphingolipid, possessing a hitherto unreported 9-methyl-C20-sphingosinemoiety, were characterized from *A. niger* EN-13, an endophytic fungus isolated from marine brown alga *Colpomenia sinuosa* along the Qingdao coastline of Shandong Province, People's Republic of China. Asperamide A displayed moderate activity against *C. albicans* with a zone of inhibition of 20 mm. A new naphthoquinoneimine derivative, namely, 5,7-dihydroxy-2-(1-(4-methoxy-6-oxo-6H-pyran-2-yl)-2-phenylethylamino)-(1,4) naphthoquinone, was also reported from *A. niger* EN-13, isolated from the inner tissue of the marine brown alga *Colpomenia sinuosa*. It displayed moderate antifungal activity against *C. albicans* with an inhibitory zone (10 mm) at 20 mg/well (6 mm). A new diphenyl ether, neoplaether, was isolated from the culture of *Neoplaconema napellum* IFB-E016, an endophytic fungus residing in the healthy leaves of *Hopea hainanensis* from Hainan Island, People's Republic of China. In vitro antifungal activity of neoplaether was examined using *A. niger*, *C. albicans*, and *Trichophyton rubrum* as test organisms, and it showed obvious activity against *C. albicans* with an MIC value of 6.2 mcg/mL (that of amphotericin co-assayed as positive control was 1.5 mcg/mL), whereas no growth inhibition to *A. niger* and *T. rubrum* could be discerned when it was tested at 100 mcg/mL. Isofusidienols A–D were produced by *Chalara* sp. (strain 6661), an endophytic fungus isolated from *Artemisia vulgaris*. All the isofusidienol compounds exhibited antifungal activity against *C. albicans* periconicin B, 6,8-dimethoxy-3-(2'-oxo-propyl)-coumarin, 2,4-dihydroxy-6-((1'E,3'E)-penta-1',3'-dienyl)-benzaldehyde was isolated from *Periconia atropurpurea*, obtained from the leaves of *Xylopia aromatica*, a native plant of the Brazilian Cerrado. These compounds were evaluated against *C. sphaerospermum* and *C. clado-sporioides*. Only compound (2,4-dihydroxy-6-((1'E,3'E)-penta-1',3'-dienyl)-benzaldehyde) exhibited strong antifungal activity against both fungi, showing detection limit of 1.0 mcg, comparable to nystatin (used as a positive control). Compound (6,8-dimethoxy-3-(2'-oxo-propyl)-coumarin) did not show any antifungal activity.

1.3.1.4 COMPOUNDS PRODUCED BY UNIDENTIFIED FUNGUS

Cabello et al. (2001) reported a novel acidic steroid arundifungin, a β-(1,3)-glucan synthesis inhibitor from F-042,833, isolated from the twig of *Olea europea* var europea, an undetermined coelomycetes, and F054,289 a sterile mycelium from the leaves of *Quercu silex*, isolated from Ontigola, Madrid, Spain. Arundifungin causes the same pattern of hallmark morphological alteration in *Aspergillus fumigates* hyphae as echinocandins, which supports the idea that arundifungin belongs to the class of glucan synthesis inhibitors. It exhibits antifungal activity against a range *Candida* sp. with MICs over 2–8 mcg/mL, whereas it is only 1 mg/mL for *A. fumigatus*. Khafrefungin was isolated from an unidentified sterile fungus (MF 6020) cultured from a Costa Rican plant sample. It inhibits fungal sphingolipid synthesis, at the step in which phosphoinositol are transferred to ceramide, resulting in the accumulation of ceramide and loss of all of the complex sphingolipids. In vitro khafrefungin inhibits the inositol phosphoceramide synthesis of *C. albicans* with an IC50 of 0.6 nanomoles. Khafrefungin inhibited the growth of *C. albicans*, *Cryptococcus neoformans*, and *Saccharomyces cerevisiae* in liquid culture with MIC of 2, 2, and 15.6 mcg/mL, respectively. Khafrefungin does not inhibit the synthesis of mammalian sphingolipids thus making this the first reported compound that is specific for the fungal pathway. Ascosterosides A and B were isolated from E99291, an endophyte from shoots of Cistus salvifolius. Both the compounds showed antifungal activity against a wide range of saprophytic, plant, and human-pathogenic fungi. Ascosteroside A was originally isolated from *Ascotricha amphitricha* as a metabolite with anti-*Candida* activity (Gorman et al., 1996), which is now known to be due to the inhibition of β-(1,3)-glucan synthesis (Onishi et al., 2000). Sphaeropsidin A was isolated from E99204, an endophyte from an asymptomatic leaf of *Quercus ilex*, and exhibited antifungal activity against yeast and filamentous fungi. Sphaeropsidins are a group of pimarane diterpenes known from the anamorphic fungi *Sphaeropsis sapinea* f. sp. cupressi and *Diplodia mutila*. 6,8-Diacetoxy-3,5-dimethylisocoumarin and 3-acetyl-6-hydroxy-4-methyl-2,3-dihydrobenzofuran were isolated from Mycelia sterila, an endophytic fungus from the Canadian thistle *Cirsium arvense* growing in Lower Saxony, Germany. The compounds were tested against three fungal test organisms namely *Mycotypha microspora*, *Eurotium repens*, and *Ustilago violacae*. Compound (6,8-diacetoxy-3,5-dimethylisocoumarin) showed moderate antifungal activity against all tested fungi, whereas the compound (3-acetyl-6-hydroxy-4-methyl-2,3-dihydrobenzofuran) as moderately antifungal against Eurotium repens. Pyrenocines A, F–H were isolated from an unidentified endophytic fungus (6760), isolated from *Trifolium dubium* from Wustrow near the Baltic Sea. Pyrenocines A and G exhibited considerable activity against *Microbotryum violaceum* (Krohn et al., 2008).

1.3.2   ANTIFUNGAL COMPOUNDS FROM ENDOPHYTIC BACTERIA

Endophytic bacteria inhabit plant internal tissues in a similar niche as phytopathogens, and they may compete with bacterial pathogens as biocontrol agents (Berg et al., 2005). Endophytic bacteria are present in most plant species and can latently or actively colonize the plant locally as well as systemically. Several authors have reported on the use of bacteria as biocontrol agents that show antibacterial and antifungal activity. However, reports of endophytic bacteria showing antifungal activity are lesser recorded. Raupach and Kloepper (1998) tested plant growth-promoting bacteria (*Bacillus* spp.) for biocontrol against cucumber fungal pathogens *Colletotrichum orbiculare*. *Burkholderia* species strains isolated from the sugarcane rhizosphere were reported to show antifungal activity against sugarcane smut (*Ustilago scitaminea*) and *Fusarium* spp. causing stalk rot. Inoculation of plants with beneficial endophytes can inhibit disease symptoms caused by viral, insect, fungal, and bacterial pathogens (Sturz et al., 2000). The beneficial effects derived from endophytic bacteria are similar to those from rhizosphere bacteria; however, it is anticipated that endophytic bacteria are more suitable as biocontrol agents because they sustainably transmit to the next generation. *Streptomyces* spp., *Pseudomonas viridiflava, Serratia marcescens*, and *Paenibacillus polymyxa* are endophytic bacteria that produce active metabolites with antimicrobial and antifungal activities (Beck et al., 2003). Moreover, endophytic bacteria with biocontrol activities have been isolated from various plant species (Table 1.3). According to recent reports, the leaf-inhabiting endophytic KB strains isolated from Arabidopsis suppressed the phytopathogen-induced disease symptoms (Hong et al., 2015, 2016a). The *Bacillus thuringiensis* KB1 strain displayed antagonistic activities against *Fusarium oxysporum* and *P. syringae* pv. tomato DC3000 in vitro and/or in planta (Hong et al., 2015). *Rhodococcus* sp. KB6 strain reduced symptoms of black rot disease in sweet potato leaves caused by the fungal pathogen *Ceratocystis fimbriata* (Hong et al., 2016b). These results support the hypothesis that endophytic bacteria act as biocontrol agents in plants. One of the strains of endophytic bacterium with a good anti-fungal ability was first isolated from *Vaccinium uliginosum*. It was identified as *Serratia marcescens* using the 16s rDNA sequence homology and through its physiological and biochemical characteristics. The antifungal activity of the isolated bacterium was tested in vitro against eleven strains of pathogenic fungi, belonging to nine different genera: *Alternaria alternata, Alternaria consortiale, A. solani, Bipolarissorokiniana, B. cinerea, Cylindrocladium colhounii, Exserohilum turcicum, Fulvia fulva, Fusarium oxysporum* f. sp. cucumerinum, *Pestalotiopsis clavispora,* and *Verticillium dahliae*. The results of the antifungal tests demonstrated that the metabolites of the strain *Serratia marcescens* have a broad activity spectrum and that they inhibit the growth of all the 11 pathogenic fungi. The inhibition rates ranged from 22.7% to 68.6%. Against the pathogens of the blueberry leaf spot *Cylindrocladium colhounii* and *Pestalotiopsis clavispora*, the inhibition rates were 57.7% and 68.6%, respectively. Reports on antifungal activities by endophytic bacteria namely *Pseudomonas aeruginosa, Pseudomonas putida* and *Bacillus megaterium* associated with *Piper nigrum* against *Phytophthora capsici* are also found. Moreover, some reports also show that the endophyric bacteria (i.e., Bacillus thuringiensis KB1 and Rhodococcus sp. KB6) associated with Arabidopsis thaliana shows antifungal activities against *Fusarium oxysporum*. Eventually, a potent antifungal strain of *Serratia marcescens* was recovered from

Bioresource Utilization and Management

Rhyncholacis penicillata and was shown to produce oocydin A (Figure 1.3), a novel anti-oomyceteous compound having the properties of a chlorinated macrocyclic, lactone. It is conceivable that the production of oocydin A by *S. marcescens* is directly related to the endophyte's relationship with its higher plant host. Currently, oocydin A is being considered for agricultural use to control the ever-threatening presence of oomyceteous fungi such as *Pythium* and *Phytophthora*. The ecomycins are produced by *P. viridiflava* (Miller, 1998). *P. viridiflava* is a member of a group of plant-associated fluorescent bacteria. It is generally associated with the leaves of many grass species and is located on and within the tissues (Miller, 1998). The ecomycins represent a family of novel lipopeptides and have molecular weights of 1153 and 1181. Besides common amino acids such as alanine, serine, threonine, and glycine, some unusual amino acids are also involved in the structure of the ecomycins, including homoserine and hydroxyaspartic acid. The ecomycins are active against such human-pathogenic fungi as *Cryptococcus neoformans* and *C. albicans*. The endophytic fungus *Paenibacillus polymyx* associated with wheat plants synthesizes the valuable compound Fusaricidin, which is a well-known antifungal agent. From the above-mentioned data, it is clear that only a few potential antifungal compounds are obtained from endophytic bacteria. However, a good number of endophytic bacteria are found to produce antifungal activities during their preliminary screening against some fungal pathogens. Some of the potent bacterial endophytes showing antimicrobial activities against some target fungal pathogen are mentioned in Table 1.2. Further, study of maize endophytes has revealed that, bacterial endophytes *Bacillus* species that naturally occur in many maize varieties may function to protect hosts by secreting antifungal "lipopeptides" that inhibit pathogens like *Fusarium* as well as inducing the upregulation of pathogenesis-related genes of host plants. The antifungal lipopeptides are grouped under "iturins" and "fengycins." In a similar study, when *R. solani* growing on agar was treated with the lipopeptide "tensin," mycelia showed retarded growth accompanied by increased branching and rosette formation as well as hyphal swelling. An endophytic bacterium, *Bacillus amyloliquefaciens* from vanilla orchids showed antifungal activity and protected plant seedlings from pathogens. Another endophytic *Bacillus* species isolated from a paddy field showed similar activity in protecting maize and horse bean from infection of *Bipolaris maydis* and *R. solani*, respectively (Wang et al., 2009). These studies show that lipopeptides secreted by endophytic bacteria especially Bacillus act as antifungal compounds against some plant-pathogenic fungi. Bacterial endophytes also produce compounds inhibitory to chitin, which is a component of fungal cell wall. It has been found that the classical inhibitors of chitin

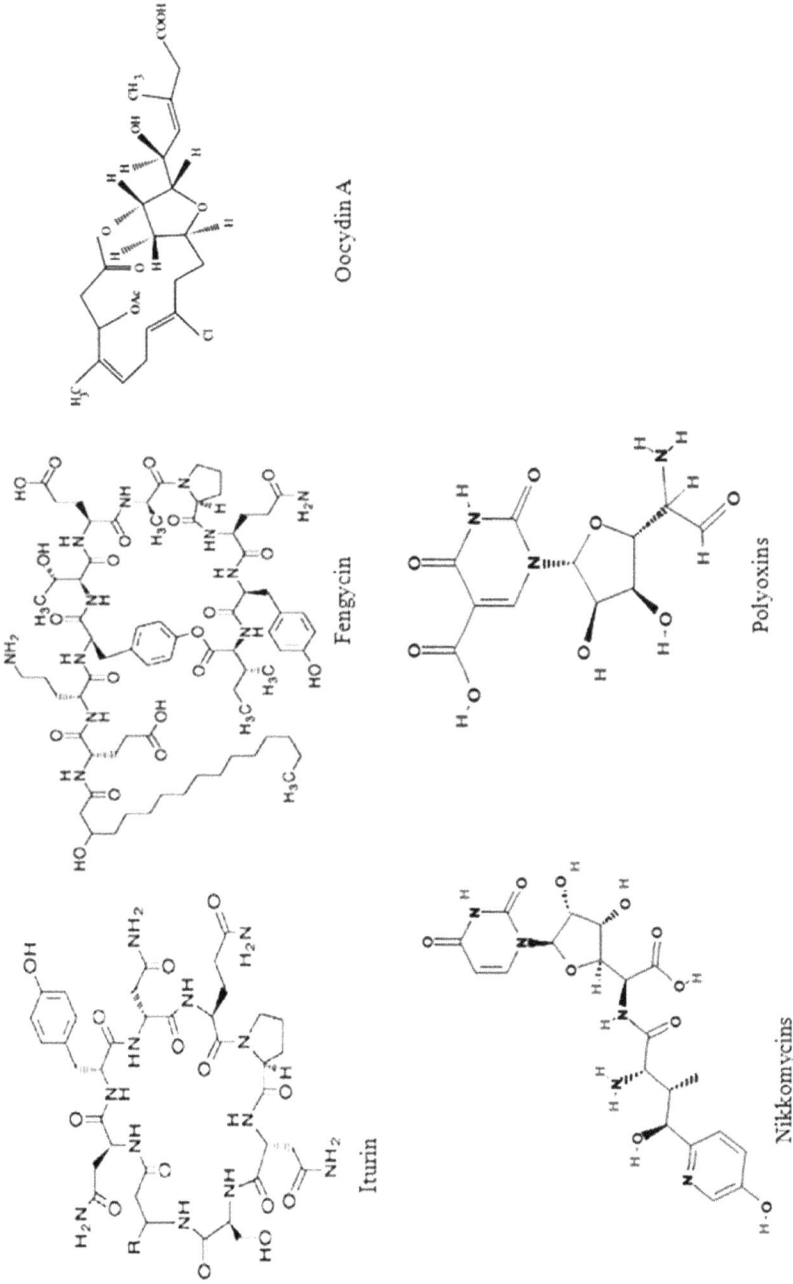

FIGURE 1.3 Structures of some antifungal metabolites obtained from endophytic bacteria.

TABLE 1.3   Some Antifungal Compounds Reported from Endophytic Microbes (Deshmukh & Verekar, 2012)

Sl. No.	Endophytic Microbes	Plant Source	Compounds Isolated
1	*Pestalotiopsis microspora*, *P. guepinii* and *Monochaetia* sp.	*Taxus baccata, Torreya taxifolia, Taxus wallichiana, Wollemia nobelis, Dendrobium speciosum*	Ambuic acid
2	*Pestalotiopsis microspora*	*Terminalia morobensis*	Pestacin, Isopestacin
3	*Pestalotiopsis jester*	*Fragraea bodenii*	Jesterone, Hydroxyjesterone
4	*Phomopsis* sp. ZSU-H76	*Excoecaria agallocha*	Cytosporone B, Cytosporone C
5	*Dothiorella* sp., strain HTF3	*Avicennia marina*	Cytosporone B
6	*Phomopsis* sp.	*Costus* sp.	Phomoxanthone A
7	*Phomopsis* sp.	*Adenocarpus foliolosus*	Phomosine A–G, 6-hydroxy-6-isopropylcyclohex-1enecarboxylic acid, 1aS,3R,4R,4aR,6S,7R,8aS)-7-chloro-3,6-dihydroxy-3,4a,8,8- tetramethyl-octahydro-1aH naphtho (1-b) oxirene-4 carboxylic acid, 5-methylmellein, 4-hydroxy-5-methylmellein, 2-quinazolin-4(3H)-one, Alternariol
8	*Phomopsis* sp. YM 311483	*Azadirachta indica*	Five lactones
9	*Phomopsis* sp.	*Erythrina crista-galli*	Phomol
10	*Phomopsis cassiae*	*Cassia spectabilis*	Ethyl 2,4-dihydroxy-5,6-dimethylbenzoate, Phomopsilactone
11	*Phomopsis* sp.	*Laurus azorica*	Cycloepoxylactone
12	*Phomopis cassia*	*Cassia spectabilis*	Two distereoisomer of 3,9,12-trihydroxycalamenenen, 3,12-dihydroxycalamenene, 3,12-dihydroxycadalene, 3,11,12-trihydroxycadalene
13	*Phoma* sp.	*Saurauia scaberrinae*	Phomodione 18
14	*Phoma* sp	*Lycium intricatum*	Pyrenophorol, dihydropyrenophorin, 4-acetylpyrenophorol, 4-acetyldihydropyrenophorin, *Cis*-dihydropyrenophorin, Tetrahydropyrenophorin, *Seco*-dihydropyrenophorin, 7-acetyl *seco*-dihydropyrenophorin, *seco*-dihydropyrenophorin-1,4-lactone

TABLE 1.3 *(Continued)*

Sl. No.	Endophytic Microbes	Plant Source	Compounds Isolated
15	*Morinia longiappendiculata*	*Santolina rosmarinifolia, Helichrysum stoechas, Calluna vulgaris, Thymus mastichina*	Moriniafungin
16	*Colletotrichum gloeosporioides*	*Artemisia mongolica*	Colletotric acid
17	*Colletotrichum gloeosporioides*	*Cryptocarya mandioccana*	*Cis*-4-hydroxy-6-deoxyscytalone, (4R)-4,8-dihydroxy-alpha-tetralone
18	*Exophiala* sp.	*Adenocarpus foliolosus*	Exochromone 24
19	*Cryptosporiopsis quercina*	*Tripterigeum wilfordii*	Cryptocandin A
20	*Cryptosporiopsis* sp	*Pinus sylvestris*	Echinocandins A, B, D, H
21	*Pezicula* sp.	*Fagus sylvatica*	Echinocandins A, B, D, H 27
22	*Cryptosporiopsis quercina*	*Phleum pratense*	Cryptocin
23	*Hormonema* sp. (ATCC 74360)	*Juniperus communis*	Enfumafungin
24	*Epichloe typhina*	*Phleum pratense*	Epichlicin
26	*Botryosphaeria rhodina*	*Bidens pilosa*	Botryorhodine A, B, C, D
27	*Dinemasporium strigosum*	*Calystegia sepium*	Dinemasone A, B, C
28	*Chaetomium globosum*	*Ginkgo biloba*	Chaetoglobosin A, C
29	*Xylaria* sp. PSU-D14	*Garcinia dulcis*	Sordaricin
30	*Xylaria* sp. F0010	*Abies holophylla*	Griseofulvin, 7,dechlorogriseofulvin
31	*Xylaria* sp.	*Palicourea marcgravii*	2-Hexyl-3-methylbutanodioic acid, Cytochalasin D
32	*Xylaria* sp.	*Ginkgo biloba*	7-Amino-4-methylcoumarin
33	*Penicillium paxilli* PSU-A71	*Garcinia atroviridis*	Penicillone, Pyrenocine A, Pyrenocine B
34	*Verticillium* sp.	*Rehmannia glutinosa*	2,6-dihydroxy-2-methyl-7-(prop-1*E*-enyl)-1-benzofuran-3 (2*H*)-one, Massariphenone, Ergosterol peroxide
35	*Nigrospora* sp. YB-141	*Azadirachta indica*	Solanapyrone C, Solanapyrone N, Solanapyrone O, Nigrosporalactone, Phomalactone

TABLE 1.3   *(Continued)*

Sl. No.	Endophytic Microbes	Plant Source	Compounds Isolated
36	*Trichoderma harzianum*	*Llexcornuta* Lindl	Trichodermin
37	*Nodulisporium* sp.	*Erica arborea*	Nodulisporin D, E, F, (3S,4S,5R)-2,4,6-trimethyloct-6-ene-3,5-diol, 5-hydroxy-2-hydroxymethyl-4H-chromen-4-one, 3-(2,3-dihydroxyphenoxy)-butanoic acid, benzene 1,2,3-triol
38	*Aspergillus clavatus, Paecilomyces* sp.	*Taxus mairei, Torreya grandis*	Brefeldin A
40	*Cladosporium* sp.	*Quercus variabilis*	Brefeldin A
41	CR377 (*Fusarium* sp.)	*Selaginella pallescens*	CR377
42	*Aspergillus niger*	*Cynodon dactylon*	Rubrofusarin B, Fonsecinone A, Asperpyrone B, Aurasperone A
43	*Coniothyrium* sp	*Sideritis chamaedryfolia*	1-Hydroxy-5-methoxy-2-nitro-naphthalene, 1,5-dimethoxy-4-nitronaphthalene, 1-hydroxy-5-methoxy-2,4-dinitronaphthalene, 1,5-di-methoxy-4,8-dinitronaphthalene, 1-hydroxy-5- methoxynaphthalene
44	*Periconia* sp.	*Taxus cuspidata*	Periconicin A, Periconicin B
45	*Nodulisporium* sp.	*Juniperus cedre*	3-Hydroxy-1-(2,6 dihydroxyphenyl)butan-1-one, 1-(2,6-dihydroxyphenyl)butan-1-one, 1-(2-hydroxy-6-methoxyphenyl)butan-1-one, 5-hydroxy-2-methyl-4H-chromen-4-one, 1,8-dimethoxynaphthalene, Nodulisporin A, Nodulisporin B, Daldinol, Nodulisporin C, (4E,6E)-2,4,6-trimethylocta-4,6-dien-3-one
46	*Penicillium* sp.	*Hopea hainanensis*	Monomethylsulochrin, Rhizoctonic acid, Asperfumoid, Physcion, 7, 8-dimethyl-iso-alloxazine, 3,5-dichloro-panisiacid
47	*Curvularia* sp.	*Ocotea corymbosa*	2-Methyl-5-methoxy-benzopyran-4-one, (2'S) 2-(propan-2'-ol)-5-hydroxy-benzopyran-4-one
48	*Aspergillus fumigatus* CY018	*Cynodon dactylon*	Asperfumoid, Physcion, Fumigaclavine C, Fumitremorgin C, Helvolic acid

TABLE 1.3 *(Continued)*

Sl. No.	Endophytic Microbes	Plant Source	Compounds Isolated
49	*Aspergillus clavatonanicus*	*Torreya mairei*	Clavatol, Patulin
50	*Colletotrichum* sp.	*Artemisia annua*	3-Beta,5-alpha-dihydroxy-6-beta-acetoxy-ergosta-7,22 dien, 3-beta,5-alphadihydroxy-6-beta-phenylacetyloxy-ergosta-7,22-diene, 3-beta-hydroxy, 5-alpha, 8-alpha-epidioxy-ergosta-6,22-diene
51	*Eupenicillium brefeldianum*	*Arisaema erubescens*	Brefeldin A
52	*Aspergillus niger EN-13*	*Colpomenia sinuosa*	Asperamide A, Asperamide B
53	*Aspergillus niger EN-13*	*Colpomenia sinuosa*	5,7-Dihydroxy-2-(1-(4-methoxy-6-oxo-6H-pyran-2-yl)-2-phenylethylamino)-(1,4) naphthoquinone
54	*Microdochium bolleyi*	*Fagonia cretica*	(12R)-12-hydroxymonocerin, (12S)-12-hydroxymonocerin, (3R,4R,10R)-4 (2–4) (4), Monocerin
55	*Neoplaconema napellum* IFBE016	*Hopea hainanensis*	Neoplaether
56	*Chalara* sp. (strain 6661)	*Artemisia vulgaris*	Isofusidienol A–D
57	*Blennoria* sp.	*Carpobrotus edulis*	Blennolide A–G, Secalonic acid B
58	*Periconia atropurpurea*	*Xylopia aromatica*	Periconicin B, 6,8-dimethoxy-3-(2′-oxo-propyl)-coumarin and 2,4-dihydroxy-6-((1′E,3′E)-penta-1′,3′-dienyl)-benzaldehyde
59	F-042,833, Undetermined Coelomycetes	*Olea europea var europea*	Arundifungin
60	F054,289, sterile mycelium	*Quercu silex*	Arundifungin
61	*Mycelia sterila*	*Cirsium arvense*	6,8-Diacetoxy-3,5-dimethyl isocoumarin, 3-acetyl-6-hydroxy-4-methyl-2,3-dihydrobenzofuran
62	*Serratia marcescens*	*Rhyncholacis penicillata*	Oocydin A
63	*Paenibacillus polymyxa*	Wheat	Fusaricidin A–D

synthesis are "nikkomycins" and "polyoxins" (Gabib, 1991). Both belong to a family of peptide-nucleoside antimycotic agents that were isolated from two different *Streptomyces* species: *S. tendae* (nikkomycin) and *S. cacaoi* var. *asoensis* (polyoxin). "Nikkomycins" exhibit activity against dimorphic fungi and are fungicidal for several clinically important fungi but show low activity against yeast and filamentous fungi (Li and Rinaldi, 1999).

1.4   CONCLUSION

There is a persistent battle between pathogens and drugs and an emerging need to discover novel antibiotics against the pathogenic microorganisms, particularly the rapidly developing drug resistant strains. The critical first step in discovering novel bioactive compounds is finding out suitable source material. Biological diversity often translates into molecular diversity, increasing the possibility of isolating new chemical entities. The ultimate aim of bioprospecting for novel compounds is to isolate compounds which are safe and efficacious for human use. Efficient screening mechanisms are crucial for targeting potential bioactive compounds. Prior knowledge of biosynthesis greatly assists in de-replicating the plethora of compounds produced by a single microorganism. Structure elucidation of the isolated chemicals and characterization of their biosynthetic pathways provides a basis for these novel compounds to be investigated in clinical trials and for commercial purposes. Utilizing traditional knowledge by studying plants that have been used to treat symptoms of disease may assist in narrowing down the plants as targets for investigating the production of novel antimycobacterial compounds. However, indiscriminate exploitation of these plant resources has rapidly declined their natural populations and threatened their existence. It has been well established now that plants serve as a vast reservoir of some untold number of microbes known as endophytes, which are defined as microbes that colonize inner healthy plant tissues without causing any disease symptoms. Some of these endophytes have produced important bioactive metabolites for wide therapeutic applications. Currently, there is growing interest to study microbial endophytes harbored in medicinal plants as many of these endophytic microbes have been reported to produce bioactive molecules similar to their respective hosts. Furthermore, based on the fact that many plant bioactive compounds are actually produced by their microbial symbionts, exploring the endophytes from these medicinal plants will assist in isolating and producing their active components. Invasive, life-threatening fungal infections are an important cause of morbidity and mortality, particularly for patients with compromised immune function. Therefore, global burden of fungal disease is significant. The therapeutic options for invasive fungal infections are quite limited. Thus, there is an urgent need for effective antifungal agents to combat the invasive fungal infections. In this context, endophytic microbes could be good candidate for obtaining antifungal metabolites to solve growing invasive fungal infection. This assumption becomes more significant considering the myriads of medicinal plants with antifungal activities, and few of these plants have been investigated from endophytic microbes so far.

KEYWORDS

- **endophytes microbes**
- **medicinal plants**
- **secondary metabolites**
- **antifungal drugs**

REFERENCES

Aravind, R.; Kumar, A.; Eapen, S. J.; Ramana, K. V. Endophytic bacterial flora in root and stem tissues of black pepper (*Piper nigrum* L.) genotype: isolation, identification and evaluation against *Phytophthora capsici. Lett Appl Microbiol.* **2009**, *48*, 58–64.

Asraful Islam, S. M.; Math, R. K.; Kim, J. M.; Yun, M. G.; Cho, J. J.; Kim, E. J.; Lee, Y. H.; Yun, H. D. Effect of plant age on endophytic bacterial diversity of balloon flower (*Platycodon grandiflorum*) root and their antimicrobial activities. *Curr Microbiol.* **2010**, *61*, 346–356.

Beck, H. C.; Hansen, A. M; Lauritsen, F. R. Novel pyrazine metabolites found in polymyxin biosynthesis by *Paenibacillus polymyxa. FEMS Microbiol Lett.* **2003**, *220*, 67–73.

Berg, G.; Krechel, A.; Ditz, M.; Sikora, R. A.; Ulrich, A.; Hallmann, J. Endophytic and ectophytic potato-associated bacterial communities differ in structure and antagonistic function against plantpathogenic fungi. *FEMS Microbiol Eco.* **2005**, *51*, 215–229.

Borges, W. S.; Borges, K. B.; Bonato, P. S.; Said, S.; Pupo, M. T. Endophytic fungi: natural products, enzymes and biotransformation. *Curr Org Chem.* **2009**, *13*, 1137–1163.

Brady, S. F. and Clardy, J. CR377, a new pentaketide antifungal agent isolated from an endophytic fungus. *J Nat Prod.* **2000**, *63*(10), 1447–1448.

Cabello, M. A.; Platas, G.; Collado, J.; Dıez, M. T.; Martin, I.; Vicente, F.; Meinz, M. et al. Arundifungin, a novel antifungal compound produced by fungi: biological activity and taxonomy of the producing organisms. *Int Microbiol.* **2001**, *4*, 93–102.

Canova, S. P.; Petta, T.; Reyes, L. F.; Zucchi, T. D.; Moraes, L. A. B.; Melo, I. S. Characterization of lipopeptides from *Paenibacillus* sp. (IIRAC30) suppressing *Rhizoctonia solani*. *World J Microbiol Biotechnol.* **2010**, *26*, 2241–2247.

Carroll, G. Fungal endophytes in stems and leaves: from latent pathogens to mutualistic symbionts. **1988**, *Ecology. 69*, 2–9.

Cho, K. M.; Hong, S. Y.; Lee, S. M.; Kim, Y. H.; Kahng, G. G.; Lim, Y. P.; Kim, H.; Yun, H. D. Endophytic bacterial communities in ginseng and their antifungal activity against pathogens. *Microb Ecol.* **2000**, *54*, 341–351.

Deshmukh, S. K. and Verekar, S. A. *Fungal endophytes: a potential source of antifungal compounds. Front Biosci* (Elite Ed). **2012**, *4*, 2045–2070.

Ding, G.; Liu, S.; Guo, L.; Zhou, Y.; Che, Y. Antifungal metabolites from the plant endophytic fungus *Pestalotiopsis foedan*. *J Nat Prod.* **2008**, *71*, 615–618.

Gabib, E. Differential inhibition of chitin synthetases 1 and 2 from *Saccharomyces cerevisiae* by polyoxin D and nikkomycins. *Antimicrob Agents Chemother.* **1991**, *35*, 170–173.

Gorman, J. A.; Chang, L. P.; Clark, J.; Gustavson, D. R.; Lam, K. S.; Mamber, S. W.; Pirnik, D.; Ricca, C.; Fernandes, P. B.; O'Sullivan, J. Ascosteroside, a new antifungal agent from *Ascotricha amphitricha*. I. Taxonomy, fermentation and biological activities. *J Antibiot.* **1996**, *40*, 547–552.

Gunatilaka, A. A. L. Natural products from plant-associated microorganisms: distribution, structural diversity, bioactivity, and implications of their occurrence. *J Nat Prod.* **2006**, *69*(3), 509–526.

Guo, B.; Wang, Y.; Sun, X.; Tang, K. Bioactive natural products from endophytes: a review. *Appl Biochem Microbiol.* **2008**, *44*, 136–142.

Harper, J. K.; Arif, A. M.; Ford, E. J.; Strobel, G. A.; Porco, J. A.; Tomer, D. P.; Oneill, K. L.; Heider, E. M.; Grant, D. M. Pestacin: a 1,3-dihydroisobenzofuran from *Pestalotiopsis microspora* possessing antioxidant and antimycotic activities. *Tetrahedron.* **2003**, *59*, 2471–2476.

Harrison, L.; Teplow, D. B.; Rinaldi, M.; Strobel, G. Pseudomycins, a family of novel peptides from *Pseudomonas syringae* possessing broad-spectrum antifungal activity. *J Gen Microbiol.* **1991**, *137*(12), 2857–2865.

Hoberg, K. A.; Cihlar, R. L.; Calderone, R. A. Characterization of cerulenin-resistant mutants of *Candida albicans*. *Infect Immun.* **1986**, *51*, 102–109.

Hong, C. E.; Jeong, H.; Jo, S. H.; Jeong, J. C.; Kwon, S. Y.; An, D.; Park, J. M. A leaf-inhabiting endophytic bacterium, *Rhodococcus* sp. KB6, enhances sweet potato resistance to black rot disease caused by *Ceratocystis fimbriata*. *J Microbiol Biotechnol.* **2016a**, *26*, 488–492.

Hong, C. E.; Jo, S. H.; Moon, J. Y.; Lee, J. S.; Kwon, S. Y.; Park, J. M. Isolation of novel leaf-inhabiting endophytic bacteria in *Arabidopsis thaliana* and their antagonistic effects on phytophathogens. *Plant Biotechnol Rep.* **2015**, *9*, 451–458.

Hong, C. E.; Kwon, S. Y.; Park, J. M. Biocontrol activity of *Paenibacillus polymyxa* AC-1 against *Pseudomonas syringae* and its interaction with *Arabidopsis thaliana*. *Microbiol Res.* **2016b**, *185*, 13–21.

Huang, Z.; Cai, X.; Shao, C.; She, Z.; Xia, X.; Chen, Y.; Yang, J.; Zhou, Z.; Lin, Y. Chemistry and weak antimicrobial activities of phomopsins produced by mangrove endophytic fungus *Phomopsis* sp. ZSU-H76. *Phytochemistry.* **2008**, *69* (7), 1604–1608.

Hussain, H.; Krohn, K.; Draeger, S.; Schulz, B. Exochromone: structurally unique chromone dimer with antifungal and algicidal activity from *Exophiala* sp. *Heterocycles.* **2007**, *74*, 331–337.

Kopcke, B.; Weber, R. W. S.; Anke, H. Galiellalactone and its biogenetic precursors as chemotaxonomic markers of the Sarcosomataceae (Ascomycota). *Phytochemistry.* **2002**, *60*, 709–714.

Korpi, A.; Jarnberg, J.; Pasanen, A. L. Microbial volatile organic compounds. *Crit Rev Toxicol.* **2009**, *39*, 139–193.

Krohn, K.; Kouam, S. F.; Cludius-Brandt, S.; Draeger, S.; Schulz, B. Bioactive nitronaphthalenes from an endophytic fungus, *Coniothyrium* sp., and their chemical synthesis. *Eur J Org Chem.* **2008**, 3615–3618.

Li, E.; Jiang, L.; Guo, L.; Zhang, H.; Che, Y. Pestalachlorides A-C, antifungal metabolites from the plant endophytic fungus *Pestalotiopsis adusta. Bioorg Med Chem.* **2008**, *16* (17), 7894–7899.

Li, J. Y.; Strobel, G. A. Jesterone and hydroxy-jesterone anti-oomycete cyclohexenenone epoxides from the endophytic fungus *Pestalotiopsis jesteri. Phytochemistry.* **2001**, *57*, 261–265.

Li, R. K.; Rinaldi, M. G. In vitro antifungal activity of nikkomycin Z in combination with fluconazole or itraconazole. *Antimicrob Agents Chemother.* **1999**, *43*, 1401–1405.

Liu, B.; Qiao, H.; Huang, L.; Buchenauer, H.; Han, Q.; Kang, Z.; Gong, Y. Biological control of take-all in wheat by endophytic *Bacillus subtilis* E1R-j and potential mode of action. *Biol Control.* **2009**, *49*, 277–285.

Liu, L.; Liu, S.; Chen, X.; Guo, L.; Che, Y. Pestalofones A-E, bioactive cyclohexanone derivatives from the plant endophytic fungus *Pestalotiopsis fici. Bioorg Med Chem.* **2009**, *17*(2), 606–613.

Ma, C. and Armstrong, A. W. Pharmacology of fungal infections. *Principles of Pharmacology: The Pathophysiologic Basis of Drug Therapy*, 4th ed. David E. Golan (Ed). Philadelphia, PA. Wolters Kluwer. **2017**, 661–673.

Ma, L.; Cao, Y. H.; Cheng, M. H.; Huang, Y.; Mo, M. H.; Wang, Y.; Yang, J. Z.; Yang, F. X. Phylogenetic diversity of bacterial endophytes of *Panax notoginseng* with antagonistic characteristics towards pathogens of root-rot disease complex. *Antonie Van Leeuwenhoek.* **2013**, *103*, 299–312.

Mayer, A. M. S.; Rodriguez, A. D.; Berlinck, R. G. S.; Fusetani, N. Marine pharmacology in 2007–2008: marine compounds with antibacterial, anticoagulant, antifungal, anti-inflammatory, antimalarial, antiprotozoal, antituberculosis, and antiviral activities; affecting the immune and nervous system, and other miscellaneous mechanisms of action. *Comp Biochem Physiol C.* **2011**, *153*, 191–222.

Miller, C. M.; Miller, R. V.; Garton-Kenny, D.; Redgrave, B.; Sears, J.; Condron, M. M.; Teplow, D. B.; Strobel, G. A. Ecomycins, unique antimycotics from *Pseudomonas viridiflava. J Appl Microbiol.* **1998**, *84*, 937–944.

Murray, F. R.; Latch, G. C. M.; Scott, D. B. Surrogate transformation of perennial rye grass, *Lolium perenne*, using genetically modified *Acremonium* endophyte. *Mol Gen Genet.* **1992**, *233*, 1–9.

National Institutes of Health. *NIAID global health research plan for HIV/AIDS, malaria and tuberculosis.* Bethesda, MD: U.S. Department of Health and Human Services, **2001**.

Onishi, J.; Meinz, M.; Thompson, J.; Curotto, J.; Dreikorn, S.; Rosenbach, M.; Douglas, C. et al. Discovery of novel antifungal (1,3)-b-D-glucan synthase inhibitors. *Antimicr Agents Chemother.* **2000**, *44*, 368–377.

Pappas, P. G.; Kauffman, C. A.; Andes, D. R. Clinical Practice Guideline for the Management of Candidiasis: 2016 Update by the Infectious Diseases Society of America. *Clin Infect Dis.* **2016**, *62*(4), e1–e50.

Perlin, D. S. Resistance to echinocandin-class antifungal drugs. *Drug Resist Updates.* **2007**, *10*, 121–130.

Petrini, O. Ecology, metabolite production and substrate utilization in endophytic fungi. *Nat Toxins.* **1992**, *1*, 185–196.

Polizzi, V.; Adams, A.; Malysheva, S. V.; De Saeger, S.; Van Peteghem, C.; Moretti A.; Picco A. M.; De Kimpe, N. Identification of volatile markers for indoor fungal growth and chemotaxonomic classification of *Aspergillus* species. *Fungal Biol.* **2012**, *116*, 941–953.

Rabkin, J. M.; Oroloff, S. L.; Corless, C. L.; Benner, K. G.; Flora, K. D.; Rosen, H. R.; Olyael, A. J. Association of fungal infection and increased mortality in liver transplant recipients. *Am J Surg.* **2000**, *179*, 426–430.

Raupach, G. S. and Kloepper, J. W. Mixtures of plant growth promoting rhizobacteria enhance biological control of multiple cucumber pathogens. *Phytopathology.* **1998**, *88*, 1158–1164.

Richardson, M. R. Changing patterns and trends in systemic fungal infections. *J. Antimicrob Chemother.* **2005**, *56*, S5–S11.

Saikkonen, K.; Faeth, S. H.; Helader, M.; Sullivan, T. J. Fungal endophytes: a continuum of interactions with host plants. *Annual Review of Ecology and Systematics.* **1998**, *29*, 319–343.

Selvin, J.; Ninawe, A. S.; Kiran, G. S.; Lipton, A. P. Sponge-microbial interactions: ecological implications and bioprospecting avenues. *Crit Rev Microbiol.* **2010**, *36*, 82–90.

Silva, G. H.; Teles, H. L.; Zanardi, L. M.; Young, M. C. M.; Eberlin, M. N.; Hadad, R.; Pfenning, L. H.; Costa-Neto, C. M.; Castro-Gamboa, I.; da Silva Bolzani, V.; Araújo, Â. R. Cadinane sesquiterpenoids of *Phomopsis cassiae*, an endophytic fungus associated with *Cassia spectabilis* (Leguminosae). *Phytochemistry.* **2006**, *67* (17), 1964–1969.

Staniek, A.; Woerdenbag, H. J.; Kayser, O. Endophytes: exploiting biodiversity for the improvement of natural product-based drug discovery. *J. Plant Interact.* **2008**, *3*, 75–93.

Strobel, G. A. Rainforest endophytes and bioactive products. *Crit Rev Biotechnol.* **2002**, *22*, 325–333.

Strobel, G. A.; Daisy, B. Bioprospecting for microbial endophytes and their natural products. *Microbiol Mol Bio Rev.* **2003**, *67*, 491–502.

Strobel, G. A.; Daisy, B.; Castillo, U.; Harper, J. Natural products from endophytic fungi. *J Nat Prod.* **2004**, *67*, 257–268.

Strobel, G. A.; Yang, X.; Sears, J.; Kramer, R.; Sidhu, R. and Hess, W. M. Taxol from *Pestalotiopsis microspora*, an endophytic fungus of *Taxus wallachiana*. *Microbiology.* **1996**, *142*, 435–440.

Sturz, A. V.; Christie, B. R.; Nowak, J. Bacterial endophytes: potential role in developing sustainable systems of crop production. *Crit Rev Plant Sci.* **2000**, *19*, 1–30.

Tan, R. X.; Zou, W. X. Endophytes: a rich source of functional metabolites. *Nat Prod Rep.* **2001**, *18*, 448–459.

Verma, S. C.; Ladha, J. K.; Tripathi, A. K. Evaluation of plant growth promoting and colonization ability of endophytic diazotrophs from deepwater rice. *J Biotechnol.* **2001**, *91*, 127–141.

Vijayakumar, E. K. S.; Roy, K.; Chatterjee, S.; Deshmukh, S. K.; Ganguli, B. N.; Kogler, H.; Fehlhaber, H. W. Arthrichitin a new cell wall active metabolite from *Arthrinium phaeospermum*. *J Org Chem.* **1996**, *61*, 6591–6593.

Wang, H.; Wen, K.; Zhao, X.; Wang, X.; Li, A.; Hong, H. The inhibitory activity of endophytic *Bacillus* sp. Strain CHM1 against plant pathogenic fungi and its plant growth-promoting effect. *Crop Prot.* **2009**, *28*, 634–639.

Wang, J.; Huang, Y.; Fang, M.; Zhang, Y.; Zheng, Z.; Zhao, Y.; Su, W. Brefeldin A, a cytotoxin produced by *Paecilomyces* sp. and *Aspergillus clavatus* isolated from *Torreya grandis*. *FEMS Immunol Med Microbiol.* **2002**, *34*, 51–57.

Wang, X. N.; Bashyal, B. P.; Wijeratne, E. M. K.; Uren, J. M.; Liu, M. X.; Gunatilaka, M. K.; Arnold, A. E.; Gunatilaka A. A. L. Smardaesidins A–G, isopimarane and 20-nor-Isopimarane diterpenoids from *Smardaea* sp., a fungal endophyte of the moss *Ceratodon purpureus*. *J Nat Prod*. **2011**, *74*, 2052–2061.

Weber, R. W. S.; Kappe, R.; Paululat, T.; Mosker, E.; Anke, H. Anti-*Candida* metabolites from endophytic fungi. *Phytochemistry*. **2007**, *68*, 886–892.

Zhang, Y. Z.; Sun, X.; Zeckner, D. J.; Sachs, R. K.; Current, W. L.; Gidda, J.; Rodriguez, M.; Chen, S. H. Synthesis and anti-fungal activities of 3-amido bearing pseudomycin analogs. *Bioorg Med Chem Lett*. **2001**, *11*, 903–907.

## CHAPTER 2

Bioactive Compounds and Herbal Therapy for Alzheimer's Disease

JAYASHREE PRUSTY and SRIKANTA JENA*

Department of Zoology, School of Life Sciences, Ravenshaw University, Cuttack 753003, Odisha, India

Corresponding author. E-mail: jenasrikanta@yahoo.co.in.

ABSTRACT

Aging is a natural biological phenomenon that comes with progressive and deleterious changes which makes the individual younger to older. Elderly people are frequently come across with many neurological disorders. Loss of neurons and synapses as a consequence of aging are the most important factors for neurodegenerative diseases. Among age-related neurodegenerative diseases Alzheimer's disease (AD) is the most prevalent one. It occurs due to aggregations of toxic proteins called amyloid-β plaques in the brain. Many ethno-pharmacologically important plants and plant products are used for their therapeutic properties to cure many diseases including neurodegenerative diseases. Bioactive compounds are the active ingredients extracted from biological sources which may have extra-nutritious values and therapeutic properties such as anti-inflammatory, anticancer, antioxidant, and anti-aging. Despite the beneficial pleiotropic effects of bioactive compounds and herbal products, they have also crucial role for neuroprotection. The manifestation of neuronal disorders particularly the age-associated diseases are more prevalence when the aging process is concomitant with other patho-physiological states such as inflammatory, hypertensive and oxidative stress conditions, and also due to genetic and many environmental factors. Although the causes of these diseases are not fully understood, the conditions may be influenced by synergistic effects of multifactorial states during elderly age. Based on these facts, the medicinal herbs and bioactive compounds having anti-aging and neuroprotective effects may be useful in a holistic approach for the treatment of age-associated diseases. However, this article targets to explore the wide range of plants, plant products and bioactive compounds for their neuroprotective properties and therapeutic values particularly for AD.

2.1   INTRODUCTION

The commencement of the age-related dementia in Alzheimer's disease (AD) is manifested with progressive neurodegeneration of the central nervous system (CNS) that leads to cognitive dysfunction and impaired memory (Bali et al., 2017; WHO, 2018). With the progression of the quality of the public health and medical care, the average life expectancy increases, so that the prevalence of AD prolonged and consistent with age. However, one of the interesting studies reveals that the prevalence of AD in the aged people of 70–79 years in United States is 4.4-fold more than India (Ganguli et al., 2000). The World Health Organization (WHO) points out that approximately 50 million people are suffering from dementia, and an additional 10 million new cases may rise in each year. Dementia is observed in 5%–8% of people who are more than 60 years of age and is one amongst the several reasons of AD, which is found in 60%–70% of cases. Further, it has been estimated that the prevalence of AD in men and that in women are about 0.4% and 0.3%, respectively, at the age of 60–69 years. This rises by 10% in men and 11.2% in women over 80 years of age. It is also estimated that the dementia cases will increase threefold from the existing number of affected people and predicted to be approximately 82 and 152 million by 2030 and 2050, respectively (Islam et al., 2017; WHO, 2018).

Alzheimer's disease is usually confused with dementia. AD is a pathophysiological condition in old age, whereas dementia is considered as a group of symptoms that lead to memory impairment and negatively impact thinking, recognition, and other several abilities (Figure 2.1). Thus, dementia is not a disease, rather a substitute term for the collective symptoms of impaired memory, confusion, poor judgmental capability, etc. Several neurological diseases and disorders are related to dementia such as symptoms of Parkinson's disease, normal pressure hydrocephalus, Wernicke–Korsakoff syndrome, and Creutzfeldt–Jakob disease (Botchway and Iyer, 2017). However, neurodegeneration and dementia are usually associated with AD in the elderly, and it damages selectively the brain tissues in particular regions of critical neural circuits of brain for memory and cognitive functions (Islam et al., 2017). Nearly a century before, most clinical and neuropathological advances have been characterized in AD. In 1907, Alois Alzheimer, a German psychiatrist, explained that the development of amyloid plaques, intracellular neurofibrillary tangles (NFTs) and neuronal loss which are the characteristic progressions of dementia; after his name, this diseased condition is called as Alzheimer's disease (1907). Several studies conducted to identify the two insoluble aggregates of peptides called amyloid-β (Aβ) and tau, which are responsible for the formation of amyloid plaques and NFTs, respectively (Jack, Jr. and Holtzman, 2013; Spires-Jones and Hyman, 2014). Certain metal ions such as aluminum (Al^{3+}), copper (Cu^{2+}), zinc (Zn^{2+}), mercury (Hg^{2+}), and iron (Fe^{2+}) are also roots in the progression of AD by mediating neurodegeneration. Another important reason of AD is due to age-related excess production of reactive oxygen species (ROS) and progressive reduction in cellular antioxidant defense mechanisms leading to oxidative stress (OS)-induced neurodegeneration. Mitochondrial impairment is another reason behind excess ROS production and onset of AD. Under such altered pathophysiological conditions, neuroinflammation is a basic outcome in CNS, which is the foremost factor that brings about the progression of AD (Heneka et al., 2015).

The current status on the availability of conventional treatment for AD includes the acetylcholinesterase (AChE) and butyrylcholinesterase (BuChE) inhibitors (rivastigmine, galantamine, donepenzil, and tacrine) and

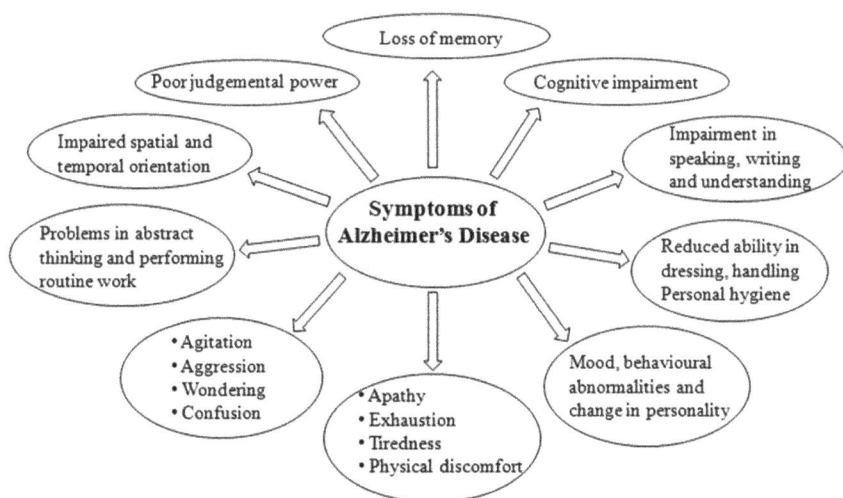

FIGURE 2.1 Symptoms of Alzheimer's disease.

the *n*-methyl-ᴅ-aspartate receptor antagonist (memantine) for the amelioration of the disease (Anand et al., 2014; Huang and Mucke, 2012). Studies
also have shown that the risk of the development of AD in postmenopausal
women is reduced on oestrogen replacement therapy. Nonsteroidal anti-
inflammatory drugs (NSAIDs) such as aspirin and ibuprofen also help in
reducing the developing AD (Anand et al., 2014; Huang and Mucke, 2012).
There are alternative strategies followed for the treatment of AD in the recent
period to improve cognition in older age and possibly delay the onset of
dementia. One of the major strategies is the approach of herbal therapy that
could be a novel pathway in the management of AD. Since 4000 years, in the
traditional Indian Ayurvedic system, several medicinal plants have been used
toward the treatment of various neurological disorders of CNS, including
memory impairment and cognitive function (Rao et al., 2012; Warrier et al.,
1993–1996). As OS is another major reason behind AD, treating the patients
with antioxidant therapy has proven to be helpful in ameliorating their
cognitive deficits (Gutzmann and Hadler, 1998). Herbal products are the
prime source of antioxidants, and the use of this phyto-antioxidant improves
cognition in elderly and possibly delays the onset of dementia. The nature
of multifactorial manifestation of AD proposed the multitargeted therapeutic
approach that might be advantageous than solitary drug treatment (Bolognesi
et al., 2009).

2.2 PATHOPHYSIOLOGY OF ALZHEIMER'S DISEASE

AD is positioned as the 7th disease leading to death in the world. Primarily
it is allied with the formation of two kinds of alternations in the brain tissues
identified as "plaques" and "tangles" (Tiraboschi et al., 2004). The formation
as well as aggregation of the abnormal β-amyloid protein between neurons
initiates the development of plaques. NFTs are composed of the insoluble tau
protein in the cortex region and further spread to the hippocampus, which
is basically responsible in forming memories. Several reports gathered
on AD over the years indicated that primarily aging with environmental
and hereditary factors is responsible in the pathogenesis of AD. The most
common factors associated with aging that frequently causes the disease are
mitochondrial dysfunction, OS, inflammation, protein glycation, metal ions,
tau protein missorting, and apolipoprotein E (ApoE) (Heppner et al., 2015;
Islam et al., 2017), Aging has the ultimate influence on the development of
AD that is further enhanced with these pathogenic factors (Figure 2.2).

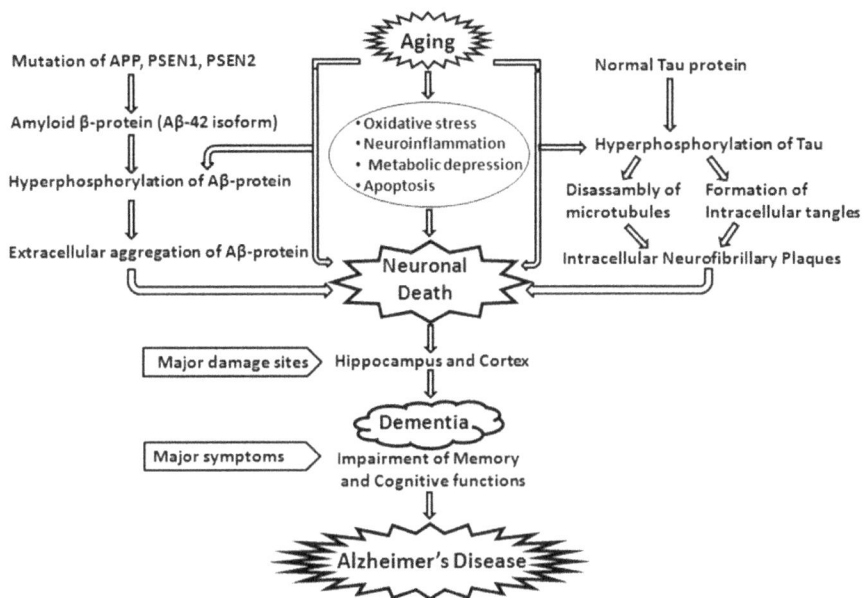

FIGURE 2.2   Causes of Alzheimer's disease.

2.2.1   *OXIDATIVE STRESS*

OS is a physiological state that is due to the altered redox status where the production of pro-oxidants is not maintained at a lower level by cellular antioxidants. Intracellular generation of free radicals (both ROS and RNS) are in excess during the advancement of aging and are considered as the foremost factor for the seeding of AD. OS precedes Aβ production and promotes its aggregation, and augments the polymerization of intracellular NFTs (Zhao and Zhao, 2013). Although there is no particular mechanism regarding alteration in redox homeostasis, but investigations on this proposed that there are several approaches to the excess generation of free radicals. These include expose to environmental toxins, pollutants, any pathological states, or deviation in the mitochondria-coupled physiochemical state. These factors overwhelm the capacity of the antioxidant and lead to oxidative damage to the biomolecules. Oxidized products of biomolecules, protein oxidation assessed by protein carbonyls and 3-nitrotyrosine, lipid peroxidation, including protein-bound marked as 4-hydroxy-2-trans-nonenal, and the oxidation of DNA and RNA were observed along with the occurrence of 8-hydroxy-2-deoxyguanosine and 8-hydroxyguanosine (Halliwell and Gutteridge 2007). These oxidized products of biomolecules may diffuse into the blood and have been considered as one of the biochemical markers for the diagnosis of AD along with other important markers (Skoumalová and Hort, 2012). Lipids play a crucial role in the structure and functions of the brain, and it has been reported that the enhanced lipid peroxidation levels in the brain tissues, cerebrospinal fluid (CSF), and plasma are linked to the OS and Aβ formation of AD patients (Butterfield and Boyd-Kimball, 2018). The formation of Aβ enhances intracellular ROS in hippocampal neurons (Harris et al., 1995) and in cortical synaptosomes (Kanski et al., 2001). Involvement of OS in AD furnishes the prospective of the supplementation of antioxidants in AD treatment.

2.2.2 *MITOCHONDRIAL DYSFUNCTION*

A mitochondrion is a power house of the cell where the maximum adenosine triphosphate (ATP) production occurs through oxidative phosphorylation by the electron transport chain (ETC) in the inner-mitochondrial membrane. In ETC, electrons pass through by the help of different carriers under redox regulation with the release of energy. Sometimes electrons are directly leaked to ETC that forms superoxide and subsequently causes OS (Halli-well and Gutteridge, 2007). It is reported that the cytochrome oxidase, a key enzyme in ETC, is reduced that promotes the increased production of ROS, thereby depleting the energy stores in AD patients (Mutisya et al., 1994). Further, Mn-superoxide dismutase (Mn-SOD), an antioxidant enzyme, is inactivated, and thus it leads to mitochondrial dysfunction, OS, and apoptosis in Aβ protein precursor/PS1 (APP/PS1) transgenic mice (Anantharaman et al., 2006). Emerging evidence indicates that Aβ perturbs ETC by decreasing the key enzyme activities and disrupting mitochondrial dynamics (Yan et al., 2013), dysfunction of mitochondrial axonal transport (Crouch et al., 2008), and mitochondrial DNA (mtDNA) mutation. Mito-chondrial dysfunction and its metabolic abnormalities have been recorded in hippocampal neurons of AD patients. Further reports depicted that the accumulation of the intracellular Aβ interferes with oxidative phosphory-lation and excess generation of ROS within mitochondria, results in the reduction in the activity of mitochondrial membrane potential, cytochrome *c* oxidase activity, and production of ATP (Hirai et al., 2001). Mitochondria, being the most active and energetic organelle, frequently endure fission and fusion reactions, and these abnormalities undergo a series of altered metabolism with enhanced free radical generation. However, fragmentation or fission is more vulnerable than fusion, and it is the potential cause behind the onset of AD by causing mitochondrial dysfunction associated with OS. Also, mutation in genes that encode mitochondrial elements as well as mutation in mtDNA is another reason behind mitochondrial dysfunction (Chen and Zhong, 2014).

2.2.3 METAL IONS

Metal ion exposure (i.e., Al^{2+}, Fe^{2+}, Cu^{2+}, Zn^{2+}, and Hg^{2+}) causes AD progression as these ions play crucial roles in the generation of free radicals. Studies indicated the possible role of iron metabolism involved in the pathogenesis of AD as the accumulation of metal ions takes place within the senile plaques of the brain of patients with AD (Robert et al., 2015), which results in the increase of OS. It has been observed that increased concentrations of Al^{3+} have also been seen in the brain tissues of AD patients (Bhattacharjee et al., 2014). Production of ROS and alteration of APP activity have been detected when Zn^{2+} binds to APP that leads to AD (Brewer, 2012). Aβ can bind Cu^{2+} with high affinity, thereby forming a cuproenzyme-like complex (Curtain et al., 2001). In this process, electrons transfer from Aβ to Cu^{2+} and convert Cu^{2+} to Cu^+, thereby forming the Aβ radical (Aβ·). In addition, Cu^+ can donate two electrons to oxygen, generating H_2O_2 (Chen and Zhong, 2014), and further producing hydroxyl radicals in a reaction called Fenton-type reaction (Lynch et al., 2000). Furthermore, Cu^{2+} has a role in the formation of ROS, and it has also a binding site in APP that further confirms the link between copper ion and AD progression (Hung, et al., 2010). As a whole, it seems that the aggregation of Aβ is generally stimulated by ROS-inducing metal ions such as Zn^{2+}, Cu^{2+}, and Fe^{2+}, which may elucidate the presence of a high concentration of metal ions found in the brain of AD patients (Sastre et al., 2015).

2.2.4 PROTEIN INVOLVEMENT

Pathophysiologically, AD is featured by the presence of the extracellular aggregation of the Aβ peptide (called Aβ plaques or senile plaque) and the accumulation of hyperphosphorylated tau protein that lead to NFTs. These unusual modified proteins cause neuronal and synaptic loss and neurotransmitter dysfunction (Huang and Mucke, 2012). Several studies have also addressed the conditions of neurons, microglia, astrocytes, and capillaries in pathogenesis and the development of senile plaques. Normally, Aβ plaques are composed of Aβ peptides surrounded by dystrophic neuritis and activated microglia (Anand et al., 2014). These extracellular deposits are the accumulation of 39–43 amino acid long products of Aβ peptides that occur in the gray matter of the brain in AD due to the sequential proteolytic processing of APP by the β-secretase and γ-secretase. After the sequencing of the gene for APP, it is identified that APP is a transmembrane protein found in almost all tissues of uncertain function. It was found that the proteolytic processing is somewhat hetero-geneous; as a result, variable lengths of Aβ are produced, particularly at the carboxyl terminus of the peptide. The peptides of Aβ are produced in two major forms of 40 and 42 amino-acid residues in length that is $A\beta_{40}$ and $A\beta_{42}$, respectively, under normal conditions of APP processing. The selective vulnerability of the brain to Aβ amyloidosis and its biological basis is not well understood; however, some clarity has been made on the levels of APP with the activities of the pro-amyloidogenic enzyme β-secretase β-amyloid precursor protein cleaving enzyme1 (BACE1), which may play the critical roles in the susceptibility of the CNS and in the pathogenesis of the disease of the elderly (Vassar et al., 2009). In a normal individual, the majority of the produced Aβ is of the shorter variety $A\beta_{40}$ than the longer variety $A\beta_{42}$. In response to a similar dose of inhibitors, the Aβ production study indicates that both $A\beta_{40}$ and $A\beta_{42}$ are produced by a single γ-secretase enzyme (Vassar, 2014). Other studies reveal that a rare familial AD develops below the age of 65 (<1% of all AD) due to autosomal-dominant AD that may arise as a result of APP mutation (Parks-Medina et al., 2016) or by the mutation of presenilin genes (PSEN1 and PSEN2) (Cruchaga et al., 2012). Another important cause of this disease is the involvement of the tau protein in AD. Tau is a microtubule-associated protein localized to the neuronal axon that becomes missorted in AD. Extracellular aggregation and hyperphosphorylation of Aβ peptides have been proposed to induce tau (microtubule-associated protein, localized in the neuronal axon) missorting in the brain of AD patients (Loewen and Feany, 2010). Phosphorylation of the tau protein takes place by several kinases including mitogen-activated protein kinase, glycogen synthase kinase 3, tau-tubulin kinase, and cyclin-dependent kinases, such as cyclin-dependent kinase 5 and stress-activated protein kinases, that leads to the formation of NFTs (O'Callaghan et al., 2014).

2.2.5 APOLIPOPROTEIN E

ApoE exists mainly as a component of lipoprotein (299 amino acids) complexes' highest expression in the liver followed by the brain. Three polymorphic forms of ApoE are apoE2 (Cys-112, Cys-158), apoE3 (Cys-112, Arg-158), and apoE4 (Arg-112, Arg-158). The position differences of these amino acids critically change the charge and structural properties of the protein, which ultimately affect the functional properties of ApoE isoforms in human (Mahley and Rall, 2000). In the brain, ApoE is mainly associated with astrocytes, followed by microglial cells and, under definite conditions, in neurons, whereas in CSF, it is predominantly associated with cholesterol and phospholipid-rich, high-density lipoprotein (HDL)-like complexes as CSF exclusively contain HDL-like lipoproteins and no low-density lipoprotein or very low density lipoprotein. However, there is no such evidence that ApoE isoforms differentially affect the CNS cholesterol or phospholipid metabolism, depicting that those structural differences in ApoE isoforms may influence neurological disorders via mechanisms that are not directly linked to isoform-specific regulation of lipid metabolism (Verghese et al., 2011). Besides progressive neurodegeneration associated with the extracellular deposition of Aβ plaques and intracellular NFTs, mutations in APP, PS1, and PS2 genes also cause rare forms of autosomal-dominant familial AD where the clinical onset of the disease occurs between the ages of 30 and 60 (Hardy and Selkoe, 2002). ApoE has been confirmed as one of the several susceptibility genes that influence AD risk, which confer the risk for sporadic, late onset AD (age >60 years), as well as autosomal-dominant familial AD (Pastor et al., 2003). There are three alleles of ApoE in human: ApoE ε2, ApoE ε3, and ApoE ε4. The most common form of ApoE in all populations is ApoE ε3 (in the range of 50%–90%), whereas ApoE ε4 and ApoE ε2 alleles are in the range of 5%–35% and 1%–5%, respectively (Mahley and Rall, 2000). However, approximately 50% of risk is associated with ApoE ε4 for AD followed by ApoE ε3 and ApoE ε2 in the development of late-onset AD compared with 20%–25% in controls. With one or two copies of ApoE ε4 allele the prevalence of late-onset AD risks is increased by approximately 3- or 12-fold, respectively. Furthermore, one or two copies of ApoE ε4 shift the age of AD onset earlier by approximately one to two decades in comparison to noncarriers in late-onset AD (Saunders et al., 2000).

2.3   AVAILABLE TREATMENTS AND CHALLENGES IN TREATMENT OF ALZHEIMER'S DISEASE

The exact diagnosis for the disease AD is very complex to manage as the cause of its occurrence is multifarious such as aging, familial history and individual genetic makeup, unbalanced metabolism, stressful lifestyle, and several other factors. Therefore, the best management is to be designed for individual patient according to their specific circumstances. The status of disease as early stage or advance stage dementia like symptoms is usually diagnosed by memory test, cognitive function test and counseling to family members and friends, or detection through neuro-imaging methods such as magnetic resonance imaging, computerized tomography scan, positron emission tomography scan and the detection of protein biomarkers in CSF. Current treatment approaches mainly focused on to maintaining the mental function, managing behavior and to slow or delay the symptomatic appearance of the disease. Several conventional medications are available to treat the symptoms of the AD by inhibiting the AChE or dual AChE–BuChE inhibitor with donepezil, galantamine or rivastigmine, respectively (Bartorelli et al., 2005). These drugs may be helpful in maintaining the memory, communication skills and improving certain behavioral problems including sleeplessness, agitation, anxiety and aggression. However, these drugs do not change the underlying cause of the disease. Furthermore, several bioactive compounds, polyphenols, herbal extracts, and food supplements have been assessed and established for their pharmacological effects, such as anti-inflammatory and antioxidant activities, and improve cognitive functions along with anti-Aβ activities that may be helpful in managing AD (Islam et al., 2017). Additionally, patients also need personal care, family support, and an enriched social environment that reduce the OS and progression of AD.

2.4   BIOACTIVE COMPOUNDS AND HERBAL THERAPY FOR ALZHEIMER'S DISEASE

Herbal medicines have several promising properties and functions to modify or delay the progress and symptoms of AD. Phytochemical studies of many herbal extracts have shown the valuable bioactive constituents, such as flavonoids, triterpenes, polyphenols, sterols, and alkaloids that show varieties of pharmacological activities, including hypolipidemic, anti-amyloidogenic, anticholinesterase, anti-inflammatory, and antioxidant effects. Currently, a new trend has been introduced in the manufacturing and marketing of drugs based on herbal medicine and their significance in health-relevant areas (Figure 2.3). These herbal products have been carefully standardized and demonstrated with their efficacy and safety for specific disease-based applications (Kennedy and Wightman, 2011; Kumar, 2006) (Table 2.1).

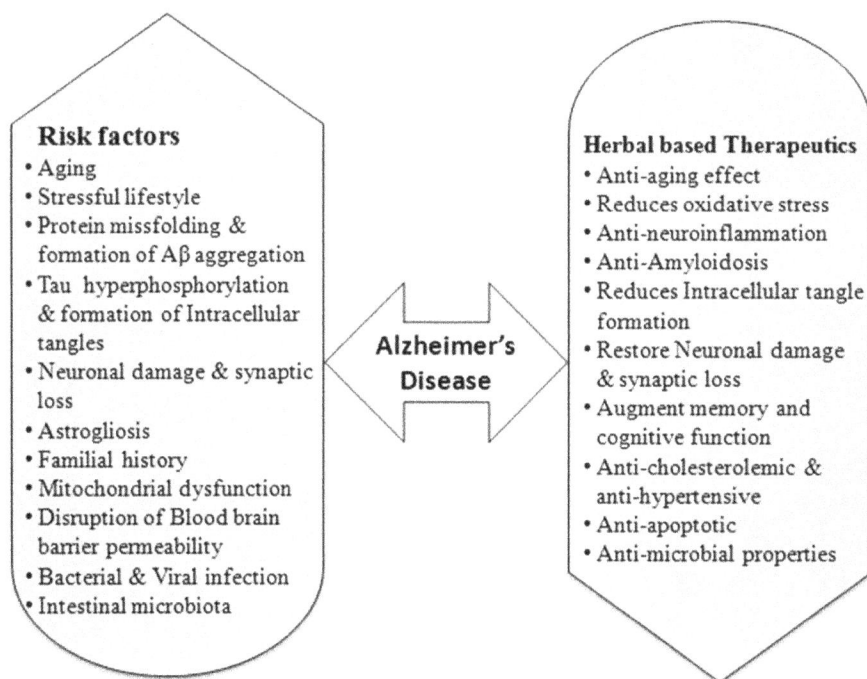

Risk factors
- Aging
- Stressful lifestyle
- Protein missfolding & formation of Aβ aggregation
- Tau hyperphosphorylation & formation of Intracellular tangles
- Neuronal damage & synaptic loss
- Astrogliosis
- Familial history
- Mitochondrial dysfunction
- Disruption of Blood brain barrier permeability
- Bacterial & Viral infection
- Intestinal microbiota

Alzheimer's Disease

Herbal based Therapeutics
- Anti-aging effect
- Reduces oxidative stress
- Anti-neuroinflammation
- Anti-Amyloidosis
- Reduces Intracellular tangle formation
- Restore Neuronal damage & synaptic loss
- Augment memory and cognitive function
- Anti-cholesterolemic & anti-hypertensive
- Anti-apoptotic
- Anti-microbial properties

FIGURE 2.3 Risk factors associated with AD and its management by herbal therapeutics.

2.4.1 BIOACTIVE COMPOUNDS FOR ALZHEIMER'S DISEASE

2.4.1.1 CURCUMIN

The yellow phenolic compound, curcumin is a major active component isolated from the rhizome of turmeric (*Curcuma longa*) and is used in Indian curries as spice, colorant and food preservative. Several studies indicate that curcumin has antioxidant, anticancer, anti-inflammatory, and anti-amyloid activities (Jena et al., 2012, 2013). Curcumin can actively scavenge ROS and also chelate various redox active metal ions such as iron (Fe^{2+}) and copper

TABLE 2.1 Different Bioactive Compounds and Herbal Extracts for the Treatment of Alzheimer's Disease

Sr. No.	Plant Source	Bioactive Compounds/Extracts	Properties and Mode of Action	References
1.	Turmeric	Curcumin	Metal ion chelator, Scavenge ROS, Antioxidant	Baum and Ng, 2004; Jena et al., 2013
2.	Lotus	Neferine	Inhibitor of AChE, BuChE and BACE1	Jung et al., 2015
3.	Red onion, grapes, apples, etc.	Quercetin	Anti-inflammatory, Anti-anxiety, Memory booster, Reduces Aβ plaque formation, AChE inhibitor	Williams et al., 2004, Menglian et al., 2018; Kumar et al., 2018
4.	Red grapes	Resveratrol	Antioxidant, Promotes Aβ clearance, Memory enhancer	Singh et al., 2013; Kumar and Khanum, 2012
5.	Ashwagandha	Root extract	Removes amyloid plaque, Enhances cognitive function	Sehgal et al., 2012; Das et al., 2015
6.	Bacopa	Leaf extract	Combat ROS by enhancing antioxidant defences system, Repair damaged neurones, Reduces amyloid plaque formation	Dhanasekaran et al, 2007
7.	Berries	Fruit	Improve spatial learning and memory, Antioxidant, Autophagy, Reduces amyloid plaque formation	Shukitt-Hale et al.; 2007, Thangthaeng et al., 2016
8.	Black pepper	Fruit	Enhances antioxidant, Anticholinesterase, Diminish amyloid plaque formation	Subedee et al., 2015
9.	Broccoli	Flower	Antioxidant	Masci et al., 2015
10.	Carica	Fruit	Free radical scavenger, Check apoptosis and DNA damage	Zhang et al., 2006
11.	Centella	Areal plant part	Reduces oxidative stress and neurodegenerative diseases	Ray and Ray, 2015
12.	Coffee	Beans	Reduces oxidative stress, decline Aβ and Tau hyperphospho-rylation, Neurobooster	Laurent et al., 2014; Basurto-Islas et al., 2014
13.	Gardenia	Fruit	Anti-inflammatory, antioxidant, learning and memory enhancer	Zhao et al., 2017; Zang et al., 2018

TABLE 2.1 *(Continued)*

Sr. No.	Plant Source	Bioactive Compounds/Extracts	Properties and Mode of Action	References
14.	Garlic	Bulb	Reduces Tau hyperphosphorylation, decrease cholesterol level, inhibit neuroinflammation, antioxidant	Chauhan, 2006; Asdaq and Inamdar, 2010; Farooqui, 2016
15.	Ginger	Rhizome	Antioxidant, inhibitor of AChE	Oboh et al., 2012
16.	Ginkgo	Leaf	Enhances autophagy, degrade Tau protein, maintain endothelial microvascular integrity, reduce Aβ deposition	Qin et al., 2018; Ahlemeyer and Krieglstein, 2003
17.	Ginseng	Root	Antioxidant, anti-inflammatory; Reduces Aβ formation, Repair neuronal damage Enhances cognitive performance	Kim et al. 2018
18.	Green Tea	Leaf	Antioxidant, Metal ion chelator, Reduces Aβ aggregation	Singh et al., 2008; Wobst et al., 2015
19.	Jatropha	Leaf	Inhibitor of AChE and BuChE	Saleem, 2016
20.	Pomegranate	Fruit	Antioxidants, Anti-inflammatory; Check abnormal Aβ formation	Neyrinck et al. 2013; Essa et al., 2015
21.	Saffron	Stigma extract	Inhibitor of AChE, reduce OS, Enhances memory	Adalier and Parker 2016; Ghaffari et al. 2015
22.	Salvia	Leaf	Inhibitor of AChE and BuChE	Akram and Nawaj, 2017
23	Virgin Olive oil	Oil extract	Anti-inflammatory, Antiamyloidegenic	Rigacci, 2015
24.	Walnut	Nut	Antioxidant, reduce DNA damage, Inhibitor of AChE	Paulose et al., 2014; Muthaiyah et al., 2011

(Cu^{2+}) (Baum and Ng, 2004). Further studies reveal that curcumin can act against Aβ aggregation, facilitate the clearance of Aβ plaques, attenuate hyperphosphorylation of tau protein, and inhibit the activity of AChE (Shen et al., 2016; Tang and Taghibiglou, 2017). OS and inflammation are thought to be the pathophysiological manifestation of AD. Some epidemiological studies showed that treatment of NSAIDs minimized risk of the development of AD (Hayden et al., 2007; Scharf and Daffner, 2007). Inflammatory response and OS is very common in AD, and it has been seen that the pro-inflammatory molecules such as lipoxygenase, cyclooxygenase-2 (COX-2), inducible nitric oxide synthase (iNOS), nuclear factor kappa-B (NF-κB), and activator protein are upregulated during the progression of the disease. Curcumin has been demonstrated as a potent inhibitor for these pro-inflammatory agents and also helpful in reducing damage due to OS (Rao, 2007). Moreover, another study also demonstrated that curcumin supplementation to the patients with AD improve the immune system and boost up the macrophages to clear the formation of amyloid proteins and improves the cognitive functions (Yao and Xue, 2014).

2.4.1.2   NEFERINE

Neferine is the most important bioactive compound found in lotus plant (*Nelumbo nucifera*) of Nelubonaceae family (Chandra and Rawat, 2015). The entire plant part; stem, seed, plumule are significant for its medicinal properties. The alkaloid derivative is called as Neferine, obtained from seed embryo which has a vast array of medicinal properties. Besides neferine, several bioactive compounds have been reported such as glycosides, fatty acids, steroids, vitamins, and other active phytochemicals (Sharma et al., 2017), and these bioactive compounds may have antimicrobial, anticancer, antidepressant, anti-inflammatory, and antioxidant effects (Baskaran et al., 2017). Lin et al. (2013) reported that neferine is one of the best inhibitors of BuChE and used as a preventive for AD. Further investigation confirmed that neferine is a potential therapeutic and preventive agent against AD, as it has a strong inhibitory property on BACE1, AChE, and BuChE, and acts as major therapeutics for AD (Jung et al., 2015). However, extensive study is required to test the therapeutic potential of extracts of *N. nucifera*, particularly neferine against the neurodegenerative diseases, hallmarked by the loss of memory and cognitive ability.

2.4.1.3   QUERECETIN

Quercetin is an important flavonoid rich in vegetables, leaves, fruits and grains such as red onions, green tea, grapes, apples, broccoli and berries and is used as a potent nutritive as well as pharmaceutical agent (Priprem et al., 2008). These prospective uses may be because of its capacity to scavenge the free radicals, reduces OS and neuronal death (Fiorani et al., 2010; Choi et al., 2014). It has also acted as an anti-inflammatory agent by inhibiting the iNOS and regulating COX-2 expression, as well as showing the anticancer effect via mechanisms that activate apoptosis and autophagy (Garcia-Mediavilla et al., 2007; Psahoulia et al., 2007). Furthermore, it has the ability to cross the blood brain barrier (Fiorani et al., 2010) and thereby exerting beneficial effects on CNS, such as anti-anxiety and booster for memory and cognitive functions by activating or inhibiting various enzyme activities and signal transduction pathways (Williams et al., 2004). Some recent findings reveals that quercetin is also helps in reducing β-amyloid plaques, neuro-inflammation, and downregulating tau phosphorylation (Menglian et al., 2018). It is also observed that quercetin act as AChE inhibitor and significantly protects the hippocampal neurons survival (Kumar et al., 2019). In another study on aged triple transgenic AD mice model demonstrated that quercetin ameliorates the histopathological hallmarks of AD and reverse the cognitive impairments without exhibiting adverse effects on nontransgenic mice (Sabogal-Guáqueta et al., 2015). However, quercetin has a potential bioactive compound for the treatment of neurodegenerative disease especially AD, and may consider this drug to reveal the underlying mechanism of different pharmacological effects in future clinical research.

2.4.1.4   RESVERATROL

Resveratrol, a phytophenol primarily found in red grapes which is helpful for the treatment of neuroinflammatory disease, cognitive impairment and AD (Krikorian et al., 2010). Many studies revealed that resveratrol supplementation is protective against ischemic injury by effectively interacting with the neurons and spinal cord (Kumar and Khanum, 2012; Witte et al., 2014). The high antioxidant activities in resveratrol can effective against the death of neurons due to nitric oxide-mediated oxidative insult (Singh et al., 2013). It has also been confirmed in cultured neuronal cells that resveratrol is a potential compound which promotes Aβ clearance and helps in the survival of the neurons (Kumar and Khanum, 2012). Another study showed resveratrol reduces OS, apoptosis and inflammation by inhibiting the activity of FOXO proteins and NF-κB through triggering sirtulin-1. It can also hinder the process of neurodegeneration in hippocampal cells and thus improving learning and memory deficiency (Vingtdeux et al., 2008). Furthermore, a report indicates that grape seed extract can minimize the DNA damage and implicated in reducing the genetic instability in a transgenic mice model (Thomas et al., 2009).

2.4.2   HERBAL THERAPY FOR ALZHEIMER'S DISEASE

2.4.2.1   ASHWAGANDHA

Accumulation of β-amyloid plaques in neurons is associated in various neurodegenerative diseases which declines the memory and cognitive functions (Prasansuklab and Tencomnao, 2013). The extract of Ashwagandha plant (*Withania somnifera*) is considered as one of the effective herbal treatments to promote cognition and memory. A study on mice model revealed that administration of Ashwagandha root extract (semipurified) significantly eliminates amyloid plaques in cortical region and hippocampal region of the mice brain and reversed the behavioral deficits by improving the cognitive functioning (Sehgal et al., 2012). Another study was performed to assess the effect of Ashwagandha extract on cognitive and psychomotor performance in healthy human subjects. The result of this study indicated that an improved reaction time in the psychomotor performance tests without showing any sedative effect (Pingali et al., 2014). Further, it was also confirmed that Ashwagandha has significant role in AD by modulating the cholinergic neurotransmission (Das et al., 2015). Although the mechanism of action of Ashwagandha in preventing the AD in humans is not clear, it promises a possibly treatment for the repair of age-related neurological disorders.

2.4.2.2   BACOPA

Brahmi (*Bacopa monniera*) is traditional Indian herbal medicine has been practised for centuries in ayurvedic system for its neuroprotective properties. This herbal drug is useful in relief anxiety and prevention of epilepsy, and acts as a neuro booster to augment learning and memory development

(Simpson et al., 2015). This herbal extract contains several alkaloids, such as saponins, D-mannitol, asbrahmine, herpestine, hersaponin, and monnierin, which mainly account for medicinal purpose. Besides, some additional active ingredients include stigmastarol, β-sitosterol, betulic acid, bacosides, and bacopasaponins (Mathur et al., 2016). It has been shown that *B. moneri* helps in reducing ROS and lipid peroxidation level along with diminish deposition of β-amyloid in the brain of the AD model (Dhanasekaran et al., 2007). In another study, it has also been confirmed that the extract of this herb helps in enhancing the antioxidant activities and reducing OS in the frontal cortex, striatum, and hippocampus (Bhattacharya et al., 2000). These properties of Brahmi make as a valuable drug for the action of neuronal repair, and the treatment of AD and other neuronal disorders.

2.4.2.3 BERRIES

Berries are small, pulpy, rounded, brightly colored, and often edible fruits. Besides nutritional values, berries such as strawberries, blueberries, and mulberries are having different constituents such as polyphenols and flavonoids with lots of antioxidant properties; these are useful in the preventing neurodegenerative diseases (Essa et al., 2012). Moreover, blueberry supplementation to aged animals modulates the level of cAMP response element binding phosphorylation and brain-derived neurotrophic factor levels, thereby improving spatial working memory (Williams et al., 2008). Strawberry contains several polyphenols such as anthocyanins, tannins and gallic acid along with vitamins A, C, and E (Fortalezas et al., 2010). It has been experimentally confirmed that strawberries can improve memory and spatial learning ability as well as hippocampus-dependent behaviors (Shukitt-Hale et al., 2007). AD is associated with cellular-advanced glycation end products that can potentially contribute neurodegenerative diseases such as AD. Due to antiglycation and antioxidant activities, some berries show their capacity in reducing Aβ accumulation and enhancing homeostatic autophagy protein and thereby attenuating AD (Sadowska-Bartosz and Bartosz, 2015; Thangthaeng et al., 2016). Blueberries also contain anthocyanin with antioxidant capacity that helps to reduce OS induced by Aβ in the NF-κB-mediated pathway and improve cognitive function (Joseph et al., 2010; Bensalem et al., 2016). Mulberry is one of the antioxidant-rich fruits containing the compound hydroxystilbenes, show neuroprotective effects by reducing ROS production, minimizing glutamate release, decreasing the cytosol calcium level by which amelio-rating Aβ-induced neurotoxicity (Chang et al., 2010).

2.4.2.4   BLACK PEPPER

Black pepper (*Piper nigrum*) is a common spice cultivated for its fruit and usually dried to use for cooking and medicinal purpose. It is native to South Asian tropical regions particularly in Sri Lanka and southern India. The herb carries a potential source of antioxidants as it can scavenge the free radicals (Subedee et al., 2015). There was a study on rat model indicates that it plays a significant role in decreasing the amyloid plaque formation and in reducing the level of cholinesterase levels. Furthermore, their study reveals that black pepper is also helpful for the treatment of AD as it augments memory and cognitive function (Subedee et al., 2015). Another report showed that piperine (1-piperoylpiperidine), an alkaloid made up of nitrogenous substance found in black pepper and long pepper (*Piper longrum*) has a beneficial effect on memory by preventing neurodegeneration in the hippocampal region of an AD animal model (Islam et al., 2017).

2.4.2.5   BROCCOLI

Broccoli (*Brassica oleracea*) is a green plant vegetable whose flowering head used as food. It has an excellent antioxidant with free radical scavenging property. A report showed that it has a capacity to reduce Aβ25–35-induced cytotoxicity, apoptosis and glutathione (GSH) depletion (Masci et al., 2015). Besides its antioxidant activity, the consumption of this vegetable is helpful in declining many age-related neurodegenerative diseases and reduces the rate of deterioration of cognition and rate of dementia (Tangney et al., 2011). Moreover, it has been reported that the vegetable intake can play a pivotal role in preventing and treatment of AD as it has potential to reverse the deleterious effects of AD (Dai et al., 2006).

2.4.2.6   CARICA

Papaya (*Carica*) is one of the beneficial fruits for patients with AD and other neurological diseases. The fermented papaya can scavenge free radicals and diminish the production of other ROS and nitric oxide (NO) (Zhang et al., 2006).

It has also helpful in reducing the inflammatory cytokines and OS-induced apoptosis and DNA damage that happens in the presence of pro-oxidants (iron, copper, aluminum, benzopyrene and methylguanidine) (Zhang et al., 2006). Furthermore, it decreases the lipid peroxidation level and the activity of SOD which augments the possible survival rate of neurons, thus provides the evidence of neuroprotective effect of the plant (Barbagallo et al., 2015).

2.4.2.7 CENTELLA

Centella asiatica, an Indian medicinal plant used in Ayurveda is effective against OS and preventing many neurological disorders including AD (Kumar and Gupta, 2003; Ray and Ray, 2015). It has been reported that aqueous extract of this plant is effective in managing the cognition deficits and OS revealed in an intra-cerebroventricular streptozotocin model of AD in rats. In this study, a significant reduction in the malondialdehyde level, enhancement in GSH content, and an increased catalase (CAT) activity in the rat brain have been observed in response to the aqueous extract of the herb (Kumar and Gupta, 2003). The key bioactive compounds found in this plant are several triterpenoid saponins, and related sapogenins (Kulkarni et al., 2012). Another active ingredient of this is triterpenoid acid, however, the mechanism of action on reduction in neurological disorders is unclear.

2.4.2.8 COFFEE

Coffee is dark-colored, bitter, slightly acidic brewed beverage drink that is prepared from roasted coffee beans. It is known for its stimulating effect in humans, predominantly due to its caffeine content (Cappelletti et al., 2015). Drinking the medically recommended amount of coffee is healthy as it attenuates the risk of progression of AD. However, it was reported that the daily intake of coffee at mid-life reduces the risk of dementia or AD (Eskelinen et al., 2009). Several studies revealed that caffeine minimizes the plasma Aβ content by reducing the β- and γ-secretase levels in hippocampus of mice brain (Islam et al., 2017). Another study showed the inhibitory effect of caffeine on OS and the pro-inflammatory process in the hippocampus region of THY-Tau22 Tg mice (Laurent et al., 2014). Caffeine has also been reported as a powerful neurobooster for its positive contribution in memory recovery and age-related cognitive functions (Van Boxtel et al., 2003). In addition to the active ingredient caffeine in coffee, eicosanoyl-5-hydroxytryptamide, a constituent in coffee, can inhibit the process of cognitive impairment and advancement of AD by downregulating the tau hyper-phosphorylation and cytoplasmic Aβ (Basurto-Islas et al., 2014).

2.4.2.9   GARDENIA

Gardenia jasminoides is a flowering plant from the family Rubiaceae and is well known for its medicinal values and also used as a natural yellow dye "crocin" extracted from its fruit practised in Asian countries (Valder, 1999). The active phytoconstituents like Monoterpenoid such as Iridoids and glucosides such as Genipin and Crocetin are commonly found in *G. jasminoides* (Yamauchi et al., 2011). These bioactive compounds perpetuate several biological activities for the protection of neurons. It behaves as an antioxidant, anti-inflammatory, antidepressant that is directly beneficial for the neurons (Zhao et al., 2017). A recent study revealed that the gardenia extract has showed good anti-inflammatory properties by inhibiting the activation of the JAK2/STAT1 signaling pathway induced by Aβ25–35, downregulating the expression of inflammatory proteins (iNOS and COX-2) and the production of pro-inflammatory mediators (PGE2 and TNF-α) in the cortex and hippocampus of the mouse model. The study further showed that *G. jasminoides* improves the learning and memory (Zang et al., 2018), and thus it can be concluded that the properties of this plant extract might have some crucial role in reversal of neurodegenaration particularly AD.

2.4.2.10   GARLIC

Garlic (*Allium sativum*) is a bulb belongs to lily family (Liliaceae) used all over the world as a spice or food additive and a medicine. It has been considered as a medicine in many cultures for thousand years in preventing and treatment of several diseases. Besides its direct free-radical scavenging activity and antioxidant property, the garlic extract may alleviate against cognitive impairment and Aβ-induced neuronal dysfunctions (Jeong et al., 2013). Garlic is useful in protecting many age-related diseases and also helpful against AD (Gupta et al., 2009). An active constituent of the "aged garlic extract" is S-allyl-L-cysteine (SAC) that is considered as the inhibitor of neuroinflammatory pathways, impeding synaptic degeneration, enhancement of antioxidant enzymes, radical scavenging, and obstructing NF-κB activation, thereby inhibiting the progression of AD (Farooqui, 2016). There is a positive correlation between cholesterol over synthesis and progression of pathophysiological events in AD, and it is thought that garlic and garlic-derived compounds reduce the cholesterol level by inhibiting 3-hydroxy-3-methylglutaryl-coenzyme A reductase, which is an essential enzyme for the biosynthesis of cholesterol (Asdaq and Inamdar, 2010). Garlic can also be helpful in reducing the hyperphosphorylation of tau proteins by downregulating the level of GSK-3, a protein kinase that can promote tau protein phosphorylation (Chauhan, 2006).

2.4.2.11   GINGER

Ginger (*Zingiber officinale*) is a rhizome used globally as a spice, flavoring agent, and traditional medicine for treating various diseases, including age-related neurodegenerative diseases (Edwards et al., 2015). Ginger root extract (GRE) has the potential for the treatment of AD as it carries anti-aging and antioxidant properties. It has been reported that GRE can pass through the blood−brain barrier and is protective against ischemia-reperfusion injury and vascular dementia in rats (Zeng et al., 2013). Another study on the effect of ginger extract inhibits the mRNA expression of pro-inflammatory genes (TNF-α, IL-1β, COX-2, macrophage inflammatory protein-1 α, monocyte chemoattractant protein-1, and interferon-inducible protein-10) on lipopolysaccharide, cytokine, and Aβ peptide-induced neuroinflammation in microglial cells which are presumed to contribute to the neuronal damage in AD (Grzanna et al., 2004). Further studies have shown that the extracts of both red ginger and white ginger inhibit AChE activity and prevent lipid peroxidation in rat brain and hence showing their usefulness in the treatment of AD (Oboh et al., 2012). The inhibition of AChE promotes the deposition of acetylcholine in synapses, which boost the cognitive function of patients with AD. Furthermore, ginger may reduce lipid peroxidation and hence help in delaying further advancement of AD (Oboh et al., 2012).

2.4.2.12   GINKGO

Ginkgo biloba (*G. biloba*) is otherwise known as the Maidenhair tree. It is the last living species of the Ginkgoaceae family with a constant morphology from the beginning, which may be around 200 years ago. It is generally found in Asian countries and predominantly used in China for its multivalent medicinal properties (Li et al., 2018). The bioactive components present in *Ginkgo* are terpene lactones and flavonoids (Solfrizzi and Panza, 2015). It has been reported that the standard *G. biloba* leaf extract, EGb761, having 24% flavon glycosides, which includes quercetin, isorharmnetin, and kaepferol, and 6% terpene lactones, particularly ginkgolides and bilobalide (Isah, 2015), is used to treat AD through different mechanism involved in regulating the appropriate brain function (Zhou et al., 2016). The bioactive components of *G. biloba* mediate lysosomal degradation of tau protein, enhancing the autophagic activity (Qin et al., 2018). EGb 761 drops the level of microglial activation, regulates endocrine function, and maintains endothelial microvascular integrity, thereby enhancing the proteolysis of hyperphosphoryated tau in AD neuropathy (Li et al., 2018). It controls $A\beta$ aggregation, which is the major factor for the initiation of AD (Ahlemeyer and Krieglstein, 2003), and also it reduces oxidative damage initiated in AD through its antioxidative property and may be used as a beneficial drug for AD (Zeng et al., 2018).

2.4.2.13 GINSENG

Ginseng (*Panax ginseng*) has been extensively used as a traditional tonic for increasing vital energy, mood elevation, and longevity in Asian countries such as China, Korea, and Japan for thousands of years (Lee et al., 2009). Ginseng extract has been considered a nonspecific agent for increasing resistance against various diseases till the identifications of ginsenosides and gintonin, the most active constituents of the plant. The first active ingredient isolated from ginseng is ginsenosides (also known as ginseng saponins), a type of triterpenoid dammarane glycosides. Besides antioxidant, anti-inflammatory, anticancer, and vasorelaxative effects, it is also very effective for the patients with AD as it improves the brain cholinergic function, thereby enhancing cognitive performance and reducing $A\beta$ formation and repairing neuronal damage (Kim et al., 2018;). In the same way, gintonin also inhibits $A\beta$-induced neurotoxicity by triggering the nonamyloidogenic pathway and enhances acetylcholine and choline through the lysophosphatidic acid receptor-mediated acetyltransferase expression in the brain. Oral supplementation of gintonin reduces brain amyloid plaque formation, thereby improving hippocampal neurogenesis and cholinergic systems and ameliorating learning and memory impairments (Kim et al., 2018).

2.4.2.14   GREEN TEA

Green tea (*Camelia sinesis*) is a flavonoid-rich beverage prepared from the dry leaves of *Camellia sinesis* and is broadly consumed worldwide. It has an extensive range of health benefits that provide a vast scope for preventing many diseases, including cardiovascular disease, inflammatory diseases, cancer, and neurodegenerative diseases (Dhouafli et al., 2018). Many bioactive compounds, particularly catechins, provide the plant with high antioxidant properties. Among the flavonoids (30% of dry weight of a leaf) in green tea (Graham, 1992), catechin having various forms of compounds such as epigallocatechin-gallate (EGCG), (−)-epigallocatechin (EGC), (−)-epicatechin (EC), and (−)-epicatechin-3-gallate (ECG) have the antioxidant potencies in the respective order of EGCG, ECG, EGC, and EC (Guo et al., 1996). It has been reported that the oral administration of the green tea extract can alleviate the activities of antioxidant enzymes (SOD, CAT, and GPx) in rat brain. In the case of human, the consumption of two cups of green tea (containing ~250 mg catechins) for 42 days elevates the total antioxidant level in plasma (Erba et al., 2003). Another study indicates that catechin reduces OS by inhibiting the enzyme xanthinoxidadase that generates the superoxide anion (Bickford et al., 2000). The metal ions, copper(II) and iron(II), enhance the necessary reactions in the process of production of ROS in the biological system. However, catechins can directly chelate these metal ions, thereby preventing the generation of free radicals (Singh et al., 2008). The suppression of ROS may be one of the key reasons of the declined translational level of APP in the hippocampus of mice in response to the prolonged supplementation of EGCG (Levites et al., 2003). EGCG also modulates apoptotic pathways to prevent OS-mediated cellular damage and promotes cell proliferation and survival through PI3k/AKT and protein kinase C-regulated pathways (Weinreb et al., 2004). Further study confirms that EGCG supplementation reduces the formation of Aβ aggregation and Tau-induced toxicity in neuronal cells (Wobst et al., 2015).

2.4.2.15   JATROPHA

The plant *Jatropha gossypiifolia* (*J. gossypiifolia*) is worldwide known as "bellyache bush," and in India it is commonly known as sibidigua. These plants belong to the angiosperm of the Euphorbiaceae family and are frequently found in America, Africa, and in the Gujarat state of India. The plant parts such as leaves, stems, roots, and seeds are commonly known for their medicinal value. The plant is extensively used to reduce hypertension, inflammation, and diabetes and also acts as antipyretic, analgesic, antimicrobial, antiseptic agents (Sabandar et al., 2013). The bark juice of this plant is used in common cold as a cough syrup (Taylor et al., 1996), and the seed oil helps to prevent leprosy and mycosis (Sabandar et al., 2013). Many bioactive compounds such as terpenoids, flavonoids, alkaloids, phenols, steroids, coumarin, and tannin are produced from this plant extract, and the key component from this plant is terpenoid (Zhang et al., 2009). One of the hypotheses regarding treatment and prevention of AD is the inhibitory activity of cholinesterase. It was reported that the plant extract of *J. gossypiifolia* possesses AChE and BuChE inhibitory activities. The leaf ethyl acetate fraction of this plant was reported for the presence of compounds having AChE and BuChE inhibitory activities and therefore this plant is used as anti-Alzheimer's agent (Saleem et al., 2016).

2.4.2.16   POMEGRANATE

Pomegranate (*Punica granatum*), a polyphenol-rich fruit, is chiefly farmed in the Middle East and Mediterranean regions, and has an effective anti-oxidants, anti-inflammatory, free radical-scavenging capacity, and can also ameliorate the neurodegeneration process (Neyrinck et al., 2013). A study reveals that pomegranate enhances the activities of major antioxidant enzymes, SOD and CAT, in the brain of the AD mouse model (Tg2576). These enzymes neutralize superoxide anions and hydrogen peroxide, which are considered one of the main reasons for neurodegeneration (Pirinccioglu et al., 2014). Intracellular redox status and neuronal OS are generally maintained via GSH and its associative enzymes such as GPx, and GSH-S transferase. Reduced levels of such antioxidant parameters in the brain of the AD mice model are increased upon the supplementation of the pome-granate extract (Ajaikumar et al., 2005). The fruit extract of pomegranate can also diminish abnormal Aβ formation, thereby checking the senile plaque progression (Essa et al., 2015).

2.4.2.17   SAFFRON

Crocus sativus or saffron is from the Iridaceae family and is famous for its flavor, aroma, and color in the spice world. Besides its potential role in food industries, it is also known for its effective anti-inflammatory, antioxidant, memory-enhancing, and antiamyloidgenic abilities. Saffron is a potentially favorable and effective remediation of AD as it prevents the accumulation of Aβ deposition. It is also reported to be beneficial as it modulates the AChE, which is involved in the formation of β-amyloid plaques and NFTs (Adalier and Parker 2016). Another study confirms that saffron is effective in improving memory and learning ability. It also declines OS by improving the antioxidant enzymatic activities in rodent models (Ghaffari et al., 2015).

2.4.2.18 SALVIA

Salvia officinalis (*S. officinalis*) is native to Europe and is known for its soothing and carminative effects. Because of these characters, like other mints, it is now adapted and grown all over the world. It has been reported that *S. officinalis* (Labiatae family) is a potential neuroprotective agent that improves the memory and cognitive function, and also act as an anti-Alzheimer's agent (Akram and Nawaj, 2017). AD is featured by the loss of cholinergic neurons, and the cholinergic synaptic function is considered to be susceptible to β-amyloid peptide toxicity (Small et al., 2001). In recent treatments, cholinesterase inhibitors have been the major promising medications approved for the treatment of AD, and the herbal extract of *S. officinalis* exhibited both AChE and BuChE inhibitory effects (Scholey et al., 2008). In a clinical study on moderate AD patients treated with 3000 mg/day of the extract for 16 weeks showed a significant cognitive improvement without any side effects when compared with the Placebo group (Akhondzadeh et al., 2003). Further, it has been observed that the herbal extract of saffron is helpful in reducing agitation behavior in AD patients. These results indicate that the extract of *S. officinalis* has potential properties in the management of mild-to-moderate AD patients.

2.4.2.19 VIRGIN OLIVE OIL

Virgin olive oil (VOO) is a common product of MeDi that primarily comprises of glycerides monounsaturated fatty acids (glycerides) and various phenolic compounds (Islam et al., 2017). The pharmacological properties of these compounds help to prevent the activation of autophagy. These compounds act as anti-inflammatory agents, stimulate neurogenesis, particularly in the hippocampal region and dopaminergic neurons, and activate anti-oxidant enzymes and the defense system in AD. It has also been reported that the presence of phenolic compounds, especially oleuropeinaglycone and oleocanthal of VOO, helps in the inhibition of the amyloidogenic Aβ construction from APP, toxicity due to the accumulation of Aβ, and the accumulation of tau, as well as the clearance of the Aβ peptide (Rigacci, 2015). The active constituents, such as polyphenols, phenolic compounds and other compounds of MeDi, considered as the possible candidates for both the prevention and treatment of AD.

2.4.2.20   WALNUT

Walnut (*Juglans regia*) is known for its multifarious properties with full of vitamins, antioxidants, different polyphenols, flavonoids, useful fatty acids, tocopherol, etc. It has been used conventionally for the remediation of many cardiac diseases, and it acts as an antimicrobial and antifungal agent; especially, it diminishes many neurodegenerative diseases in individuals if consumed daily. Several studies showed that walnut extracts can be effective against OS and cytotoxicity that because of its free radical scavenging capacity, protecting DNA damage and reducing the formation of Aβ fibrils. It is also proved that its beneficial properties can facilitate the expression of neuronal transcription factors, resulting in prevention from the deterioration of neurons (Muthaiyah et al., 2011; Paulose et al., 2014). The polyphenolic constitutions of walnut show the antioxidant activities and inhibitory effects of AChE, which make it a potential medicinal source against neurodegenerative diseases, particularly for AD (Muthaiyah et al., 2011).

2.5   FUTURE PROSPECTIVE

The occurrence of AD primarily due to the formation of misfolded proteins and aggregation of these malformed proteins associated with Aβ plaques buildups an NFT assembly in the cortex of the brain. The progression of this disease is leading to more complicated neuropathology through increased inflammation and OS, and many neurological insults. The cause behind the initiation of AD is a multistep process, so it is possible to treat AD by focusing on different steps during therapeutic interventions. Although few recognized drugs and psychometric-based therapies are available for AD, these are mainly focused on reducing the progression of AD without approaching the actual underlying causes of the disease. Therefore, multiple approaches with combinatorial treatment of the available drugs and psychometric therapy along with the supplementation of potential bioactive compounds and herbal extracts are needed for the better management of the disease. The disease must be scrutinized properly, and for early detection, specific proteins and development of monoclonal antibodies may be useful against AD. In addition to the conventional method of treating AD with drugs for inhibitions of cholinergic neurons, herbal extracts and their bioactive compounds may be used for the treatment of age-related diseases, particularly AD. Furthermore, prolonged adherence to the intake of bioactive phytochemicals from diverse sources of plants may deliver the pleiotropic beneficial effect against age-related neurodegenerative diseases and might also be helpful in the prevention of progression of AD. Apart from these treatments, patients also need personal care, family support, and an enriched social environment that reduce OS and progression of AD. Therefore, a holistic approach is needed with "standard drugs-," herbal remedies-," and "nondrug"-based combinatorial therapies along with early, accurate diagnosis and evaluation of response to therapy.

KEYWORDS

- aging
- neurodegenerative diseases
- Alzheimer's diseases
- bioactive compounds
- herbal therapy

REFERENCES

Adalier, N.; Parker, H. Vitamin E, turmeric and saffron in treatment of Alzheimer's disease. *Antioxidants*. **2016,** *5*(4), 40. doi:10.3390/antiox5040040.

Ahlemeyer, B.; Krieglstein, J. Pharmacological studies supporting the therapeutic use of *Ginkgo biloba* extract for Alzheimer's disease. *Pharmacopsychiatry*. **2003,** *36*, S8– S14.

Ajaikumar, K.; Asheef, M.; Babu, B.; Padikkala, J.The inhibition of gastric mucosal injury by *Punicagranatum*L.pomegrane) methanolicextract. *J. Ethnopharmacol*. **2005,** *96*(1), 171–176.

Akhondzadeh, S.; Noroozian, M.; Mohammadi, M.; Ohadinia, S.; Jamshidi, A. H.; Khani, M. *Salvia officinalis* extract in the treatment of patients with mild to moderate Alzheimer's disease: a double blind, randomized and placebo controlled trial. *J. Clin. Pharm. Ther.* **2003**, *28*, 53–59.

Akram, M.; Nawaz, A. Effects of medicinal plants on Alzheimer's disease and memory deficits. *Neural Regen Res.* **2017**, *12*(4), 660–670.

Alzheimer, A. Uber eineeigenartigeErkrankung der Hirnrinde. *AllgZeitschrift Psychiatr.* **1907**, *64*, 146–148.

Anand, R.; Gill, K. D.; Mahdi, A. A. Therapeutics of Alzheimer's diseases: Past, present and future. *Neuropharmacology.* **2014**, *76*, 27.

Anantharaman, M.; Tangpong, J.; Keller, J. N.; Murphy, M. P.; Markesbery, W. R.; Kiningham, K. K.; St Clair, D. K. Beta-amyloid mediated nitration of manganese superoxide dismutase: implication for oxidative stress in a APPNLH/NLH X PS-1P264L/P264L double knock-in mouse model of Alzheimer's disease. *Am. J. Pathol.* **2006**, *168*, 1608–1618.

Asdaq, S.; Inamdar, M. Potential of garlic and its active constituent, S-allyl cysteine, as antihypertensive and cardioprotective in presence of captopril. *Phytomedicine.* **2010**, *17*(13), 1016–1026.

Bali, P.; Lahiri, D. K.; Banik, A.; Nehru, B.; Anand, A. Potential for Stem Cells Therapy in Alzheimer's Disease: Do Neurotrophic Factors Play Critical Role? *Curr. Alzheimer Res.* **2017**, *14*(2), 208–220.

Barbagallo, M.; Marotta, F.; Dominguez, L. J. Oxidative stress in patients with Alzheimer's disease: effect of extracts of fermented papaya powder. *Mediat. Inflamm.* **2015**, *2015* (2015), 1–6.

Bartorelli, L.; Girald, C.; Saccardo, M.; Cammarata, S.; Bottini, G.; Fasanaro, A. M.; Trequattrini, A. Effects of switching from an AChE inhibitor to a dual AChE-BuChE inhibitor in patients with Alzheimer's disease. *Curr. Med. Res. Opin.* **2005**, *21*(11), 1809–1817.

Baskaran, R.; Priya, L. B.; Kalaiselvi, P.; Poornima, P.; Huang, C.Y.; Padma, V. V. Neferine from *Nelumbonucifera* modulates oxidative stress and cytokines production during hypoxia in human peripheral blood mononuclear cells. *Biomed. Pharmacother.* **2017**, *93*, 730–736.

Basurto-Islas, G.; Blanchard, J.; Tung, Y. C.; Fernandez, J. R.; Voronkov, M.; Stock, M.; Zhang, S.; Stock, J. B.; Iqbal, K. Therapeutic benefits of a component of coffee in a rat model of Alzheimer's disease. *Neurobiol. Aging.* **2014**, *35*(12), 2701–2712.

Baum, L.; Ng, A. Curcumin interaction with copper and iron suggests one possible mechanism of action in Alzheimer's disease animal models. *J. Alzheimer's Dis.* **2004**, *6*, 367–377.

Bensalem, J.; Dal-Pan, A.; Gillard, E.; Calon, F.; Pallet, V. Protective effects of berry polyphenols against age-related cognitive impairment. *Nutr. Aging* **2016**, *3*(2–4), 89–106.

Bhattacharjee, S.; Zhao, Y.; Hill, J. M.; Percy, M. E.; Lukiw, W. J. Aluminum and its potential contribution to Alzheimer's disease (AD). *Front. Aging Neurosci.* **2014**, *6*(62), 1–3.

Bhattacharya, S.; Bhattacharya, A., Kumar, A.; Ghosal, S. Antioxidant activity of *Bacopa monniera* in rat frontal cortex, striatum and hippocampus. *Phytother. Res.* **2000**, *14*, 174.

Bickford, P. C.; Gould, T.; Briederick, L.; Chadman, K.; Pollock, A.; Young, D.; Shukitt-Hale, B.; Joseph, J. Antioxidant-rich diets improve cerebellar physiology and motor learning in aged rats. *Brain Res.* **2000**, *866*, 211–217.

Bolognesi, M. L.; Cavalli, A.; Melchiorre, C. Memoquin: a multi-target-directed ligand as an innovative therapeutic opportunity for Alzheimer's disease. *Neurotherapeutics.* **2009**, *6*(1), 152–162.

Botchway, B. O. A.; Iyer, I. C. Alzheimer's disease-The past, the present and the future. *Sci. J. Clin. Med.* **2017,** *6*(1), 1–19.

Brewer, G. J. Copper excess, zinc deficiency, and cognition loss in Alzheimer's disease. *BioFactors.* **2012,** *38*(2), 107–113.

Butterfield, D. A.; Boyd-Kimball, D. Oxidative stress, amyloid- peptide, and altered key molecular pathways in the pathogenesis and progression of Alzheimer's disease. *J. Alzheimer's Dis.* **2018,** *62,* 1345–1367.

Cappelletti, S.; Piacentino, D.; Daria, P.; Sani, G.; Aromatario, M. Caffeine: cognitive and physical performance enhancer or psychoactive drug. *Curr. Neuropharmacol.* **2015,** *13*(1), 71–88.

Chandra, S.; Rawat, D. S. Medicinal plants of the family Caryophyllaceae: a review of ethnomedicinal uses and pharmacological properties. *Integr. Med. Res.* **2015,** *4,* 123–131.

Chang, R. C; Chao, J.; Yu, M.; Wang, M. Neuroprotective effects of oxyresveratrol from fruit against neurodegeneration in Alzheimer's disease. In: Ramassamy, C.; Bastianetto, S. (Eds.). *Recent Advances on Nutrition and the Prevention of Alzheimer's Disease.* Kerala, India: Transworld Research Network. **2010,** 155–168.

Chauhan, N. B. Effect of aged garlic extract on APP processing and tau phosphorylation in Alzheimer's transgenic model Tg2576. *J. Ethnopharmacol.* **2006,** *108*(3), 385–394.

Chen, Z.; Zhong, C. Oxidative stress in Alzheimer's disease. *Neurosci. Bull.* **2014,** *30*(2), 271–281.

Choi, S. M.; Kim, B. C.; Cho, Y. H.; Choi, K. H.; Chang, J.; Park, M. S.; Kim, M. K.; Cho, K. H.; Kim, J. K. Effects of flavonoid compounds on beta-amyloid-peptide-induced neuronal death in cultured mouse cortical neurons. *Chonnam Med J.* **2014,** *50*, 45–51.

Crouch, P. J.; Harding, S. M.; White, A. R.; Camakaris, J.; Bush, A. I.; Masters, C. L. Mechanisms of A beta mediated neurodegeneration in Alzheimer's disease. *Int. J. Biochem. Cell Biol.* **2008,** *40,* 181–198.

Cruchaga, C.; Chakraverty, S.; Mayo, K.; Vallania, F. L.; Mitra, R. D.; Faber, K.; Williamson, J.; Bird, T.; Diaz-Arrastia, R.; Foroud, T. M. Rare variants in APP, PSEN1 and PSEN2 increase risk for AD in late-onset Alzheimer's disease families. *PLoS One* **2012,** *7*(2), e31039.

Curtain, C. C.; Ali, F.; Volitakis, I.; Cherny, R. A.; Norton, R. S.; Beyreuther, K.; Barrow, C. J.; Masters, C. L.; Bush, A. L.; Barnham, K. J. Alzheimer's disease amyloid-beta binds copper and zinc to generate an allosterically ordered membrane-penetrating structure containing superoxide dismutase-like subunits. *J. Biol. Chem.* **2001,** *276,* 20466–20473.

Dai, Q.; Borenstein, A. R.; Wu, Y.; Jackson, J. C.; Larson, E. B. Fruit and vegetable juices and Alzheimer's disease: the Kame Project. *Am. J. Med.* **2006,** *119*(9), 751–759.

Das, T. K.; Abdul, H. M. R.; Das, T.; Shad, K. F. Potential of Glycowithanolides from Withaniasomnifera (Ashwagandha) as therapeutic agents for the treatment of Alzheimer's disease. *World J. Pharma. Res.* **2015,** *4*(6), 2277–7105.

Dhanasekaran, M.; Tharakan, B.; Holcomb, L. A.; Hitt, A. R.; Young, K. A.; Manyam, B. V. Neuroprotective mechanisms of ayurvedic antidementia botanical *Bacopa monniera. Phytother. Res.* **2007,** *21,* 965–969.

Dhouafli, Z.; Cuanalo-Contreras, K.; Hayouni, E. A.; Mays, C. E.; Soto, C.; Moreno-Gonzalez, I. Inhibition of protein misfolding and aggregation by natural phenolic compounds. *Cell. Mol. Life Sci.* **2018,** *75,* 3521–3538, doi:10.1007/s00018-018-2872-2.

Edwards, S. E.; da Costa Rocha, I.; Williamson, E. M.; Heinrich, M. *Ginger* Zingiber officinale *Roscoe.* Chennai: John Wiley & Sons: Chennai **2015,** p. 164.

Erba, D.; Riso, P.; Criscuoli, F.; Testolin, G. Malondialdehyde production in Jurkat T cells subjected to oxidative stress. *Nutrition.* **2003,** *19,* 545–548.

Eskelinen, M. H.; Ngandu, T.; Tuomilehto, J.; Soininen, H.; Kivipelto, M. Midlife coffee and tea drinking and the risk of late-life dementia: a population-based CAIDE study. *J. Alzheimer's Dis.* **2009**, *16*(1), 85–91.

Essa, M. M.; Subash, S.; Akbar, M.; Al-Adawi, S.; Guillemin, G. J. Long-term dietary supplementation of pomegranates, figs and dates alleviate neuroinflammation in a transgenic mouse model of Alzheimer's disease. *PLoS One* **2015**, *10*(3), e0120964.

Essa, M. M.; Vijayan, R. K.; Castellano-Gonzalez, G.; Memon, M. A.; Braidy, N.; Guillemin, G. J. Neuroprotective effect of natural products against Alzheimer's disease. *Neurochem. Res.* **2012**, *37*(9), 1829–1842.

Farooqui, A. A. (ed.). Treatment of Alzheimer disease with phytochemicals other than curcumin. In: *Therapeutic Potentials of Curcumin for Alzheimer Disease.* Springer, Cham **2016**, 335–369, doi: 10.1007/978-3-319-15889-1.

Fiorani, M.; Guidarelli, A.; Blasa, M.; Azzolini, C.; Candiracci, M.; Piatti, E.; Cantoni, O. Mitochondria accumulate large amounts of quercetin: prevention of mitochondrial damage and release upon oxidation of the extramitochondrial fraction of the flavonoid. *J. Nutr. Biochem.* **2010**, *21*, 397–404.

Fortalezas, S.; Tavares, L.; Pimpão, R.; Tyagi, M.; Pontes, V.; Alves, P. M.; McDougall, G.; Stewart, D.; Ferreira, R. B.; Santos, C. N. Antioxidant properties and neuroprotective capacity of strawberry tree fruit (*Arbutus unedo*). *Nutrients* **2010**, *2*(2), 214–229.

Ganguli, M.; Chandra, V.; Kamboh, M. I.; Johnston, J. M.; Dodge, H. H.; Thelma, B. K.; Juyal, R. C.; Pandav, R.; Belle, S. H.; DeKosky, S. T. Apolipoprotein E polymorphism and Alzheimer disease: The Indo-US Cross-National Dementia Study. *Arch. Neurol.* **2000**, *57*, 824–830.

Garcia-Mediavilla, V.; Crespo, I.; Collado, P. S.; Esteller, A.; Sanchez-Campos, S.; Tunon, M. J.; Gonzalez-Gallego, J. The anti-inflammatory flavones quercetin and kaempferol cause inhibition of inducible nitric oxide synthase, cyclooxygenase-2 and reactive C-protein, and downregulation of the nuclear factor kappa-B pathway in Change Liver cells. *Eur. J. Pharmacol.* **2007**, *557*, 221–229.

Ghaffari, S. H.; Hatami, H.; Dehghan, G. Saffron ethanolic extract attenuates oxidative stress, spatial learning, and memory impairments induced by local injection of ethidium bromide. *Res. Pharm. Sci.* **2015**, *10*, 222–232.

Graham, H. N. Green tea composition, consumption, and polyphenol chemistry. *Prev. Med.* **1992**, *21*, 334–350.

Grzanna, R.; Phan, P.; Polotsky, A.; Lindmark, L.; Frondoza, C. G. Ginger extract inhibits β-amyloid peptide-induced cytokine and chemokine expression in cultured THP-1 monocytes. *J. Altern. Complement. Med.* **2004**, *10*(6), 1009–1013.

Guo, Q.; Zhao, B.; Li, M.; Shen, S.; Xin, W. Studies on protective mechanisms of four components of green tea polyphenols against lipid peroxidation in synaptosomes. *Biochim. Biophys. Acta* **1996**, *1304*, 210–222.

Gupta, V. B.; Indi, S.; Rao, K. Garlic extract exhibits antiamyloidogenic activity on amyloid-beta fibrillogenesis: relevance to Alzheimer's disease. *Phytother. Res.* **2009**, *23*(1), 111–115.

Gutzmann, H.; Hadler, D. Sustained efficacy and safety of idebenone in the treatment of Alzheimer's disease: update on a 2-year double-blind multicentre study. *J. Neural. Transm. Suppl.* **1998**, *54*, 301–310.

Halliwell, B.; Gutteridge, J. M. C. Free Radicals in Biology and Medicine, 4th ed. Oxford University Press: Oxford, 2007.

Hardy, J.; Selkoe, D. J. The amyloid hypothesis of Alzheimer's disease: progress and problems on the road to therapeutics. *Science.* **2002**, *297*, 353–356.

Harris, M. E.; Hensley, K.; Butterfield, D. A.; Leedle, R. A.; Carney, J. M. Direct evidence of oxidative injury produced by the Alzheimer's amyloid beta peptide (1–40) in cultured hippocampal neurons. *Exp. Neurol.* **1995,** *131,* 193–202.

Hayden, K. M.; Zandi, P. P.; Khachaturian, A. S.; Szekely, C. A.; Fotuhi, M.; Norton, M. C.; Tschanz, J. T.; Pieper, C. F.; Corcoran, C.; Lyketsos, C. G.; Breitner, J. C.; Welsh-Bohmer, K. A. Does NSAID use modify cognitive trajectories in the elderly? The Cache County study. *Neurology* **2007,** *69,* 275–282.

Heneka, M. T.; Carson, M. J.; El Khoury, J.; Landreth, G. E.; Brosseron, F.; Feinstein, D. L.; Jacobs, A. H.; Wyss-Coray, T.; Vitorica, J.; Ransohoff, R. M. Neuroinflammation in Alzheimer's disease. *Lancet Neurol.* **2015,** *14*(4), 388–405.

Heppner, F. L.; Ransohoff, R. M.; Becher, B. Immune attack: the role of inflammation in Alzheimer disease. *Nat. Rev. Neurosci.* **2015,** *16*(6), 358–372.

Hirai, K.; Aliev, G.; Nunomura, A.; Fujioka, H.; Russell, R. L.; Atwood, C. S.; Johnson, A. B.; Kress, Y.; Vinters, H. V.; Tabaton, M.; Shimohama, S.; Cash, A, D.; Siedlak, S. L.; Harris, P. L.; Jones, P. K.; Petersen, R. B.; Perry, G.; Smith, M. A. Mitochondrial abnormalities in Alzheimer's disease. *J. Neurosci.* **2001,** *21,* 3017–3023.

Huang, Y.; Mucke, L. Alzhimer mechanisms and therapeutics strategies. *Cel l* **2012,** *48,* 1204.

Hung, Y. H.; Bush, A. I.; Cherny, R. A. Copper in the brain and Alzheimer's disease. *JBIC J. Biol. Inorg. Chem.* **2010,** *15*(1), 61–76.

Isah, T. Rethinking *Ginkgo biloba* L.: medicinal uses and conservation. *Pharmacogn. Rev.* **2015,** *9,* 140–148.

Islam, M. A.; Khandker, S. S.; Alam, F.; Md. Khalil, M. I.; Kamal, M. A.; Gan, S. H. Alzheimer's disease and natural products: future regimens emerging from nature. *Curr. Top. Med. Chem.* **2017,** *17,* 1408–1428.

Jack, Jr. C. R.; Holtzman, D. M. Biomarker modeling of Alzheimer's disease. *Neuron.* **2013,** *80*(6), 1347–1358.

Jena, S.; Anand, C.; Chainy, G. B. N.; Dandapat, J. Induction of oxidative stress and inhibition of superoxide dismutase expression in rat cerebral cortex and cerebellum by PTU-induced hypothyroidism and its reversal by curcumin. *Neurol. Sci.* **2012,** *33*(4), 869–873.

Jena, S.; Dandapat, J.; Chainy, G. B. N. Curcumin differentially regulates the expression of superoxide dismutase in cerebral cortex and cerebellum of L-thyroxine (T_4)-induced hyperthyroid rat brain. *Neurol. Sci.* **2013,** *34*(4), 505–510.

Jeong, J. H.; Jeong, H. R.; Jo, Y. N.; Kim, H. J.; Shin, J. H.; Heo, H. J. Ameliorating effects of aged garlic extracts against Aβ-induced neurotoxicity and cognitive impairment. *BMC Complement. Altern. Med.* **2013,** *13*(1), 1–11.

Joseph, J. A.; Bielinski, D. F.; Fisher, D. R. Blueberry treatment antagonizes C-2 ceramide-induced stress signaling in muscarinic receptor-transfected COS-7 cells. *J. Agric. Food Chem.* **2010,** *58*(6), 3380–3392.

Jung, H. A.; Karki, S.; Kim, J. H.; Choi, J. S. BACE1 and cholinesterase inhibitory activities of *Nelumbonucifera* embryos. *Arch. Pharm. Res.* **2015,** *38,* 1178–1187.

Kanski, J.; Varadarajan, S.; Aksenova, M.; Butterfield, D. A. Role of glycine-33 and methionine-35 in Alzheimer's amyloid-peptide 1–42-associated oxidative stress and neurotoxicity. *Biochim. Biophys. Acta.* **2001,** *1586,* 190–198.

Kaur, U.; Banerjee, P.; Bir, A.; Sinha, M.; Biswas, A.; Chakrabarti, S. Reactive oxygen species, redox signaling and neuroinflammation in Alzheimer's disease: The NF-κB connection. *Curr. Top. Med. Chem.* **2015,** *15*(5), 446–457.

Kennedy, D. O.; Wightman, E. L. Herbal extracts and phytochemicals: plant secondary metabolites and the enhancement of human brain function. *Adv. Nutr.* **2011,** *2,* 32–50.

Kim, H. J.; Jung, S. W.; Kim, S. Y.; Cho, I. H.; Kim, H. C.; Rhim, H.; Kim, M.; Nah, S. Y. Panax ginseng as an adjuvant treatment for Alzheimer's disease. *J. Ginseng Res.* **2018,** *42*(4), 401–411.

Krikorian, R.; Nash, T. A.; Shidler, M. D.; Shukitt-Hale, B.; Joseph, J. A. Concord grape juice supplementation improves memory function in older adults with mild cognitive impairment. *Br. J. Nutr.* **2010,** *103*(5), 730–734.

Kulkarni, R.; Girish, K.J.; Kumar, A. Nootropic herbs (MedhyaRasayana) in Ayurveda: An update. *Pharmacogn. Rev.* **2012,** *6*(12), 147–153.

Kumar, A.; Mehta, V.; Raj, U.; Varadwaj, P. K.; Udayabanu, M.; Yennamalli, R. M.; Singh, T. R. Computational and in-vitro validation of natural molecules as potential acetylcholinesterase inhibitors and neuroprotective agents. *Curr. Alzheimer Res.* **2019,** *16*(2), 116–127, doi:10.2174/1567205016666181212155147.

Kumar, G. P.; Khanum, F. Neuroprotective potential of phytochemicals.*Pharmacogn. Rev.* **2012,** *6*(12), 81–90.

Kumar, V. Potential medicinal plants for CNS disorders: an overview. *Phytother. Res.* **2006,** *20,* 1023–1035.

Kumar, V. M..; Gupta, Y. Effect of Centellaasiatica on cognition and oxidative stress in an intra-cerebroventricular streptozotocin model of Alzheimer's disease in rats. *Clin. Exp. Pharmacol. Physiol.* **2003,** *30*(5–6), 336–342.

Laurent, C.; Eddarkaoui, S.; Derisbourg, M.; Leboucher, A.; Demeyer, D.; Carrier, S.; Schneider, M.; Hamdane, M.; Müller, C. E.; Buée, L. Beneficial effects of caffeine in a transgenic model of Alzheimer's disease-like tau pathology. *Neurobiol. Aging* **2014,** *35*(9), 2079–2090.

Lee, M. S.; Yang, E. J.; Kim, J. I.; Ernst, E. Ginseng for cognitive function in Alzheimer's disease: a systematic review. *J Alzheimers Dis.* **2009,** *18,* 339–344.

Levites, Y.; Amit, T.; Mandel, S.; Youdim, M. B. Neuroprotection and neurorescue against Abeta toxicity and PKC-dependent release of nonamyloidogenic soluble precursor protein by green tea polyphenol (−)-epigallocatechin-3-gallate. *FASEB J.* **2003,** *17,* 952–954.

Li, H.; Sun, X.; Yu, Fan; Xu, L.; Miu, J.; Xiao, P. In *Silico* investigation of the pharmacological mechanisms of beneficial effects of *Ginkgo biloba* L. on Alzheimer's disease. *Nutrients* **2018,** *10*(5), 589.

Lin, Z.; Wang, H.; Fu, Q.; An, H.; Liang, Y.; Zhang, B.; Hashi, Y.; Chen, S. Simultaneous separation, identification and activity evaluation of three butyrylcholinesterase inhibitors from Plumulanelumbinis using on-line HPLC-UV coupled with ESI–IT–TOF-MS and BChE biochemical detection. *Talanta,* **2013,** *110,* 180–189.

Loewen, C. A.; Feany, M. B. The unfolded protein response protects from tau neurotoxicity *in vivo. PLoS One* **2010,** *5*(9), e13084.

Lynch, T.; Cherny, R. A.; Bush, A. I. Oxidative processes in Alzheimer's disease: the role of abeta-metal interactions. *Exp. Gerontol.* **2000,** *35,* 445–451.

Mahley, R. W.; Rall, S. C. Jr. Apolipoprotein E: far more than a lipid transport protein. *Annu. Rev. Genomics Hum. Genet.* **2000,** *1,* 507–537.

Masci, A.; Mattioli, R.; Costantino, P.; Baima, S.; Morelli, G.; Punzi, P.; Giordano, C.; Pinto, A.; Donini, L. M.; d'Erme, M. Neuroprotective effect of Brassica oleracea sprouts crude juice in a cellular model of Alzheimer's disease. *Oxid. Med. Cell. Longev.* **2015,** 781938.

Mathur, D.; Goyal, K.; Koul, V.; Anand, A. The Molecular Links of Re-Emerging Therapy: A Review of Evidence of Brahmi (Bacopamonniera). *Front Pharmacol.* **2016,** *7,* 44.

Menglian, L. V.; Yang, S.; Cai, L.; Qin, L. Q.; Li, B. Y.; Wan, Z. Effects of quercetin intervention on cognition function in APP/PS1 mice was affected by vitamin D status. *Mol. Nutr. Food Res.* **2018,** doi: 10.1002/mnfr.201800621.

Muthaiyah, B.; Essa, M. M.; Chauhan, V.; Chauhan, A. Protective effects of walnut extract against amyloid beta peptide-induced cell death and oxidative stress in PC12 cells. *Neurochem. Res.* **2011,** *36*(11), 2096–2103.

Mutisya, E. M.; Bowling, A. C.; Beal, M. F. Cortical cytochrome oxidase activity is reduced in Alzheimer's disease. *J. Neurochem.* **1994,** *63*, 2179–2184.

Neyrinck, A. M.; Van Hée, V. F.; Bindels, L. B.; De Backer, F.; Cani, P. D.; Delzenne, N. M. Polyphenol-rich extract of pomegranate peel alleviates tissue inflammation and hypercholesterolaemia in high-fat diet-induced obese mice: potential implication of the gut microbiota. *Br. J. Nutr.* **2013,** *10*(5), 802–809.

O'Callaghan, C.; Fanning, L. J.; Barry, O. P. p38δ MAPK: emerging roles of a neglected isoform. *Int. J. Cell Biol.* **2014,** 2014, 272689.

Oboh, G.; Ademiluyi, A. O.; Akinyemi, A. J. Inhibition of acetylcholinesterase activities and some pro-oxidant induced lipid peroxidation in rat brain by two varieties of ginger (*Zingiber officinale*). *Exp. Toxicol. Pathol.* **2012,** *64*(4), 315–319.

Parks-Medina, E.; Redman, R.; Reed, S.; Brabetz, B. The Mutation of Amyloid Precursor Protein and the Development of Alzheimer's Disease. New York University Ottendorfer Series 2016.

Pastor, P.; Roe, C. M.; Villegas, A.; Bedoya, G.; Chakraverty, S.; Garcia, G.; Tirado, V.; Norton, J.; Rios, S.; Kosik, K. S.; et al. Apolipoprotein Eepsilon4 modifies Alzheimer's disease onset in an E280A PS1 kindred. *Ann. Neurol.* **2003,** *54*, 163–169.

Pingali, U.; Pilli, R.; Fatima, N. Effect of standardized aqueous extract of Withania somnifera on tests of cognitive and psychomotor performance in healthy human participants. *Pharmacognosy Res.* **2014,** *6*(1), 12–18.

Pirinccioglu, M.; Kizil, G.; Kizil, M.; Kanay, Z.; Ketani, A. The protective role of pomegranate juice against carbon tetrachloride induced oxidative stress in rats. *Toxicol. Ind. Health* **2014,** *30*(10), 910–918.

Poulose, S. M., Miller, M. G., Shukitt-Hale., B. Role of Walnuts in Maintaining Brain Health with Age. *J. Nutr.* **2014,** 144(4), 561S–566S.

Prasansuklab, A.; Tencomnao, T. Amyloidosis in Alzheimer's disease: The toxicity of amyloid beta (Aβ), mechanisms of its accumulation and implications of medicinal plants for therapy. *Evid. Based Complement Alternat. Med.* **2013,** *2013*, 413808.

Priprem, A.; Watanatorn, J.; Sutthiparinyanont, S.; Phachonpai, W.; Muchimapura, S. Anxiety and cognitive effects of quercetin liposomes in rats. *Nanomed. Nanotechnol. Biol. Med.* **2008,** *4*(1), 70–78.

Psahoulia, F. H.; Moumtzi, S.; Roberts, M. L.; Sasazuki, T.; Shirasawa, S.; Pintzas, A. Quercetin mediates preferential degradation of oncogenic Ras and causes autophagy in Ha-RAS transformed human colon cells. *Carcinogenesis* **2007,** *28*, 1021–1031.

Qin, Y.; Zhang, Y.; Tomic, I.; Hao, W.; Menger, M. D.; Liu, C.; Fassbender, K.; Liu, Y. *Ginkgo biloba* Extract EGb 761 and its specific components elicit protective protein clearance through the autophagy-lysosomal pathway in tau-transgenic mice and cultured neurons. *J. Alzheimer's Dis.* **2018,** DOI: 10.3233/JAD-180426.

Rao, C. V. Regulation of COX and LOX by curcumin. *Adv. Exp. Med. Biol.* **2007,** *595*, 213–226.

Rao, R. V.; Descamps, O.; John, V.; Bredesen, D. E. Ayurvedic medicinal plants for Alzheimer's disease: a review. *Alzheimers Res. Ther.* **2012,** *4*, 22.

Ray, S.; Ray, A. Medhya Rasayanas in brain function and disease. *Med. Chem.* **2015,** *5*, 505–511.

Rigacci, S. Olive oil phenols as promising multi-targeting agents against Alzheimer's disease In Springer: Heidelberg, **2015**, 1–20.

Robert, A.; Liu, Y.; Nguyen, M.; Meunier, B. Regulation of copper and iron homeostasis by metal chelators: A possible chemotherapy for Alzheimer's disease. *Acc. Chem. Res.* **2015**, *48*(5), 1332–1339.

Sabandar, C. W.; Ahmat, N.; Jaafar, F. M.; Sahidin, I. Medicinal property, phytochemistry and pharmacology of several Jatropha species (Euphorbiaceae): a review. *Phytochemistry* **2013**, *85*, 7–29.

Sabogal-Guáqueta, A. M.; Mũnoz-Manco, J. I.; Ramirez-Pineda, J. R.; Lamprea-Rodriguez, M.; Osorio, E.; Cardona-G´omez, G. P. The flavonoid quercetin ameliorates Alzheimer's disease pathology and protects cognitive and emotional function in aged triple transgenic Alzheimer's disease model mice. *Neuropharmacology* **2015**, *93*, 134–145.

Sadowska-Bartosz, I.; Bartosz, G. Prevention of protein glycation by natural compounds. *Molecules* **2015**, *20*(2), 3309–3334.

Saleem, H.; Ahmad, I.; Shahid, M. N.; Gill, M. S.; Nadeem, M. F.; Mahmood, W.; Rashid, I. In vitro acetylcholinesterase and butyrylcholinesterase inhibitory potentials of Jatropha gossypifolia plant extracts. *Acta. Pol. Pharm.* **2016**, *73*(2), 419–423.

Sastre, M.; Ritchie, C. W.; Hajji, N. Metal ions in Alzheimer's disease brain. *JSM Alzheimer's Dis. Relat. Dement.* **2015**, *2*(1), 1014.

Saunders, A. M. Apolipoprotein E and Alzheimer disease: an update on genetic and functional analyses. *J. Neuropathol. Exp. Neurol.* **2000**, *59*, 751–758.

Scharf, J. M.; Daffner, K. R. NSAIDs in the prevention of dementia: a Cache-22. *Neurology* **2007**, *69*, 235–236.

Scholey, A. B.; Tildesley, N. T.; Ballard, C. G.; Wesnes, K. A.; Tasker, A.; Perry, E. K.; Kennedy, D. O. An extract of Salvia (sage) with anticholinesterase properties improves memory and attention in healthy older volunteers. *Psychopharmacology (Berl).* **2008**, *198*(1), 127–139.

Sehgal, N.; Gupta, A.; Valli, R. K.; Joshi, S. D.; Mills, J. T.; Hamel, E.; Khanna, P.; Jain, S. C.; Thakur, S. S.; Ravindranath, V. Withania somnifera reverses Alzheimer's disease pathology by enhancing low-density lipoprotein receptor-related protein in liver. *Proc. Natl. Acad. Sci. USA* **2012**, *109*(9), 3510–3515.

Sharma, B. R.; Gautam, L. N.; Adhikari, D.; Karki, R. A Comprehensive review on chemical profiling of *Nelumbo nucifera*: Potential for drug development. *Phytother. Res.* **2017**, *31,* 3–26.

Shen, L.; Liu, C. C.; An, C. Y.; Ji, H. F. How does curcumin work with poor bioavailability? Clues from experimental and theoretical studies. *Sci. Rep.* **2016**, *6*, 20872.

Shukitt-Hale, B.; Carey, A. N.; Jenkins, D.; Rabin, B. M.; Joseph, J. A. Beneficial effects of fruit extracts on neuronal function and behaviour in a rodent model of accelerated aging. *Neurobiol. Aging* **2007**, *28*(8), 1187–1194.

Simpson, T.; Pase, M.; Stough, C. Bacopa monnieri as an Antioxidant Therapy to Reduce Oxidative Stress in the Aging Brain. *Evid. Based Complement Alternat. Med.* **2015**, *2015*, 615384.

Singh, M.; Arseneault, M.; Sanderson, T.; Murthy, V.; Ramassamy, C. Challenges for research on polyphenols from foods in Alzheimer's disease: bioavailability, metabolism, and cellular and molecular mechanisms. *J. Agric. Food Chem.* **2008**, *56*, 4855–4873.

Singh, N.; Agrawal, M.; Doré, S. Neuroprotective properties and mechanisms of resveratrol in *in vitro* and *in vivo* experimental cerebral stroke models. *ACS Chem. Neurosci.* **2013**, *4*(8), 1151–1162.

Skoumalová, A.; Hort, J. Blood markers of oxidative stress in Alzheimer's disease. *J. Cell. Mol. Med.* **2012,** *16,* 2291–2300.

Small, D. H.; Mok, S. S.; Bornstein, J. C. Alzheimer's disease and Abeta toxicity: from top to bottom. *Nat. Rev. Neurosci.* **2001,** *2,* 595–598.

Solfrizzi, V.; Panza, F. Plant-based nutraceutical interventions against cognitive impairment and dementia: meta-analytic evidence of efficacy of a standardized Gingko biloba extract. *J. Alzheimers.* **2015,** *43,* 605–611.

Spires-Jones, T. L.; Hyman, B. T. The intersection of amyloid beta and tau at synapses in Alzheimer's disease. *Neuron* **2014,** *82*(4) 756–771.

Subedee, L.; Suresh, R.; Jayanthi, M.; Kalabharathi, H.; Satish, A.; Pushpa, V. Preventive role of Indian black pepper in animal models of Alzheimer's disease. *J. Clin. Diagn. Res.* **2015,** *9*(4), 1–4.

Tang, M.; Taghibiglou, C. The mechanisms of action of curcumin in Alzheimer's disease. *J. Alzheimers Dis.* **2017,** 58, 1003–1016.

Tangney, C. C.; Kwasny, M. J.; Li, H.; Wilson, R. S.; Evans, D. A.; Morris, M. C. Adherence to a Mediterranean-type dietary pattern and cognitive decline in a community population. *Am. J. Clin. Nutr.* **2011,** *9* (3), 601–607.

Taylor, R. S. L.; Hudson, J. B.; Manandhar, N. P.; Towers, G. H. N. Antiviral activities of medicinal plants of southern Nepal. *J. Ethnopharmacol.* **1996,** *53,* 97–104.

Thangthaeng, N.; Poulose, S. M.; Miller, M. G.; Shukitt-Hale, B. Preserving brain function in aging: the anti-glycative potential of berry fruit. *NeuroMol. Med.* **2016,** *18*(3), 465–473.

Thomas, P.; Wang, Y. J.; Zhong, J. H.; Kosaraju, S.; O'Callaghan, N. J.; Zhou, X. F.; Fenech, M. Grape seed polyphenols and curcumin reduce genomic instability events in a transgenic mouse model for Alzheimer's disease. *Mutat. Res. Fundam. Mol. Mech. Mutagen.* **2009,** *661*(1), 25–34.

Tiraboschi, P.; Hansen, L. A.; Thal, L. J.; Corey-Bloom, J.The importance of neuritic plaques and tangles to the development and evolution of AD. *Neurology,* **2004,** *62*(11), 1984–1989.

Valder, P. *Garden Plants of China.* Glebe, New South Wales: Florilegium **1999,** p. 289. ISBN 1-876314-02-8.

Van Boxtel, M.; Schmitt, J.; Bosma, H.; Jolles, J. The effect of habitual caffeine use on cognitive change: a longitudinal perspective. *Pharmacol. Biochem. Behav.* **2003,** *75*(4), 921–927.

Vassar, R. BACE1 inhibitor drugs in clinical trials for Alzheimer's disease. *Alzheimer's Res. Ther.* **2014,** *6,* 89.

Vassar, R.; Kovacs, D. M.; Yan, R.; Wong, P. C. The β-secretase enzyme BACE in health and Alzheimer's disease: regulation, cell biology, function, and therapeutic potential. *J. Neurosci.* **2009,** *29*(41), 12787–12794.

Verghese, P. B.; Castellano, J. M.; Holtzman, D. M. Roles of apolipoprotein E in Alzheimer's disease and other neurological disorders. *Lancet Neurol.* **2011,** *10*(3), 241–252.

Vingtdeux, V.; Dreses-Werringloer, U.; Zhao, H.; Davies, P.; Marambaud, P. Therapeutic potential of resveratrol in Alzheimer's disease. *BMC Neurosci.* **2008,** *9* (2), S6.

Warrier, P. K.; Nambiar, V. P. K.; Ramankutty, C. Indian medicinal plants: A compendium of 500 species (Vol. I–V), Universities Press, New Delhi. 1993–1996.

Weinreb, O.; Mandel, S.; Amit, T.; Youdim, M. B. Neurological mechanisms of green tea polyphenols in Alzheimer's and Parkinson's diseases. *J. Nutr. Biochem.* **2004,** *15,* 506–516.

Williams, C. M.; El Mohsen, M. A.; Vauzour, D.; Rendeiro, C.; Butler, L. T.; Ellis, J. A.; Whiteman, M.; Spencer, J. P. Blueberry induced changes in spatial working memory correlate with changes in hippocampal CREB phosphorylation and brain derived neurotrophic factor (BDNF) levels. *Free Radical Biol. Med.* **2008**, *45*, 295–305.

Williams, R. J.; Spencer, J. P.; Rice-Evans, C. Flavonoids: antioxidants or signalling molecules? *Free Radic. Biol. Med.* **2004**, *36*, 838–849.

Witte, A. V.; Kerti, L.; Margulies, D. S.; Flöel, A. Effects of resveratrol on memory performance, hippocampal functional connectivity, and glucose metabolism in healthy older adults. *J. Neurosci.* **2014**, *34*(23), 7862–7870.

Wobst, H. J.; Sharma, A.; Diamond, M. I.; Wanker, E. E.; Bieschke, J. The green tea polyphenol (−)-epigallocatechin gallate prevents the aggregation of tau protein into toxic oligomers at substoichiometric ratios. *FEBS Lett.* **2015**, *589*(1), 77–83.

World Health Organization (WHO). Dementia [online]. [Fact Sheet, 17 December, 2018]. Available at: http://www.who.int/mediacentre/factsheets/fs362/en/, 2018.

Yamauchi, M.; Tsuruma, K.; Imai, S.; Nakanishi, T.; Umigai, N.; Shimazawa, M.; Hara, H. Crocetin prevents retinal degeneration induced by oxidative and endoplasmic reticulum stresses via inhibition of caspase activity. *Eur. J. Pharmacol.* **2011**, *650*(1), 110–119.

Yan, M. H.; Wang, X.; Zhu, X. Mitochondrial defects and oxidative stress in Alzheimer's disease and Parkinson disease. *Free Radic. Biol. Med.* **2013**, *62*, 90–101.

Yao, E. C.; Xue, L. Therapeutic effects of curcumin on Alzheimer's disease. *Adv. Alzheimer's Dis.* **2014**, *3*, 145–159.

Zang, C. X.; Bao, X. Q.; Li, L.; Yang, H. Y.; Wang, L.; Yu, Y.; Wang, X. L.; Yao, X. S.; Zhang, D. The protective effects of *Gardenia jasminoides* (Fructus Gardenia) on amyloid-β-induced mouse cognitive impairment and neurotoxicity. *Am. J. Chin. Med.* **2018**, *46*(2), 1–17.

Zempel, H.; Mandelkow, E. Lost after translation: missorting of tau protein and consequences for Alzheimer's disease. *Trends Neurosci.* **2014**, *37*(12), 721–732.

Zeng, G. F.; Zhang, Z. Y.; Lu, L.; Xiao, D. Q.; Zong, S. H.; He, J. M. Protective effects of ginger root extract on Alzheimer's disease induced behavioral dysfunction in rats. *Rejuvenation Res.* **2013**, *16*(2), 124–133.

Zeng, K.; Li, M.; Hu, J.; Mahaman, Y. A. R.; Bao, J.; Huang, F.; Xia, Y.; Liu, X.; Wang, Q.; Wang, J. Z.; Yang, Y.; Liu, R.; Wang, X. *Ginkgo biloba* extract EGb761 attenuates hyperhomocysteinemia-induced AD like tau hyperphosphorylation and cognitive impairment in rats. *Curr. Alzheimer Res.* **2018**, *15*(11) 89–99.

Zhang, J.; Mori, A.; Chen, Q.; Zhao, B. Fermented papaya preparation attenuates β-amyloid precursor protein: β-amyloid-mediated copper neurotoxicity in β-amyloid precursor protein and β-amyloid precursor protein Swedish mutation overexpressing SH-SY5Y cells. *Neuroscience* **2006**, *143*(1), 63–72.

Zhang, X. P.; Zhang, M. L.; Su, X. H.; Huo, C. H.; Gu, Y. C.; Shi, Q. W. Chemical constituents of the plants from genus *Jatropha*. *Chem. Biodivers.* **2009**, *6*(12), 2166–2183.

Zhao, C.; Zhang, H.; Li, H.; Lv, C.; Liu, X.; Li, Z.; Xin, W.; Wang, Y.; Zhang, W. Geniposide ameliorates cognitive deficits by attenuating the cholinergic defect and amyloidosis in middle aged Alzheimer model mice. *Neuropharmacology.* **2017**, *116*, 18–29.

Zhao, Y.; Zhao, B. Oxidative stress and the pathogenesis of Alzheimer's disease. *Oxid. Med. Cell. Longev.* **2013**, 316523.

Zhou, X.; Cui, G.; Tseng, H. H. L.; Lee, S. M. Y.; Leung, G. P. H.; Chan, S. W.; Kwan, Y. W.; Hoi, M. P. M. Vascular contributions to cognitive impairment and treatments with traditional Chinese medicine. *Evid. Based. Complement. Alternat. Med.* **2016**, 9627258. doi: 10.1155/2016/9627258.

CHAPTER 3

Wild Cucurbits: An Ethnomedicinally Important Plant Species for Aboriginals of the Similipal Biosphere Reserve, Odisha, India

YASASWINEE ROUT[1], SANJEET KUMAR[1*], GITISHREE DAS[2], and JAYANTA KUMAR PATRA[2*]

[1]*Ambika Prasad Research Foundation, Bhubaneswar 751006, Odisha, India*

[2]*Research Institute of Biotechnology and Medical Converged Science, Dongguk University, Seoul, Gyeonggi-do, Republic of Korea*

Corresponding author. E-mail: jkpatra.cet@gmail.com; sanjeet.biotech@gmail.com

ABSTRACT

Fecundity-oriented research and breeding among cultivated plants in agriculture have resulted in narrowing down the genetic base of indigenous food plants. Cucurbits are the most common plants that grow either in wild or under cultivated conditions. They have enormous genetic diversity, leading to vegetative and reproductive characters. They grow in almost all vegetative regions of the world. Many reports revealed that the Cucurbits possess certain bioactive compounds that are responsible for their therapeutic values against diseases and disorders. Therefore, in the present study, an enumeration survey was carried out to document the wild Cucurbits (WCs) as prime genetic resources available in and around the protected areas of Northern part (Similipal Biosphere Reserve), Odisha State, India, during 2012-2016. A total of 11 species belonging to 8 genera of WCs were enumerated that were used as food and medicines. Out of 11, *Cucumis melo*, *Trichosanthus cucumerina*, and *Coccinea grandis* were very common in all parts of the study area. *Solena amplexicaulis* and *Mukia madrespatna* were rare in these localities. The medicinal values and active compounds present in the enumerated Cucurbits were gathered and have been documented. The present chapter highlights the importance of WCs in the advance research of pharmacology.

3.1　INTRODUCTION

Forests play a vital role in the sustainability of habitats. It is highly complex, and the constantly changing environment made up of numbers of living and nonliving things. Natural forests are found in such parts of the globe where the factors for plant growth have been ideal for several centuries (Abalaka et al., 2011; Kumar et al., 2017). The resources of forest are useful in maintaining the ecological balance, providing food and medicines, contributing raw materials to many life-line industries, protecting the wild animals (Ava, 1988; Chopra, 1993; Easterling et al., 2007; Sakals et al., 2006; Sills et al., 2011). In tropical forests, there are different types of wild, edible plants, containing essential nutrients that are useful to supplement the food for human life (Bortolotto et al., 2015; Romojaro et al., 2013). Forest resources are basically collected by tribal communities in and around the forests. The resources are obtained not only from wild plants but also from wild animals (Chopra, 1993; Delang et al., 2006; Deshmukh et al., 2011; Jefferey et al., 2003). The Food and Agricultural Organization (FAO) of the United Nations has estimated that India is covered with 19.5% of forest of the world (Food and Agricultural Organization, 2012). India is rich in floral diversities and about 45,000 plant species are found throughout the regions (Mohammed et al., 2004). In Eastern Ghats, there are about 2600 plant species (Nayak et al., 2013) that are spread from West Bengal, Odisha, Andhra Pradesh, Karnataka to Tamil Nadu. Many of them are having ethnic uses among the locals. Many of these forests have hubs of different types of medicinal plants, which are also reflected in earlier studies (Panda et al., 2013; Reddy et al., 2002).

　　The major forests of Odisha are the part of Eastern Ghats. In Odisha, as per the state record, about 37.34% of area is covered with forest. As per the Indian forest report, out of 30 districts, Mayurbhanj is covered with 1340 km^2 of dense forest, followed by Sundargarh (1046 km^2), Kandhamal (660 km^2), Sambalpur (533 km^2), and Rayagada (453 km^2) (Kumar et al., 2012; Misra et al., 2011, 2013; Panda et al., 2013). From the beginning of civilization till date, there has been the interacting relationship between human beings, forests, and their flora and fauna. The impacts of human activities and natural disasters have drastically reduced the diversity of genus, plant and animal species, ecosystems, and the landscapes. This has threatened human welfare since biodiversity acts as the potential source of foods, fibers, and medicines for livelihood. The most startling factor is that the forest resources are depleting in posthaste. In view of this, the actuality of innovative approaches was realized by the United Nations Educational, Scientific and Cultural Organization (UNESCO) for living and working in harmony with nature. The aims and objectives of the UNESCO were reflected in the form of the biosphere reserves. The Government of India has declared 18 biosphere reserves, out of which the Similipal Biosphere Reserve (SBR) stands 15th in the list. It also comes under the world network of Biosphere Reserve based on the "Man and the Biosphere Program" of the UNESCO list that was established in the year 2008 (Biosphere Reserve in India, 2007; Ministry of Environment and Forests, 2012; Kumar et al., 2017; Tripathy et al., 2013, 2014).

SBR (Figure 3.1) is situated in the central part of Mayurbhanj district in the state of Odisha, India (Bhakta et al., 2014; Kumar et al., 2013, 2014; Tripathy et al., 2013). It lies between 21° 10¢ to 22° 12¢ N latitude and 85° 58¢ to 86° 42¢ E longitude, ranging between 300 and 1180 m above sea level (Das et al., 2008; Tripathy et al., 2013, 2014). The Biosphere Reserve is also rich in fauna too with about 42 species of mammals, including wild elephant, tiger; about 30 species of reptiles; and about 242 species of birds (Rout et al., 2008). The flora of SBR has mixed vegetation, such as Orissa Semi Evergreen Forest, Tropical Moist Broad-leaf Forest, Tropical Moist Deciduous Forest, Dry Deciduous Hill Forest, High Level Sal Forest with Grassland and Savanna. It is enriched with about 1075 species of plants including 8 endangered, 8 vulnerable, 34 species of rare plants (Kumar et al., 2017; Misra et al., 2011). Out of available 96 orchid species, 2 endemic species are found in SBR (Mishra et al., 2010, 2011). Among the tribal races Ho, Kolha, Santhal, Bathudi, Bhumija, Mahali, Saunti, Munda, Gonda, and Pauri Bhuiyan have their scattered dwellings in the core, buffer, and the adjoining areas of SBR. The common occupation of these tribal communities is the collection of forest products, hunting, and traditional farming (Tripathy et al., 2013; Tripathy et al., 2014a; Tripathy et al., 2014b; Kumar et al., 2017). During their day-to-day activities for their livelihood, they collect wild food and medicines. Different tribal communities have different approaches with regard to the collection of wild plant resources and their utilization as food and medicines (Kumar et al., 2012, 2013, 2017; Misra et al., 2013). They not only collect the wild food from plants but also collect the plants or their parts to cure and protect against diseases.

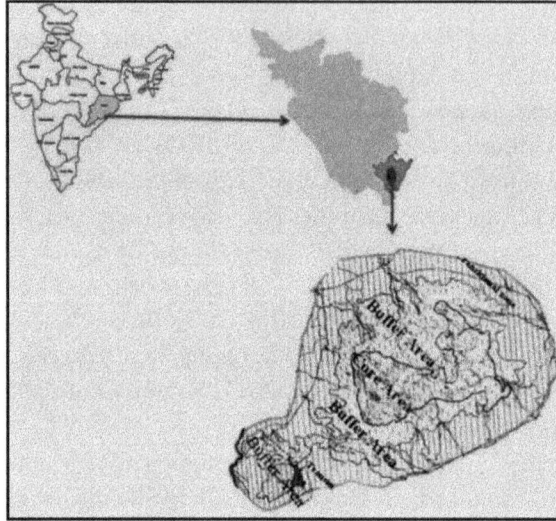

FIGURE 3.1   Geographical location of study area (Similipal Biosphere Reserve) Odisha, India.

In this context, wild Cucurbits (WCs) are important as per their availability and palatability. Most common Cucurbits grow either in wild or under cultivated conditions (Tripathy et al., 2013, 2014a, 2014b). They have tremendous genetic diversity, extending to vegetative and reproductive characters. They grow in tropical, subtropical, deserts, and temperate regions of the world. In an ancient Chinese medicinal book *BanCao Gang Me*, the importance and uses of Cucurbits are described. The plant species of this group produced pepo fruits that were often used as vegetables in most of the early civilizations (Kumar et al., 2017). Many early therapeutic literature and practices show their uses as medicines. Many reports revealed that Cucurbits possess certain bioactive compounds that are responsible for their therapeutic values against diseases and disorders. Some published reports show that the most common Cucurbits such as *Cucumis sativa* is used to reduce blood pressure and *Luffa cylindrica* seeds flour are consumed for child growth. It was noted that *Coccinia grandis* (*C. grandis*) is used against ring worm, small pox, scabies, skin eruption, ulcers, while *Momordica charantia* (*M. charantia*) is a common plant of the state used to suppress neural response

(Kumar et al., 2017; Tripathy et al., 2013, 2014a, 2014b, 2014c). The major common WCs available in SBR are *C. grandis*, *Cucumis trigonus*, *Diplocyclos palmatus*, *Luffa aegyptica*, *M. charantia*, *Momordica dioica*, *Mukia maderaspatana*, *Solena amplexicaulis*, *Trichosanthes tricuspidata*, *Trichosanthes cucumerina*, among others (Kumar et al., 2012, 2017; Tripathy et al., 2013) (Figure 3.2). Keeping all these in views, the present review presents the detail medicinal and food values of the selected wild Cucurbits available in the study area along with their ethnobotany and pharmaceutical potential. These plants were selected based on their availability and use by the local and tribal inhabitants.

3.2 ETHNIC COMMUNITIES OF THE PROTECTED AREA OF NORTHERN ODISHA (SIMILIPAL BIOSPHERE RESERVE)

SBR is covered with a dense forest. Its frosted field tracks are the homeland of tribal communities. The tribes present in and around SBR are HO, Kolho, Santhal, Bathudi, Bhumijo, Mahali, Saunti, Munda, Gond, Huria, Bhuiyan, etc. (Kumar et al., 2012, 2017; Tripathy et al., 2013). Among the tribal groups, Hill-Kharia and Mankirdias are considered as the primitive tribal groups (Kumar et al., 2012, 2017; Tripathy et al., 2013). SBR comprises about 61 villages within the buffer zones and about 1200 villages inside the peripheral zone. The forest inhabitant communities have maintained their primitive culture, traditional skills, and rituals. The main occupations of the tribal groups are hunting and traditional farming, along with the collection of forest products. In the present study, the Ho, Bathudi, Kolho, Munda, and Santhal tribes were contacted for their traditional knowledge on wild plants and use of these plants or their parts in curing different diseases (Kumar et al., 2013, 2017; Tripathy et al., 2013).

Rice, maize, and millets are the main crops for the cultivation of tribal communities of SBR (Kumar et al., 2017). In addition, they also grow vegetables such as pumpkin, brinjal, pea, and onion in their home gardens (Kumar et al., 2012; Tripathy et al., 2013). Rice is the principal food for them. Apart from their main occupation, they collect different forest products from the forests. They depend upon wild medicinal plants available in the nearby forest to cure different diseases (Kumar et al., 2013; Tripathy et al., 2013, 2014; Vinod et al., 2010). Detailed information on some WCs found abundantly in and around SBR, and their use by the local people has been documented in this review. The plants were selected as per the availability and rate of consumption and storage by the rural and tribal communities (Figure 3.2). This plant species of the region were identified following Flora's Books (Haines, 1925; Kumar et al., 2012, 2017; Saxena et al., 1995; Tripathy et al., 2013). The selected plant species were enumerated by the Bentham and Hooker system, and taxonomical characterization was done using morphological characteristics, followed by Flora's Book and published articles (Haines, 1925; Ikediobi et al., 1983; Saxena et al., 1995).

FIGURE 3.2 Common Wild Cucurbits available in Similipal Biosphere Reserve, Odisha, India.

3.3 ENUMERATED WILD CUCURBITS OF SBR

Coccinia grandis (**L.**) **Voigt.:** Climbing herb having attractive white flower. It is a common weed in SBR. Its stem is angular and glabrous. Leaves are ovate, entric, and lobed. Frequently several large circular glistening glands are seen near the base. Tendrils are simple. Fruits are oblong and narrowed apically. It is green when young and turn red when ripen. Seeds are oblong and compressed (Kumar et al., 2017; Tripathy et al., 2013).

Cucumis hardwickii **Royle:** It is a slender with scabrid stems. Leaves are of 12 cm in size, entire or shallowly 5-lobed, and scabrous. Flowers are yellow in color. It possesses multiple fruiting habits; its fruits are dark green in color, and oblong or subglobose. Seeds are many and compressed (Kumar et al., 2017).

Cucumis melo **L.:** It is a prostrate procumbent herb. Its stem is thickened toward the base, scabrid, and hispid with hairs. Leaves are orbicular and ovate with shallow rounded subangular lobes. Leaf margins are denticulate. Flowers are yellow in color, and both are present in the same plant. Fruits are spherical and ovoid, and obtuse at ends. Fruits appear as green striped when young and turn yellow when ripen (Kumar et al., 2017; Tripathy et al., 2013).

Cucumis sativus **L.:** Herbaceous vine with pubescent stems and unbranched tendrils. Leaves are alternate and simple, with 3–7 palmate lobes and serrated margins. Flowers are yellow in color. Hairless cylindrical fruits are warty, yellow to green (Kumar et al., 2017; Sahu et al., 2015; Tripathy et al., 2013).

Diplocyclos palmatus (**L.**) **Jeffrey.:** It is a climbing herb with smooth stem. Its leaves are orbicular and ovate, deeply lobed, slightly scarbid above, and almost smooth beneath. Leaf margins are minutely denticulate. Flowers are small, yellowish, and both clustered in the same axils. Fruits are globose, smooth, and green with white strips when young and turn red with stripes when ripen. Seeds are embedded in the blue green pulp. They are pyriform, surrounded by a thick ring on either side which appear like "*Shivaling*," so it is locally known as "Shivilingi" (Bhakta et al., 2014; Kumar et al., 2017; Tripathy et al., 2014a).

Gymnopetalum cochinchinensis (Lour.) Kurz: A slender creeper with a scabrid and hairy stem. Leaves reniform to triangular, 5-angled or lobed half-way, edges crenate-dentate. Flowers are white, ovate, and oblong. Fruits are 10-ribbed beaked and orange (Kumar et al., 2017).

Gynostemma pedata **Blume:** Slender climber herb; stem pubercelous, tendrils simple; leaves pedately 3–5 foiate. Flowers are axillary and whitish-yellow yellow in color; fruits are globose with compressed seeds (Kumar et al., 2017; Rout et al., 2008).

Luffa acutangula **(L.) Roxb.:** It is an annual climber with an angular stem. Its leaves are lobed-angled and deeply cordate. Tendrils are trifid and subhispid. Flowers are yellow. Male and female are present in the same axils. Fruits are clavate and oblong with 10-angled and apex obtuse. Seeds are black, ovate, compressed, and 9–10 mm long (Kumar et al., 2017; Tripathy et al., 2013).

Luffa aegyptiaca **Mill.:** It is an annual climber with yellow flowers. Leaves alternate and large. Fruits green, cylindrical and smooth (Tripathy et al., 2013; Kumar et al., 2017).

Momordica charantia **L.:** It is monaecious, softly hairy climbing annual herb. Leaves are lobed, 3–7foliate with entire margin. Tendrils are simple. Flowers yellow. Fruits ovoid or fusiform with tapering end. Fruits green when young and gradually turn to yellow when ripen and finally turn to red. Seeds are compressed, ovate, sculptured on surfaces (Tripathy et al., 2013; Kumar et al., 2017).

Momordica dioica **Roxb. ex Willd.:** It is a slender and glabrous. Stem angular. Leaves simple or lobed, ovate, deeply cordate and often sinuately denticulate. Flowers solitary, yellow and dioecious. Fruits ellipsoid, ovoid and covered with soft fleshy spines. Seeds are ellipsoid and closely inverted with an aril-like integument (Tripathy et al., 2013; Kumar et al., 2017).

Mukia maderaspatana **(L.) Roem.:** It is spiny, branched and monoceious with red ripen berry in groups of three fruits. It as scarbous climbing herb. Leaves ovate or deltoid, angular and acute. Flowers are yellow and small. Fruits are scarlet, globose, green when young and turned red when ripen (Tripathy et al., 2013; Kumar et al., 2017).

Solena amplexicaulis **Lam.:** It usually found on the foot hills at SBR. It is prostrate or climbing herb. Stem is angled and smooth. Leaves are poly-morphic, ovate, lobed, cordate, and hastate at base. Tendrils simple. Flowers white. Fruits are ellipsoid with red pulp. Seeds slightly compressed and white. Roots are in tuberous form (Tripathy et al., 2013; Kumar et al., 2017).

Thladiantha cordifolia **(Blume) Cogn.:** Climber with ovate-heart-shaped leaves and bell-shaped golden yellow flowers. Leaves are round and bristly haired. Fruit is ellipsoidal, about 3.5 cm long, longitudinally.

Trichosanthes cucumerina **L.:** It is a climber having tri-branched tendrils along beautiful white flowers. Stems often angled. Leaves long-petiole, orbicular-reniform or broadly ovate, or lobed, deeply cordate, denticulate from the mucronate nerve endings. Leaves are also pubescent or somewhat scabrous beneath, puberulous above. Flowers monoecious, male and female arising from the same axil. Fruits spindle-shaped rostate, green with white strips when young and red without strips when ripen (Vinod et al., 2010; Tripathy et al., 2013, 2014a, 2014b; Kumar et al., 2017).

***Trichosanthes tricuspidata* Lour.:** It is a large climber having attractive fruits like red bulb along bright white flowers. Stem is suffruticose, branches long pendent. Leaves are broadly ovate, simple or deeply palmately 3–5 lobed, cordate, denticulate, with large green glands near base, upper surface smooth and bright green when fresh but very scabrous when old, lower surface paler with cystoliths on the nerves when dry. Flowers: white and dioecious. Fruit: brightscarlet, globose, on axillary short stout peduncle. Seeds embedded in dark green pulp, oblong, flattened, slightly narrowed at base (Kumar et al., 2017; Tripathy et al., 2013, 2014).

3.4   ETHNOBOTANICAL DATA ON SELECTED WILD CUCURBITS FOUND IN THE REGION

Field work was conducted with the rural and tribal communities of SBR and its adjoining areas during 2009–2017, and the data on the potential of the WCs were documented (Figure 3.3). The methodological frameworks for the ethnobotanical study were done as per the standard techniques of exploration and germplasm collection (Christian et al., 2004; Hawkes et al., 1980), qualitative and quantitative ethno-biological approaches in the field, interviews, elicitation methods, and data collection followed by authentication (Martin, 1995). Intensive and extensive field surveys were done in different landscapes and micro-ecological niches across forest types, adjoining valleys, homesteads, kitchen gardens, farmland, fallow lands, etc. in the core, buffer, and peripheral regions of SBR, covering randomly selected 20 villages of the locality. The field surveys including weekly markets (*haat*) of peripheral regions of SBR were undertaken during different seasons of the study periods. Each survey took 1–2 weeks. The standard participatory rural appraisal method (Cunningham, 2001) was adopted for sampling and data collection to incorporate the indigenous knowledge. Opinions of tribal people were taken regarding the uses of experimental plant species through questionnaires.

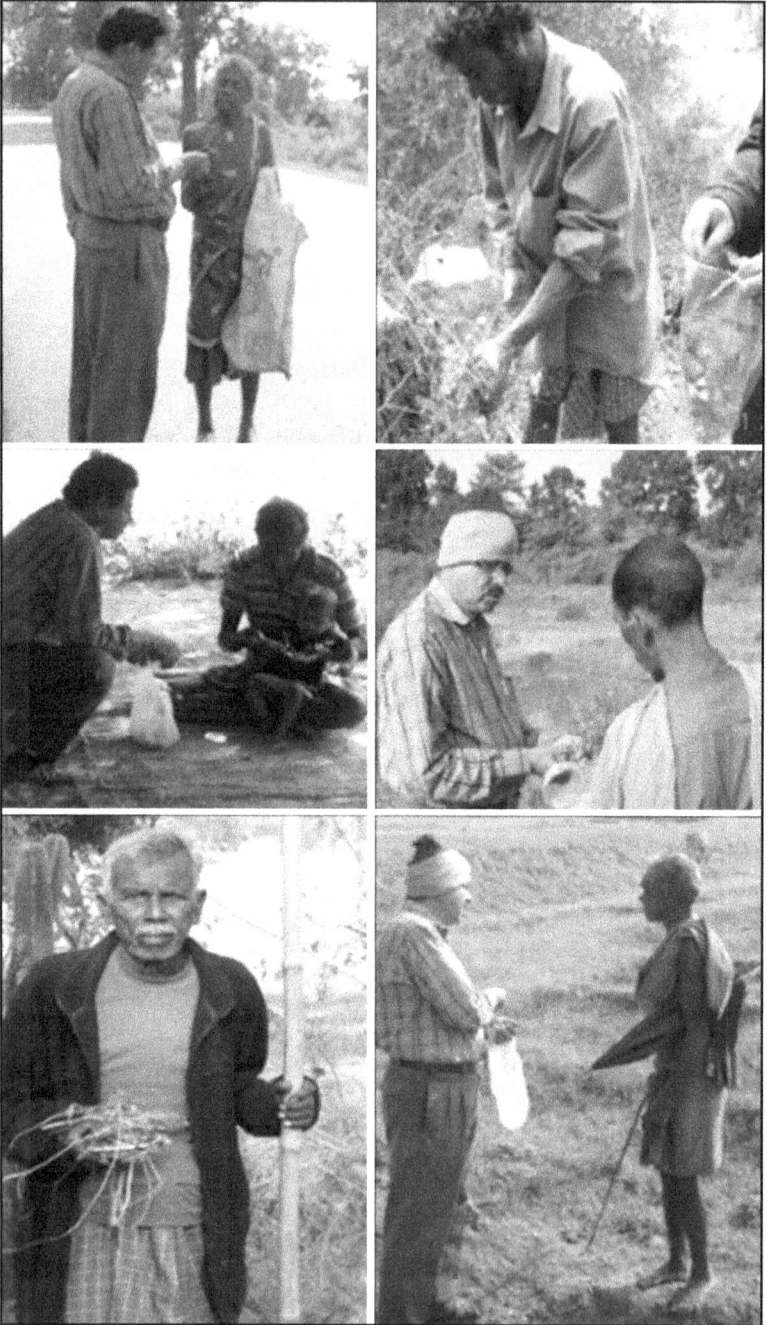

FIGURE 3.3   Collection of ethnobotanical and medicinal values of Cucurbits available in peripheral areas of Similipal Biosphere Reserve, Odisha, India.

3.5   CUCURBITACEAE: A FAMILY OF GOURD, THEIR BOTANY, ORIGIN, AND DISTRIBUTION

About 118 genera and 825 species belong to the Cucurbitaceae family worldwide (Jeffrey, 1962; Kumar et al., 2017). They are believed to be among the primitive plant families and are mostly utilized as vegetables by human beings (Tripathy et al., 2017). Researches indicate that the fruits of many Cucurbits have been used since ancient civilization till modern era as a source of food or alternative food (Kumar et al., 2017). In primitive, people in different parts of the world have observed the nutritional importance of Cucurbits and started to use their own techniques to remove bitter components from vegetative parts of Cucurbits to make them palatable (Anilkumar et al., 2015). The earliest records of edible Cucurbits used by humans have come from Mexico (Bisognin, 2002). The earlier evidence further traced out that the people of China started the domestication of some Cucurbits, which reflect their origin too (Bisognin, 2002).

The family Cucurbitaceae is predominantly distributed around the Tropics and Sub-Tropical Regions (Kumar and Tripathy, 2017; Tupe et al., 2013). They are sensitive to frost. Most of the species of this family are annual or perennial vines (Haines, 1925; Kumar et al., 2017). Some are also woody lianas, shrubs with watery sap, thorny shrubs, scandent, or prostrate. Roots are fibrous and tuberous (Tripathy et al., 2013). Most of the species have yellow or white flowers. The stems are hairy and penta-angular. Tendrils are present at 90° to the leaf petioles at nodes. Tendrils are solitary, lateral, simple or branched, spirally twisted, and rarely absent. Leaves are ex-stipulate, alternate simple, palmi-nerved, palmately lobed, or palmately compound. Lamina is a variable among the members in the same species or even in the individual plant. Flowers are usually unisexual. They are either small or large. Male and female flowers are found in different plants or in the same plant. Flowers are paniculate, racemose, or subumbel-late, and rarely solitary. The fruit is often modified berry called a pepo. It is indehiscent, pendulous, or ascending, often compressed, rarely winged. Seeds are usually many (Kumar et al., 2017; Saxena et al., 1995; Tripathy et al., 2013).

3.6   ETHNOBOTANICAL IMPORTANCE OF CUCURBITS

Many species of the family possess high traditional therapeutic values among the rural and tribal communities (Kumar et al., 2017; Tripathy et al., 2013). Numerous studies have been conducted on the taxonomy, ethnobotany, economic values, medicinal properties, pharmacological values, and domestication of species of this family. However, very less or few scientific reports are available on the wild species belonging to this family (Kumar et al., 2012, 2017; Tripathy et al., 2013). Most of wild species have nutritional as well as therapeutic values. They are mostly grown wild in and around the rural and tribal villages and forest edges. Tribal communities mostly use the plant parts of these WCs as juice, paste, and other formulations (Tripathy et al., 2013, 2014a, 2014b). Rout et al. (2008) reported the ethnobotanical values of M. charantia used by the local inhabitants of the Koraput district of Odisha. Murthy et al. (2013) observed the traditional medicinal systems using WC from the Eastern Ghats of Odisha. Mallik and Akhter (2012) documented the therapeutic values of C. grandis leaf to cure cough and cold and M. charantia fruit to enhance the appetite of domestic animals. Sadangi et al. (2005) documented that Trichosanthes species are used to cure ear problems in India. Panda et al. (2013) reported that C. grandis is used to cure jaundice and M. dioica is used in diabetes, while Pekamwar et al. (2013) documented that the raw fruits of C. grandis are eaten to reduce cough. Nowadays such knowledge on WC is gradually declining (Kumar et al., 2017). Details are listed in Table 3.1.

3.7   BIOACTIVE COMPOUNDS PRESENT IN WILD CUCURBITS

Floral wealth is a rich source of secondary metabolites (Kumar et al., 2017; Tripathy et al., 2014). However, the nature and composition of these compounds vary from species to species (Sassidharan et al., 2011). Some of these compounds form a part of the secondary metabolites present in plants. The bioactive compounds present in plants most often protect the plants against various infections forming plant defense systems (War et al., 2012). These compounds are also the base of medicinal properties of the plant species. The members of the family Cucurbitaceae are also quite rich in various bioactive compounds (Castillejo et al., 2016; Kumar et al., 2017). The potential of some of these compounds has been studied. However, not much of information/reports available on the bioactive compounds present WC. Therefore, in the present study, steps have been taken for screening the active constitutes present in plant parts of some WCs. WCs contain common compounds called Cucurbitacin (Kumar et al., 2017; Tripathy et al., 2014). Details are listed in Table 3.2.

TABLE 3.1 Ethno-medicinal Values of the Selected Wild Cucurbits

Plant Name	Parts	Uses	Sources
Cucumis melo	Seeds	Seeds are diuretic promotes the production of urine	Tahseen et al., 2013
C. melo	Leaves	Used to treat the skin infection	Hussain et al., 2010
C. melo	Fruits	Used in stomach ache	Mathur et al., 2011
C. melo	Leaves	Used to treat the skin infection	Hussain et al., 2010
Diplocyclos palmatus	Leaves	Paste and juice is used for the treatment of scorpion bite, used against fever. Leaf juice is given for fever	Tahseen et al., 2013
D. palmatus	Fruits	Skin infection	Chaturvedi et al., 2014
D. palmatus	Seeds	Seed extract is taken during dysentery	Kamble et al., 2008
D. palmatus	Seeds	Powder of seeds and roots is given twice a day in empty stomach to induce fertility	Gupta et al., 2010
D. palmatus	Leaves	Leaf paste is mixed with fruit powder in equal quantities is taken with a cup of local wine prepared by *Madhuca indica* flowers at morning without any food for three months to cure hysteria	Neelima et al., 2011
D. palmatus	Leaves	Leaves are used against inflammation	Chopra et al., 1956
D. palmatus	Seeds	Used in asthma, cholera and promotes fertility in women	Singh et al., 2012
D. palmatus	Leaf	Paste is applied externally to reduce rheumatic pain	Subramanyam et al., 2009
D. palmatus	Leaf	Used in burning and diarrhoea	Radha et al., 2012
D. palmatus	Fruits	Used in hysteria and jaundice	Radha et al., 2012
D. palmatus	Seed	Powder in quarter spoon is administrated with milk to break sterility	Patel et al., 2013
D. palmatus	Leaves	Young leaves are used in joint pain	Ganeshan et al., 2006
D. palmatus	Fruits	Fruits are used to reduce fever	Vadnere et al., 2013
D. palmatus	Leaf	Leaves are used for fertility	Venkateshwarlu et al., 2010
D. palmatus	Roots	Roots are used in teeth decay	Patel et al., 2013
D. palmatus	Leaf	Paste is applied externally on join to cure pain in the morning once in three days for one month	Ganeshan et al., 2006
D. palmatus	Seed	Seed powder is put in water and ½ cup full of water is taken, seed powder mixed in ½ cup milk is taken by women	Kamble et al., 2008

TABLE 3.1 *(Continued)*

Plant Name	Parts	Uses	Sources
D. palmatus	Leaves	Paste and juice is used for the treatment of scorpion bite, used against fever. Leaf juice is given for fever	Tahseen et al., 2013
D. palmatus	Fruits	Skin infection	Tripathy et al., 2013
D. palmatus	Seeds	Seed extract is taken during dysentery	Kamble et al., 2010
D. palmatus	Seeds	Powder of seeds and roots is given twice a day in empty stomach to induce fertility	Gupta et al., 2010
D. palmatus	Seeds	Seeds are given to women for conception	Patel et al., 2013
D. palmatus	Leaves	Used in colic, fever, paralysis of tongue, piles	Patel et al., 2013
Trichosanthes cucumerina	Fruits	Fruits are used against stomach worm	Tahseen et al., 2013
T. cucumerina	Leaves	Used in snake bite	Jain et al., 2011
T. cucumerina	Root	Used to reduce colic pain	Jain et al., 2011
T. cucumerina	Seed	Seed powder mixed is given for the treatment of scropian bite	Jain et al., 2011
T. tricuspidata	Roots	Used against skin infections	Tahseen et al., 2013
T. tricuspidata	Leaf	Juice of leaves extract applied to relieve joint pain	Bharath et al., 2010
T. tricuspidata	Leaves	Used in ear pain	Sadangi et al., 2005
T. tricuspidata	Fruits	Fruits are smoked in asthma	Jadhav et al., 2011
T. tricuspidata	Fruits	Used in asthma and ear pain	Ganeshan et al., 2006
T. tricuspidata	Root	Powdered roots and stems are taken with hot water twice a day for the treatment of dysentery by Nyshi tribe	Shrivastava et al., 2013
T. tricuspidata	Flowers	Used for blood-related problems Flowers are taken during night once in 2 days for one month	Kamble et al., 2010
T. tricuspidata	Seed	Seed powder mixed with water is given once a day for conjugative 5 days after manse	Kamble et al., 2010

TABLE 3.2 Bioactive Compounds Isolated from the Wild Cucurbits

Plant Name	Bioactive Compound[s]	Supporting Literature[s]
Cucumis melo	Phenolic Glycosidess	Dhaiman et al., 2012
C. melo	Cytokines	Arora et al., 2011
C. sativus	Isovitexin and Saponin	Dhaiman et al., 2012
D. palmatus	Bryonin, Punicic acid, Glucomannon,	Joshi, 2010
L. siceraria	Flavones-C Glycosidess Lagenin 22-Deoxy Cucurbitacin-D	Kumar et al., 2012
L. cylendrica	Lucysides, Lucyin A	Pratap et al., 2012
L. cylindrical	Lucosides	George et al., 2011
M. charantia	Momorcharins, Momordin	Paul et al., 2010
Trichosanthes cucmerina	Trimethoxy 3',4'-methylene dioxyiso flavene	Reddy et al., 2010
T. cucumerina	Cucurbitacin, 23–24 Dihydro-iso Sterol 2/3-sito sterol Stigmasterol	Sandhya et al., 2010
T. cucumerina	Tricosanthin	Bakare et al., 2010
T. cucumerina	Sio Cucurbitacin, 2 bita Sitosterol stigma sterol	Sandhya et al., 2010
T. cucumerina	Triterpenoid-Saponins	Murthy et al., 2013
T. cucumerina	Cucurbitacin	Patel et al., 2013
T. cucumerina	Trichosanthin	Edeoga et al., 2010
T. tricospidata	Cucurbitacin K 2-*o*-beta-gluco-pyranoside	Dhaiman et al., 2012
Trichosanthes spp.	Trichosanthin	Kage et al., 2009

KEYWORDS

- **wild cucurbits**
- **biodiversity**
- **ethnomedicine**
- **richness**
- **bioactive compounds**
- **neutraceuticals**

ACKNOWLEDGMENTS

Authors are thankful to Dr. Prakash Kumar Tripathy; Dr. Padan Kumar Jena, Department of Botany, Ravesnhaw University, Cuttack, India and forest officials of Similipal Biosphere Reserve, Odisha, India. Authors are also thankful to the local community. JK Patra and G Das are thankful to Dongguk University, Republic of Korea for support.

REFERENCES

Abalaka, M. E.; Inabo, H. I.; Onaolapo, J. A.; Olonitolo, O. S. Antioxidant capabilities of extracts and antimicrobial activities of chromatographic fractions of *Momordica charantia* L. (Cucurbitaceae). *Ferm. Tech. Bioeng.* **2011**, *1*, 1–6.

Anilakumar, K. R.; Kumar, G. P.; Ilaiyaraja, N. Nutritional, pharmacological and medicinal properties of *Momordica charantia. Int. J. Nut. Food Sci.* **2015**, *4*(1), 75–83.

Arora, R.; Kaur, M.; Gill, N. S. Antioxidant activity and pharmacological evaluation of *Cucumis melo* var *agrestis* methanolic seed extract. *Res. J. Phyto. Chem.* **2011**, *5*(3), 146–155.

Ava, W. Small-scale utilization of rotten by a semi community in West Malaysia. *Econ. Bot.* **1988**, *42*(1), 105–119.

Bakare, R. I.; Magbagbeola, O. A.; Akinwande, A. I.; Okunowo, O. W. Nutritional and chemical evaluation of *Momordica charantia. J. Med. Plants Res.* **2010**, *4*(2), 2189–2193.

Bhakta, S.; Pattanaik, L.; Dutta, P.; Sahu, E.; Bastia, A. K. Diversity of corticolours algae from Similipal Biospher Reserve, Mayurbhanj, Odisha. *Phykos.* **2014**, *44*(1), 9–16.

Bharath, R. K.; Narayanam, B. S.; Tribal medicinal studies on Sriharikota Island, Andhra Pradesh. *Ethno. Leaflets* **2010**, *14*, 95–107.

Bisognin, D. A. Origin and evolution of cultivated Cucurbits. *Sci. Rural.* **2002**, *32*(4), 103–103.

Bortolotto, I. M.; Amorozo, M. C.; Neto, G. G.; Olderland, J.; Damasceno, G. A. Knowledge and use of wild edible plants in rural communities along Paragyay River, Pantanal, Brazil. *J. Ethnobiol. Ethnomed.* **2015**, *11*, 46–49.

Biosphere Reserve in India. *Guidelines and Performance*. New Delhi: Ministry of Environment and Forests, **2007**, pp. 1–35.

Castillejo, N.; Martinez, H. G. B.; Monaco, K.; Gomez, P. A.; Aguayo, E.; Artes, E.; Artes, H. F. Preservation of bioactive compounds of a green vegetables smoothie using short time-high temperature mild thermal treatment. *Food Sci. Technol. Int.* **2016**, *27*, 1–6.

Chaturvedi, Y.; Saxena, M. Ethnomedicinal study of plants with special reference to bacterial diseases in Kondar and Saur tribes of district Chhatarpur, Madhaya Pradesh, India. *World J. Pharm. Pharm. Sci.* **2014**, *3*(7), 1899–1904.

Chopra, K. The value of non-timber forest products: estimation for tropical deciduous forest in India. *Econ. Bot.* **1993**, *47*(3), 251–257.

Chopra, R. N.; Chopra, S. L.; Chopra, I. C. *Glossary of Indian Medicinal Plant*. New Delhi: CSIR, **1956**, pp. 1–42.

Christian, R. V.; Brigitte, V. L. Tools and methods for data collection in ethnobotanical studies of home gardens. *Field Method.* **2004**, *16*(3), 285–306.

Cunningham, A. B. *Applied Ethnobotany: People, Wild Plant Use and Conservation*. London: Earthscan Publications. **2001**.

Das, S.; Das, B. P. Similipal Biosphere: genesis and history. *Orissa Rev.* **2008**, *1*, 1–9.

Delang, C. O. The role of wild food plants in poverty alleviation and biodiversity conservation in tropical countries. *Prog. Dev. Stud.* **2006**, *6*(4), 275–286.

Deshmukh, B. S.; Waghmode, A. Role of wild edible fruits as a food resource: traditional knowledge. *Int J. Pharm. Life Sci.* **2011**, *2*(7), 919–924.

Dhaiman, K.; Gupta, A.; Sharma, D. K.; Gill, N. S.; Goyal, A. A review on the medicinally important plants of the family Cucurbitaceae. *Asian J. Clin. Nutr.* **2012**, *4*(1), 16–26.

Easterling, W. E.; Aggarwal, P. K.; Batima, P.; Brander, K. M.; Erda, L. Food, fiber and forest products. In: Parry, M. L.; Canvziani, O. F.; Palutikof, J. P.; Linden P. J. V.; Hanson C. E. (Eds.). *Climate Change: Impacts, Adaptation and Vulnerability.* Cambridge: Cambridge University Press, **2007**, pp. 273–286.

Edeoga, H. O.; Osuagwu, G. G. E.; Omosun, G.; Mbaebie, B. O.; Osuagwu, A. N.; Pharmaceutical and therapeutic potential of some wild Cucurbitaceae species from South-East Nigeria. *Recent Res. Sci. Technol.* **2010**, *2*(1), 63–68.

Eisenber, D. M.; Kessler, R. C.; Foster, C.; Norlock, F. E.; Calkins, D. R.; Delbanco, T. L. Unconventional medicine in the United States: prevalence, costs and pattern of use. *New Engl. J. Med.* **1993**, *328*(4), 246–252.

Food and Agriculture Organization. *Food and Agriculture Organization Report 2011-2012.* Rome: FAO, **2012**, pp. 1–51.

Ganeshan, S.; Venkateshan, G.; Banumathy, N. Medicinal plants used by ethnic group Thottianaickans of Semmalai hills (reserved forest), Tiruchirappalli district, Tamil Nadu. *Indian J. Tradit. Know.* **2006**, *5*(2), 245–252.

George, V. V. S.; Surekha, S. P. Photochemical and biological screening of *Luffa cylendrica* Linn. Fruit. *Int. J. Pharm. Res.* **2011**, *3*(3), 1582–1585.

Gupta, R.; Vairale, M. G.; Deshmukh, R. R.; Chaudhary, P. R.; Wate, S. R. Ethnomedicinal uses of some plants used by *Gond* tribe of Bhandara district, Maharashtra. *Indian J. Tradit. Know.* **2010**, *9*(4), 713–717.

Haines, H. H. *The Botany of Bihar and Orissa.* London: Adlard and Son and West Newman, **1925**, pp. 1115–1124.

Hawkes, J. G. *Crop Genetic Resources: Field Collection Manual.* International Board of Plant Genetic Resources, Rome, Italy and EUCARPIA. England: University of Birmingham, **1980**.

Hussain, K.; Nisar, M. F.; Majeed, A.; Nawaz, K.; Bhatti, K. H. Ethnomedicinal survey for important plants of Jalalpur Jattan, District Gujrat, Punjab, Pakistan. *Ethnobotan. Leaflets.* **2010**, *14*, 807–825.

Ikediobi, C. O.; Gboanual, L. C. Identification of Yam (*Dioscorea* spp.) species and cultivars by use of electrophoretic patterns of soluble tuber proteins. *Biotropica.* **1983**, *15*(1), 65–67.

Jadhav, P. D.; Mahadkar, S. D.; Valvi, S. R. Documentation and ethnobotanical survey of wild edible plants from Kolhapur district. *Recent Res. Sci. Technol.* **2011**, *3*(12), 58–63.

Jain, A.; Katewa, S. S.; Sharma, S. K.; Praveen, G.; Jain, V. Snake lore and indigenous snakebite remedies practiced by some tribals of Rajasthan. *Indian J. Tradit. Know.* **2011**, *10*(2), 258–268.

Jefferey, R.; Sunder, N.; Mishra, A.; Peter, N.; Tharakn, P. J. A move from minor to major: competing discourses of non-timber forest products in India. In: Greenough, P. (Ed.). *Nature in the Global South: Environmental Projects in South and South East Asia.* Durham, NC: Duke University Press, **2003**, pp. 9–103.

Jeffrey, C. Notes on Cucurbitaceae, including a proposed new classification of the family. *Kew Bullet.* **1962**, *15*(3), 337–371.

Joshi, S. G. *D. Palmatus* Jeff. *Medicinal Plants.* New Delhi: Oxford University Press, **2010**, pp. 161–161.

Kage, D. N.; Seetharam, Y. N.; Malashetty, B. *In vitro* antibacterial property and phytochemical profile of *Trichosanthes cucumerina* L var. *cucumerina*. *Adv. Nat. Appl. Sci.* **2009**, *3*(3), 438–441.

Kamble, S. Y.; More, T. N.; Patil, S. R.; Pawar, S. G.; Bindurani, R.; Bodhankar, S. L. Plants used by the tribes of Northwest Maharashtra for the treatment of gastrointestinal disorders. *Indian J. Tradit. Know.* **2008**, *7*(2), 321–325.

Kamble, S. Y.; Patil, S. R.; Sawant, P. S.; Sawant, S.; Pawar, S. G.; Singh, E. A. Studies of plants used in traditional medicine by Bhilla tribe of Maharashtra. *Indian J. Tradit. Know.* **2010**, *9*(3), 591–598.

Kumar, S.; Behera, S. P.; Jena, P. K. Validation of tribal claims on *Dioscorea pentaphylla* L. through phytochemical screening and evaluation of antibacterial activity. *Plant Sci. Res.* **2013**, *35*, 55–61.

Kumar, S.; Das, G.; Shin, H. S.; Kumar, P.; Patra, J. K. Evaluation of medicinal values of *Gymnopetalum chinense* (Lour.) Merr.: a lesser known cucurbits from Eastern Ghats of India. *Brazal. Arch. Biol. Technol.* **2017**, *60*, e17160580.

Kumar, S.; Jena, P. K.; Tripathy, P. K. Study of wild edible plants among tribal groups of Similipal Biosphere Reserve Forest, Odisha, India; with special reference to *Dioscorea* species. *Int. J. Biol. Technol.* **2012**, *3*(1), 11–19.

Kumar, S.; Mahanti, P.; Das, G.; Patra, J. K. Country liquors of Similipal Biosphere Reserve, Odisha India: a staple fermented food of the tribal communities. *EC. Microbiol.* **2017**, *9*(3), 140–145.

Kumar, S.; Tripathy, P. K. *Wild Cucurbits: Source of Traditional Therapeutic Systems and Medicines*. Germany: Lambert Academic Publishing, **2017**.

Mallik, J.; Akhter, R. Phytochemical screening and *in-vitro* evaluation of reducing power, cytotoxicity and anti-fungal activities of ethanol extracts of *Cucumis sativus*. *Int. J. Pharm. Biol. Arch.* **2012**, *3*(3), 555–560.

Martin, J. M.; Ethnobotany: a method manual. London: Champman and Hall, **1995**, pp. 1–150.

Mathur, A.; Singh, R.; Yousuf, S.; Bhardwaj, A.; Verma, S. K.; Babu, P.; Gupta, V.; Prasad G. B. K. S.; Dua, V. K. Antifungal activity of some plant extracts against clinical pathogens. *Adv. Appl. Sci. Res.* **2011**, *2*(2), 260–264.

Ministry of Environment and Forests. *Annual Report*. **2012–2013**. New Delhi: Ministry of Environment and Forests, pp. 1–400.

Mishra, B. K. Conservation and management effectiveness of Similipal Biosphere Reserve, Orissa, India. *Indian Forest.* **2010**, *136*(10), 1310–1326.

Misra RC, Sahoo HK, Pani DR, Bhandari DC. Genetic resources of wild tuberous food plants traditionally used in Similipal Biosphere Reserve, Odisha, India. *Genet. Resour. Crop Evol.* **2013**, doi:10.1007/s10722-013-9971-6.

Misra, R. C.; Sahoo, H. K.; Mohapatra, A. K.; Reddy, R. N. Addition to the flora of Similipal Biosphere Reserve, Odisha, India. *J. Bombay Nat. Hist. Soc.* **2011**, *108*(1), 69–76.

Mohammed, S.; Kasera, P. K.; Shukla, J. K. Unexploited plants of potential medicinal value from the Indian Thar desert. *Nat. Prod. Rad.* **2004**, *3*(2), 69–74.

Murthy, K. S. R.; Ravindranath, D.; Sandhya, R. S.; Pullaiah, T. Ethnobotany and distribution of wild and cultivated genetic resources of Cucurbitaceae in the Eastern Ghats of Peninsular India. *Topclass J. Herb. Med.* **2013**, *2*(6), 149–158.

Nayaka, S.; Reddy, A. M.; Devi, A.; Ayyappadasan, G.; Uperti, D. K. Eastern Ghats biodiversity reserve with unexplored lichen wealth. *Curr. Sci.* **2013**, *104*(7), 821–825.

Neelima, L.; Prasad, G. P.; Sudarsanam, G.; Pratap, G. P.; Jyothi, B. Ehnobotanical studies in Raipur forest division of Nellore district in Andhra Pradesh. *Life Sci. Leaflet.* **2011**, *11*, 333–345.

Panda, P. C.; Mohapatra, A. K.; Acharya, P. K.; Debata, A. K. Plant diversity in tropical deciduous forest of Eastern Ghats India: a land-scape level assessment. *Int. J. Biodivers. Conserv.* **2013**, *5*(10), 625–639.

Panda, S. K.; Rout, S. D.; Mishra, N.; Panda, T. Phytotherapy and traditional knowledge of tribal communities of Mayurbhanj district, Odisha, India. *Indian J. Pharmacol. Phytother.* **2011**, *3*(7), 101–113.

Patel, E.; Krishnamurthy, R. A review on potency of some Cucurbitaceae plants against Hepatitis and antimicrobial activities. *Indian J. Fundam. Appl. Life Sci.* **2013**, *3*(2), 13–18.

Patel, P. K.; Parekh, P. P.; Sorathia, K. D. Studies on ethnomedicinal aspects of family Cucurbitaceae in North Gujarat. *Unique J. Pharm. Biol. Sci.* **2013**, *1*(1), 34–36.

Paul, A.; Raychaudhuri, S. S. Medicinal and molecular identification of two *Momordica charantia* varieties: a review. *Electron. J. Biol.* **2010**, *6*(2), 43–51.

Pekamwar, S. S.; Kalyankar, T. M.; Kokate, S. S. Pharmacological activities of *Coccinia grandis. J. Appl. Pharm. Sci.* **2013**, *3*(5), 114–119.

Pratap, S.; Kumar, A.; Sharma, N. K.; Jha, K. K. *Luffa cylindrica*: an important medicinal plant. *J. Nat. Prod. Resour.* **2012**, *2*(1), 127–134.

Radha, P.; Sultana, R.; Murthy, B. L. N. Medicinally important climbing plants from Eastern Ghats-Andhra Pradesh. Eastern Ghats: EPTRI. *ENVIS Newsl.* **2012**, *18*(2), 1–8.

Reddy, C. S.; Murthy, M. S. R.; Dutt, C. B. S. *Vegetation Diversity and Endemic in Eastern Ghats, India.* Hyderabad: EPTRI, **2002**, pp. 109–134.

Reddy, J. L.; Jose, B.; Anjana, J. C.; Ruveena, T. N. Evaluation of antibacterial activity of *Trichosanthes cucumerina* and *Cassia didymoboya* frees leaves. *Int. J. Pharm. Pharm. Sci.* **2010**, *2*(4), 153–155.

Romojaro, A.; Botella, M. A.; Obon, C.; Pretel, M. T. Nutritional and antioxidant properties of wild edible plants and their use as potential ingredients in the modern diet. *Int. J. Food Sci. Nutr.* **2013**, *64*(8), 944–952.

Rout, S. D. Anthropogenic threats and biodiversity conservation in Similipal Biosphere Reserve, Odisha, India. *Tiger Paper.* **2008**, *35*(3), 22–26.

Rout, S.; Rout, S.; Kumar. S. S.; Satapathy, S.; Patnaik, D. An ethnobotanical survey of medicinal plants in Semiliguda of Koraput district, Odisha, India. *Res. J. Recent Sci.* **2013**, *2*(8), 20–30.

Sadangi, N.; Padhy, R. N.; Sahu, R. K. A contribution o medico-ethnobotany of Kalahandi district, Orissa on ear and mouth disease. *Anc. Sci. Life.* **2005**, *24*(3), 160–163.

Sahu T, Sahu J. *Cucumis sativus* (cucumber): a review on its pharmacological activity. *J. Appl. Pharm. Res.* **2015**, *3*(1), 4–9.

Sakals, M. E.; Johan, L. I.; David, J. W.; Roy, C. S.; Gordon, E. G. The role of forest in reducing hydro-geo morphic hazards. *For. Snow Landsc. Res.* **2006**, *80*(1), 11–22.

Sandhya, S.; Chandracekhar, J.; Banji. D.; Rao, K. N. V. Pharmacological study on the leaf of *Tricosanthes cucumerina* Linn. *Arch. Appl. Sci. Res.* **2010**, *5*, 414–421.

Sassidharan, S.; Chen, Y.; Saravanan, D.; Sundram, K. M.; Latha, L. Y. Extraction, isolation and characterization of bioactive compounds from Plants extracts. *Afr. J. Tradit. Complement. Altern. Med.* **2011**, *8*(1), 1–10.

Saxena, H. O.; Brahmam, M. *The Flora of Orissa.* Orissa. Bhubaneswar: Forest Development Corporation Ltd. and Regional Research Laboratory. **1995**, 1940–1956.

Shrivastava, A.; Roy, S. Cucurbitaceae: an ethnomedicinally important vegetable family. *J. Med. Plants Stud.* **2013**, *1*(4), 16–20.

Sills, E.; Shanley, P.; Paumgarten, F.; Beer, J. D.; Pierce, A. Evolving prospective on non-timber forest products. In: Shackleton, C. (ed.). *Non-timber Forest Products in Global Context.* Heidelberg: Springer-Verlag, **2011**.

Singh, A.; Dubey, N. K. An ethnobotanical study of medicinal plants in Sonebhadra District of Uttar, Pradesh, India with reference to their infection by foliar fungi. *J. Med. Plants Res.* **2012**, *6*(14), 2727–2746.

Subramanyam, R.; Steven, G. N. Valorizing the "Irulas" traditional knowledge of medicinal plants in the Kodiakkarai Reserve Forest, India. *J. Ethnobiol. Ethnomed.* **2009**, doi: 10.1186/1746-4269-5-1.

Tahseen, M. A.; Mishra, G. Ethnobotany and diuretic activity of some selected Indian medicinal plants: a scientific review. *Pharm. Innov. J.* **2013**, *2*(3), 109–121.

Tripathy, P. K.; Kumar, S.; Jena, P. K. Diversity and ethno-botanical assessment of some wild cucurbits of Similipal Biosphere Reserve forest, Odisha, India. *PPMNC Recent Adv. Plant Biotechnol.* **2013**, 77–83.

Tripathy, P. K.; Kumar, S.; Jena, P. K. Nutritional and medicinal values of selected wild cucurbits available in Similipal Biosphere Reserve forest, Odisha. *Int. J. Pharm. Sci. Res.* **2014a**, *5*(10), 5430–5437.

Tripathy, P. K.; Kumar, S.; Ofoeze, M. A.; Gouda, S.; Singh, N. R.; Jena, P. K. Validation of traditional therapeutic claims through phytochemical screening and antibacterial assessment: a study on mahakaal (*Trichosanthes tricuspidata* L.) from Similipal Biosphere Reserve forest, Odisha, India. *Algerian J. Nat. Prod.* **2014b**, *2*(3), 85–97.

Tripathy. P. K.; Kumar, S.; Jena, P. K. Assessment of food, ethno-botanical and antibacterial activity of *Trichosanthes cucumerina* L. *Int J. Pharm. Sci. Res.* **2014c**, *5*(7), 2919–2916.

Tupe, S. D.; Patil, P. D.; Thoke, R. B.; Aparadh, V. T. Phytochemical screening in some Cucurbitaceae members. *Int. Res. J. Pharm. Appl. Sci.* **2013**, *3*(1), 49–51.

Vadnere, G. P.; Pathan, A. R.; Kulkarni, B. U.; Singhai, A. K. *Diplocyclos palmatus*: a phytopharmacological review. *Int. J. Res. Pharm. Chem.* **2013**, *3*(1), 157–159.

Venkateshwarlu, G.; Shantha, T. R.; Shiddamallayya, N.; Ramarao, V.; Kishore, K. R.; Giri, S. K.; Sridhar, B. N.; Pavankumar, S. Physicochemical and preliminary phytochemical studies on the fruits of "*Shivalingi*" (*Diplocyclos palmatus* (Linn.) Jeffrey). *Int. J. Ayurvedic Med.* **2010**, *2*(1), 20–26.

Vinod, S. S.; Sekhar, C. K.; Aradhana, R.; Nath, V. S. An updated review on *Tricosanthes cucumerina* L. *Int J. Pharm. Sci. Rev. Res.* **2010**, *1*(2), 56–60.

War, A. R.; Paulraj, M. G.; Ahmad, T.; Buhroo, A. A.; Hussain, B.; Ignacimuthu, S.; Sharma, H. H. Mechanisms of plant defense against insect herbivores. *Plant Signal Behav.* **2012**, *7*(10), 1306–1320.

CHAPTER 4

The Miracle Plant—*Moringa*

KARABI BISWAS and SANKAR NARAYAN SINHA*

Environmental Microbiology Research Laboratory, Department of Botany, University of Kalyani, Kalyani, West Bengal 741235, India

Corresponding author. E-mail: sinhasn62@yahoo.co.in.

ABSTRACT

Moringa are sustainable, extensive, easily available, and inexpensive plants. These plants are sources of various bioactive components such as alkaloids, flavonoids, carbohydrates, minerals, phenolic acids, phytosterols, natural sugars, organic acids, proteins, and vitamins that can be found from different vegetative structures such as leaves, seeds, stems, and pod husks. These bioactive compounds have potential applications in various fields, namely medicines, functional food preparations, water purification, and biodiesel production. This chapter describes the informative source of concerning bioactive molecules in the species of the Moringaceae family, including *Moringa oleifera*, and an overview of patents required to protect *Moringa*-derived products.

4.1 INTRODUCTION

Out of 13 species of the genus *Moringa*, *Moringa oleifera* (*M.* oleifera) is the best-known plant. *Moringa* was highly valued in the ancient world. The Greeks, Romans, and Egyptians extracted edible oil from the *M. oleifera* seeds and used it for the preparation of perfume and skin lotion. As for example, in the 19th century, *Moringa* plantation in the West Indies yielded oil that was exported to Europe for perfumes and lubricants for machinery. *Moringa* pods have long been used for food by the people in the Indian subcontinent. The edible leaves are eaten throughout West Africa and in parts of Asia (Alhusnan and Alkahtani, 2016; Salaheldeen et al., 2014). These plants are well known as nutritional, medicinal, and water-purifying agents (Salaheldeen et al., 2015). However, some experiments have revealed that bioactive compounds from *Moringa* plants could be used for the innovation of functional food products and for other industrial food applications (Oyeyinka and Oyeyinka, 2018). Due to the presence of potent bioactive constituents in *Moringa* plants, these plants might be used in various food technologies as antioxidant factors, antimicrobial agents, and food fortificants, among other nutritional and technological applications (Devisetti et al., 2016).

4.2   DESCRIPTION

M. oleifera is a fast-growing, deciduous tree that can reach a height up to 10–12 m (32–40 ft) with a diameter of trunk 45 cm (1.5 ft). The bark is whitish gray and is surrounded by a thick cork. Young stems have purple- or greenish-white-colored, hairy bark. *M. oleifera* has an open crown of weak, pendulous branches with feathery tripinnate leaves. The flowers have fragrance and surrounded by five unequal, yellowish-white thinly veined petals. The length and breadth of the flower are 1.2–1.5 cm (1/2") and 2.0 cm (3/4"), respectively. They grow on the slender, hairy stalks of drooping or spreading flower clusters of 10–25 cm long. Flowering starts within the first six months after plantation. In seasonally cool regions, flowering occurs only once in a year between April and June. In more persistent seasonal temperatures and with fixed rainfall, flowering occurs two times or even all year round (Parrotta, 1993). The fruit is a hanging, three-sided brown capsule of 20–45 cm in size that holds dark brown, globular seeds with a diameter near about 1 cm. The seeds have three whitish papery wings and are dispersed by wind and water (Parrotta, 1993). During cultivation, it is usually cut back annually to 1–2 m (3–6 ft) and allowed for regrowth; hence, the leaves and pods remain within the arm's reach.

For growth, suitable rainfall between 250 and 3000 mm is needed annually. In drought conditions, the plant may lose its leaves. This, however, does not indicate that the plant is dead. It recovers on the onset of rains. It grows best at altitudes up to 600 m, but it may grow at altitudes of up to 1000 m. It may survive at a temperature ranging from 25 °C to 40 °C, but the plant is found to tolerate temperatures of 48 °C and light frosts. *Moringa* grows well in neutral to slightly acidic soils, and the best growth is noted in well-drained-loamy to clay-loamy soils. It can tolerate clay soils, but in the waterlogged condition, it does not grow. Moreover, the *Moringa* tree has no known major pests that indicate that it can quickly be established as a perennial tree crop, which can be added to different agroforestry systems. Young leaves and drumsticks are the delicious parts of many eastern cuisines with a unique "green" flavor. The *Moringa* root can also be consumed and gives a flavor resembling horseradish. Each and every part of the *Moringa* tree showed beneficial properties that can serve humanity. A large number of people in societies throughout the world have made use of these properties (Pandey et al. 2011).

4.3 USES OF *MORINGA*

Every part of this tree can be used for different purposes. *Moringa* is full of nutrients and vitamins and is good for our food as well as that for our animals (Dou and Kister, 2016). *Moringa* is used to clean dirty water and is an important source of medicines. It gives a lot of leafy material that is useful at the time of using alley-cropping systems.

4.3.1 *FOOD*

All food products prepared from the *Moringa* plant possesses extremely high nutritional value. One can consume the leaves, particularly young pods, shoots, flowers, roots, and, in some species, even the bark. Leaves are low in carbohydrates and fats and rich in minerals, iron, and vitamin B (Vaknin and Mishal, 2017). Leaves are an important source of vitamins and minerals when served as raw, cooked, or dried (Price, 2000). They are especially useful as a human food as they appear at the end of the dry season when few other sources of green leafy vegetables are available. Flowers may be cooked and mixed with other foods or fried in batter. Fuglie (1999) discussed that the 8 g serving of dried leaf powder will ensure supplements of 14% of protein, 23% of iron, 40% of calcium, and nearly all forms of vitamin A in children below 13 years of age, equivalent to what a child needs in a day. Leaves of 100 g could fulfill over a one-third of the daily need of calcium in a woman and required quantities of copper iron, sulfur, protein, and vitamin B. Pods are cooked like other green beans and have a similar flavor to *Asparagus*. This can be used in soups and stews. The seeds contain 35% oil and are used for cooking purposes. The gum that is found in the bark can be used to season food. According to Dachana et al. (2010), the use of dried *Moringa* leaves (10%) in cookie formulations significantly increased the iron, calcium, protein, and β-carotene contents, but maintained sensory qualities such as crumb color, texture, mouthfeel, and flavor. These outcomes corroborated the results of Sengev et al. (2013), in which the addition of the *Moringa* leaf powder to a wheat bread formulation significantly increased its nutrimental composition, especially magnesium, calcium, and β-carotene. Moreover, apart from the mineral content, vitamin A content increases in wheat bread when the *Moringa* seed powder is added to the formulation, thus making *Moringa* seed powder a potent, fortified food product which could help to maintain proper vision (Bolarinwa et al., 2017). As noted by Omosuli et al. (2017), high-oleic-acid vegetable oils, such as *Moringa* oil, are found to be very stable even in a frying condition because these oils produce less conjugated dienes and trienes during such processes than do polyunsaturated fatty-acid-rich oils (Abdulkarim et al., 2007). The root bark must be completely removed as it is rich in alkaloids, namely moringine, a toxic compound similar to ephedrine. According to Fuglie (1999), the root bark should be completely removed as it contains harmful substances, and vinegar is added to the grinded root usually.

4.3.2   ANIMAL FODDER

Cattle, pigs, sheep, goats, and poultry use the leaves, bark, and young stems of *Moringa*. If trees are to be used for animal fodder, it is better to prune them at a height of 4 m, but if they are not, they should be pruned at a height of 6 m so that harvesting for consumption by human beings can be smoothly carried out. Diets of livestock are improved by the supplementation of *Moringa* products. In many parts of the world, leaves and twigs are used as fodder for cattle, sheep, goats, and camels (Mahatab et al., 1987). Flowers of this plant are an important source of pollen for honey bees (Rajan, 1986).

4.3.3   WATER PURIFIER

M. oleifera leaves, seeds, and flowers are found to be highly efficient in waste-water treatment (Figure 4.1). It has been found that *M. oleifera* is the ideal natural coagulant yet discovered, which can be a substitute of synthetic coagulants; in fact, it is used widely all around the world (Ali et al., 2009). Coelho et al. (2009) suggested that the lectin protein is present in *M. oleifera* that is used for the flocculation/coagulation of the waste water. Further, *M. oleifera* contains high molecular weight, cationic dimeric protein that destabilizes the particles contained in the water and flocculates the colloids through a process of adsorption and neutralization, followed by sedimentation (Ndabigengesere et al., 1995). The significant reduction of these pollution-indicating parameters of the tannery effluent demonstrated its high applicability to treat the tannery wastes along with other wastewater treatment (Sinha et al., 2016). The amino acid sequence exhibited high contents of arginine, glutamine, and proline, with a total of 60 residues (Gassenschmidt et al., 1995). This cationic protein is generally known as the *Moringa oleifera* cationic protein (also known as Flo), which inhibits bacterial cell growth and settles negatively charged particles in a solution. Plant 2S albumin proteins are basic sources of nitrogen and carbon, which are involved in plant defense. A 30 kDa temperature-resistant coagulant lectin (cMoL) protein, which functions in the pH range of 4.0–9.0 and stables at 100 °C, is also identified from *Moringa* seeds and possesses potent water-coagulant properties (Santos et al., 2010).

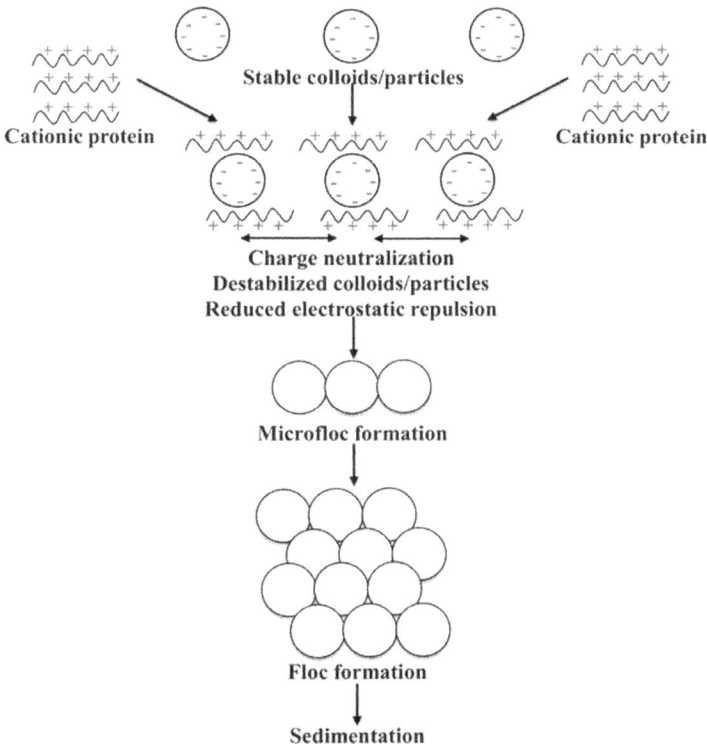

FIGURE 4.1   Mechanism of water purification by *Moringa*.

4.3.4   AGRICULTURE

Moringa-alley cropping reduces soil acidity as reported in Undie et al. (2013). The juice from fresh *Moringa* leaves can be used to produce an effective plant growth hormone, increasing yields by 25%–30% for nearly any crop: bell pepper, chili, coffee, onions, maize, melon, sorghum, soya, tea, and others. Among the active substances is zeatin—a plant hormone from the cytokine group. This foliar spray should be sprayed along with other fertilizers, watering, and strong agricultural practices.

4.3.5   FUEL AND OTHER USES

Biodiesel is a renewable and eco-friendly alternative to conventional nonrenewable fossil fuel. Biodiesel is the long-chain alkyl (ethyl, methyl, or propyl) esters prepared by chemically reacting lipids of vegetable oil and animal fat. The vegetable oils obtained from *Moringa* had been successfully used in biodiesel preparation (Tenenbaum, 2008). *M. oleifera* seeds contain 33%–41% (w/w) oil, known as "ben oil," because of the contents of behenic acid (C22, docosanoic acid, and 7% w/w), which possesses significant resistance to oxidative degradation (Rashid et al., 2008). Due to the presence of the significant amount of monounsaturated fatty acids in the form of oleic acid (C18:1, 72.2%), *Moringa* seed oil is a potential candidate for biodiesel production. The bark fiber is used in making rope, mats, and the wood produce a blue dye; a good-quality paper was made by chipping the wood. The trees also produce viscose resin produce from the tree, which are used in the textile industries.

4.4   MEDICINAL VALUE OF *MORINGA*

Moringa has the folkloric value from the ancient time in India (Fuglie, 1999) for its antitumor, anti-inflammatory, diuretic, abortifacient, antispasmodic, emmenagogue, and ecbolic properties. Indian worriers were using leaves of *M. oleifera* to reduce pain and enhance their energy during wars (Mahmood et al., 2010) (Figure 4.2). The roots, leaves, and seeds are great significance in Ayurveda, and the utilization of the stem bark, stem exudates, leaves, roots, root bark, flowers and seeds treating a broad range of ailments has been considered in ancient Sanskrit books on medicine (Ramachandran et al., 1980). The presence of vitamins A and C in leaves is considered great value in scurvy and respiratory ailments and they are also used as emetic. The juice extracted from the leaves has strong antibacterial and antimalarial properties (Gbeassor et al., 1990). Various parts of the *M. oleifera* tree, including, roots, leaves, flowers, bark, fruits, and seeds are traditionally used in different therapeutic applications, including hysteria (a psychological disorder), abdominal tumors, prostate problems, scurvy, paralysis, helminthic bladder, sores, and other skin problems. The medicinal values and therapeutic potential of *M. oleifera* have been extensively reviewed (Farooq et al., 2012; Mbikay, 2012). The pharmacological and physiological properties of the, roots, leaves, bark, flower sap, and seeds of *M. oleifera* have been described by Stohs and Hartman (2015). The leaf, flower, and fruit extracts of *M. oleifera* showed a high degree of safety without any ill effects on human beings. The phytochemicals of *M. oleifera* have shown anti-inflammatory, anthelmintic, antidyslipidemic, antihyperglycemic, antioxidant, antimicrobial, anti-ulcer, hepatoprotective, and antiproliferative properties. The leaf extracts of *M. oleifera*, rich in kaempferol phenolic compounds and quercetin, have been reported to be used as the human tumor (KB) cell line model. Different fatty acids such as linoleic acid, oleic acid, linolenic acid, behenic acid, and antibiotic (pterygospermin) are found in *Moringa* seeds. Some phytochemicals such as tannins, terpenoids, saponin, phytate, and flavonoids are found in *Moringa* seeds. The leaf extract of *M. oleifera* showed remarkable morphological changes and lower cell viability along with elevated internucleosomal DNA fragmentation and reactive oxygen species (ROS) generation in the KB cells (Sreelatha et al., 2011). *M. oleifera* leaves are also a rich source of flavonoids and phenolics and showed potent antioxidant properties both in in vitro and in vivo systems. With 70% ethanol, the highest extractability of total flavonoids and that of total phenolics are 6.20 g isoquercitrin equivalents/100 g extract and 13.23 g chlorogenic acid equivalents/100 g extract, respectively. This extract has exhibited a high 2,2-diphenyl-1-picrylhydrazyl-scavenging activity (EC_{50} = 62.94 lg/mL) and the maximum ferric-reducing power (51.50 mM $FeSO_4$ equivalents per 100 g extract). Moreover, at a concentration of 100 g/mL, the extract is observed to reduce a relative amount of intracellular ROS significantly (Vongsak et al., 2013). An alkaloid *N*, α-ʟ-rhamnopyranosyl vincos amide, extracted from leaves of *M. oleifera*, was found to protect against isoproterenol-induced cardiac toxicity in rats.

Moringa leaves have phytosterols (Jain et al., 2010) that help to reduce cholesterol uptake by the intestines (Lin et al., 2010). It seems that this could result in a decrease in cholesterol levels and an increase in excreted cholesterol seen in study rodents (Mehta et al., 2003; Jain et al., 2010). The high-fiber content in this plant (12% w/w) may also play a role in the hypocholesterolemic

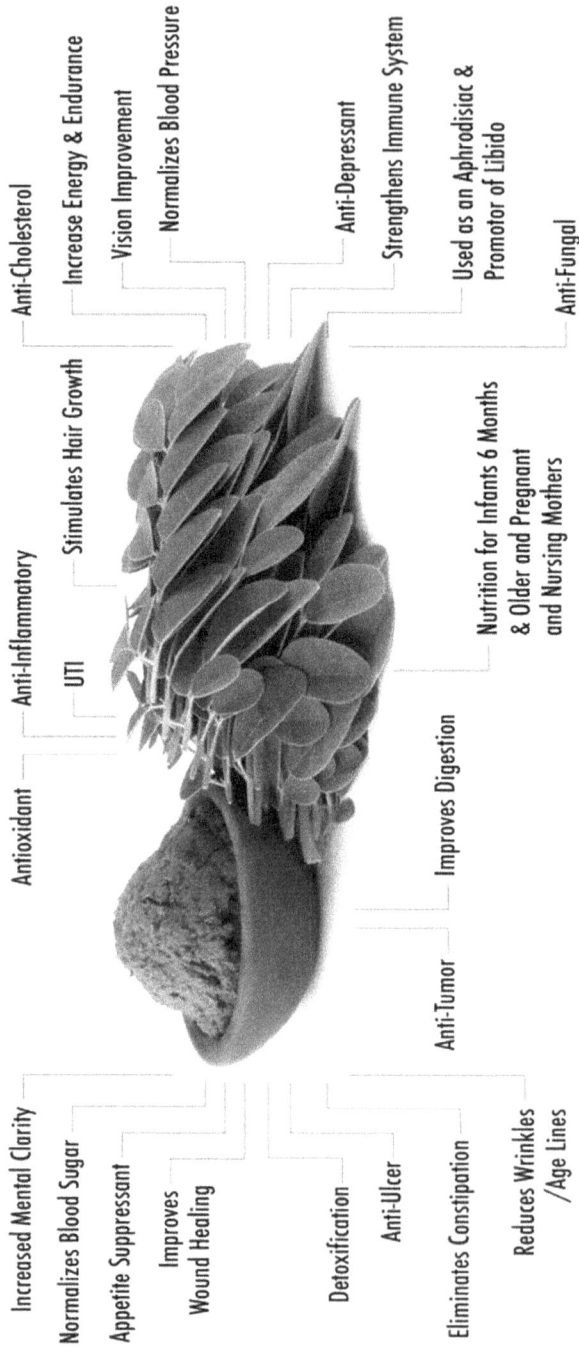

FIGURE 4.2 Traditional use of *Moringa.*

effect due to enhanced excretion/gastric emptying (Bortolotti et al., 2008). Moringine, an alkaloid from *Moringa*, is closely associated with ephedrine, inducing the relaxation of the bronchioles (Kirtikar and Basu, 1975). *Moringa* seeds and kernels have been utilized for the treatment of bronchial asthma, by reducing the acuteness of the symptoms, as a result to improve respiratory problems in patients (Agrawal and Mehta, 2008). Study in Wistar and Goto-Kakizaki rats with type 2 diabetes exhibits a positive result of *Moringa* leaves, with a significant reduction in blood glucose levels (Ndong et al., 2007). Besides, in high-fatty-diet mouse models, the plant extracts showed improved glucose tolerance (Jaja-Chimedza et al., 2018). The declining blood sugar levels within 3 hours after ingestion have been shown (Mittal et al., 2007) and the inhibitory activity of the extracts against the α-glucosidase enzyme (Natsir et al., 2018). Polyphenols such as kaempferol glycosides, 3-glycoside, and others (Ndong et al., 2007) probably show an impact in this antidiabetic activity. Studies revealed that the antioxidant properties of *Moringa* can rescue apoptosis of beta cells, thereby blocking cell damage that leads to curing properties against diabetes (Al-Malki and El Rabey, 2015). Further studies may lead to the commercialization of *Moringa* extracts for the conventional treatment of diabetes.

4.5   PHYTOCONTITUENTS IN *MORINGA*

For the presence of the bioactive compounds, *Moringa* shows these activities. Table 4.1 shows several bioactive compounds obtained from different vegetative parts of *Moringa* plants and the functional activity of these compounds.

4.6   VALUE PRODUCTS OF *MORINGA* IN INDIA

Moringa seed oil is a light oil contain vitamins A, B, C, E, unsaturated fatty acids and palmitoleic, oleic and linoleic acids provides great moisturizing and nourishing qualities that spreads and absorbs easily into the skin. Due to its antioxidant, antiseptic and anti-inflammatory properties, helps diminish the appearance of fine lines and wrinkles and used to purify and heal the skin. Moringa seed oil is applied for spa and aromatherapy, fragrance and Perfumery, cosmeceutica and also for, massage. Moringa tea is prepared from leaves mixed with ginger/lemon /mint/tulsi used for reduce body weight and nourishing and detoxifying nutrient rich superfood (Tetteh et al., 2010). Moring leaves supplemented with dry, alfalfa, ginger, ground nut, spirulina,

TABLE 4.1 Phytoconstituents of Different Parts of Moringa

Biological Activity	Bioactive Compound	Class of Phytochemical	Isolated Parts	References
Antimicrobial Activity	1,2-Benzenedicarboxylic acid, diethyl ester	Fatty acid	Ethyl acetate Leaf extract	Abbas, 2013
	1,2-Benzenedicarboxylic acid, bis-2-ethylhexyl ester	Fatty acid	Methanol leaf extract	Adegun and Aye, 2013
	Octadecanoic acid	Fatty acid	Ethanol leaf extract	Adegun and Aye, 2013
	Indole	Hydrocarbon	Steam distillation of leaf essential oil	Abbas, 2013
Antiviral activity	1,2,3-Cyclopentanetriol	Fatty acid	Ethanol leaf extract	Adegun and Aye, 2013
Anticancer activity	Hexadecanoic acid, ethyl ester	Fatty acid	Hexane leaf oil extract	Abou-Elezz et al., 2011
	O-Cymene	Terpenoids	Petroleum ether root extract	Abou-Elezz et al., 2011
Antioxidant activity	Benzyl isothiocyanate	Glycosides	Ethanol root extracts	Abiodun et al., 2015
	Ergosta-5,22-dien-3-ol, (3β,22E)	Lactic acid	Ethanol leaf extract	Abbas, 2013
	Linalool oxide	Alcohol	Ethyl acetate leaf extract	Adeniji and Lawal, 2013
	γ-Tocopherol	Fatty acid	Hexane essential oil extract	Abou-Elezz et al., 2011
Pesticides	*n*-Hexadecanoic acid	Fatty acid	Ethanol leaf extract	Adegun and Aye, 2013
	Pentacosane	Hydrocarbon	Water leaf extract	Adeniji and Lawal, 2013
	Upiol	Alcohol	Ether root extract	Abiodun et al.,2015
Flavouring agents	Nonanol	Alcohol	Aqueous leaf extract	Abbas, 2013
	21 (Z)-3-Octenol	Alcohol	Methanol flower extract	Adegun and Aye, 2013
	2,3-Butanedione	Ketone	Ethyl acetate leaf extract	Adeniji and Lawal, 2013
	Gamma-butyrolactone	Ketone	Methanol root extract	Abiodun et al., 2015

TABLE 4.1 *(Continued)*

Biological Activity	Bioactive Compound	Class of Phytochemical	Isolated Parts	References
	2-Methyl furan	Aldehyde	Steam distillation of leaf	Abbas, 2013
	3,7-Dimethyl octa 2,6-dienyl-3 chlorobenzoate	Sugar	Aqueous leaf extract	Abbas, 2013
Fragrance agent	Dimethoate	Organophosphate	Methanol leaf extract	Adegun and Aye, 2013
	β-Damascenone	Ketone	Aqueous root extract	Abiodun et al., 2015
	Pyrocatechol	Alcohol	Ethyl acetate leaf extract	Adeniji and Lawal, 2013
	Pentacosane	Hydrocarbon	Aqueous methanol leaf extract	Abbas, 2013

and colostrums to produce Mogo-Colostrum organic energy bar is used a supplements for nutrient. Oil cake prepared from Moringa seed are used as an agent for water purification. From various parts of *Moringa* plant growth promoter has been screened with other extracts named as moringa bio-booster (De-heer, 2011). Ellis et al. (2011) developed high nutritious biscuits from cassava and sweet potato flour fortified with moringa leaf powder which is suitable for gluten intolerants. Some even produced ingredients by Indian company registered as an intellectual property. The company Sabinsa formulated a standardized ingredient from dried *M. oleifera* leaves with a standardized minimum level of antioxidant activity. Few Indian companies, however, manufacture final products based on moringa, such as the company Grenera (Ministry of Foreign Affairs, 2017).

4.7   CONCLUSION

This chapter highlighted the bioactive compounds in *Moringa* plants and the recent approaches regarding the functional applications and influence of these biocompounds on functional characteristics in all aspect. It was observed that the major food products based on *Moringa* plants presented high dietary fiber and low fat contents, which suggests that this plant can be used in the formulation of hypocaloric foodstuffs.

KEYWORDS

- miracle plant
- *Moringa oleifera*
- bioactive components
- patent
- biodiesel

REFERENCES

Abbas, T. E. The use of *Moringa oleifera* in poultry diets. *Tur. J. Vet. Ani. Sci.* **2013**, *37*, 492–496.

Abdulkarim, S.M.; Lai, O.M.; Muhammad, S.K.S.; Long, K.; Ghazali, H.M. Frying quality and stability of high-oleic *Moringa oleifera* seed oil in comparison with other vegetable oils. *Food Chem.* **2007,** *105,* 1382–1389.

Abiodun, B. S.; Adedeji, A. S.; Taiwo, O.; Gbenga, A. Effects of *Moringa oleifera* root extract on the performance and serum biochemistry of *Escherichia coli* challenged broiler chicks. *J. Agri. Sci.* **2015,** *60*(4), 505–513.

Abou-Elezz, F. M. K.; Sarmiento-Franco, L.; Santos-Ricalde, R; Solorio-Sanchez, F. Nutritional effects of dietary inclusion of *Leucaena leucocephala* and *Moringa oleifera* leaf meal on Rhode Island Red hens' performance. *C. J. Agri. Sci.* **2011,** *45,* 163–169.

Adegun, M. K.; Aye, P. A. Growth performance and economic analysis of West African Dwarf Rams fed *Moringa oleifera* and cotton seed cake as protein supplements to Panicum maximum. *Am. J. Food Nutri.* **2011,** *3*(2), 58–63.

Adeniji, A. A.; Lawal, M. Effects of replacing groundnut cake with *Moringa oleifera* leaf meal in the diets of grower rabbits. *Int. J. Mol. Vet. Res.* **2011,** *2*(3), 8–13.

Agrawal, B.; Mehta, A. Antiasthmatic activity of *Moringa oleifera* Lam: A clinical study. *Indian J. Pharmacol.* **2008,** *40,* 28–31.

Alhusnan, L. A.; Alkahtani, M. D. Impact of *Moringa* aqueous extract on pathogenic bacteria and fungi in vitro. *Ann. Agric. Sci.* **2016,** *61*(2), 247–50.

Ali, E. N.; Muyibi, S. A.; Salleh, H. M.; Salleh, M. R. M.; Alam M. Z. *Moringa oleifera* seeds as natural coagulant for water treatment. In *Proceedings of the 13th International Water Technology Conference*, IWTC'09, Hurghada, Egypt. **2009,** pp. 163–168.

Al-Malki, A. L.; El Rabey, H. A. The antidiabetic effect of low doses of *Moringa oleifera* Lam. Seeds on Streptozotocin induced diabetes and diabetic nephropathy in male rats. *Biomed. Res. Int.,* **2015,** 1–13.

Bolarinwa, A.; Adeyeri, J. B.; Okeke, T. C. Compaction and consolidation characteristics of lateritic soil of a selected site in Ikole Ekiti, southwest Nigeria. *Nigerian J. Technol.* **2017,** *36*(2), 339–345.

Bortolotti, M.; Levorato, M.; Lugli, A.; Mazzero, G. Effect of a balanced mixture of dietary fibers on gastric emptying, intestinal transit and body weight. *Ann. Nutr. Metab.* **2008,** *52,* 221–226.

CBI Ministry of Foreign Affairs. CBI is the Centre for the Promotion of Imports from Developing Countries, **2017.** https://www.cbi.eu

Coelho, J. S.; Santos, N. D. L.; Napoleão, T. H.; Gomes, F. S.; Ferreira, R. S.; Zingali, R. B.; Coelho, L. C. B. B.; Leite, S. P.; Navarro, D. M. A. F.; Paiva, P. M. G. Effect of *Moringa oleifera* lectin on development and mortality of *Aedes aegypti* larvae. *Chemosphere.* **2009,** *77*(7), 934–938.

Dachana, K. B.; Rajiv, J.; Indrani, D.; Prakash, J. Effect of dried moringa (*Moringa oleifera* lam) leaves on rheological, microstructural, nutritional, textural and organoleptic characteristics of cookies. *J. Food Qual.,* **2010,** *33*(5), 660–677.

De-Heer, N. E. A. Formulation and sensory evaluation of herb tea from *Moringa oleifera, Hibiscus sabdariffa* and *Cymbopogon citratus*. M.Sc. Thesis (Ghana: Kwame Nkrumah University of Science and Technology), **2011.**

Devisetti, R.; Sreerama, Y. N.; Bhattacharya, S. Processing effects on bioactive components and functional properties of moringa leaves: development of a snack and quality evaluation. *Food Sci. Technol.* **2016,** *53*(1), 649–657.

Dou, H.; Kister, J. Research and development on Moringa Oleifera–Comparison between academic research and patents. *World Pat. Info.* **2016,** *47,* 21–33.

Ellis, W. O.; Oduro, I.; Owusu, D. Development of crackers from cassava and sweet potato flours using *Moringa oleifera* and Ipomoea batatas leaves as fortificant, *Am. J. Food Nutr.* **2011**, *4*(3):115–117.

Farooq, M.; Hussain, M.; Wahid, A.; Siddique, K. H. M. Drought stress in plants: an overview. In *Plant responses to drought stress*. Springer, Berlin, Heidelberg, **2012**, pp. 1–33.

Fuglie, L. J. The miracle tree: *Moringa oleifera*, natural nutrition for the tropics, **1999**.

Gassenschmidt, U.; Jany, K. D.; Bernhard, T.; Niebergall, H. Isolation and characterization of a flocculating protein from *Moringa oleifera* Lam. *Biochimica. Biophysica. Acta (BBA)-General Subjects* **1995**, *1243*(3), 477–481.

Gbeassor, M.; Kedjagni, A. Y.; Koumaglo, K.; De Souza, C.; Agbou, K.; Aklikokou, K. *In-vitro* antimalarial activity of six medicinal plants. *Phyto. Res.* **1990**, *4*(3), 115–117.

Jain, P. J.; Patil, S. D.; Haswani, N. G.; Girase, M. V.; Surana, S. J. Hypolipidemic activity of *Moringa oleifera* Lam., Moringaceae, on high fat died-induced hyperlipidemia in albino rats. *Braz. J. Pharmacogn.* **2010**, *20*, 969–973.

Jaja-Chimedza, A.; Zhang, L.; Wolff, K.; Graf, B. L.; Kuhn, P.; Moskal, K.; Raskin, I. A dietary isothiocynate-enriched moringa (*Moringa oleifera*) seed extract improves glucose tolerance in a high-fat-diet mouse model and modulates the gut microbiome. *J. Funct. Foods,* **2018**, *47*, 376–385.

Kirtikar, K. R.; Basu, B. D. Indian Medicinal Plants. In: Singh B. and M. P. Singh (Eds.) Dehradun, **1975**, 676–683.

Lin, X.; Racette, S. B.; Lefevre, M.; Spearie, C. A.; Most, M.; Ma, L.; Ostlund, R. E. Jr. The effects of phytosterols present in natural food matrices on cholesterol metabolism and LDL-cholesterol: a controlled feeding trial. *Eur. J. Clin. Nutr.* **2010**, *64*, 1481–1487.

Mahatab, S. N.; Ali, A.; Asaduzzaman, A. H. M. Nutritional potential of *sajna* leaves in goats. *Livestock Advisor.* **1987**, *12*(12), 9–12.

Mahmood, K. T.; Mugal, T.; Haq, I. U. *Moringa oleifera*: A natural gift: a review. *J. Pharma. Sci. Res.* **2010**, *2*(II), 775–781.

Mbikay, M. Therapeutic potential of *Moringa oleifera* leaves in chronic hyperglycemia and dyslipidemia: a review. *Front. Pharmacol.* **2012**, *3*, 24.

Mehta, K.; Balaraman, R.; Amin, A. H.; Bafna, P. A.; Gulati, O. D. Effect of fruits of *Moringa oleifera* on the lipid profile of normal and hypercholesterolaemic rabbits. *J. Ethnopharmacol.* **2003**, *86*, 191–195.

Mehta, K.; Balaraman, R.; Amin, A. H.; Bafna, P. A.; Gulati, O. D. Effect of fruits of *Moringa oleifera* on the lipid profile of normal and hypercholesterolaemic rabbits. *J. Ethnopharmacol.* **2003** *86*, 191–195.

Mittal, M.; Mittal, P.; Agarwal, A. C. Pharmacognostic and phytochemical investigation of antidiabetic activity of *Moringa oleifera* lam leaf. *Ind. Pharm.* **2007**, *6*, 70–72.

Natsir, H.; Wahab, A. W.; Laga, A.; Arif, A. R. Inhibitory activities of *Moringa oleifera* leaf extract against α-glucosidase enzyme in vitro. *J Physics: Conf. Sers.* **2018**, *979*, 12–19.

Ndabigengesere, A.; Narasiah, K. S,; Talbot, B. G. Active agents and mechanism of coagulation of turbid waters using *Moringa oleifera*. *Water Res.* **1995**, *29*(2), 703–710.

Ndong, M.; Uehara, M. Katsumata, S; Suzuki, K. Effects of oral administration of *Moringa oleifera* Lam on glucose tolerance in gotokakizaki and wistar rats. *J. Clin. Biochem. Nutr.* **2007**, *40*, 229–233.

Omosuli, S. V.; Oloye, D. A.; Ibrahim, T. A. Effect of drying methods on the physicochemical properties and fatty acid composition of Moringa seeds oil. *Progs. Food Nutri. Sci.* **2017**, 27–32.

Oyeyinka, A. T.; Oyeyinka, S. A.*Moringa oleifera* as a food fortificant: Recent trends and prospects. *J. Saudi. Soci. Agri. Sci.* **2018**, *17*(2), 127–136.

Parrotta, J. A. *Moringa Oleifera* Lam: Resedá, Horseradish Tree, Moringaceae, Horseradish-tree Family. International Institute of Tropical Forestry, US Department of Agriculture, Forest Service **1993**.

Price, M. L. The *Moringa* Tree. Echo Technical note. Echo, Florida, USA, **2000**, 12.

Rajan, B.K.C. Apiculture and farm forestry in semi-arid tracts of Karnataka. *My Forest.* **1986**, *22*: 41–49.

Ramachandran, C.; Peter, K. V. G.; Opalakrishnan, P. K. Drumstic (*Moringa oleifera*): a multipurpose Indian vegetables. *Econ. Bot.* **1980, ** *34*(3), 276–283.

Rashid, H.; Shafi, S.; Booy, R.; Bashir, H. E.; Ali, K.; Zambon, M. C.; Haworth, E. Influenza and respiratory syncytial virus infections in British Hajj pilgrims. *Emerg. Health Threats J.* **2008**, *1*(1), 7072.

Salaheldeen, M.; Aroua, M. K.; Mariod, A. A.; Cheng, S. F.; Abdelrahman, M. A. An evaluation of *Moringa peregrina* seeds as a source for bio-fuel. *Indus. Crops Prod.* **2014**, *61*, 49–61.

Salaheldeen, M.; Aroua, M. K.; Mariod, A. A.; Cheng, S. F.; Abdelrahman, M. A.; Atabani, A. E. Physicochemical characterization and thermal behavior of biodiesel and biodiesel–diesel blends derived from crude Moringa peregrina seed oil. *Energ. Convers. Manag.* **2015**, *92*, 535–542.

Santos, M. J.; Kanekar, N.; Aruin, A. S. The role of anticipatory postural adjustments in compensatory control of posture: 2. Biomechanical analysis. *J. Electromyogr. Kinesiol.* **2010**, *20*(3), 398–405.

Sengev, A. I.; Abu, J. O.; Gernah, D. I. Effect of *Moringa oleifera* leaf powder supplementation on some quality characteristics of wheat bread. *Food Nutri. Sci.* **2013**, *4*(3), 270–275.

Sinha, S.N.; Paul D.; Biswas K. Effects of *Moringa oleifera* Lam. and *Azadirachta indica* A. Juss. Leaf extract in treatment of tannery effluent. *Our Nature.* **2016, ** *14*(1), 47–53.

Sreelatha, S.; Jeyachitra, A.; Padma, P. R. Antiproliferation and induction of apoptosis by *Moringa oleifera* leaf extract on human cancer cells. *Food. Chem. Toxicol.* **2011**, *49*(6), 1270–1275.

Stohs, S. J.; Hartman, M. J. Review of the safety and efficacy of *Moringa oleifera*. *Phytother. Res.* **2015**, *29*(6), 796–804.

Tenenbaum, D. J. Food vs. fuel: diversion of crops could cause more hunger. *Environ. Health. Persp.* **2008**, *116*(6), A254–A254.

Tetteh, O. N. A.; Oduro, I. N.; Ellis, W. O.; Appaw, W. A comparative evaluation of the effect of blanching and dehydration methods on the proximate composition of *Moringa oleifera* and *Moringa stenopetala*. Sunyani Polytechnic Lecture Series VI Conference Proceedings, **2010**, 223–233.

Undie, U. L.; Kekong, M. A.; Ojikpong, T. Moringa (*Moringa oleifera* Lam.) leaves effect on soil pH and garden egg (*Solanum aethiopicum* L.) yield in two Nigeria agro-ecologies. *Eur. J. Agric. Res.* **2013**, *1*(1), 17–25.

Vaknin, Y.; Mishal, A. The potential of the tropical "miracle tree" *Moringa oleifera* and its desert relative *Moringa peregrina* as edible seed-oil and protein crops under Mediterranean conditions. *Sci. Horticul.* **2017**, *225*, 431–437.

Vongsak, B.; Sithisarn, P.; Mangmool, S.; Thongpraditchote S.; Wongkrajang, Y.; Gritsanapan,W. Maximizing total phenolics, total flavonoids contents and antioxidant activity of *Moringa oleifera* Leaf extract by the appropriate extraction method. *Ind. Crops Prod.* **2013**, *44*, 566–571.

CHAPTER 5

Biomedical Application of Polymeric Biomaterial: Polyhydroxybutyrate

S. MOHAPATRA[1,5], D. MOHANTY[1], S. MOHAPATRA,[2] S. SHARMA[3],
S. DIKSHIT[4], I. KOHLI[5], D. P. SAMANTARAY[1*], R. KUMAR[6], and
MEGHAVI KATHPALIA[5]

[1]Department of Microbiology, OUAT, Bhubaneswar 751003, Odisha, India

[2]Department of Mechanical Engineering, IIT, Roorkee, India

[3]Department of Economics, OUAT, Bhubaneswar 751003, Odisha, India

[4]Department of Biotechnology, IIT, Bombay, India

[5]Department of Microbial Technology, Amity University, Uttar Pradesh, Noida, India

[6]Department of Environmental Sciences, H.N.B. Garhwal University (A Central University), Srinagar Garhwal, Uttarakhand 246174, India

*Corresponding author. E-mail: dpsamantaray@yahoo.com

ABSTRACT

The biomedical field is found to have a shift from nonbiodegradable polymers' application to ecofriendly biopolymers due to its easy degradability and compatibility with biological matters. Because of all these attractive features, polyhydroxyalkanoates (PHAs) are the choice of interest of all biologists. This microbial-origin polymeric biomaterial has expanded its efficacy and efficiency as an excellent biomedical candidate having various applications in the field of biomedical applications. Additionally, the presence of 3-hydroxybutyrate (3HB) has been observed in components of blood in the eukaryotic cell. Furthermore, these days, polyhydroxybutyrate (PHB) is used for several biomedical uses such as drug delivery, skin grafting, physical fixation substantiates, and inert medical meshes extra. More specifically,

PHB, due to its physiochemical properties that are suitable for a large extent of medical applications, is a proper biomaterial for tissue engineering such as preparation of scaffolding and fabrication of nontoxic, biocompatible resorbable medical apparatus such as plates, sutures, stents, bone-fixation devices, tissue repairs, and screws. Generally, the lower rate of the resorbing nature is worthful while focusing toward tissue engineering as this concept provides ample time to heal properly. In vivo and in vitro studies revealed that most of the monomers from PHAs are biocompatible with the human body that enhances the use of these materials for the use of drug delivery. After finishing all the served work, it degrades properly with zero toxicity. The new area of research strives toward various biopolymer applications, especially target clinical use, as well as biomedical appliance. Here we have compiled the how PHA is an excellent candidate in the biopolymer field, precisely, its advanced uses in drug-delivery applications and tissue engineering.

5.1   INTRODUCTION

There is an extensive range of degradable polymers exist currently which have potential as biomaterials and can be used as a good prospect for an eco-friendly substitute of plastics made up of petroleum products in various applications. In medical applications, polyurethanes and derivatives from polyethylene glycol (PEG) have been used for a long period of time as the golden parameters for polymers, but now they are being replaced by several blends and polymers from natural resources which contain superior biodegradability and biocompatibility. Biological originated biodegradable plastics are the most demanding polymers compared to conventional nondegradable plastics as it does not demand its remotion after any types of surgery or any types of implantation in the organ of the body. Due to these exclusive features, polyhydroxyalkanoates (PHAs) cover most of the biomedical clinical market nowadays, finding application in targeted drug delivery, skin grafting, tissue engineering, etc. (Liu et al., 2012; Tran et al., 2009). There are a lot of problems being placed by the applications of polymers, that is, nonbiodegradable in nature in the biomedical field along with their surgical removal after surgery/recovery. Microbial bioplastics are found to be completely biodegradable, with zero toxicity, and never cause any toxicity or hazardous effect during surgery. Hence, they are considered as a worthy material in biomedical fields (Chanprateep, 2010). PHAs, being new in the field, have gained a lot of attention. The ongoing research virtues the enlargement of the bio-polymers' application and seems to be the future of biomedical applications. Especially in clinical uses, the device and its uses should be biologically compatible, portraying that they are unable to create serious immune reactions at the time they are entered to the blood or soft tissues of the host. Modern medicine advancements have been impacted to a great extent by the utilization of polymers as biomaterials. Particularly, polymeric biomaterials which are biodegradable in nature provide a substantial advantage as they can be dissimilated and eliminate subsequently while completion of targeted work (Bonartsev et al., 2007; Boskhomdzhiev et al., 2010). Generally, PHAs serve as carbonosomes—packages of a carbon- and energy-storing component found in the cytoplasm of several microorganism species. An environment having plentiful carbon but a limited amount of other nutrients (e.g., nitrogen, phosphorus) is very suitable for the production of PHAs in organisms. Lemoine first discovered polyhydroxybutyrate (PHB), which is one among the PHA family present in *Bacillus megaterium,* in 1927. Since then, various inventions have created interest in the worthy utilization of PHAs in scientific research, medical research, households, and industrial packages (such as smart packaging, indicator packaging, etc.), and in other applications. Microorganisms have made PHAs as a fatty ester which is used by them easily as when nutrients are required in depletion condition. Several findings have revealed that PHAs are promptly damaged by a sort of microorganisms at the time it is kept in a natural ecosystem, like the aqueous or soil ambiance; hence, presenting that PHAs are a biological; based and family of polymers that are biodegradable (Boyandin et al., 2012; Brigham et al., 2011; Budde et al., 2011). PHB is a biodegradable and biocompatible polymer which is generally cast off for various medical appliances and projects like physical fixation supports, medical meshes of inert material, used a shuttle for targeted drug release. Introduction of advance characteristics prompt using this type of materials for bioimplants, scaffolds, and biodegradable and comparable latex and other medical instruments, like plates, stents, screws, orthopedic fixating apparatus, and sutures, and for accelerating tissue regeneration. Moreover this extensive scope with attractive physicochemical aspects proven as a best material of choice for medical applications (Bonartsev et al., 2007). There is much experimental evidence to prove that the mammalian system, including the human body, tolerated PHA very well, and therefore PHA is studied for use in drug delivery devices. Numerous novel applications will emerge for PHA in medicine after the refinement of fabrication procedures. PHA application as a scaffolding material will keep on growing with the constant interest in tissue engineering. Nontoxicity and biocompatibility are the general needs to be utilized in the human system. Complete nontoxicity, as well as biodegradability and hence the biocompatibility, has been demonstrated by the analysis and research on biocompatibility, biodegradation, and the evaluation of in vitro and in vivo cyto-compatibility and toxicity assays in vitro and in vivo on PHB. Monomers generated the PHB are those which are established as intermediaries in metabolic paths in every higher organism. 3-Hydroxybutyric acid is the key product for biodegradation in PHB, and the same intermediaries are seen in human bodies (Shishatskaya et al., 2008).

5.2   MICROORGANISMS ASSOCIATED WITH THE PRODUCTION OF PHA

Poly(3-hydroxybutyrate) (P(3HB)) is the one which is most generally utilized PHA polymers and a French scientist, Lemoigne described it first in 1925. Afterward, several bacterial strains among the archaebacteria, that is, Gram-positive and Gram-negative bacteria and photosynthetic microorganisms, with the inclusion of cyanobacteria, were identified as accumulating P(3HB), both anaerobically and aerobically. P(3HB)'s role as a polymer of bacterial storage possesses the mostly similar functions to starch, and glycogen was acknowledged constantly in 1973. Macrae and Wilkinson found that when the glucose to nitrogen proportion in the culture medium stayed high then the P(3HB) homopolymer accumulation is initiated by *B. megaterium* and without carbon and energy sources the resulting intracellular degradation known to be the mobilization of P(3HB). At the point when the other monomer types were discovered, the thought of 3HB monomer being the main constituent of this polymer was changed following a time of its acknowledgment and named as bacterial storage material. PHA for stress survival is a vital factor for microorganisms. In nutrient-deficit conditions, PHA, being a reservoir of both carbon and energy for sporulating as well as nonsporulating bacteria, promotes long-term bacteria survival. Also, PHA-harboring bacteria have demonstrated more stress tolerance against ephemeral environmental assaults, for example, heat radiation, osmotic shock, and ultraviolet (UV) irradiation. PHA biosynthetic pathways are complexly interlinked with the bacterium's central metabolic pathways, which involve *de novo* fatty acids synthesis, glycolysis, amino acid metabolic pathway, citric acid cycle, Calvin Cycle, β-oxidation, and the serine pathway (Figure 5.1).

5.3   PHAS AND THEIR CLASSIFICATION

A huge rate of waste accumulation and emission of greenhouse gas have been resulted due to the boundless use of conventional plastics. Hence recent technologies are being formed for bio-green materials' development, which gives rise to various negligible side effects on the environment. A bioplastic, PHA, due to its similar physical characteristics as synthetic plastics, has been attracting major interests. In contrast to synthetic plastics, PHA is yield from sustainable resources and is degraded by microorganisms aerobically that produces CO_2 gas and H_2O upon disposal. The selections of reasonable carbon sources, appropriate bacterial strains, recovery processes, and productive fermentation are major aspects which thought for PHA commercialization. The beauty of these polymers is that it is available in various forms like monomer can extend up to 4 to more than 16 monomers with this most exciting features and it can come up with various combinations of the monomer as well (Figure 5.2). The chain length carbon three to carbon six lying in between scl PHAs, the polymer having six to fifteen carbon monomers are denoted as mcl PHAs, more than 15 carbon number polymer is denoted as lcl PHAs. Till date, more than 150 monomers of PHAs have been reported.

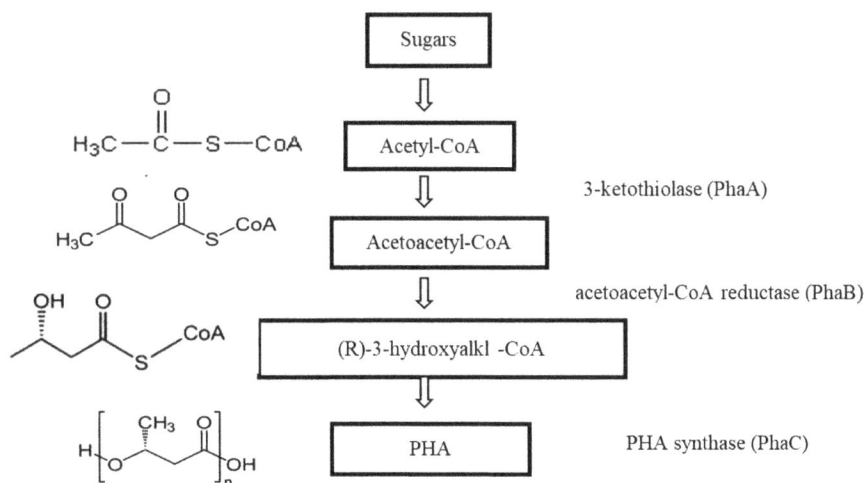

FIGURE 5.1 Common pathway for the production of PHA.

5.4 PHA'S BIOSYNTHESIS AND ITS REGULATION IN BACTERIA

When there are imbalanced nutrient supplies, PHAs naturally accumulate for bacteria for the storage of carbon and energy. When the growth of bacteria is restricted by oxygen, phosphorous, or nitrogen and a carbon source is consumed and present in abundance, these polyesters are agglomerated.

Although the most well-known cause of the limited growth of bacteria is nitrogen, for certain bacteria, for example, *Azotobacter* spp., oxygen is the most effective cause to limit their growth.

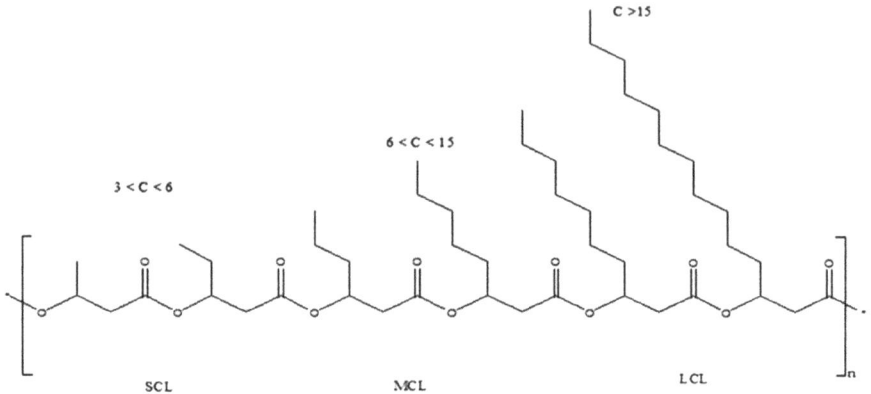

FIGURE 5.2   Classification of polyhydroxyalkanoates.

In bacterial cells, PHAs are accumulated at intracellular granules sites that are water-insoluble; they store surplus nutrients inside the bacterial cells as their general physiological flexibility is unaffected. Due to the polymerization of soluble intermediates into insoluble ones, the cell does not change its osmotic state. Hence, the leakage of these relevant compounds out of the cell is prevented, and nutrients remain accessible at a low maintenance cost. The PHA granule surface is coated by using phospholipids layering and proteins. Phasins, kinds of proteins, are important compounds present in the granule interface and influence the size and number of PHA granules. Gene expression of phasins can occur due to closely packed granules in bacterial cells.

5.5   BIODEGRADABILITY AND BIOCOMPATIBILITY OF PHA

Biodegradability in various environmental conditions is a significant property of biological PHA materials. Numerous microorganisms like fungi and bacteria in sludge, sea water, and soil give rise to extracellular PHA-damaging enzymes which hydrolyze the solid PHA into water-solvent monomers and oligomers, and after that, they utilize the resultant products, for example, the nutrients inside the cells. The enzymatic and hydrolytic degradation processes of P(3HB-*co*-4HB) films were analyzed by supervising the time-dependent changes in molecular weights and weight loss. PHA depolymerase and lipases were found to hydrolyze P(3HB-*co*-4HB) films (Sudesh et al., 2000). Piskin found that enzymes that exist in vivo catalyze the degradation, as in vivo deprivation of PHB is quicker than in vitro hydrolysis at body temperature. PHA implants and other therapeutic devices are observed to be degraded at the spot of implantation in animals. It has been discovered that lipase activities occurred in the rodent gastrointestinal are close to the PHA implant, suggesting the involvement of lipases in the metabolism of PHA in vivo.

Biodegradation, as well as biocompatibility, is very much necessary for PHA to be included in drug delivery and other many biomedical applications. For several medical applications, materials should be biologically compatible. PHA films' surface properties are discovered to be favorable for the attachment and proliferation of tissue cells (Misra et al., 2006), portraying that PHA is generally appropriate for scaffolding materials in tissue engineering. NIH 3T3 fibroblast cells are believed to be adhered to and proliferated on PHA membranes (Shishatskaya and Volova, 2004). Mesenchymal stem cells were observed to adhere to and proliferate on various substrates of PHA, with a terpolymer, poly(hydroxybutyrate-*co*-hydroxyvalerate-*co*-hydroxyhexanoate) (P[HB-*co*-HV-*co*-HHx]) (Ji et al., 2009; Wei et al., 2009). PHA matrices are also analyzed for hemocompatibility response by incubating the mammalian blood cells with the polymer films. Sevastianov et al. (2003) and Brigham and Sinskey (2012) depicted that PHB or P(HB-*co*-HV) when interacts with blood neither influences the responses of platelet nor activates the complement system. Juni and Nakano observed the in vivo biocompatibility of PHB through the infusion of microspheres (100 μm) into the muscle of a rodent's thigh. Transient intense inflammation was recognized, which was halted seven days after injection. During the four-week postinjection study period, the microspheres were again found encapsulated by the connective tissue.

5.6   SCAFFOLDS IN TISSUE ENGINEERING

PHB is a high-demand biopolymer among all types of microorganism biopolymers due to its biocompatibility, biodegradability, guidance, support to cell growth, and organization of cells that allow tissues' growth. Hence, it is used as a worthy candidate in the preparation of various items used in medical fields such as surgery, transplantology, medicine, pharmacology, and in tissue-engineering scaffolds. Cartilage and bone-generation concepts, which are based on the autogenous cell/tissue transplantation, are one of the most promising techniques used in biomedical and tissue-engineering research. The ultimate shape of new cartilage and bone, which provides adequate mechanical support for tissue engineering and cells to maintain the differentiated function, is defined by the scaffold architecture of the three-dimensional (3D) construct. Several factors play a chief role in scaffold tissue engineering, such as 3D and highly porous with an interconnected pore network for cell growth and flow transport of nutrients and metabolic waste that are biocompatible and bioresorbable with an adjustable degradation and resorption rate to match the cell/tissue growth in vitro and/or in vivo along with suitable surface chemistry for cell attachment, differentiation and proliferation, and mechanical properties to match with those of the tissues at the area of implantation (Hutmacher, 2000). Success in this field requires the use of biomaterials in tissue and organ regeneration. The principal classes of degradable polymers which could be used as the scaffold in tissue engineering generally include poly(α-hydroxy acids), such as poly(glycolic acid) (PGA), poly(lactic acid) (PLA), and a range of their copolymers. Gunatillake and Adhikari, (2003) simplified that polymers can be used as sutures and fixtures for fracture-fixation devices and scaffolds for cell transplantation. Polymer crystallinity is found to be one of the factors in the interactions of PHA with cartilage chondrocytes. Maturational differentiation of chondrocytes was found affected by the amount of PHB in a PHB-P(HB-*co*-HHx) blend present on a surface. Scaffolds that were produced from the unblended P(HB-*co*-HHx) were seen to be effectual in cartilage repair. Matrices that are fabricated from P(HB-*co*-HV) that is implanted into the cartilage defects in rabbits demonstrated a better healing response than fabrication from collagen infused with calcium phosphate. PHA copolymer scaffolds, when tested in rats, showed mild tissue response. However, PHB implants, due to their rigidity and provision of mechanical stimulus to tissues surrounding the implant, have shown more tissue response. The rate of bio-absorption of the implants declined in correlation with the 3HB content (Kose et al., 2005; Wang et al., 2008; Zheng et al., 2005). Another study has pointed out that PHA matrices permit proliferation of the neural stem cells. The presence of the P(3HB-*co*-3HHx) porosity allows P(HB-*co*-HHx) to penetrate stem cells into the polymer matrix. Osteoblasts were also seen to deposit, proliferate, and adhere to calcium on PHA substrates (Wang et al., 2004; Xu et al., 2010). Another strategy for tissue repair was reported by Ellis and colleagues, who discovered laser-perforated and biodegradable PHA scaffold films. The films of statistical copolymers of P(HB-*co*-HV) have pores exhibiting micrometer dimensions. Hence, by the time the cells were seeded into the film surface, they could attach and proliferate on the upper surface, as well as through the pores and into the region of the damaged tissue (Ellis et al., 2011). In Ellis et al.'s study, an increased surface amorphicity was achieved at the pore edges, which helped cell adhesion and could encourage growth and migration of cells for regenerative medicine.

5.7   PHB IN DRUG DELIVERY SYSTEMS

Biologically degradable polymers with a drug entrapping can be put inside the human body and made use for the deliverance of local drug with the controlled ejection of a drug over months (Lenz & Marchessault, 2005). Drugs can be microencapsulated very well in a copolymer or PHA homopolymer. Microcapsule- or microsphere-founded delivery systems are widely being used for numerous drug delivery such as antibiotics, hormones, anesthetics, anticancer agents, steroids, anti-inflammatory agents, and vaccines (Orts et al., 2008). In controlled-release drug delivery systems, a harmless carrier material containing the needful biomedical, mechanical, and physical properties, along with degradability in biological media, is needed (Diaz et al., 2014; Sin et al., 2013; Turesin et al., 2001).

The suitableness of PHA's inclusion in the delivery of drugs and in other biomedical applications will rely on its biological degradation attributes as well as biocompatibility. PHAs are hydrophobic and biocompatible and hence can be changed into poriferous matrices, films, nanoparticles, microspheres, and microcapsules. PHBs are reported to be used as biodegradable carriers for long-term dosage of medicines, hormones, and drugs (Diaz et al., 2014; Sin et al., 2013; Turesin et al., 2001). Hemocompatibility is tested in PHA matrices through the inspection of the reaction of mammalian blood after incubation with the polymer films. When PHB or P(HB-*co*-HV) came in liaison with blood, it neither affected responses of platelet nor did activate the complementary system. Still, significantly the purification procedures of polymer had to be complied to lessen the measure of the bacterial cell wall material that is consorted with the purified PHA (Brigham and Sinskey, 2012; Sevastianov et al., 2003).

Recently, there is an emerging interest in novel-drug-delivery-system development through the use of nanotechnology. Nanoparticles can present a broad array of drugs to different body areas for a long period and symbolize a bright drug delivery system of targeted and regulated release; hence, they have formed a key field of research in drug delivery. Its surface properties can be transformed for the delivery of the target drug. A large variety of drugs can also be delivered through the use of nanoparticles through numerous routes. Nanoparticles are often used to produce hydrophobic drugs, hydrophilic drugs, vaccines, biological macromolecules, proteins, and so forth (Yamamoto et al., 2005).

Xiong and colleagues developed statistical copolymers of P(HB-*co*-HHx), PLA, and nanoparticles of PHB and attached them with lipid-soluble colorant rhodamine B isothiocyanate (RBITC) like a model compound. Nanoparticles have a better loading efficiency, to the extent of 75%, with statistical copolymers of PHB homo- and P(HB-HHx), and the release of a drug for some time, at least 20 days, whereas the denotation of PLA nanoparticles only persists up to 15 days. Chaturvedi et al. took the PHB mixture along with the cellulose acetate phthalate (CAP) in various compositions for drug concentrated with 5-fluorouracil, which is an anticancer drug, and then went for the investigation of the simulated colon delivery of the said drug. The blend, due to its pH sensitivity, created a larger in vitro release in alkaline pH than acidic pH, which suggests its potentiality for the delivery of colon (Chaturvedi et al., 2011; Xiong et al., 2010). Yao et al. developed a receptor-mediated drug-delivery system in which the RBITC model drug was targeted to cancer cells or macrophages by incorporating P(HB-*co*-HHx) and colligating with a recombinant polyhydroxyalkanoate granule binding protein (phasin) derived from *Cupriavidus necator* (Yao et al., 2008).

5.8 POLYHYDROXYALKANOATES AS *IN VIVO* OR SURGICAL APPLICATION

Wang et al. delineated the use of PHAs as scaffolds for human bone marrow stromal cells as a new scheme for repairing nerve injury. A comparison was found between a statistical terpolyester of the composition P(HB-*co*-HV-*co*-HHx) and poly(lactic acid) and P(HB-*co*-HHx) regarding the transformation of the human bone marrow stromal cells to nerve cells. The terpolyester had firmer cell proliferation, differentiation, and adhesion compared to the other two polymers. Zhao et al. propounded that a 3D bioplotter was printed on the mesopotous bioactive glass and 3D scaffolds of a composite which are made up of P(HB-*co*-HHx) with the aim of delivering the materials used for the regeneration of bone. These are largely porous and huge scaffolds which have shown enhanced human bone marrow stromal cells adhesion, good bioactivity, and stimulated bone regeneration in in vivo experiments (Wang et al., 2010; Zhao et al., 2014). Sutures of PHB and P(HB-*co*-HV) were seen to have the mechanical strength for the use in facial wounds, and, therefore, they were intramuscularly administered on animals for examination. The ambient tissue did react to the PHAs by a transient post-traumatic inflammation, and also led to the formation of fibrous capsules with up to 200 μm thickness, which thinned with prolonged exposure. If the sutures were implanted for periods of up to 1 year, they stimulated no suppurative inflammation or necrosis (Shishatskaya et al., 2004).

In order to be effectual, especially in wound closures, the polymeric material should have the exceptional tensile strength to be used in sutures. Muscle-fascial wounds are being found to be healed by PHB and P(HB-*co*-HV) sutures (Shishatskaya et al., 2002, 2004). Post oral surgery in dogs, P(HB-*co*-HV) films helped in wound healing. Poly(4-hydroxybutyrate) (P4HB) is a very common PHA type which is used for surgical-material fabrication. Being a suture material, oriented P4HB fibers (545 MPa) are firmer than polypropylene sutures (410–460 MPa). Also, Young's modulus of P4HB sutures is much lower than the other monofilament sutures available in markets. Tepha Inc. in Cambridge, MA, USA, manufactures several medical devices from PHA. The famous TephaFLEX® suture fabricated from P4HB is the first suture to get approval from the US Food and Drug Administration (FDA). Tepha Inc. (Lexington, MA, USA) is also engaged in the production of surgical films and meshes formed from PHA.

5.9 TREATMENT OF VARIOUS DISEASES

The degraded materials of PHB, that is, 4-hydroxy butyrate (HB) units are found to be pharmacologically active compounds and hence are worthy for treatment of several diseases such as narcolepsy, atypical psychoses, neurosis, Parkinson's, radiation exposure, circulatory collapse, drug addiction and withdrawal, catatonic schizophrenia, chronic schizophrenia, alcohol withdrawal syndrome, chronic brain syndrome, cancer, and many other neuropharmacological illnesses. HB (4-hydroxy butyrate) units are found to be useful in narcolepsy treatment, a disorder of heavy sleep noticed in human beings starting at early adulthood, resulting in sudden sleep attacks, paralysis, and sometimes in a temporary loss of muscle tone. HB behaves as a neurotransmitter and, therefore, can be used in the central nervous system of mammals as it is closely related to gamma aminobutyric acid (GABA, a muscle tone regulator); chemically, it affects the GABA receptor and decreases the narcolepsy. GABA generally regulates the muscle tone (Watanabe et al., 2002).

5.10   CONCLUSION

These days plastic has immense importance in pharmaceutical, household, and industries. The applications of plastic have a magical role because of its diverse properties. Each of its monomer shows different properties, and its combinational monomeric polymer can be used in different biomedical applications such as bio-implants, drug coatings, and preparation of bio-instruments like sutures. Plastic can also be used as adhesive agents. The increasing applications of PHAs require a higher supply for future generations. These increasing demands of PHAs due to bio-degradable, bio-compartible, nonimmunogenic and biological origin will push more industries for the production of PHAs.

KEYWORDS

- **biodegradability**
- **biocompatibility**
- **biomaterial**
- **PHB**
- **resorbable**

REFERENCES

Bonartsev, A.; Myshkina, V.; Nikolaeva, D.; Furina, E.; Makhina, T.; Livshits, V.; Boskhomdzhiev, A.; Ivanov, E.; Iordanskii, A.; Bonartseva, G. Biosynthesis, biodegradation, and application of poly(3-hydroxybutyrate) and its copolymers-natural polyesters produced by diazotrophic bacteria. *Communicating Current Research and Educational Topics and Trends in Applied Microbiology* 2007, 1, 295–307.

Boskhomdzhiev, A.P.; Bonartsev, A.P.; Makhina, T.K.; Myshkina, V.L.; Ivanov, E.A.; Bagrov, D.V.; Filatova, E.V.; Iordanskii, A.L.; Bonartseva, G.A. Biodegradation kinetics of poly(3-hydroxybutyrate)-based biopolymer systems. *Biochemistry Supplement Series B: Biomedical Chemistry* 2010, 4, 177–183.

Boyandin, A.N.; Prudnikova, S.V.; Filipenko, M.L.; Khrapov, E.A.; Vasilev, A.D.; Volova, T.G. Biodegradation of polyhydroxyalkanoates by soil microbial communities of different structures and detection of PHA degrading microorganisms. *Applied Biochemistry Microbiology* 2012, 35–44.

Brigham, C.J.; Sinskey, A.J. Applications of polyhydroxyalkanoates in the medical industry. *International Journal of Biotechnology for Wellness Industries* 2012, 1(1), 53–60.

Brigham, C.J.; Kurosawa, K.; Rha, C.K.; Sinskey, A.J. Bacterial carbon storage to value added products. *Journal of Microbial and Biochemical Technology* 2011, 83, S3–002.

Budde, C.F.; Riedel, S.L.; Willis, L.B.; Rha, C.; Sinskey, A.J. Production of poly(3-Hydroxybutyrate-co-3-Hydroxyhexanoate) from plant oil by engineered Ralstonia eutropha strains. *Applied and Environmental Microbiology* 2011, 77(9), 2847–2854.

Chanprateep, S. Current trends in biodegradable polyhydroxyalkanoates. *Journal of Bioscience Bioengineering* 2010, 110, 621.

Chaturvedi, K.; Kulkarni, A.R.; Aminabhavi, T.M. Blend microspheres of poly(3-hydroxy-butyrate) and cellulose acetate phthalate for colon delivery of 5-Fluorouracil. *Industrial & Engineering Chemistry Research* 2011, 50, 10414.

Diaz, A.; Katsarava, R.; Puiggali, J. Synthesis, properties and applications of biodegradable polymers derived from diols and dicarboxylic acids: From polyesters to poly(ester amides). *International Journal of Molecular Sciences* 2014, 15(5), 7064–7123.

Ellis, G.; Cano, P.; Jadraque, M.; Martín, M.; López, L.; Núñez, T.; de la Pena, E.; Marco, C.; Garrido, L. Laser microperforated biodegradable microbial polyhydroxyalkanoate substrates for tissue repair strategies: An infrared microspectroscopy study. *Analytical and Bioanalytical Chemistry* 2011, 399, 2379–2388.

Gunatillake, P.A.; Adhikari, R. Biodegradable synthetic polymers for tissue engineering. *European Cells & Materials* 2003, 5, 1–16.

Hutmacher, D.W. Scaffolds in tissue engineering bone and cartilage. *Biomaterials* 2000, 21, 2529–2543.

Ji, G.Z.; Wei, X.; Chen, G.Q. Growth of human umbilical cord Wharton's jelly-derived mesenchymal stem cells on the terpolyester poly(3-hydroxybutyrate-co-3-hydroxyvalerate-co-3-hydroxyhexanoate). *Journal of Biomaterials Science Polymer* 2009, 20(3), 325–339.

Kose, G.T.; Korkusuz, F.; Ozkul, A.; Soysal, Y.; Ozdemir, T.; Yildiz, C.; Hasirci, V. Tissue engineered cartilage on collagen and PHBV matrices. *Biomaterials* 2005, 26, 5187–5197.

Lenz, R.W.; Marchessault, R.H. Bacterial polyesters: Biosynthesis, biodegradable plastics and biotechnology. *Biomacromolecules* 2005, 6(1), 1–8.

Liu, Q.Y.; Jiang, L.; Shi, R.; Zhang, L.Q. Synthesis, preparation, *in vitro* degradation, and application of novel degradable bioelastomers—A review. *Progress in Polymer Science* 2012, 37, 715–765.

Misra, S.K.; Valappil, S.P.; Roy, I.; Boccaccini, A.R. Polyhydroxyalkanoate (PHA)/inorganic phase composites for tissue engineering applications. *Biomacromolecules* 2006, 7(8), 2249–2258.

Orts, W.J.; Nobes, G.A.R.; Kawada, J.; Nguyen, S.; Yu, G.E.; Ravenelle, F. Poly (hydroxyalkanoates): Biorefinery polymers with a whole range of applications. The work of Robert H. Marchessault. *Canadian Journal of Chemistry* 2008, 86(6), 628–640.

Sevastianov, V.I.; Perova, N.V.; Shishatskaya, E.I.; Kalacheva, G.S.; Volova, T.G. Production of purified polyhydroxyalkanoates (PHAs) for applications in contact with blood. *Journal of Biomaterials Science Polymer* 2003, 14, 1029–1042.

Shishatskaya, E.I.; Volova, T.G.; Efremov, S.N.; Puzyr, A.P.; Mogilnaya, O.A. Tissue response to biodegradable suture threads made of polyhydroxyalkanoates. *Biomedical Engineering* 2002, 36(4), 210–217.

Shishatskaya, E.I.; Volova, T.G.; Puzyr, A.P.; Mogilnaya, O.A.; Efremov, S.N. Tissue response to the implantation of biodegradable polyhydroxyalkanoate sutures. *Journal of Materials Science: Materials in Medicine* 2004, 15(6), 719–728.

Shishatskaya, E.I.; Voinova, O.N.; Goreva, A.V.; Mogilnaya, O.A.; Volova, T.G. Biocompatibility of polyhydroxybutyrate microspheres: *in vitro* and *in vivo* evaluation. *Journal of Materials Science: Materials in Medicine* 2008, 19(6), 2493–2502.

Sin, L.T.; Rahmat, A.R.; Rahman, W.A.W.A. *3-Applications of poly(lactic acid)*. In: *Handbook of Biopolymers and Biodegradable Plastics*; S. Ebnesajjad, Ed; William Andrew Publishing, Boston, MA, USA, 2013, pp. 55–69.

Sudesh, K.; Abe, H.; Doi, Y. Synthesis, structure and properties of polyhydroxyalkanoates: Biological polyesters. *Progress in Polymer Science* 2000, 25, 1503–1555.

Tran, R.T.; Zhang, Y.; Gyawali, D.; Yang, J. Recent development on citric acid derived biodegradable elastomers. *Recent Patents on Biomedical Engineering* 2009, E2, 216–227.

Turesin, F.; Gursel, I.; Hasirci, V. Biodegradable polyhydroxy alkanoate implants for osteo-myclitis therapy: *in vitro* antibiotic release. *Journal of Biomaterials Science, Polymer Edition* 2001, 12(2), 195–207.

Wang, Y.; Wu, Q.; Chen, G.Q. Attachment, proliferation and differentiation of osteoblasts on random biopolyester poly(3-hydroxybutyrate-*co*-3-hydroxyhexanoate) scaffolds. *Biomaterials* 2004, 25, 669–675.

Wang, Y.; Bian, Y.; Wu, Q.; Chen, G.Q. Evaluation of three dimensional scaffolds prepared from poly(3-hydroxybutyrate-*co*-3-hydroxyhexanoate) for growth of allogeneic chondrocytes for cartilage repair in rabbits. *Biomaterials* 2008, 29, 2858–2868.

Wang, L.; Wang, Z.H.; Shen, C.Y.; You, M.L.; Xiao, J.F.; Chen, G.Q. Differentiation of human bone marrow mesenchymal stem cells grown in terpolyesters of 3-hydroxyal-kanoates scaffolds into nerve cells. *Biomaterials* 2010, 31, 1691.

Watanabe, M.; Maemura, K.; Kanbara, K.; Tamayama, T.; Hayasaki, H. GABA and GABA receptors in the central nervous system and other organs. *International Review of Cytology* 2002, 213, 1–47.

Wei, X.; Hu, Y.J.; Xie, W.P.; Lin, R.L.; Chen, G.Q. Influence of poly(3-hydroxybutyrate-co-4-hydroxybutyrate-*co*-3-hydroxyhexanoate) on growth and osteogenic differentiation of human bone marrow-derived mesenchymal stem cells. *Journal of Biomedical Materials Research* 2009, 90(3), 894–905.

Xiong, Y.C.; Yao, Y.C.; Zhan, X.Y.; Chen, G.Q.; Application of polyhydroxyalkanoates nanoparticles as intracellular sustained drug-release vectors. *Journal of Biomaterials Science, Polymer* 2010, 21, 127.

Xu, X.Y.; Li, X.T.; Peng, S.W.; Xiao, J.-F.; Liu, C.; Fang, G.; Chen, K.C.; Chen, G.-Q. The behaviour of neural stem cells on polyhydroxyalkanoate nanofiber scaffolds. *Biomaterials* 2010, 31(14), 3967–3975.

Yamamoto, H.; Kuno, Y.; Sugimoto, S.; Takeuchi, H.; Kawashima, Y. Surface-modified PLGA nanosphere with chitosan improved pulmonary delivery of calcitonin by mucoadhesion and opening of the intercellular tight junctions. *Journal of Controlled Release* 2005, 102(2), 373–381.

Yao, Y.C.; Zhan, X.Y.; Zhang, J.; Zou, X.-H.; Wang, Z.-Hi.; Xiong, Y.-C.; Chen, J.; Chen, G.-Q. A specific drug targeting system based on polyhydroxyalkanoate granule binding protein PhaP fused with targeted cell ligands. *Biomaterial* 2008, 29(36), 4823–4830.

Zhao, S.; Zhu, M.; Zhang, J.; Zhang, Y.; Liu, Z.; Zhu, Y.; Zhang, C. Three dimensionally printed mesoporous bioactive glass and poly(3-hydroxybutyrate-*co*-3-hydroxyhexanoate) composite scaffolds for bone regeneration. *Journal of Materials Chemistry* 2014, B2, 6106.

Zheng, Z.; Bei, F.F.; Tan, H.l.; Chen, G.Q. Effects of crystallization of polyhydroxyalkanoate blend on surface physicochemical properties and interactions with rabbit articular cartilage chondrocytes. *Biomaterials* 2005, 26, 3537–3548.

CHAPTER 6

Thermostable Cyclodextrin Glycosyltransferase (CGTase): Recent Advances in Pharmaceutical Application of Cyclodextrins and Its Enzymatic Production

AMRITA SWAIN, LUNA SAMANTA, and DHANANJAY SOREN[*]

Department of Zoology, Ravenshaw University, Cuttack, Odisha 753003, India

Corresponding author. E-mail: dsoren@ravenshawuniversity.ac.in.

ABSTRACT

Cyclodextrin glycosyltransferase (CGTase; EC 2.4.1.19) is an extracellular enzyme which converts starch into nonreducing, cyclic malto-oligosaccharides called cyclodextrins (CDs). It is a hydrolytic enzyme that carries out reversible intermolecular as well as intramolecular transglycosylation reaction toward cyclization, coupling, and disproportionation of malto-oligosaccharides. CD is a nonreducing closed-ring malto-oligosaccharide with glucose monomers linked with each other by $\alpha 1 \rightarrow 4$ glycosidic bonds. The most common form of CDs are consist with Six, seven, and eight glucose residues, i.e., α-, β-, γ-cyclodextrin, respectively. CGTase exhibits an axis of cyclization either with phenylalanine (Phe) or tyrosine (Tyr), essentially for the formation of CDs. The CGTase enzyme is efficient of transglycosylation and is specifically dependent on the centrally located Tyr residue for cyclization. Similarly, Phe and Arginine (Arg) are known for CD binding residues, whereas, lysine (Lys), and asparagine (Asp) are known for linear substrate binding residues of CGTase enzyme. The kinetics of coupling reaction exhibit the binding of both the donor (CD) and acceptor (monosaccharide) to the active binding cleft of the CGTase enzyme module before processing of CDs. Environmental conditions are regulatory aspects for growth of the microbial population and metabolic production. Enhanced CD production could be achieved through providing appropriate conditions. Induction of more thermostable CGTase enzymes are required to increase the CD production by ignoring the traditional problem of increase in temperature with inactivating the enzyme. Fermentation conditions such as the concentration of nutrients, temperature, and compositions of the carbon and nitrogen sources determine the optimum production of CGTase. Similarly, the CD–drug inclusion complex can be utilized for management of drug toxicity and regulated drug delivery.

6.1 INTRODUCTION

In modern days, cyclodextrins (CDs) perform as a recognized guest molecule in the area of molecular complexation. The primary cause of its acceptance is based on its cost-effectiveness and easy availability. The key enzyme CGTase required for its production is most often isolated from different microorganisms and plants. CDs are one of the most flexible supports in pharmaceutical technology, being a vital requisite for the formulation of a wide range of delivery devices from the most classical dosage forms to the newest drug carrier molecules. There are other applications of CD in food, textile, cosmetics, as well as in the biotechnological and environmental fields, which drag interest toward this molecule.

CDs were first isolated by Villiers (1981) as degradation products of starch by *Bacillus amylobacter*, which were known as cellulosine for their similarity to cellulose, available in two forms of dextrins—α and β. The oligosaccharide nature of CDs produced by the enzymatic breakdown of potato starch by *Bacillus macerans* had been well characterized, and termed as Schardinger dextrins (Cova et al., 2018; Maheriya, 2017; Marques, 2010). The First International Cyclodextrin Symposium on CDs was organized by Szejtli in 1981, in Budapest. Gradually, the research world took interest in this molecule due to its structural peculiarity and broad industrial application in various fields (Buschmann and Schollmeyer, 2002).

Molecular complexation is the association of two or more molecules during formulation of drug inclusion, which enhances its dissolution and bioavailability. CDs are primarily cyclic, cage-like glucose residues produced from enzymatic hydrolysis of starch with help of the enzyme CGTase. This review elaborates the journey of CD from the molecular structure to its application in different fields such as pharmaceuticals, food, cosmetics, flavor, and its production strategies through a proficient enzyme.

6.2 STRUCTURE AND PHYSICOCHEMICAL PROPERTIES

CDs are cyclic oligosaccharides of glucose residues consisting of 6, 7, or 8 D-glucose units linked by α-1,4 glycosidic bonds, classified as α, β, and γ, respectively (Figure 6.1; Table 6.1). They are synthesized from starch by the cyclization reaction of a bacterial enzyme, CGTase. The major CDs are crystalline in nature and are homogeneous nonhygroscopic substances with no reducing groups. They are resistant toward acid-catalyzed hydrolysis, unlike that of linear sugars, and with increasing cavity size, the ring-opening rate increases gradually. A rise in temperature and the concentration of acid increases the rate of acid hydrolysis (Das et al., 2013). The torus shape of CDs are achieved by the unique positioning of glucose residues in the CD rings in which secondary hydroxyl groups (C2 and C3) are present on one side and primary hydroxyl groups (C6) are arranged on another side on the edges. The hydrophobic C3 and C5 hydrogen and ether-linked oxygen are found in the internal site and the hydrophilic hydroxyl groups at the external site of the CD molecule. The disruption of exterior hydrogen bonds altering the 2- or 3-hydroxyl group permits more interactions with water molecules, resulting in altered solubility (Nitalikar et al., 2012). CDs upon crystallization yield two types of crystal packing, i.e., channel-packing structure and cage-packing structure, depending on the available guest molecules. These crystal structures convey that CDs exhibit the expected "round" structure, with all glucopyranose units in the $^4C_1$ chair conformation when they are in the complex. It contains a hydrophobic internal surface and a hydrophilic external surface, which is a special feature that enables CDs to form inclusion complexes with most of the organic and inorganic compounds. These peculiarities of CDs allow them to change the physical and chemical properties of encapsulated guest compounds. Therefore, CDs are becoming increasingly popular day by day for extensive use in industrial sectors.

6.3 INCLUSION COMPLEX FORMATION

An appropriate guest molecule substitutes the water from the CD cavity, leading to the formation of the CD-inclusion complex. Inclusion complex formation does not initiate the breakdown or formation of any covalent bond; rather, a probable alteration of the temperature-dependent solubility, electrochemical properties, chemical reactivity, and spectral properties play a major role in inclusion complex formation (Loftsson and Brewster, 2010; Radi and Eissa, 2010). The solubility of inclusion complex and

CDs are inversely proportional to each other. Inclusion complex with guest molecule of low water solubility usually results with decrease in solubility of CD, where solubility of inclusion complex is generally less than that of the CD, and greater than the guest molecule (Das et al., 2013). A microenvironment is provided by the lipophilic cavity of CD molecules in which an appropriately sized nonpolar moiety can be accommodated to form

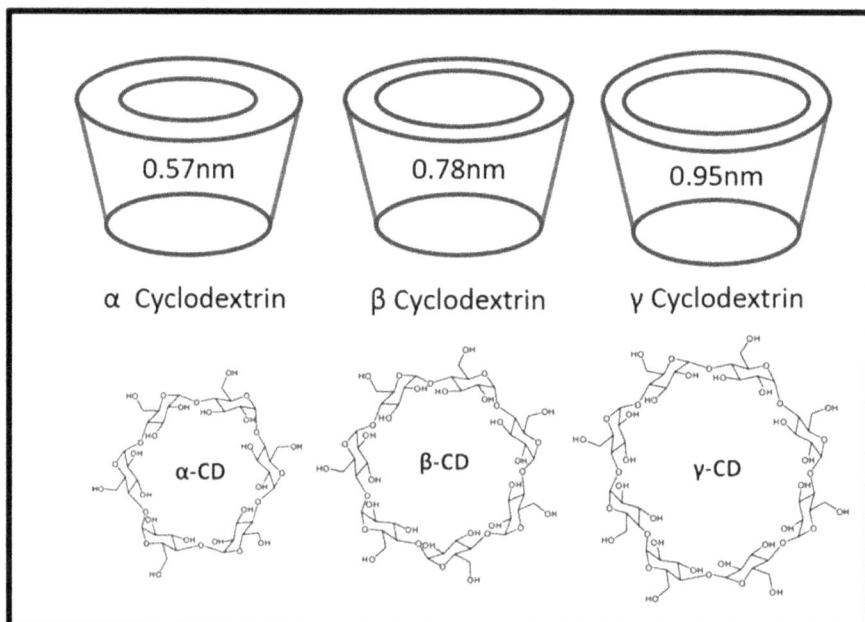

FIGURE 6.1   Structure of cyclodextrin (Narayanan et al., 2017).

TABLE 6.1   Details of CD Classes and Their Properties (Katageri et al. 2012; Miranda et al. 2011)

Cyclodextrin Type	α-CD	β-CD	γ-CD
No. of glucopyranose units	6	7	8
Molecular weight	972	1135	1297
Cavity volume (nm^3)	0.174	0.262	0.472
External diameter (Å)	14.6	15.4	17.5
Internal diameter (Å)	4.7–5.3	6.0–6.5	7.5–8.3
Shape of crystals	Hexagonal lattice	Monocyclic parallelograms	Quadratic prism
pK$_a$	12.33	12.2	12.08
Water solubility	14.5	1.85	23.2

inclusion complexes (Figure 6.2) (Loftsson and Brewster, 1996). CD as a host binds to guest molecules and creates a temporary vibrant equilibrium by developing a "host–guest" complex with the locally interacting surface molecule, depending on the appropriate size of the guest molecule. This whole phenomenon is established by binding strength and the driving force to drag the visitor molecule toward the CD cavity. The aqueous solvent is the optimum choice of complex formation, both in the solution and crystalline forms. Similarly, the formation of an inclusion complex can be accomplished through different conformational stages and charge-transfer interactions in the presence of any nonaqueous solution or cosolvent, which includes the involvement of van der Waals interactions, hydrogen bonding, hydrophobic interactions, electrostatic interactions for liberating enthalpy-rich water molecules from the cavity to make it hydrophobic (Jambhekar and Breen 2016; Loftsson and Brewster, 2010). The reason for extensive acceptance of CD is its quite diversified list of guest compounds starting from different aliphatic or aromatic groups to organic acids, alcohols, aldehydes, ketones, fatty acids, polar molecules such as amines, halogens, and oxyacids along with multiple reactive hydroxyl groups, which favor chemical modification of CD boosting in its functionality (Schmid, 2001).

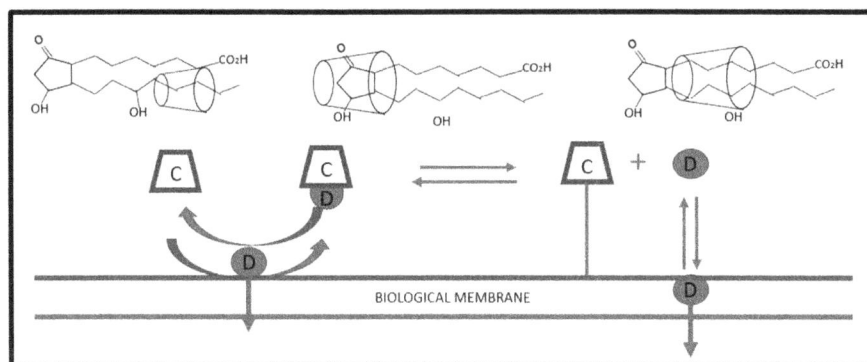

FIGURE 6.2   CD-drug complex formation and drug release to membrane (Savjani et al., 2012; Loftsson and Stefánsson 1997).

6.4   DETERMINATION OF CYCLODEXTRIN COMPLEXES

The different instrumental techniques available to characterize complex formation include nuclear magnetic resonance, high-performance liquid chromatography, phase solubility, X-ray powder diffraction, Fourier transform infrared spectroscopy, circular dichroism, differential scanning calorimetry, Fourier transform Raman spectroscopy, thermogravimetric analysis, ultra-violet–visible spectroscopy, differential solubility (Das et al., 2013), and nanodifferential scanning fluorimetry (Sonnendecker and Zimmermann, 2019). The most generalized type of CD complexes available is 1:1 drug/CD (D/CD) complex wherein one molecule of drug forms a complex with one molecule of CD.

This is represented by equation as:

$$D + CD \xrightleftharpoons{K_{1:1}} D/CD$$

Higher-order D/CD complex can also be formed having a ratio of 1:2, when an additional CD molecule forms a complex with the existing 1:1 complex (Kategeri and Sheikh, 2012).

The equation is

$$D/CD + CD \xrightleftharpoons{K_{1:2}} D/CD_2$$

The inclusion complex formation of CD with a guest molecule relies on two principal factors (Figure 6.3). The first is steric interaction, which depends on the relative size of the CD to the guest molecule or functional groups within the guest, and the second critical factor is the thermodynamic interactions between different components of the system (i.e., CD, guest, solvent) governed by a favorable net driving force which helps in the formation of the complex. The internal diameter of the cavity and its volume are determined by the number of glucose units assembled to form the types of CDs. These dimensions give the ability to α-CD to complex with low molecular weight molecules or compounds with aliphatic side chains, β-CDs to complex with aromatics, and heterocycles, and γ-CD to accomodate larger molecules such as macrocycles and steroids. However, the molecular heights of all CD types remain the same.

6.5 MECHANISMS OF GUEST RELEASE FROM CYCLODEXTRIN COMPLEXES

The CD–guest complexation is a dynamic process of noncovalent interactions where continuous association and dissociation take place between the molecule and the CD within the cavity. In the case of a 1:1 complex, the interaction is as follows:

$$CD + G \xrightleftharpoons{} CD\text{-}G; \ K = K_R/K_D.$$

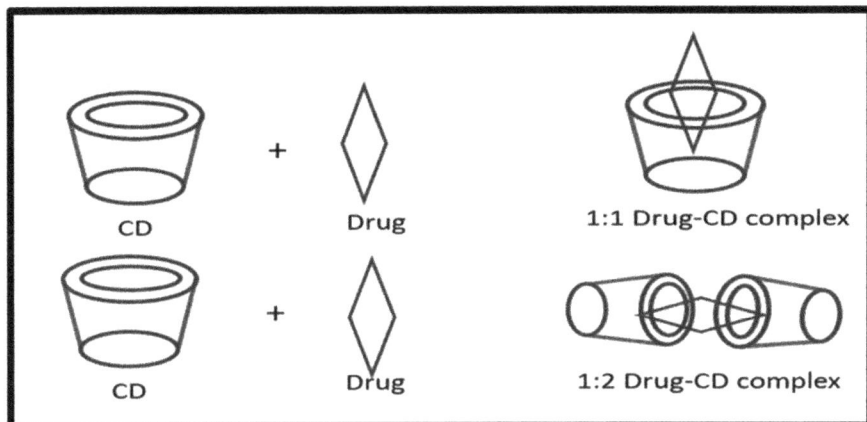

FIGURE 6.3   Drug-CD complex formation (Savjani et al., 2012).

The above equation explains CD and the guest molecule (G) forming CD-G as the inclusion complex, where K defines the vital equilibrium constant, with K_R and K_D as the recombination and dissociation constant, respectively. The size of the guest molecule regulates the formation and breakdown of the inclusion complex. Larger the size of the guest molecule, slower is the rate of formation and dissociation of the inclusion complex, where local ionization decreases the rates of complex formation and dissociation. In spite of dilution appearing to be a major drug-release mechanism, other such mechanisms are there such as drug uptake by tissues, competitive displacements of the drug from the complex, ionic strength, binding to protein, and temperature may also be considered for the stability and dissociation of the CD-drug complex.

6.6   DRUGS AND CYCLODEXTRIN

Encapsulation of drug molecules within CD as guest molecules has become an area of new interest for efficient drug delivery function. The drug solubility characteristics of a specific CD is often studied by phase solubility (PS) analysis, which deciphers the ability of forming complex with both natural and modified CD. There are two PS profiles named as A-type and B-type, in which A-type explains the concentration of dissolved drug increases with the amount of added CD where β- and γ-CD possess a comparatively limited solubility with precipitate from the solution. The most accepted types of CD complexes are one with 1:1 D/CD complex and a higher order D/CD complex having 1:2 ratio where the former complex forms with one drug to one CD molecule and in the later case, one additional CD molecule forms complex with an existing 1:1 complex, respectively, with the release of the drug to the membrane (Figures 6.2 and 6.3) (Kategeri and Sheikh, 2012; Savjani et al., 2012).

6.6.1   TECHNIQUES FOR DRUG–CYCLODEXTRIN COMPLEX FORMATION

Different types of techniques have been used for the preparation of inclusion complexes, as described below.

Physical bending/kneading: The inclusion complexes of liquid or oil guest molecules can be prepared by simply grinding the drug with the CD in mortar or in a rapid mass granulator. The guest molecule and CD with minimum volume of water are prepared into a paste, which is then evaporated under vacuum at room temperature, filtered and stored in a desiccator in powdered form until further evaluation.

Co-precipitation: CD is dissolved in a polar aqueous solution, and the guest molecule is gradually added to form a complex by precipitation. Sometimes, heating enhances the solubility of both CD and guest molecule when a guest molecule can withstand the temperature and the precipitate obtained is cooled, separated by centrifugation, decanting or filtration, and finally processed with subsequent washing. The major limitation of this procedure is during scaling-up and interference of organic solvents.

Dry mixing: It involves simple mixing of CD with an oil or liquid guest to form an inclusion complex. Although it does not require any washing step for scaling-up, the main causative limitation observed was incomplete complexation which resulted due to inadequate mixing.

Gas–liquid method: Passing a vapor through a hot or cold CD solution will complex many solvents or other chemicals present. Further, upon filtration, the complex is separated and collected, and through steam distillation the volatiles are removed.

Spray-drying: When a monophasic solution of drug and CD is prepared with a suitable solvent system, complexation is attained by constant stirring to an equilibrium state from where the solvents can easily be removed by spray-drying. This technique is applied to thermostable compounds, as the CD and guest molecules are dissolved in deionized water and dried using a spray dryer (Cheirsilp and Rakmai, 2016).

Lyophilization or freeze-drying: This method is suitable for heat-labile guest molecules and possible for CD–drug complex formation. Initially the

CD and guest are mixed with buffer to form a homogeneous suspension and then it is freeze dried to get a powder by vacuum drying.

Solid dispersion/co-evaporated dispersion: In this technique, the solutions of both CD and drug are suspended in water and ethanol, respectively, and mixed in an appropriate medium with constant stirring till the equilibrium is reached, resulting in evaporation under vacuum.

Melting: It is the process of simply melting the guest and adding it to CD powder with a higher concentration of guest molecule, and then the excess is removed cautiously by a solvent of weak complex or by vacuum sublimation (Nitalikar et al., 2012).

Neutralization: A discrete mixture of drug and CDs are prepared in NaOH/NH$_4$OH with mixing and neutralizing of pH to 7.5 and subsequently adding HCl till the precipitation of complexes, followed by the filtration of precipitate, cured by washing to make it chlorine-free complex formulation.

Extrusion: It involves CD, guest, and water, which go into the extruder for continuous heating and mixing in a regulated manner and subsequently drying to form the complex.

Microwave irradiation: It involves using of certain moles of CD and guest dissolved in water and organic solvent in a flask, followed by heating at 60 °C for 1–2 min. The residual complex parts are filtered, vacuum-dried, and desiccated for further use (Chaudhary and Patel, 2013; Patil et al., 2010).

6.6.2 EFFECTS OF CD PROPERTIES ON DRUG FORMULATION

Drug solubility and dispersion: The CD has an important role in formulation of poorly water-soluble drugs by improving the apparent drug solubility and dissolution through inclusion complexation or solid dispersion (Table 6.2). It also acts as a hydrophilic carrier for drugs with low molecular characteristics for complexation, or as a tablet-dissolution enhancer for drugs with high dose such as hormones, peptides and proteins, and immunological factors (Tasic et al., 1992; Tian et al., 2013).

Bioavailability: Drug solubility, dissolution, and permeability are the three major factors that can make one drug bioavailable by decreasing the limitations. The drug also has to cross specific biological barriers such as the mucosa, skin, or the eye cornea without disturbing the lipid bilayer. CD and drug complexation can achieve these characteristics if they are used in a regulated manner (Challa et al., 2005; Loftsson et al., 2007). CDs can enhance membrane fluidity by their cholesterol-removing ability and can stimulate membrane invagination by loss of bending resistance, resulting in cell lysis.

Unstable drugs when complexed with CD usually show less contact time at the absorption site during drug delivery.

TABLE 6.2 Modification of the Drug Release Site and/or Time Profile by CDs (Uekama et al., 2006)

Release Pattern	Aim	Use of CD
Immediate release	Enhanced dissolution and absorption of poorly water soluble drugs	HP-β-, DM-β-, SB-β-, and branched-β-CDs
Prolonged release	Sustained release of water soluble drugs	Ethylated β-CDs, acylated β-CDs
Modified release	More balanced oral bioavailability with prolonged therapeutic effects	Simultaneous use of different CDs and/or other excipients
Delayed, pH-dependent release	(Enteric) Acid protection of drugs	CME-β-CD
Site-specific release	Colon targeting	Drug/CD conjugate

Drug safety: CD-drug inclusion bodies not only increase the efficiency of drugs even used in lower dosage, but it also downgrade the side effects, irritation levels, and toxicity to the human system by preventing the direct contact to membranes. The antiviral activity of ganciclovir has been enhanced by β-CD on the human cytomegalovirus clinical strain, with increased drug potency and reduced toxicity (Rasheed et al., 2008).

Drug stability: Encapsulation of CD with drug can formulate a stable compound by modulating against various physical, chemical and thermal factors like hydrolysis, dehydration, oxidation, temperature, heat, light, metal salts, relative humidity thereby increasing its shelf life (Tian et al., 2013; Tiwari et al., 2010). CD stabilizes the drug by altering with the nature and effect of the included functional group on drug. The two antihistamine drugs Trimeprazine (Lutka and Koziara, 2000) and Promethazine (Lutka, 2002) have been encapsulated with CDs with improved photo stability for treating allergic responses.

6.7 PHARMACEUTICAL APPLICATIONS

CD complexation property has resolved many difficulties of targeted drug delivery in pharmaceutics (Table 6.3). It plays a vital role in the drug-delivery system as a potential candidate with ability to alter the physical, chemical, and biological properties of guest molecules with enhanced bio-adaptability, thereby enabling the multifunctional route of administration exclusively for

TABLE 6.3 Selective Drugs Formulated with CD in India and Other Countries (Jambhekar and Breen 2016; Cyclodextrin news, 2013; Chordiya and Senthlkumaran 2012; Miranda et al., 2011; Loftsson and Brewster 2010)

CD Type	Drug Name	Country	Commercial Name	Formulation Form/ Administration Route	Pharmaceutical Company
β-CD	Aceclofenac	India	Aceclofenac-β-cyclodextrin	Tablet	Taj Pharma
β-CD	Betahistine	India	Betahist	Tablet	Geno Pharma
β-CD	Flunarizine	India	Fluner	Tablet	Geno pharma
β-CD	Norfloxacin Tinidazole	India	Entronor −TZ/Noroxin	Tablet	
β-CD	Piroxicam	India	Cycladol/Pyrodex/Medicam	Tablet	Ranbaxy/SunPharma MMC
β-CD	Refocoxib	India	Rofizgel	Tablet	Wockhardt
HP(2-hydroxy prolyl) β-CD	Voriconazole	India	Vorzu	Tablet	Ranbaxy
α-CD	Alprostadil	Europe/Japan/USA	Caverjectdual/Provastatin/Rigidur	Intravenous solution	Pfizer
α-CD	Cefotiam hexetil hydrochloride	Japan	Pansporin T	Oral tablets	Takeda
α-CD	OP-1206	Japan	Opalmon	Tablet	
β-CD	Piroxicam	Europe	Brexin	Oral tablets/Suppository	Chiesi
β-CD	Nicotin	Europe	Nicorette	Sublingual tablet	Pfizer
β-CD	Omeprazole	Europe	Omebeta	Tablet	betapharm
β-CD	Cephalosporin	Japan	Meiact	Tablet	Mejji seika
β-CD	Cetirzene	Germany	Cetrizin	Chewable tablet	Losan pharma
HP β-CD	Itraconazole	Europe/USA	Sporanox	Oral and intravenous solution	Janssen

TABLE 6.3 *(Continued)*

CD Type	Drug Name	Country	Commercial Name	Formulation Form/ Administration Route	Pharmaceutical Company
HP β-CD	Indometacin	Europe	Indocid	Eye drop solution	Chauvin
HP β-CD	Cisapride	Europe	Propulsid	Suppository	Janssen
HP β-CD	Hydrocortisone	Europe	Dexocort	Buccal	
SBE-β-CD (sulfobutyl Ether-β-CD)	Voriconazole	Europe/USA/Japan	Vfend	Intravenous solution	Pfizer
SBE β-CD	Ziprasidone maleate	Europe/USA	Zeodon/Zeldox	Intramuscular solution	Pfizer
Methyl-β-CD	Chloramphenicol	Europe	Chlorocil	Opthalmic solution	Oftalder
Methyl-β-CD	17β-Estradiol	Europe	Aerodiol	Nasal drop	Servier
HP γ-CD	Diclofenac sodium	Europe	Voltarenophtha	Eye drop solution	Novartis
HP γ-CD	Tc-99 Teboroxime	USA	Cardiotec	Intravenous solution	Bracco

rectal, ocular, nasal, dermal, and transdermal delivery systems (Singh et al., 2002). Anticancer drugs as guest within CDs are universally accepted for their excellent bioavailability. The chemical structure and pharmacological properties augment the rate of drug absorption, bioavailability, clearance, volume of distribution, and plasma half-life with suitable dose size and frequency, simultaneously to meet patient needs. Most candidate drugs in clinical development fail to reach the clinical market due to some of the above reasons. Solubility in aqueous medium has been the principal determining factors in drug formulation and development. Drugs with in nature have been designed to permeate the biological membranes via passive diffusion. Class II and Class IV drugs are absorbed through biological membrane by techniques such as solid dispersion, particle size reduction, melt extrusion, salt formation, spray drying, and complexation, where drug solutions are made in the form of microemulsions, liposomes, with nonaqueous solvents, which increase the drug's solubility and bioavailability without decreasing their lipophilicity. CDs are used worldwide by pharmaceutical industries as a carrier complex for their peculiar structure and versatility in complex formation with any type of drug (Loftsson et al., 2002). Among all CD derivatives, hydroxypropyl-β-cyclodextrin (HP β-CD) and sulfobutyl ether-β-cyclodextrin (SBE β-CD) are the two promising CD derivatives that are proved to be more efficient than the traditional one; these are used in a broad range of drugs, especially in oral and intravenous dosage forms with higher solubility and less toxicity (Brewster and Loftson, 2007). CD-based nanoparticles are very much promising due to their desired drug size, better drug-loading, extended circulation in blood, and good surface morphology for tumor-targeting and cancer therapy (Erdogar and Bilensoy, 2018). CD-associated therapeutics are also reliable in other medical conditions such as cholesterol efflux from membranes, cardiovascular diseases, antipathogenic in the case of HIV and herpes simplex virus, and neural disorders in Alzheimer's disease with the help of bioavailable drugs (Leclercq, 2016). Table 6.3 illustrates a number of specific drugs with their administration route formulated in India and other countries.

6.7.1 CYCLODEXTRIN-BASED DRUG DELIVERY

6.7.1.1 OPHTHALMIC DRUG DELIVERY

The outer lipophilic layer of the eye cornea needs a lipophilic drug that can permeate through the cornea into the eye. But regular ophthalmic drugs like suspensions, gels, oily drops, ointments, and solid inserts lead to unnecessary negative effects like eye irritation and blurred vision. CD-drug complexes replace the irritants and can be easily absorbed through biological membranes of eye cornea and skin by extracting or complexing with some nonpolar components such as cholesterol and phospholipids from the membrane without causing any discomfort (Rasheed et al., 2008).

6.7.1.2   NASAL DRUG DELIVERY

Nasal formulations with CD complexation satisfy high aqueous solubility of hydrophobic drugs particularly for peptides with better dispersal rate, i.e., methylated CD could induce bioavailability up to 100% (Merkus et al., 1991; Nitalikar et al., 2012). Besides that, drugs for Alzheimer's disease and neurological disorder prefer CD complex of HP β-CD for intranasal delivery as it has neuroprotective to β amyloid toxicity with strong absorbing capacity (Jacob and Nair, 2018).

6.7.1.3   ORAL DRUG DELIVERY

This type of drug delivery can be explained based on the duration of release, mode, and site of action, and can be divided into (1) immediate release, (2) prolonged release, (3) modified release, and (4) delayed release (Nitalikar et al., 2012). In addition to increasing solubility, oral CD-encapsulated drugs also enhance buccal and sublingual bioavailability, gastrointestinal stability, and permeability (Challa et al., 2005). Fast-acting drugs such as antipyretics, analgesics, and coronary vasodilators are from immediate-release groups. Except that, polar CD molecules are used to enhance the bioavailability of some nonsteroidal anti-inflammatory drugs, anti-epileptics, steroids, barbiturates, antidiabetics, vasodilators, cardiac glycosides, benzodiazepines, etc. (Nitalikar et al., 2012).

6.7.1.4   SUBLINGUAL DRUG DELIVERY

Sublingual drug delivery is almost the same as that of oral type, with the difference being that in the former, the drug has to be released from the CD complex before it gets diffused within a short time span (Rasheed et al., 2008).

6.7.1.5   DERMAL DRUG DELIVERY

The most difficult layer for drug penetration to the epidermis is the stratum corneum, and traditional drugs use fatty acids or alcohol for optimum absorption, which later on can cause skin irritation and low solubility. Therefore, CD can minimize the side effects in topical application as well as improve the solubility, stability, and sustained drug release for dermally applied drugs (Rasheed et al., 2008).

6.7.1.6   PULMONARY DRUG DELIVERY

Because of the large surface area, low enzymatic activity, and sufficient blood supply, limiting drug degradation and first-pass metabolism like through the gastrointestinal (GI) tract makes an altered option for local treatment of pulmonary diseases. The CD–drug complex can transform solution to the powder form, lowering irritation, unpleasant smell, and bad taste, increasing drug solubility, dissolution, and stability (Rasheed et al., 2008).

6.7.1.7   PARENTERAL DRUG DELIVERY

α-CD and some polar derivatives of β-CD such as HP-β- and SBE-β-CDs are used in parenteral formulations for better solubility and minimizing toxicity, whereas γ-CD-forming aggregates cannot be applied for these drugs (Challa et al., 2005; Nitalikar et al., 2012).

6.7.1.8   RECTAL DRUG DELIVERY

Rectal absorption is the main limitation as the rectal fluid is viscous as compared to GI fluid and found in minimum volume. The CD carriage makes the drug nonpolar in nature, thus avoiding direct contact with the rectal membrane with quick release and well dispersal rate (Nitalikar et al., 2012). There are certain lipophilic drugs for rectal delivery which are anti-inflammatory agents like flurbiprofen (Hirayama and Uekema, 1999; Uekama et al., 2006).

6.7.1.9   BRAIN DRUG TARGETING

CD complexation may improve target drugs like antitumor agents, steroids, and Ca^{2+} channel antagonists to brain crossing the blood brain barrier providing greater solubility and stability (Tiwari et al., 2010).

6.7.2   NOVEL DRUG DELIVERY SYSTEMS

6.7.2.1   LIPOSOMES

Liposomes are the vesicles which can improve the application of complexed CD and drug by accommodating drugs to the hydrophilic and lipid bilayer site as per their polar or nonpolar nature, respectively. The difficulties seen before with intravenous application, like increased kidney toxicity and fast release of drugs into urine, lowering the bioavailability, could be diminished with liposomal drug targeting (Rasheed et al., 2008; Tiwari et al., 2010).

6.7.2.2   MICROSPHERES AND MICROCAPSULES

The presence of CD in microspheres or microcapsules can enhance the potency and efficacy of drugs with a better drug-loading capacity (Rasheed et al., 2008). During microsphere preparation, HP β-CD provides a stable environment for the bovine serum albumin and lysozyme by amplifying the polarity and encapsulating hydrophobic groups in the CD cavity (Chordiya and Senthilkumaran, 2012). Microcapsule crosslinked with β-CD acts as a release modifier by decreasing the retardation of hydrophilic drugs through semipermeable membrane, facilitating the release of drugs in a regulated manner (Challa et al., 2005; Tiwari et al., 2010).

6.7.2.3   OSMOTIC PUMP TABLET

Osmotic tablets are used in some steroid drugs and oral drugs which are poorly water soluble. An osmogent is the principal component that exerts pressure to release drugs in a controlled manner along with the central active agent and other excipients, coated with semipermeable membrane and a drilled delivery passage. Its limitation is that it requires a drug in a solution, which may be prepared by many solubilizing techniques (Rasheed et al., 2008).

6.7.2.4 GENE AND OLIGONUCLEOTIDE

The nonviral vector development for gene delivery is accepted in medical science because of the immunogenic and toxic response of viral vectors for which nucleotide gene delivery technologies are on the rise. But there are certain drawbacks such as endonuclease susceptibility with degradation of products, lowered blood release and cell membrane transport capacity, nonspecific communication with outer and inner cellular cations due to its polyanionic nature, and being immunogenic and toxic (Challa et al., 2005). CD complexation is well adapted to nucleotide accommodation to overcome the above limitations.

6.7.2.5 PEPTIDES AND PROTEINS

Biotechnology has made it possible to produce a generous amount of proteins which are therapeutically active and may cure some rare diseases, but being proteinaceous in nature the drug–delivery system is quite a tough job because of the problems arising such as less absorbance to cell membrane, unstable enzymatic and chemical activity, fast release from blood, altered dose response, and immunogenicity. This situation can be avoided by CD encapsulation, as it induces its bioavailability by modulating its chemical and biological properties and acts as potential carriers for the delivery of proteins (Challa et al., 2005).

6.7.2.6 NANOPARTICLES

Although nanoparticles have proven themselves as better drug-target vehicles in terms of stability than liposomes, they have some limitations related to drug-loading of polymeric nanoparticles and inefficient encapsulation. CD provides an improvement in nanoparticle modeling by a greater loading capacity by increasing the hydrophobic sites (Challa et al., 2005; Rasheed et al., 2008).

6.8 OTHER INDUSTRIAL APPLICATIONS

6.8.1 FOOD

CD complexation has various roles in improving the food industry. The encapsulation in the hydrophobic cavity can store and stabilize flavor, fragrance, color, lipophilic food components, light, heat, and oxygen-sensitive products, as well as preventing the degradation of essential oils and vitamins. It can act as a taste enhancer in alcoholic drinks and a bitter-taste stabilizer in citrus fruits, and can help in cholesterol removal from animal products (Astray et al., 2009; Das et al., 2013).

6.8.2   ENVIRONMENT

The association capacity of CD with numerous types of guest molecules can also entrap environmental pollutants in the CD cavity from water and soil and can remove suspended particulate matters from the atmosphere. Similarly, it can entrap and remove toxic industrial effluents by forming the inclusion complex for safe disposal (Cheirsilp and Rakmai 2016; Singh et al., 2002).

6.8.3   COSMETICS, TOILETRIES, AND PERSONAL CARE

The stabilization of color and flavor of lipsticks, fragrance of perfumes, improvement of shelf life of sunscreen lotion and skin creams, and solubility of nonpolar cosmetic products in the present cosmetic world much needed the CD-inclusion complex formation (Buschmann and Schollmeyer, 2002; Cheirsilp and Rakmai, 2016). Eventually, it can increase the lifespan of personal care and beauty products and lower the unhealthy odor from sanitary napkins, diapers, and clothes. It can also help solubilize insoluble triclosan, which is used as a topical antiseptic and disinfectant in silica-based toothpaste (Das et al., 2013; Singh et al., 2002). CD complexes in talcum powder stabilize fragrances and prevent their loss due to evaporation and oxidation over a long manufacturing period with improved antimicrobial efficacy.

6.8.4   PACKING AND TEXTILE INDUSTRY

The quality of fabric had been demonstrated to be improved by binding CD derivatives with the fiber alternative monochlorotriazinyl (MCT) produced by Wacker Chemie AG (Munich, Germany), the world's largest producer of γ-CD, which imparts superb textile finishing to cottons, woolens, and blended materials (Singh et al., 2002). CDs are also incorporated to reduce sweat, smoke, and clothing odors. They are also used for enhancing dye uptake to avoid color fade and also contribute in forming the inclusion complex with oily antimicrobial and volatile agents coated on a hydrophilic sheet to a natural resin binder, used for wrapping fresh products (Amrit et al., 2011; Singh et al., 2002).

6.8.5 BIOTECHNOLOGICAL FIELD

CD complexation has upregulated the substrate solubility without damaging the microbial cells or enzymes; as a result, the enzymatic conversion has accelerated with lowered toxicity and a better fermentation yield, independent of any concentration (Das et al., 2013).

6.9 PRODUCTION OF CYCLODEXTRIN

Since CD can resolve many complications in different areas, its production must be accelerated. Most of the studies reveal that there are two basic aspects regarding CD production: (1) solvent extraction and (2) nonsolvent process, in which we can extract a particular CD using an organic complexing agent and the other generates a mixture of CD without any complexing agent. They can be separated later by chromatographic procedures (Das et al., 2013; Li et al., 2007). However, in both the processes, there is a need for the enzyme CGTase (EC 2.4.1.19) for enzymatic conversion from starch or starch derivatives after liquefaction. CGTase is a member of the α-amylase family (family 13) of glycosyl hydrolases. The end product of the reaction often produces a mixture of α-, β-, and γ-CD, consisting of six, seven, and eight glucose units, respectively (Figure 6.1), and trace amounts of large-ring CDs with more than nine glucose units. The specific amount and ratio of CDs in α, β, and γ forms are determined by both CGTase and the reaction conditions, including reaction time, temperature, and presence of solvent (Li et al., 2007). The preparation process of CD was primarily initiated with the culturing of CGTase-producing bacteria with the separation and purification of the enzyme by fermentation, followed by the enzymatic conversion of prehydrolyzed starch to the mixture of cyclic and acyclic dextrins (Table 6.4). The concluding steps include separation of CDs from the mixture and their purification and crystallization (Das et al., 2013). The CGTase enzyme acts upon amylose and amylopectin of starch, which can be used as raw materials for CD production. A larger percentage of amylopectin are present

TABLE 6.4 List of Thermostable CGTase Produced from Microbial Sources

Name of the Microbes	Enzyme type/Maximum Producing CD Type	Maximum pH Range	Maximum Temperature (°C)	References
Thermococcus sp.	α-CGTase	5–5.5	120	Tachibana et al., 1999
Pyrococcus furiosus DSM 3638	β-CGTase	5	95–100	Lee et al., 2007
Thermoanaerobacter sp.	β-CGTase	7.0–8.0	95	Avci and Donmez 2009
Thermoanaerobacter sp. ATCC 53627	β-CGTase	5–6.7	90–95	Starne et al., 1991; Norman and Jorgensen, 1992;
T. thermosulfurigenes EM1	α-CGTase	6	85	Knegtel et al., 1996
Bacillus sp. Strain MK6	α-CGTase/β-CGTase	4–10	30–80	Noi et al., 2008
Bacillus sp. TPR71HNA6	β-CGTase	6.04	80	Kashipeta, 2015
Bacillus licheniformis Sk 13.002	β-CGTase	6.0	75	Letsididi et al., 2011
Paenibacillus macerans(Modified)	α-CGTase	8.0	65	Wang, et al., 2016
Bacillus flexus SV 1	β-CGTase	8.0	60	Reddy, et al., 2017
Microbacterium terrae KNR 9	β-CGTase	6.0	60	Rajput et al., 2016
Bacillus circulans.	β-CGTase	6.5	60	Li et al., 2014
Bacillus licheniformis MCM-B 1010	β-CGTase	8.5	60	Thombre and Kanekar 2013

in potato starch (~70%–75%), which has been adapted more for effective production (Das et al., 2013).

6.10 CONCLUSION

A broad range of microbes are able to use starch as a carbon and energy source for growth. Maximum starch-degrading enzymes are predominantly extracellular enzymes with multiple reaction specificities to yield a wide variety of products (Bart et al., 2000). Examples of CGTase-producing bacteria are *Bacillus circulans, Paenibacillus macerans, Klebsiella pneumoniae, Bacillus megaterium, Bacillus stearothermophilus*, *Thermoanaerobacter* sp., *Bacillus amyloliquefaciens, Bacillus lentus,* alkalophilic *Bacillus* sp., *Micrococcus*, etc. (Das et al., 2013).

CGTase-producing bacteria can be found in various places, such as soil, waste-disposal sites, plantation sites or vegetation habitats, hot springs, and even in deep sea water and mud. *Bacillus licheniformis* MCM-B 1010 isolated from Lonar lake, India, was employed in the production of CD (Thombre and Kanekar, 2013). Similarly, CGTase was isolated from *Bacillus agaradhaerens* strain LS-3C, isolated from an Ethiopian soda lake, which was purified up to 43-fold by starch adsorption with a yield of 50% (Martins and Kaul, 2002). CGTase-producing alkalophilic bacterial screening was found from a collected sample of hyper saline soda lakes of Wadi Natrun Valley, Egypt (Ibrahim et al., 2012). The major limitations of enzymatic production of CDs are mixture of α, β, and γ-CD, and susceptibility to product inhibition by these dextrins (Bart et al., 2000). The reaction also requires a higher range of temperature in which the enzyme cannot function to its fullest and a decrease in production takes place. This difficulty can be solved through using a temperature-resistant enzyme that can be produced by thermostable bacteria. *Thermoanaerobacter* and *Thermoanaerobacterium* are two isolated thermostable bacteria which can produce the desired enzyme (Bart et al., 2000). These CGTases can sustain higher temperatures and low pH values, with no need of α-amylase treatment, with an enhanced reaction rate and minimum cost and time (Bart et al., 2000). CDs, as a result of their complexation ability and other peculiar characteristics, continue to have different applications in different areas of drug delivery and pharmaceutics. If a more stable CGTase enzyme can be discovered for improvement of CD production, then its price can drop significantly and more industrial applications may be carried out. Thus, usage of CDs will increase rapidly in the coming years.

KEYWORDS

- cyclodextrin glycosyltransferase (CGTase)
- thermostable
- cyclization

REFERENCES

Amrit, U. R. B.; Agrawal, P. B.; Warmoeskerken, M. M. C. G. Application of β-cyclodextrins in textiles. *Autex Research Journal* 2011, 11(4), 94–101.

Astray, G.; Gonzalez-Barreiro, C.; Mejuto, J. C.; Rial-Otero, R.; Simal-Ga´ndara, J. Review on the use of cyclodextrins in foods. *Food Hydrocolloids* 2009, 23, 1631–1640.

Bart, A. V. V.; Joost, C. M. U.; Bauke, W. D.; Lubbert, D. Engineering of cyclodextrin glycosyl transferase reaction and product specificity. *Biochimica et Biophysica Acta* 2000, 1543, 336–360.

Brewster, M. E.; Loftsson, T. Cyclodextrins as pharmaceutical solubilizers. *Advanced Drug Delivery Reviews* 2007, 59, 645–666.

Buschmann, H. J.; Schollmeyer, E. Applications of cyclodextrins in cosmetic products: A review. *Journal of Cosmetic Science* 2002, 53, 185–191.

Challa, R.; Ahuja, A.; Ali, J.; Khar, R. Cyclodextrins in drug delivery: An updated review. *AAPS Pharm Sci Tech* 2005, 6(2), 329–357.

Chaudhary, V. B.; Patel J. K. Cyclodextrin inclusion complex to enhance solubility of poorly water soluble drugs: A review. *International Journal of Pharmaceutical Sciences and Research* 2013, 4(1), 68–76.

Cheirsilp, B.; Rakmai, J. Inclusion complex formation of cyclodextrin with its guest and their applications. *Biology, Engineering and Medicine* 2016, 2(1), 2–6.

Chordiya M. A.; Senthilkumaran K. Cyclodextrin in drug delivery: A review. *Research & Reviews: Journal of Pharmacy and Pharmaceutical Sciences* 2012, 1(1), 219–229.

Cova, T. F.; Murtinho, D.; Pais, A. A. C. C.; Valente, A. J. M. Combining cellulose and cyclodextrins: Fascinating designs for materials and pharmaceutics. *Frontiers in Chemistry* 2018, 6, 271.

Das, S. K.; Rajabalaya, R.; David, S.; Gani, N.; Khanam, J.; Nanda, A. Cyclodextrins the molecular container. *Research Journal of Pharmaceutical, Biological and Chemical Sciences* 2013, 4(2), 1694–1720.

Erdoğar, N.; Bilensoy, E. *Cyclodextrin-based nanosystems in targeted cancer therapy*. In: *Environmental Chemistry for a Sustainable World*; Fourmentin, S.; Crini, G.; Lichtfouse, E., Eds; Cyclodextrin Applications in Medicine, Food, Environment and Liquid Crystals. Springer, 2018, Vol. 17, pp. 59–80.

Hirayama, F.; Uekama, K. Cyclodextrin-based controlled drug release system. *Advanced Drug Delivery Reviews* 1999, 36, 125–141.

Ibrahim, A. S. S.; Al-Salamah Ali, A.; El-Tayeb, M. A.; El-Badawi, Y. B.; Garabed, A. A novel cyclodextrin glycosyltransferase from alkaliphilic *Amphibacillus* sp. NPST-10:

Purification and properties. *International Journal of Molecular Sciences* 2012, 13, 10505–10522.

Jacob, Shery.; Nair, A. B. Cyclodextrin complexes: perspective from drug delivery and formulation. *Drug Development Research* 2018, 79(5), 201–217.

Jambhekar, S. S.; Breen, P. Cyclodextrins in pharmaceutical formulations I: Structure and physicochemical properties, formation of complexes, and types of complex. *Drug Discovery Today* 2016, 21(2), 356–362.

Kategeri, A. R.; Sheikh, M. A. Cyclodextrin a gift to pharmaceutical industry. *International Research Journal of Pharmacy* 2012, 3(1), 52–56.

Leclercq, L. Interactions between cyclodextrins and cellular components: Towards greener medical applications? *Beilstein Journal of Organic Chemistry* 2016, 12, 2644–2662.

Li, Z.; Wang, M.; Wang, F.; Gu, Z.; Du, G.; Wu, J.; Chen, J. γ-Cyclodextrin: A review on enzymatic production and applications. *Applied Microbiology and Biotechnology* 2007, 77(2), 245–255.

Loftsson, T.; Brewster, M. E. Pharmaceutical applications of cyclodextrins, drug solubilisation and stabilization. *Journal of Pharmaceutical Sciences* 1996, 85(10), 1017–1025.

Loftsson, T.; Brewster, M. E. Pharmaceutical applications of Cyclodextrins: Basic science and product development. *Journal of Pharmacy Pharmacology* 2010, 62(11), 1607–1621.

Loftsson, T.; Dominique D. Cyclodextrins and their pharmaceutical applications. *International Journal of Pharmaceutics* 2007, 329, 1–11.

Loftsson, T.; Magnúsdóttir, A.; Másson, M.; Sigurjónsdóttir, J. F. Self-association and cyclodextrin solubilization of drugs. *Journal of Pharmaceutical Sciences* 2002, 91(11), 2307–2316.

Lutka, A. Investigation of interaction of promethazine with cyclodextrins in aqueous solution. *Acta Poloniae Pharmaceutica* 2002, 59, 45–51.

Lutka, A.; Koziara, J. Interaction of trimeprazine with cyclodextrins in aqueous solution. *Acta Poloniae Pharmaceutica* 2000, 57, 369–374.

Maheriya, P. M. Cyclodextrin: A promising candidate in enhancing oral bioavailability of poorly water soluble drugs. *MOJ Bioequivalence and Bioavailability* 2017, 3(3), 60–63.

Marques, H. M. C. A review on cyclodextrin encapsulation of essential oils and volatiles. *Flavour and Fragrance Journal* 2010, 25, 313–326.

Martins, R. F.; Kaul, R. H. A new cyclodextrin glycosyltransferase from an alkaliphilic *Bacillus agaradhaerens* isolate: Purification and characterization *Enzyme and Microbial Technology* 2002, 30(1), 116–124.

Merkus, F. W.; Verhoef, J.; Romeijn, S. G.; Schipper, N. G. Absorption enhancing effect of cyclodextrins in intranasally administered insulin in rats. *Pharmaceutical Research* 1991, 8, 588–592.

Nitalikar, M. M.; Sakarkar, D. M.; Jain, P. V. The Cyclodextrins: A review. *Journal Current Pharmaceutical Research* 2012, 10(1), 1–6.

Patil, J. S.; Kadam, D. V.; Marapur, S. C.; Kamalapur, M. V. Inclusion complex system; a novel technique to improve the solubility and bioavailability of poorly soluble drugs: A review. *International Journal of Pharmaceutical Sciences and Research* 2010, 2(2), 29–34.

Radi, A. E.; Eissa, S. Electrochemistry of cyclodextrin inclusion complexes of pharmaceutical compounds. *The Open Chemical and Biomedical Methods Journal* 2010, 3, 74–85.

Rasheed, A.; Ashok Kumar, C. K.; Sravanthi, V. V. N. S. S. Cyclodextrins as drug carrier molecule: A review. *Scientia Pharmaceutica* 2008, 76, 567–598.

Savjani, K. T.; Gajjar, A. K.; Savjani, J. K. Drug solubility: Importance and enhancement techniques. *ISRN Pharmaceutics* 2012, 1–10.

Schmid, R. Recent advances in the description of the structure of water, the hydrophobic effect, and the like-dissolves-like rule. *Chemical Monthly* 2001, 132, 1295–1326.

Singh, M.; Banerjee, U. C.; Sharma, R. Biotechnological applications of cyclodextrins. *Biotechnology Advances* 2002, 20, 341–359.

Sonnendecker, C.; Zimmermann, W. Domain shuffling of cyclodextrin glucano transferases for tailored product specificity and thermal stability. *FEBS Open Bio* 2019, 9, 338–395.

Tasic, L. M.; Jovanovic, M. D.; Djuric, Z. R. The influence of beta cyclodextrin on the solubility and dissolution rate of paracetamol solid dispersions. *Journal of Pharmacy and Pharmacology* 1992, 44, 52–55.

Thombre, R. S.; Kanekar, P. P. Synthesis of β-cyclodextrin by cyclodextrin glycosyl transferase produced by *Bacillus licheniformis*, MCM-B 1010. *Journal of Microbiology and Biotechnology Research* 2013, 3(1), 57–60.

Tian, Y. Q.; Zhou, X.; Jin, Z. Y. Use of cyclodextrins in food, pharmaceutical and cosmetic industries. In: *Cyclodextrin Chemistry*; Jin, Z. Y., Ed; World Scientific, 2013, pp. 215–233.

Tiwari, J.; Tiwari, R.; Rai, A. K. Cyclodextrins in delivery systems: applications. *Journal of Pharmacy and Bioallied Sciences* 2010, 2(2), 72–79.

Uekama, K.; Hirayama, F.; Arima, H. Recent aspect of cyclodextrin-based drug delivery system. *Journal of Inclusion Phenomena* 2006, 56, 3.

## CHAPTER 7

Mangrove Plants in Therapeutic Management of Diabetes: An Update

SWAGAT KUMAR DAS*, DIBYAJYOTI SAMANTARAY, and
SWATISMITA BEHERA

Department of Biotechnology, College of Engineering and Technology, Biju Patnaik University of Technology, Techno Campus, Ghatikia, Bhubaneswar, Odisha 751003, India

*Corresponding author. E-mail: das.swagat@gmail.com

ABSTRACT

Diabetes mellitus (DM) is the most common endocrine and metabolic disorder clinically manifested with high blood glucose level. It affects all ages of people and has become one of the major causes of morbidity and mortality in the 21st century. Although several synthetic antidiabetic drugs have been developed, they possess several side effects. In recent years, there has been an exponential growth in the use of herbal medicines for antidiabetic therapy, because of their natural origin and less side effects. Mangrove plants growing in the ecologically hostile conditions prevailing at the interface between land and sea are endowed with unique phytochemicals rich in bioactive compounds responsible for curing several ailments, including diabetes. Several studies have reported that extracts from mangrove plants can alleviate diabetic and its oxidative stress-associated complications. The present review highlights the mangrove plants with antidiabetic potentials and their phytochemical constituents and active compounds, along with their mode of action, evaluated through various in vitro and in vivo studies.

7.1 INTRODUCTION

Diabetes mellitus (DM) is a metabolic disorder associated with impairment in insulin secretion and insulin action, as well as aberrations in intermediary metabolism of carbohydrates, proteins, and lipids (DeFronzo, 2004). Several reports indicate that diabetes is a major worldwide health problem and likely to increase further in the future, both in developed and developing countries, including India. The global prevalence of diabetes is estimated to be 4.4% of the world's population in 2030, affecting about 366 million people across the globe (WHO, 2001).

Persistent hyperglycemia, which is commonly associated with diabetes, is responsible for the development of many common complications that include polydipsia, polyuria, weight loss, polyphagia, poor wound-healing, gingivitis, and blurred vision. In a few cases, ketoacidosis may develop under chronic hyperglycemic conditions leading to stupor, coma, and, in the absence of effective treatment, death (WHO, 1999). The elevated blood glucose increases the mitochondrial reactive oxygen species level and stimulates the activation of several stress-signaling pathways like polyol pathway, phosphokinase-C (PKC) pathway, advanced glycation end products (AGEs) pathway, and hexosamine pathway. These pathways are altogether responsible for development of several macrovascular and microvascular post-diabetic complications in diabetic patients, such as diabetic neuropathy, nephropathy, and retinopathy.

Therefore, management of diabetes in recent times poses a big challenge. Though there are several synthetic drugs available to control diabetes, they are not efficient enough to control late-diabetic complications and are also associated with side effects (Das et al., 2016). Recently, there has been a growing interest in herbal medicines for the management of diabetes, both in developing and developed countries, due to their natural origin and less side effects (Modak et al., 2007). In this context, mangrove plants growing in the ecologically stressful conditions at the interface of land and sea can be useful in the treatment of diabetes (Bandaranayake, 2002). Recently, mangroves from India and across the globe have been investigated for their beneficial use in treating diabetes and their associated complications (Bandaranayake, 1998, 2002; Das et al., 2016). However, studies on the details of their mechanism of action and on phytochemicals responsible for antidiabetic properties are limited. Therefore, the present chapter provides an overview of diabetes and its pathogenesis, the limitations of current chemical-based therapeutics, and the effects of various mangrove plants in therapeutic management of DM.

7.2 PATHOGENESIS OF DIABETES MELLITUS

Diabetes is a multipathogenic disorder, and many factors like food habit, life-style, drug intake, environment, and genetic factors can play a significant role in its development. Recent studies suggest that a complex interaction between inflammation, endoplasmic reticulum stress, oxidative stress, mitochondrial dysfunction, and autophagy dysregulation plays an important role in insulin resistance. Diabetes is a polygenic disorder, with obesity-related insulin resistance playing a major role in its onset and progression. It is characterized by excessive hepatic glucose production, decreased insulin secretion from pancreatic beta cells, and insulin resistance in peripheral tissue such as muscle adipose and liver (Ahmed, 2006). There are convincing data to indicate a genetic component associated with insulin resistance (Kumar et al., 1992). The pathogenesis of insulin resistance and type 2 diabetes is summarized in Figure 7.1. Gestational diabetes is another category of diabetes that occurs during pregnancy, which may improve or disappear after delivery. Even though it might be transient, gestational diabetes can damage the health of the fetus or mother, and about 40% of women with gestational diabetes develop type 2 diabetes later in life (Mayfield, 1998).

7.3 CONVENTIONAL ANTIDIABETIC DRUGS AND THEIR MECHANISM OF ACTION

Type 2 diabetes is controlled and managed by a combination of diet restriction, weight reduction programs, and oral hypoglycemic drugs (Evans and Rushakoff, 2007). When hyperglycemia becomes severe, patients are usually switched to insulin injections, with or without oral agents, to improve insulin action. Oral hypoglycemia agents exert their glucose-lowering effects via different mechanisms (Figure 7.2). These mechanisms of action include reduction of hepatic glucose production, enhancement of insulin secretion by pancreatic beta cells, improvement of insulin sensitivity, and inhibition of intestinal glucose digestion and absorption (Table 7.1). However, current antidiabetic medications have toxic side effects such as nausea, diarrhea, hypoglycemia at higher doses, liver problems, lactic acidosis, and weight gain (Bastaki, 2005; Evans and Rushakoff, 2007). Further, despite the intensive use of current antidiabetic agents, about 50% type 2 diabetic patients exhibit poor glycemic control (Nathan, 1993). Hence, exploration of new antidiabetic agents is the need of the hour.

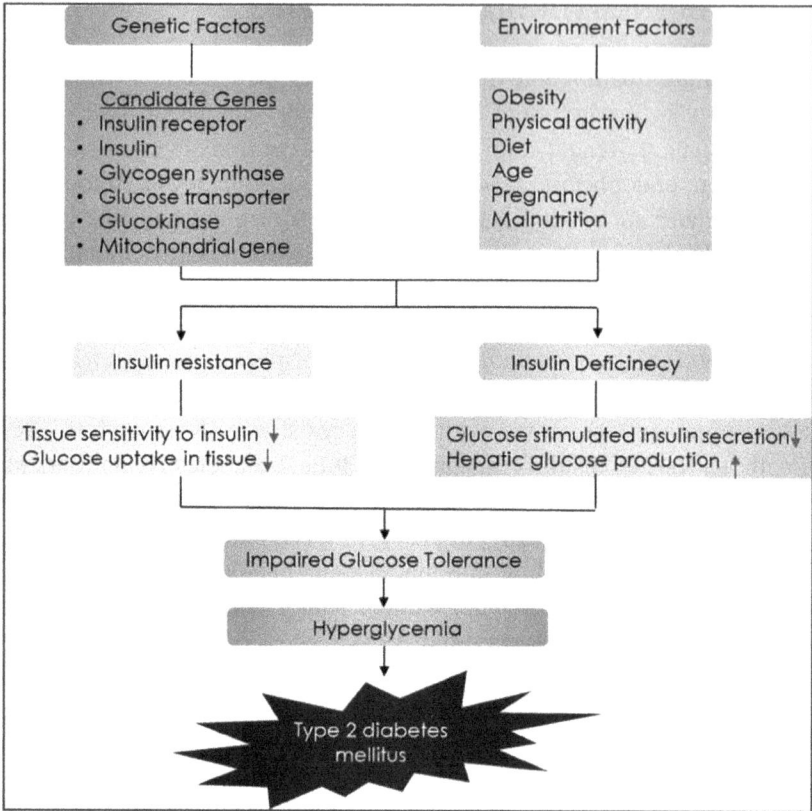

FIGURE 7.1   Pathogenesis of type 2 DM or NIDDM (De Fronzo, 2004).

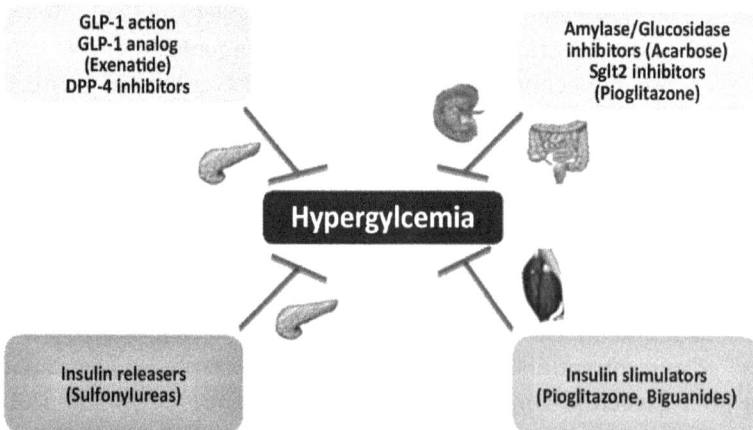

FIGURE 7.2   Current therapies in treatment of DM.

TABLE 7.1 Oral Hypoglycemic Drugs, Their Mode of Action and Side Effects

Oral Hypoglycemic Drugs	Mechanism of Action	Possible Side Effects
Amylin mimetics	Stimulate the release of insulin	Hypoglycemia; nausea or vomiting; headache; redness, and irritation at injection site
Incretin mimetics	Stimulate the release of insulin	Nausea or vomiting; headache; dizziness; kidney damage or failure
Meglitinides	Stimulate the release of insulin	Hypoglycemia; weight gain
Sulfonylureas	Stimulates the pancreas to release more insulin	Hypoglycemia; weight gain; nausea; skin rash
Dipeptidy peptidase-4 (DPP-4) inhibitors	Stimulate the release of insulin; inhibit the release of glucose from the liver	Upper respiratory tract infection; inflammation of the pancreas
Biguanides	Inhibit the release of glucose from the liver	Nausea; diarrhea; rarely, lactic acidosis
Thiazolidinediones	Improve sensitivity to insulin; inhibit the release of glucose from the liver	Heart failure; heart attack; stroke; liver disease
Carbohydrate hydrolyzing enzyme inhibitors	Slow the breakdown of starches and some sugars	Stomach pain; gas; diarrhoea
Bile acid sequestrants	Works with other diabetes medications to lower blood glucose	Constipation, nausea, diarrhea, gas, heartburn, headache

7.4   MODE OF ANTIDIABETIC ACTIVITY OF MANGROVE PLANT EXTRACTS

Traditionally, more than 100 mangroves and their associated species were being used for the treatment of diabetes. However, only a few have been evaluated and reported scientifically (Bandaranayake, 2002). Based on the antidiabetic research carried upon mangrove plants, it has been suggested that these plants can exhibit their antidiabetic action in several ways, such as: (1) imitating insulin activity, (2) decreasing intestinal glucose absorption, (3) stimulating the secretion of insulin, (4) increasing glucose uptake, and (5) inhibiting negative insulin signals through suppression of dipeptidyl peptidase-4 (DPP-IV) and protein tyrosine phosphatises (PTPase) activity (Figure 7.3). These mangrove plants were also found to be effective in the management of post-diabetic complications through decreasing AGEs and exerting antioxidative effects on oxidative stress-associated diabetic complications (Das et al., 2016).

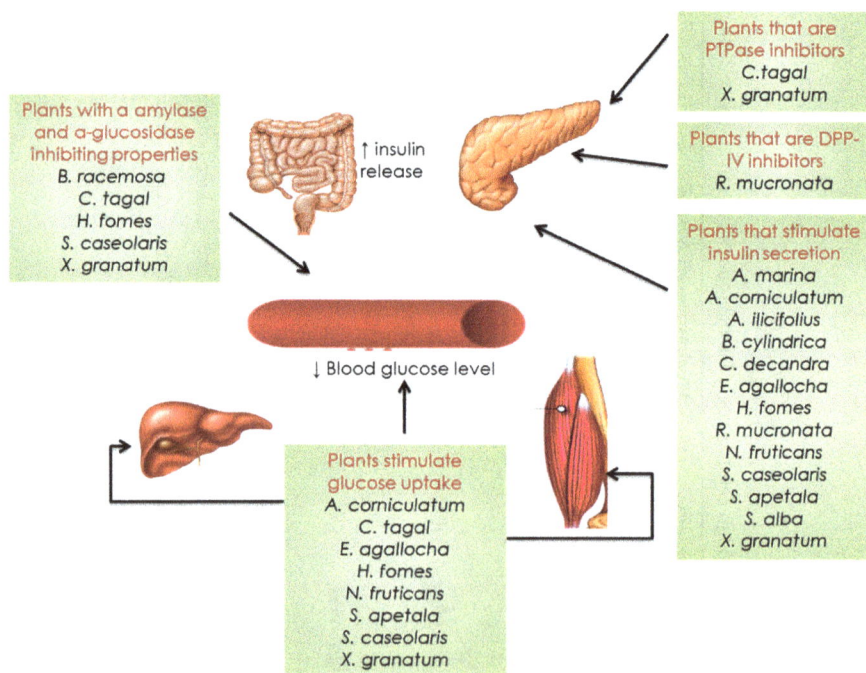

FIGURE 7.3   Mode of antidiabetic activity of mangrove plants.

7.5 RESEARCH ON ANTIDIABETIC MANGROVE PLANTS

Mangroves plants have recently generated great interest for their antidiabetic potential. A survey of literature revealed that many mangrove species have been scientifically evaluated and reported for their antidiabetic potentials. A few of the mangrove plants studied for their antidiabetic potentials are discussed below.

7.5.1 AEGICERAS CORNICULATUM

The ethanolic leaf extract of *Aegiceras corniculatum* (*A. corniculatum*) (Family: Myrsinaceae) in alloxanized diabetic rats at a dose of 100 mg/kg exhibited regulation of the blood glucose level. Further, an increase in body weight and liver hexokinase, along with a decrease in the activities of glucose-6-phosphatase, fructose 1,6-bisphosphatase, and glycosylated hemoglobin (Hb), was also observed in diabetic-induced rats upon treatment with ethanol leaf extracts of *A. corniculatum* (Gurudeeban et al., 2012). In another study, the aqueous leaf extract of *A. corniculatum* exhibited hypoglycemic effect in streptozotocin (STZ)-induced diabetic. The antidiabetic effect may be attributed to the presence of different secondary metabolites in the extract. The aqueous leaf extract of this plant exerts its antidiabetic effect by healing the pancreatic β cells, thereby increasing the serum insulin level and decreasing the serum glucose level (Geegi and Manoharan, 2018).

7.5.2 ACANTHUS ILICIFOLIUS

Ethanol root extract of *Acanthus ilicifolius* (*A. ilicifolius*) (Family: Acanthaceae) at doses of 200 and 400 mg/kg body weight significantly reduced the blood sugar level in normal, glucose-fed hyperglycemic- and alloxan-induced diabetic rats. The antidiabetic properties of the root extracts of *A. ilicifolius* can be attributed to the regeneration of β cells in the diabetic rats (Venkataiah et al., 2013). In another study, the methanolic leaf extract of this plant reduced blood glucose levels in glucose-loaded mice, in a dose-dependent manner (Ahmed et al., 2014).

7.5.3 AVICENNIA sp.

The ethanol leaf extract of *Avicennia marina* (*A. marina*) (Family: Avicenniaceae) has been shown to exhibit antihyperglycemic activity, as evident from in vivo studies in alloxanized diabetic rats. The extracts at 250 and 500 mg/kg doses significantly reduced the blood glucose level, along with increasing total Hb, total protein, and serum insulin levels. The possible mechanism underlying the antihyperglycemic action of *A. marina* is attributed to the stimulation of surviving β cells releasing more insulin (Babuselvam et al., 2013). In another study, the methanol extract of pneumatophores of *A. marina* exhibited antihyperglycemic action by inhibiting AGEs (Mahera et al., 2011). Hamzevi et al. (2017) have also reported the antidiabetic and antioxidant properties of aqueous extracts of *A. marina* in alloxan-induced diabetic rats.

Avicennia alba, another medicinally important mangrove, has been reported for its antidiabetic effects. The stem and leaf methanolic extract of this plant reduced the blood glucose level in diabetic mice at 100, 200, and 400 mg/kg doses. The hematological parameters like glycosylated Hb were found normal compared to those of normal, untreated rats. Glibenclamide was also equally effective against diabetes (Michael et al., 2018).

In another study, Das et al. (2018) have reported the antidiabetic activities of ethanolic leaf and bark extracts of *Avicennia officinalis*. The ethanolic leaf and bark extracts of this plant could inhibit the carbohydrate-metabolizing enzymes, namely, α-glucosidase and α-amylase enzymes, in a dose-dependent manner.

7.5.4   *BARRINGTONIA RACEMOSA*

Barringtonia racemosa (*B. racemosa*) (Family: Lecythidaceae), a medicinally important mangrove plant, has been reported for its antidiabetic potential by various researchers. The hexane, ethanol, methanol extracts of seeds of *B. racemosa* were also reported for α-glucosidase- and α-amylase-inhibition properties (Gowri et al., 2007). The presence of pentacyclic triterpenoid bartogenic acid in the methanol extracts of this plant is responsible for its antidiabetic activity. In another study, the flavonoid extracts obtained from fruit kernels of *B. racemosa* have shown an antidiabetic effect in diabetic rats. The extract exerts a glucose-lowering effect along with an increase in β-cell granulation effect in diabetic rats (Musman et al., 2017). The methanolic and dichloromethane leaf extracts of this plant could lower the blood glucose level in diabetic rats in a dose-dependent manner. The antidiabetic potential of this plant may be due to an increase in glucose utilization by peripheral tissues and increase in β-cell granulation (Umaru et al., 2018, 2019).

7.5.5 *BRUGUIERA* sp.

The ethanolic leaf extract of *Bruguiera cylindrica* (*B. cylindrica*) (Family: Rhizophoraceae) exhibited antidiabetic activity, as demonstrated by an in vitro yeast glucose uptake assay (Pitchaipillai and Ponniah, 2016). In another study, the ethanolic leaf extract of *B. cylindrica,* upon oral administration at a dose of 0.15 g/kg body weight, showed antihyperglycemic activity in alloxan-induced diabetic mice. The ethanolic extract of *B. cylindrica* leaves may exert a stimulating effect on the β cells, leading to increased insulin secretion and a decrease in the blood sugar level (Shyam and Kadalmani, 2014).

In an in vivo study, the ethanolic bark extract of *Bruguiera gymnorrhiza* displayed an antihyperglycemic effect in streptozotocin-induced diabetic rats. Significant reduction in the blood glucose level was reported in the STZ-induced diabetic rats treated with ethanolic bark extracts (400 mg/kg), which was comparable to that of a standard drug, glibenclamide (0.5 mg/kg body weight) (Karimulla and Kumar, 2011).

7.5.6 *CERIOPS* sp.

Two species of the genus of *Ceriops* (Family: Rhizophoraceae), *Ceriops decandra* (*C. decandra*) and *Ceriops tagal* (*C. tagal*), have been reported for antidiabetic potential. The ethanol leaf extract of *C. decandra* at a dose of 120 mg/kg lowered the serum glucose level in alloxan-induced diabetic mice (Nabeel et al., 2010). Increase in insulin secretion, body weight, Hb levels and decrease in HbA1c levels were observed on diabetic rats upon ethanolic leaf extracts treatment of this plant. Stimulation of surviving β cells to release more insulin might be the mechanism underlying the antihyperglycemic action of *C. decandra* (Nabeel et al., 2010).

The ethanolic leaf extract of *C. tagal* at a dose of 250 mg/kg improved the glucose tolerance of the normoglycemic rats and lowered the blood glucose levels in STZ-induced diabetic rats (Tiwari et al., 2008). The *n*-hexane-soluble fraction of ethanolic leaf extracts of *C. tagal* also exhibited antidiabetic activity by stimulating the glucose uptake in L6 muscle cells in a dose-dependent manner (Tamrakar et al., 2008). The hydroalcoholic bark extracts of *C. tagal* were also reported to exhibit in vitro antihyperglycemic activity by inhibiting the α-glucosidase enzyme (Lawag et al., 2012).

7.5.7   *EXCOECARIA AGALLOCHA*

Excoecaria agallocha (*E. agallocha*) (Family: Euphorbiaceae) methanol bark extract reduces the serum glucose levels at doses of 200 and 400 mg/kg (Rahman et al., 2010). The ethanolic leaf extracts of *E. agallocha* also exhibited antidiabetic activity, which may be attributed to the presence of bioactive principles like flavonoids, triterpenoids, alkaloids, phenolics, etc. (Thirumurugan et al., 2009). In another study, soxhlated ethanolic leaf extracts of *E. agallocha* at 250 and 500 mg/kg body weight showed antidiabetic potentials in STZ-induced diabetic rats. The activity of the *E. agallocha* may be attributed to the ability of the extract to elicit antioxidant enzymes (Kiran et al., 2018). The leaf extract of this plant is rich in polyphenols and flavonoids, which contribute to the antioxidant potential of the leaf extracts of this plant.

7.5.8   *HERITIERA FOMES*

The methanol bark extract of *Heritiera fomes* (Family: Sterculiaceae) exhibited antidiabetic properties, as evident from an in vivo study on the mice model. The bark extract of this plant at a dose of 250 mg extract per kilogram of body weight significantly lowered serum glucose levels by 49.2%, in comparison to 43.5% by glibenclamide (Ali et al., 2011).

7.5.9   *KANDELIA CANDEL*

The ethanol leaf extracts of *Kandelia candel* (*K. candel*) (Family: Rhizophoraceae) demonstrated antidiabetic potential by decreasing the blood glucose level in both sucrose-loaded model (SLM) and STZ-induced diabetic rat model at a dose of 500 mg/kg (Lakshmi et al., 2013). In another study, Shettar and Vedamurthy (2017) reported the antidiabetic potential of the methanolic extract of *K. candel.* The study showed that among the chloroform, ethyl acetate, methanol, ethanol, and water-soxhlated extracts of the *K. candel* leaf, the methanol leaf extract exhibited antidiabetic properties, as evaluated by an α-amylase inhibition assay and yeast glucose uptake assay.

7.5.10   *NYPA FRUTICANS*

The methanolic leaf and stem extracts of *Nypa fruticans* (*N. fruticans*) (Arecaceae) showed potent blood glucose-lowering property in glucose-induced hyperglycemic mice at a dose of 500 mg/kg. The antidiabetic effect may be attributed due to the potential of *N. fruticans* extracts to stimulate pancreatic β-cell function or by increasing peripheral utilization of glucose (Reza et al., 2011). The seed extracts of *N. fruticans* also reported antidiabetic potential. The hydroethanolic extract of the seed mesocarp of *N. fruticans,* having a rich source of dietary polyphenols, exhibited an inhibitory effect on the last phase of carbohydrate digestion (Martin et al., 2017). In another study, Yusoff et al. (2017) reported the antidiabetic potential of aqueous extract of vinegar made from *N. fruticans.* The aqueous extract of vinegar, upon being administered to STZ-induced diabetic rats, improved postprandial glucose levels in diabetic rats. The antidiabetic activity of the vinegar may be attributed to its insulin-stimulatory and hepatoprotective effects.

7.5.11 *RHIZOPHORA* sp.

Among different species of genus *Rhizophora* (Family: Rhizophoraceae), *Rhizophora apiculata* (*R. apiculata*), *Rhizophora annamalayana* (*R. anna-malayana*), and *Rhizophora mucronata* (*R. mucronata*) showed promising antidiabetic activities. Nabeel et al. (2012) have reported the antidiabetic potential of the three mangrove plants *R. annamalayana, R. apiculata,* and *R. mucronata.* Oral administration of aqueous leaf extracts of *R. annamalayana, R. apiculata,* and *R. mucronata* at a 60 mg/kg dose ameliorated the different parameters such as blood glucose, plasma insulin, body weight, total Hb, glycosylated Hb, liver glycogen, plasma and tissue lipids, cholesterol, triglycerides, free fatty acids, and phospholipids toward normalcy in the alloxan-induced diabetic rats. The ethanolic root extracts of *R. apiculata* showed antihyperglycemic activity at a 250 mg/kg dose in the experimental rat model. The chloroform and aqueous fractions of *R. apiculata* showed antidiabetic potential in an STZ model at a 100 mg/kg dose (Lakshmi et al., 2006). The ethanolic leaf extracts of *R. apiculata* also showed antihyperglycemic activity in normal, glucose-fed, and STZ-diabetic rats (Sur et al., 2004, 2015). The study showed that among the chloroform, ethyl acetate, methanol, ethanol, and water-soxhlated extracts of *R. apiculata,* the aqueous leaf extract exhibited antidiabetic properties, as evaluated by an α-amylase inhibition assay and yeast glucose uptake assay (Shettar and Vedamurthy, 2017).

The aqueous (Gaffar et al., 2011; Haque et al., 2013) and hydroalcoholic (Lawag et al., 2012) bark extracts of *R. mucronata* also showed in vitro antihyperglycemic activities. The leaves of *R. mucronata* extracted with

80% methanol exhibited antidiabetic activity and could reduce the blood glucose level in STZ-induced diabetic rats at doses of 50 and 100 mg/kg. The antidiabetic activity of the extract may be attributed to the antiradical action of the extract (Sur et al., 2015).

The fresh juice and ethanolic leaf extract of *R. mucronata* showed anti-hyperglycemic activity in STZ-induced diabetic rats at a dose of 200 mg/kg body weight and significantly reduced the blood glucose level (Adhikari et al., 2016, 2017).

7.5.12   *SONNERATIA* sp.

In genus *Sonneratia* (Family: Lythraceae), antidiabetic activity has been reported in three species, namely, *Sonneratia alba* (*Sonneratia alba*), *Sonneratia apetala* (*S. apetala*), and *Sonneratia caseolaris* (*S. caseolaris*). The methanolic fruit extract of *S. caseolaris* exhibited antidiabetic potential by inhibiting the α-glucosidase enzyme (Tiwari et al., 2010) and reducing the serum glucose concentrations in glucose-loaded mice (Hasan et al., 2013; Rahmatullah et al., 2012). The methanolic leaf extracts of *S. alba* significantly reduced sugar levels upon administration to the diabetic mice (Morada et al., 2011). In another study, the seeds and pericarps of *S. apetala* fruits also exhibited antidiabetic activity in STZ-induced diabetic mice. The antidiabetic property of *S. apetala* may be attributed to its insulin-mimetic activity, increased glucose utilization, islets of Langerhans regeneration, and enhanced transport of blood glucose (Hossain et al., 2013).

7.5.13   *XYLOCARPUS* sp.

In vitro and in vivo antidiabetic potentials have been reported in *Xylocarpus granatum* (*X. granatum*) and *Xylocarpus moluccensis* (*X. moluccensis*) that belong to the family Meliaceae. The ethanolic extract of epicarp of *X. granatum* showed blood glucose lowering effect and improvement in insulin resistance in STZ-treated after the oral treatment of the extracts at 250 mg/kg dose for three weeks (Srivastava et al., 2011). The leaf and bark extracts of *X. granatum* also possessed α-glucosidase enzyme properties and also found to enhance glucose uptake in *ex situ* yeast model (Das et al., 2016). The ethyl acetate-soluble fraction of the epicarp of *X. moluccensis* (EAXm) was also reported for its antidiabetic and antidyslipidemic activities in diabetic rats. The antihyperglycemic activity of this plant extract may be due to

TABLE 7.2 Mangrove Plants with Antidiabetic Potentials

Mangrove Species (Family)	Parts/Solvent	Mechanism of Action	References
A. corniculatum (*Myrsinaceae*)	Leaf/Ethanol Leaf/Aqueous	▪ ↑ Glucose utilization by direct stimulation of glucose uptake or via enhanced insulin secretion ▪ Healing of pancreatic β-cells that produced insulin ▪ ↑ Insulin secretion	Gurudeeban et al., 2012; Geegi and Manoharan, 2018
A. ilicifolius (*Acanthaceae*)	Root/Ethanol Leaf/Methanol	▪ Regeneration of β cells of pancreas	Venkataiah et al., 2013; Ahmed et al., 2014
A. marina (*Avicenniaceae*)	Leaf/eEhanol Pneumatophore/Methanol Leaf/ aqueous	▪ Stimulation of β cells to release more insulin ▪ Antiglycation activity	Mahera et al., 2011; Babuselvam et al., 2013; Hamzevi et al., 2017
A. alba	Leaf, stem/Methanol	▪ ↓Elevated blood glucose level	Michael et al., 2018
A. officinalis	Leaf, bark/Ethanol	▪ α-Glucosidase and α-amylase inhibitory property	Das et al., 2018
B. racemosa (*Lecythidaceae*)	Seed/Hexane, ethanol, methanol Fruit Leaf/Methanol, dichloromethane	▪ α-Glucosidase and α-amylase inhibitory property ▪ Stimulating glucose utilization by peripheral tissues ▪ Increase in the β-cell granulation	Gowri et al., 2007; Umaru et al., 2018; Umaru et al., 2019; Musman et al. 2017
B. cylindrica (*Rhizophoraceae*)	Leaf/Ethanol,	▪ Yeast glucose uptake ▪ Stimulation of β cells to release more insulin ▪ Stimulating effects on glucose utilization	Revathi et al. 2015 Pitchaipillai and Ponniah, 2016; Shyam and Kadalmani, 2014
B. gymnorrhiza (*Rhizophoraceae*)	Ethanol bark extract	▪ Presence of antidiabetic principles	Karimulla and Kumar, 2011
C. decandra (*Rhizophoraceae*)	Leaf /Ethanol	▪ Stimulation β-cells to release more insulin ▪ Increased hexokinase activity resulting in increased glycolysis and utilization of glucose	Nabeel et al. 2010

TABLE 7.2 (Continued)

Mangrove Species (Family)	Parts/Solvent	Mechanism of Action	References
C. tagal (*Rhizophoraceae*)	Leaf/Ethanol Bark/Hydroalcoholic	• Inhibition against PTPase enzyme • Stimulation of glucose uptake • α-Glucosidase inhibitory property	Tiwari et al. 2008; Tamrakar et al., 2008; Lawag et al. 2012
E. agallocha (*Euphorbiaceae*)	Bark/Methanol Leaf/Ethanol	• ↑ Pancreatic secretion of insulin • ↑ Uptake of glucose	Rahman et al., 2010; Thirumurugan et al. 2009; Kiran et al., 2018
H. fomes (*Sterculiaceae*)	Methanol bark extract	• Enhancement of pancreatic secretion of insulin • ↑ The glucose uptake • Inhibition in glucose absorption in gut	Ali et al., 2011
K. candel (*Rhizophoraceae*)	Leaf/Ethanol Leaf/Methanol	• Decrease blood glucose level • Alpha amylase enzyme inhibition • ↑ Glucose uptake	Lakshmi et al., 2013; Shettar and Vedamurthy, 2017
N. fruiticans (*Arecaceaea*)	Leaf and stem/Methanol	• ↑ Glucose utilization either by direct stimulation of glucose uptake or via the mediation of enhanced insulin secretion	Reza et al. 2011; Martin et al. 2017; Yusoff et al. 2017
R. annamalayana (*Rhizophoraceae*)	Ethanol bark extract	• Improved level of insulin secretion and its action • Insulin mimetic activity	Ali et al. 2011; Nabeel et al. 2012
R. apiculata (*Rhizophoraceae*)	Root/Ethanol Leaf/Ethanol Leaf/Aqueous	• Improved level of insulin secretion and its action • Insulin mimetic activity • β-cell protection • Alpha amylase enzyme inhibition • ↑ Glucose uptake	Lakshmi et al., 2006; Sur et al., 2004, 2015; Nabeel et al., 2012; Shettar and Vedamurthy, 2017

TABLE 7.2 (Continued)

Mangrove Species (Family)	Parts/Solvent	Mechanism of Action	References
R. mucronata (Rhizophoraceae)	Bark/Hydroalcoholic	• Improved level of insulin secretion and its action;	Lawag et al., 2012; Nabeel et al., 2012; Gaffar et al., 2011; Haque et al., 2013; Sur et al., 2015; Adhikari et al., 2016, 2017
	Leaf/Methanol	• Insulin mimetic activity;	
		• α-Glucosidase inhibitory	
S. caseolaris (Lythraceae)	Methanol fruit extracts	• Intestinal α-glucosidase inhibitory activity;	Tiwari et al., 2010; Rahmatullah et al., 2012;
		• Increased pancreatic secretion of insulin;	Hasan et al., 2013
		• ↑Glucose uptake from serum;	
		• ↓Glucose absorption from gut.	
S. alba (Lythraceae)	Seeds and pericarps	• Enhanced insulin releasing activity;	Hossain et al., 2013; Patra et al., 2014
		• Insulin mimetic activity;	
		• Modifying glucose utilization	
		• Stimulating the regeneration of islets of langerhans in pancreas	
		• Enhance transport of blood glucose to the peripheral tissue	
S. alba (Lythraceae)	Methanol leaf extracts	• Modifying glucose utilization	Morada et al., 2011
X. granatum (Meliaceae)	• Epicarp of fruit/Ethanol extract	• Stimulation on β cells and help in the release of insulin	Srivastava et al., 2011
	• Bark/Ethanol	• Elevation in insulin sensitivity to glucose	Das et al., 2016
		• ↓Protein tyrosine phosphatase activity may be helping in insulin action	Das et al., 2019a, b
		• α-Glucosidase inhibitory activity	

TABLE 7.2 *(Continued)*

Mangrove Species (Family)	Parts/Solvent	Mechanism of Action	References
X. moluccensis (*Meliaceae*)	Ethyl acetate soluble fraction of epicarp of fruit	▪ Insulin mimetic or insulin secretagogue activity ▪ Insulin resistance reversal activity ▪ α-Glucosidase inhibitory activity	Srivastava et al., 2014
R. mucronata (*Rhizophoraceae*)	Hydro methanolic leaf extract	▪ Restricts oxidative stress in liver tissue ▪ Reduce level of lipid peroxides and higher level of glutathione in liver ▪ Decrease blood glucose level	Sur et al., 2015

insulin mimetic, insulin secretagogue, and insulin resistance reversal activity (Srivastava et al., 2014). Recently, the ethanol bark extract of *X. granatum* has been reported for its antihyperglycemic potential in the STZ-induced diabetic mice model. The study demonstrated that the ethanol bark extract of this plant could normalize the elevated blood sugar level and ameliorate the different oxidative stress-mediated complications as well (Das et al., 2019a). Further, a bioactivity-guided study demonstrated that xyloccensin-I present in this ethanol bark extract might be the active principle that possesses both antioxidant and antidiabetic potential (Das et al., 2019b).

7.6 CONCLUSION AND FUTURE PERSPECTIVES

The medicinal potential of herbal drugs from mangrove plants for diabetic management is significant, and they have negligible side effects compared to synthetic antidiabetic drugs. There is an increasing demand by patients for natural products with antidiabetic activity. This chapter provides detailed information about the mangrove plants that are responsible for the hypogly-cemic activities. These mangrove plant extracts are also seen to ameliorate post-diabetic complications due to their strong antioxidant potential. Though mangrove plants exhibit potential antidiabetic activities, studies on them are very limited. Even in most of the cases, neither detail investigations have been undertaken to identify the bioactive compound responsible for antidiabetic activity nor the mode of the action of the compounds was studied. Therefore, an in-depth study is required to identify the lead compound responsible for antidiabetic activities, leading to its drug development through proper laboratory and clinical investigation to control diabetes and its associated complications.

KEYWORDS

- mangrove
- phytochemicals
- diabetes
- antidiabetic principle

REFERENCES

Adhikari, A.; Ray, M.; Das, A.K.; Sur, T.K. Antidiabetic and antioxidant activity of *Rhizophora mucronata* leaves (Indian sundarban mangrove): An *in vitro* and *in vivo* study. *Ayu* 2016, 37(1), 76–81.

Adhikari, A.; Das, A.K.; Ray, M.; Sur, T.K.; Biswas, S. Comparative evaluation of antidiabetic activity of fresh juice and ethanolic extract of *Rhizophora mucronata* leaves in animal model. *International Journal of Basic & Clinical Pharmacology* 2017, 6(9), 2193–2198.

Ahmed, I. Diabetes mellitus, Vol. 4. Division of endocrinology, Diabetes and metabolic Disease, Department of Medicine, Jefferson Medical College, Thomas Jefferson University, Philadelphia, PA, 2006, pp. 237–246.

Ahmed, Md.N.; Sultana, T.; Azam, Md.N.K.; Rahmatullah, Md. A preliminary antihypergly-cemic and antinociceptive activity evaluation of a mangrove species *Acanthus ilicifolius* L. leaves in mice. *AJTM* 2014, 9(6), 143–149.

Ali, M.; Nahar, K.; Sintaha, M.; Khaleque, H.N.; Jahan, F.I.; Biswas, K.R.; Swarna, A.; Monalisa, M.N.; Jahan, R.; Mohammed, R. An evaluation of antihyperglycemic and antinociceptive effects of methanol extract of *Heritiera fomes* Buch-Ham. Sterculiaceae. barks in Swiss Albino mice. *Advances in Natural and Applied Sciences* 2011, 5, 116–121.

Babuselvam, M.; Abideen, S.; Gunasekaran, T.; Beula, J.M.; Dhinakarraj, M. Bioactivity of *Avicennia marina* and *Rhizophora mucronata* for the management of diabetes mellitus. *World Journal of Pharmaceutical Research* 2013, 3, 11–18.

Bandaranayake, W.M. Traditional and medicinal uses of mangroves. *Mangroves Salt Marshes* 1998, 12, 133–148.

Bandaranayake, W.M. Bioactivities, bioactive compounds and chemical constituents of mangrove plants. *Wetlands Ecology and Management* 2002, 10, 421–452.

Bastaki, S. Diabetes mellitus and its treatment. *International Journal of Diabetes and Metabolism* 2005, 13, 111–134.

Das, S.K.; Samantaray, D.; Patra, J.K.; Samanta, L.; Thatoi, H. Antidiabetic potential of mangrove plants: A review. *Frontiers in Life Science* 2016, 9, 75–88.

Das, S.K.; Prusty, A.; Samantaray, D.; Hasan, M.; Jena, S.; Patra, J.K.; Samanta, L.; Thatoi, H.N. Effect of *Xylocarpus granatum* bark extract on amelioration of hyperglycaemia and oxidative stress associated complications in STZ-induced diabetic mice. *eCAM* 2019a, https://doi.org/10.1155/2019/8493190.

Das, S.K.; Samantaray, D.; Sahoo, S.K.; Pradhan, S.K.; Samanta, L.; Thatoi, H.N. Bioactivity guided isolation of antidiabetic and antioxidant compound from *Xylocarpus granatum* J. Koenig bark. *3 Biotech* 2019b, 9(198), 1–9.

DeFronzo, R.A. Pathogenesis of type 2 diabetes mellitus. *Medical Clinics of North America* 2004, 88, 787–835.

Evans, J.L.; Rushakoff, R.J. Oral pharmacological agents for type 2 diabetes: Sulfonylureas, meglitinides, metformin, thiazolidinediones, α-glucosidase inhibitors, and emerging approaches. In: *Diabetes and Carbohydrate Metabolism*; Degroot, L.; Goldfine, I.D.; Rushakoff, R.J., Eds.; Endotext.com, 2007.

Gaffar, M.U.; Morshed, M.A.; Uddin, A.; Roy, S.; Hannan, J.M.A. Study the efficacy of *Rhizophora mucornata* poir leaves for diabetes therapy in long Evans rats. *International Journal of Biomolecules and Biomedicine* 2011, 1, 20–26.

Geegi, P.G.; Manoharan, N. Phytochemical screening and hypoglycemic effect of leaves of *Aegiceras corniculatum (L.)* blanco (Black mangrove) in streptozotocin-induced diabetic rats. *WJPPS* 2018, 7(4), 1412–1427.

Gowri, P.M.; Tiwari, A.K.; Ali, A.Z.; Rao, J.M. Inhibition of α-glucosidase and amylase by bartogenic acid isolated from *Barringtonia racemosa* Roxb. seeds. *Phytotherapy Research* 2007, 21, 796–799.

Gurudeeban, S.; Satyavani, S.; Ramanathan, T.; Balasubramanian, T. Antidiabetic effect of a black mangrove species *Aegiceras corniculatum* in alloxan-induced diabetic rats. *Journal of Advanced Pharmaceutical Technology and Research* 2012, 3, 52–56.

Hamzevi, A.; Sadoughi, S.D.; Rahbarian, R. The effect of aqueous extract of *Avicennia marina* (Forsk.) Vierh. leaves on liver enzymes' activity, oxidative stress parameters and liver histopathology in male diabetic rats. *FEYZ* 2017, 21(4), 305–316.

Haque, M.; Ahmed, A.; Nasrin, S.; Rahman, M.M.; Raisuzzaman, S. Revelation of mechanism of action of *Rhizophora mucornata* Poir. bark extracts for its antidiabetic activity by gut perfusion and six segment method in long evan rats. *International Research Journal of Pharmacy* 2013, 4, 1–4.

Hasan, M.N.; Sultana, N.; Akhter, M.S.; Billah, M.M.; Islamp, K.K. Hypoglycemic effect of methanolic extract from fruits of *Sonneratia caseolaris*—a mangrove plant from Bagerhat region, The Sundarbans, Bangladesh. *Journal of Innovation and Development Strategy* 2013, 7, 1–6.

Hossain, S.J.; Basar, M.H.; Rokeya, B.; Arif, K.M.T.; Sultana, M.S.; Rahman, M.H. Evaluation of antioxidant, antidiabetic and antibacterial activities of the fruit of *Sonneratia apetala* Buch Ham. *Oriental Pharmacy and Experimental Medicine* 2013, 13, 95–102.

Karimulla, S.; Kumar, B.P. Antidiabetic and Anti hyperlipidemic activity of bark of *Bruguiera gymnorrhiza* on streptozotocin induced diabetic rats. *Asian Journal of Pharmaceutical Science and Technology* 2011, 1, 4–7.

Kiran, G.; Sharma, G.N.; Shrivastava, B.; Babu, S. Protective role of *E. agallocha* L. against STZ induced diabetic complication. *The Pharma Innovation Journal* 2018, 7(9), 17–26.

Kumar, V.; Cotran, R.S.; Robbins, J. *Basic Pathology*, 5th ed. A Division of Harcourt Brace and Company, Philadelphia, London, 1992, pp. 570–573.

Lakshmi, V.; Gupta, P.; Tiwari, P.; Srivastava, A.K. Antihyperglycemic activity of *Rhizophora apiculata* Bl. in rats. *Natural Product Research* 2006, 20, 1295–1299.

Lakshmi, V.; Kumar, R.; Srivastava, A.K. Antidiabetic effect of the leaves of *Kandelia candel* Linn. *Natural Products* 2013, 9, 319–321

Lawag, I.L.; Aguinaldo, A.M.; Naheed, S.; Mosihuzzaman, M. α-Glucosidase inhibitory activity of selected Philippine plants. *Journal of Ethnopharmacology* 2012, 144, 217–219.

Mahera, S.A.; Ahmad, V.U.; Saifullah, S.M.; Mohammad, F..V.; Ambreen, K. Steroids and triterpenoids from grey mangrove *Avicennia marina. Pakistan Journal of Botany* 2011, 43, 1417–1422.

Martin, F.; Boris, N.S.; Kengne, S.R.; Chia, T.E.; Guy, T.N.; Gabin, A.K.B.; Ngondil, J.L.; Innocent, G. Antioxidant and postprandial glucose-lowering potential of the hydroethanolic extract of Nypa fruticans seed mesocarp. *Biology and Medicine (Aligarh)* 2017, 9(4), 1–6.

Mayfield, J. *Diagnosis and Classification of Diabetes Mellitus; New Criteria*. Bowen Research Center, Indian University Indianapolis, IN, 1988.

Michael, R.A.Z.; Ramireddy, D.S.; Rao, B.G. Antidiabetic activity of *Avicennia alba* in Streptozotocin induced Type-2 diabetic rats. *IJRPPS* 2018, 3(1), 253–258.

Modak, M.; Dixit, P.; Londhe, J.; Ghaskadbi, S.; Paul, A.; Devasagayam, T. Indian herbs and herbal drugs used for the treatment of diabetes. *Journal of Clinical Biochemistry and Nutrition* 2007, 40(3), 163–173.

Morada, N.J; Metillo, E.B.; Uy, M.M; Oclarit, J.M. Anti-diabetic polysaccharide from mangrove plant, *Sonneratia alba* Sm. *International Conference on Asia Agriculture and Animal* 2011, 3, 197–200.

Musman, M.; Audina, E.; Ratu, F.I.R.; Erlidawati, E.; Safrida, S. Assessment of type 2 anti-diabetes on bound flavonoids of *Barringtonia racemosa* (L.) Spreng. Kernel in glucose-induced diabetic rats. *AJPT* 2017, 12(3), 48–61.

Nabeel, M.A.; Kathiresan K.; Manivannan, S. Antidiabetic activity of the mangrove species *Ceriops decandra* in alloxan-induced diabetic rats. *Journal of Diabetes* 2010, 2, 97–103.

Nabeel, A.M.; Kathiresan, K.; Manoharan, C.; Subramanian, M. Insulin-like antigen of mangrove leaves and its anti-diabetic activity in alloxan-induced diabetic rats. *Natural Product Research* 2012, 26, 1161–1166.

Nathan, D.M. Long-term complications of diabetes mellitus. *The New England Journal of Medicine* 1993, 328, 1676–1685.

Pitchaipillai, R.; Ponniah, T. *In vitro* antidiabetic activity of ethanolic leaf extract of Bruguiera cylindrica L.—Glucose uptake by yeast cells method. *International Journal of Biological and Biomedical Journal* 2016, 2(4), 171–175.

Rahman, M.; Siddika, A.; Bhadra, B.; Rahman, S.; Agarwala, B.; Chowdhury, M.H.; Mohammed, R. Antihyperglycemic activity studies on methanol extract of *Petrea volubilis* L. Verbenaceae leaves and *Excoecaria agallocha* L. Euphorbiaceae stems. *Advances in Natural and Applied Sciences* 2010, 4, 361–364.

Rahmatullah, M.; Azam, M.N.K; Pramanik, S.; Sania, R.S.; Jahan, R. Antihyperglycemic activity evaluation of rhizomes of *Curcuma zedoaria* Christm. Roscoe and fruits of *Sonneratia caseolaris* L. Engl. *International Journal of PharmTech Research* 2012, 4, 125–129.

Reza, H.; Haq, W.M.; Das, A.K.; Rahman, S.; Jahan, R.; Rahmatullah, M. Anti- hyperglycemic and antinociceptive activity of methanol leaf and stem extract of *Nypa fruticans* Wurmb. *Pakistan Journal of Pharmaceutical Sciences* 2011, 24, 485–488.

Shettar, A.K.; Vedamurthy, A.B. Evaluation of antidiabetic potential of *Kandelia candel* and *Rhizophora apiculata*—An *in vitro* approach. *International Journal of Pharmaceutical Sciences Research* 2017, 8(6), 2551–2559.

Shyam, K.P.; Kadalmani, B. Antidiabetic activity of *Bruguiera cylindrica* (Linn.) leaf in Alloxan induced diabetic rats. *International Journal of Current Research in Biosciences and Plant Biology* 2014, 1, 56–60.

Srivastava, A.K.; Srivastava, S.; Srivastava, S.P.; Raina, D.; Ahmad, R.; Srivastava, M.N.; Raghubir, R.; Lakshmi, V. Antihyperglycemic and antidyslipidemic activity in ethanolic extract of a marine mangrove *Xylocarpus granatum. Journal of Pharmaceutical and Biomedical Sciences* 2011, 9, 1–12.

Srivastava, A.K.; Tiwari, P.; Srivastava, S.P.; Srivastava, R.; Mishra, A.; Rahuja, N.; Pandeti, S.; Tamrakar, A.K.; Narender, T.; Srivastava, M.N.; Lakshmi, V. Antihyperglycaemic and antidyslipidemic activities in ethyl acetate fraction of fruits of marine mangrove *Xylocarpus moluccensis. International Journal of Pharmacy and Pharmaceutical Sciences* 2014, 6, 809–826.

Sur, T.K.; Hazra, A.; Hazra, A.K.; Bhattacharya, D. Antiradical and antidiabetic properties of standardized extract of Sunderban mangrove *Rhizophora Mucronata. Pharmacognosy Magazine* 2015, 11(42), 389–394.

Sur, T.K.; Seal, T.; Pandit, S.; Bhattacharya, D. Hypoglycaemic activities of a mangrove plant *Rhizophora apiculata* Blume. *Natural Product Sciences* 2004, 10, 11–15.

Tamrakar, A.K.; Kumar, R.; Sharma, R.; Balapure, A.K.; Lakshmi, V.; Srivastava, A.K. Stimulatory effect of *Ceriops tagal* on hexose uptake in L6 muscle cells in culture. *Natural Product Research* 2008, 22, 592–599.

Thirumurugan, G.; Vijayakumar, T.M.; Poovi, G.; Senthilkumar, K.; Sivaraman, K.; Dhanaraju, M.D. Evaluation of antidiabetic activity of *Excoecaria agallocha* L. in alloxan induced diabetic mice. *Natural Products* 2009, 1–5.

Tiwari, A.K.; Viswanadh, V.; Gowri, P.M.; Ali, A.Z.; Radhakrishnan, S.V.S.; Agawane, S.B.; Madhusudana, K.; Rao, J.M. Oleanolic acid an α-Glucosidase inhibitory and antihyperglycemic active compound from the fruits of *Sonneratia caseolaris*. *Open Access Journal of Medicinal and Aromatic Plants* 2010, 1, 19–23.

Tiwari, P.; Tamrakar, A.K.; Ahmad, R.; Srivastava, M.N.; Kumar, R.; Lakshmi, V.; Srivastava, A.K. Antihyperglycaemic activity of *Ceriops tagal* in normoglycaemic and streptozotocin-induced diabetic rats. *Medicinal Chemistry Research* 2008, 17, 74–84.

Umaru, I.J.; Umaru, H.A.; Umaru, K.I. Potential of Barringtonia Racemosa (L.) Dichloro-methane extract on Streptozotocin (STZ)-induced type 2 diabetic rats. *Current Research in Clinical Diabetes & Obesity Journal* 2019, 9(4), 1–6.

Umaru, I.J.; Boyi, R.H.; Miyel, M.H.; Kukoyi, A.J.; Umaru, K.I.; Umaru, H.A.; Madaki, K. Antidiabetic potentials of leaves extract of Barringtonia Racemosa (L) in Alloxan-induced Albino rats. *AJPP* 2018, 5, 1–4.

Venkataiah, G.; Ahmed, I.M.; Reddy, D.S.; Rejeena, M. Anti-diabetic activity of *Acanthus ilicifolius* root extract in alloxan induced diabetic rats. *Indo American Journal of Pharmaceutical Research* 2013, 3, 9007–9012.

WHO. *Definition, Diagnosis and Classification of Diabetes Mellitus and Its Complications*. Report of a WHO Consultation. Part 1, Diagnosis and Classification of Diabetes Mellitus. 1999, WHO, Geneva.

WHO. *Legal Status of Traditional Medicines and Complementary/Alternative Medicine, A Worldwide Review*, 2001, World Health Organization.

Yusoff, N.A.; Lim, V.; Al-Hindi, B.; Razak, K.N.A.; Widyawati, T.; Anggraini, D.R.; Ahmad, M.; Asmawi, M. Z. *Nypa fruticans* Wurmb. Vinegar's aqueous extract stimulates insulin secretion and exerts hepatoprotective effect on STZ-induced diabetic rats. *Nutrients* 2017, 9(925), 1–12.

PART II
Utilization of Bioresources in Biofuel

CHAPTER 8

Microalgae as a Viable Bioresource for Sustainable Biofuel Production Through Biorefinery Technologies

BUNUSHREE BEHERA[1], SUMAN NAYAK[1], S. RANGABHASHIYAM[2], R. JAYABALAN[3], and P. BALASUBRAMANIAN[1,*]

[1]*Agricultural and Environmental Biotechnology Group, Department of Biotechnology and Medical Engineering, National Institute of Technology Rourkela, Odisha 769 008, India*

[2]*Department of Biotechnology, School of Chemical and Biotechnology, SASTRA University, Thanjavur, Tamil Nadu 613401, India*

[3]*Department of Life Science, National Institute of Technology Rourkela, Odisha 769 008, India*

Corresponding author. E-mail: biobala@nitrkl.ac.in

ABSTRACT

Microalgae are being recognized globally as the sustainable feedstock to replace the finite fossil fuels. However, the utilization of algae as the "fuel-only option" is questionable in view of their potential economic impacts, being incompetent with the current market-based fuels. Hence, it is critically essential to comprehend the underlying biochemical principles to integrate the water–food–energy–environment nexus that microalgae inhabit. The present chapter outlines the emerging concept of microalgal biorefinery with an insight on tackling the technological and economic issues pertinent to the algal biofuel sector. The sequential and integrated framework for processing the algal biomass into several energy-based and value-added products has been delineated to aid in the financial sustainability of the production unit. The idea of synergistic integration of the industrial sectors with the algal biorefinery to efficiently utilize the multitude of resources toward sustainable biofuel production has also been discussed in detail. The resulting industrial symbiosis could offer the tangible benefits of waste-resource utilization for energy generation, with intangible advantages like the reduction of pollutant load and net-zero carbon emissions. Further, this would generate employment options for the rural population in addition to the availability of cheaper food/feed supplements and organic fertilizers. Adoption of innovative technologies for creating novel products and capabilities, with minimized environmental impacts, could ensure the realization of algal technology as a bio-based industry from the view of sustainable development.

8.1 INTRODUCTION

Urbanization and industrialization had horrendous impacts on natural resources, leading to fossil-fuel depletion, increase in atmospheric carbon dioxide (CO_2) levels, global warming, and climate change (Mohan et al., 2016). This phenomenon has shifted the focus of scientists toward the exploration of cleaner and greener renewable sources of bioenergy (Behera et al., 2019a). The first-generation biofuels from sugarcane, corn, starch based plant resources, and animal products, though popular in the previous decade, were criticized globally due to their potential competitiveness with food/feed products (Mohr and Raman, 2013). To circumvent these issues, second-generation lignocellulosic biomass and waste-energy crops were studied as a source of raw material for biofuel. However, the high pretreatment costs and energy associated with the process of reducing the lignin fraction to expose the entrapped cellulose have limited their use at the field scale. In lieu of the above-mentioned problems, the third-generation biofuels from microalgae have attracted attention as a sustainable feedstock for fulfilling the energy needs (Behera et al., 2019a).

Microalgae are considered as photoautotrophs that can metabolize nutrients from resources like wastewater, as well as marine/brackish water, and can sequester carbon either from the atmosphere or from the industrial emissions (Behera et al., 2019b; Rangabhashiyam et al., 2017). Microalgae with simultaneous waste remediation can assimilate, metabolize, and transform nutrients into intracellular bio-macromolecules like lipids, carbohydrates, and proteins (Trivedi et al., 2015). Microalgal biomass (mostly lipids) are traditionally chemically converted into biodiesel that is expected to act as an alternative to conventional diesel (Sharma and Singh, 2017). The commercial application of microalgae for the production of biofuel is limited, due to the higher capital and operating costs. The current costs of algal biofuel production are comparatively high. Pertinent to the emerging issues of financial incompetence, algal researchers and scientists have come up with the idea of microalgal biorefinery, which, instead of focusing on a single end product, employs the cascade for the methodical, parallel synthesis of a variety of commercial products (bioenergy) and by-products (value-added chemicals).

Though much literature exists related to microalgal biorefinery, the sequential and integrated multicascade framework for exploring the bioenergy and value-added products is limited. Most literature related to biorefinery has discussed the different energy products and value-added chemicals that can be derived from microalgal biomass (Mohan et al., 2016; Moreno-Garcia et al., 2017; Subhadra, 2011). However, very few studies have addressed the economic aspects of making the algal biorefinery feasible.

Thus, in an attempt to explore the immense potential of microalgae, the present study provides a comprehensive review of the bioprocess techniques for transforming microalgal biomass into energy-based and value-added products. The biochemical nature and algal biology have been delineated to aid in providing the idea of diversified uses of microalgae in the commercial market. The economic hurdles associated with the traditional utilization of algal biomass have been discussed in detail. Integration of the zero waste algal biorefinery model framework with existing industries has been elucidated to minimize the biogenic wastes, thereby reducing environmental pollution. The socioeconomic benefits of establishing a sustainable, circular, bio-based economy have been highlighted. Further, the challenges associated with the implementation of the microalgal biorefinery model and the plausible solutions have been detailed. Annexing the existing microalgal cultivation strategies with cheaper waste resources from the industries, with minimal energy consumption for multivariate product formation, will help in establishing a sustainable microalgal bio-economy in the near future.

8.2 EXPLORATION OF MICROALGAL BIOMASS FOR NUMEROUS PRODUCTS

Microalgae are best known for their diversified chemical composition, because of which they can be utilized for producing energy and other commodities. These microorganisms possess the inherent potential to assimilate glucose, starch, cellulose, and various other kinds of carbohydrates up to 50% of their dry weight, due to their high photoconversion efficiency (Hwang et al., 2016). The lipid content of microalgae usually varies from 20% to 60%

of their total weight under stress conditions or specific culture conditions (Yen et al., 2013). Protein is one of the crucial constituents, accounting for 50%–70% of the total algal weight (Pignolet et al., 2013).

Microalgal carbohydrates are mainly made up of monosaccharides like glucose, galactose, mannose, fructose, and xylose in different ratios, along with the components of starch, cellulose, and hemicellulose (Markou et al., 2012). The biomolecules in the form of glucose and starch can be hydrolyzed and converted into fermentable bioethanol, or anaerobically converted into biohydrogen or biogas (Behera et al., 2019a). Several other microalgal polysaccharides, due to their specialized recalcitrant properties, act as protective structural components of the cell (Hwang et al., 2016). These compounds can be extracted and used as bioactive compounds with diverse applications in food, clinical drugs, and supplements (Trivedi et al., 2015). They can also be used to produce a range of value-added chemicals that could be used as cosmetics, plasticizers, thickening agents, etc.

Proteins constitute 50%–70% of the dry weight of microalgae (Yen et al., 2013). They constitute an essential product or by-product in a microalgal biorefinery and can be used as a supplementary nutrient feed/nutraceutical or as animal fodder (Zhu and Hiltunen, 2016). Proteins from the residual biomass after oil/carbohydrate extraction can also be utilized as a source of fertilizer for soil conditioning (De Bhowmick et al., 2018). Proteins can be extracted using suitable organic solvents under optimal operating conditions of pH, ionic strength of salt, and concentration of solvent, without much loss in their functional properties (Maurya et al., 2016a). Protein-rich algae can also be utilized for synthesizing bioplastics (Rahman and Miller, 2017).

Microalgae have increasingly attracted our attention due to their metabolic capacity to accumulate large quantities of lipids (20%–60% of their dry weight) (Rangabhashiyam et al., 2017). The lipid content depends on the culture conditions. A high carbon-to-nitrogen ratio (C/N) favors lipid accumulation. Several researchers have also proposed that under limited nitrogen conditions, microalgae can accumulate 70%–80% of triglycerides (Karemore et al., 2013; Mehrabadi et al., 2015). These fractions via biochemical conversion can be processed into biodiesel, bio-oil, and biocrude oil (Behera et al., 2019a; Trivedi et al., 2015). The selection of species depends on the end product to be targeted. Table 8.1 shows the biochemical composition of the microalgae widely utilized in biotechnology.

Apart from bio-macromolecules, microalgae also harbor significant amounts of pigments, which can be extracted using a suitable solvent system and thereby converted into value-added chemicals (Chew et al., 2017).

TABLE 8.1 Biochemical Composition of Commonly Used Microalgae

Microalgae	Macromolecules (%)			Pigments mg/g (dry weight)						References
	Protein	Carbohydrate	Lipids	Lutein	Fucoxanthin	Astaxanthin	Zeaxanthin	Cantaxanthin	β carotene	
C. sorokiniana	40–50	25–35	32.3	0.8	–	–	–	–	0.1	Banskota et al., 2018
N. oleoabundans	32	24.3	22.2–31.8	3.4	–	–	0.1	0.5	0.6	Banskota et al., 2018; Rashidi & Trindade, 2018
S. obliquus	50–56	10–17	12–14	1.6	–	–	1.6	0.3	0.4	Banskota et al., 2018; Behera et al., 2019a
C. vulgaris	51–58	12–27	14–22	–	–	–	–	–	–	Behera et al., 2019a
B. braunii	11–24	20–76	41–86	0.7	–	0.2	0.1	–	0.2	Blifernez-Klassen et al., 2018; Banskota et al., 2018
P. tricornutum	52–59	10–15	44.8	2.1	24.3	–	0.1	0.6	1.6	Banskota et al., 2018; Benavides et al., 2018
T. chui	15–30	–	6–13	0.6	–	0.1	–	0.4	1.0	Banskota et al., 2018; Behera et al., 2019a

Pigments are usually utilized as food colorants. β-carotenoids are often utilized as a source of vitamins and find applications in pharmaceutical industries (Zhu, 2015). Further, genetic engineering and metabolic flux engineering strategies can be utilized for altering the biochemical characteristics of microalgae for obtaining the desired products (Hathwaik et al., 2017). Fu et al. (2016) demonstrated the use of computational metabolic modeling tools like expression-dependent gene effect (EDGE) [developed by Wagner et al., (2013), Dalhousie University, Canada], OptKNOCK (Genomatica, Inc., USA) [developed by Burgard et al., (2003) at Pennsylvania State University, USA] and OptORF (Gevo, Inc., USA) [developed by Kim and Reed, (2010) at the University of Wisconsin, USA] for studying the effects of gene alterations. Behera et al. (2019c) have also discussed the insilico process of enhancing lipid biogenesis in microalgae via manipulating the fatty acid synthesis pathway. Understanding the microalgal ecology and optimizing bioprocess engineering principles would aid in unleashing the multiproduct ability of microalgae, as shown in Figure 8.1. A recent study by Behera et al. (2019a) has summarized the bioprocess engineering strategies for obtaining multiple products from microalgae.

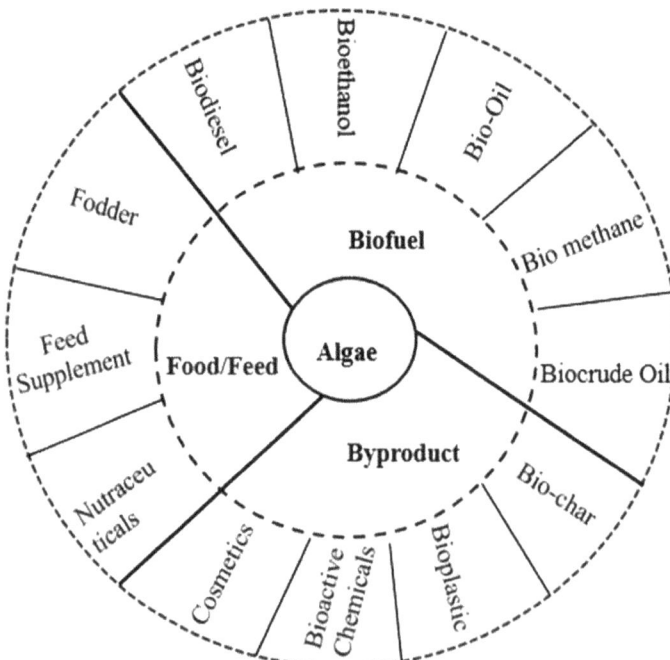

FIGURE 8.1 Potential applications of microalgal biomass.

8.3 CONVENTIONAL APPLICATIONS OF MICROALGAE

Over decades, microalgae have been widely used as feedstock in the areas of food, fodder, and fertilizers. With the rising costs of fodders worldwide, microalgal biomass are being utilized to replace a proportion of the animal feed. Becker (2007) have reported that due to the high intracellular content of proteins (about 60% of dry weight) and the presence of the required essential amino acids, they could be used as the nutritional supplement in food and fodder industries. A study by Yan et al. (2012) used fermented *Chlorella* biomass at a dose of 0.1%, which resulted in the improvement of gut microflora, thereby enhancing the intestinal digestibility. The microalgae *Scenedesmus almeriensis* was utilized by Vizcaino et al. (2014) as a feed for sea bream, thereby confirming its use as a healthy nutritional replacement in fish meal. A similar study by Kaspar et al. (2014) proved the suitability of *Chaetoceros calcitrans* as a continuous feed in hatcheries.

A very recent study by Lerat et al. (2018) discussed in detail the traditional utilization of microalgae in different food/fodder industries. Commercially, microalgal products for human nutrition are available in the forms of tablets, capsules, and liquids. They are also being used as ingredients in foods and beverages (Zhu and Hiltunen, 2016). Spolaore et al. (2006) reported that the strains belonging to *Arthrospira* sp., *Chlorella* sp., *Dunaliella salina*, and *Aphanizomenon flosaquae* are commonly used on the commercial scale for nutritional supplements. High-value molecules from microalgae like polyunsaturated fatty acids (PUFAs) are often added into infant products as supplements (Hudek et al., 2014). Out of several PUFAs, currently only docosahexaenoic acid (DHA) is widely being utilized, with others remaining unexplored. In the case of cosmetics, microalgal extracts are being used in different skin and hair care products (Mourelle et al., 2017). Pigments like β carotene, zeaxanthin, astaxanthin, etc., produced from different microalgae like *Haematococcus pluvialis*, *Botryococcus braunii*, *Dunaliella salina*, etc. are widely used as food colorants (Zhu et al., 2016). Microalgal biomass have also been reported to find applications as organic fertilizers for the slow release of nitrogen, phosphorus, and potassium to well attune plant growth and metabolism, resulting in improved fruit quality and carotenoid content (Maurya et al., 2016a). Global companies like Seambiotic Ltd., Fuji Chemical Industry Co., and Cyanotech Corporation are dealing with the use of algal biomass for cosmetics and nutritious feed (Khan et al., 2018).

Owing to the evidential problems of energy crisis, global warming, and climate change, microalgal biofuel, especially algal biodiesel, has widely been researched over the previous years. Microalgae have cutting-edge advantages

over traditional energy crops, including higher lipid content, flexibility of cultivation on nonarable and wastewater sources, and the capacity to biologically
sequester huge amounts of CO_2 (Rangabhashiyam et al., 2017). Thus, they are
undoubtedly regarded as the promising feedstock for biofuel production and
a potential sustainable alternative to conventional fossil fuels. The quality of
algal biodiesel and fatty acid methyl esters composition greatly depends on
the microalgal strain utilized (Milano et al., 2016). Further, biochemical and
thermochemical processing of microalgal carbohydrates into bioethanol by
fermentation (Ho et al., 2013; Lee et al., 2015), biohydrogen via dark fermentation (Batista et al., 2014; He et al., 2016), biogas by anaerobic digestion
(Passos et al., 2014; Uggetti et al., 2017), and bio-oil via pyrolysis (Tag et al.,
2016; Vardon et al., 2012) are increasingly explored. Despite heavy research,
algal biofuel could not be translated to the field scale due to the low net
energy ratio (NER) owing to the energy-intensive processes involved. Several
researchers have reported that the fuel-only option from microalgae is neither
technically nor economically feasible (Gerber et al., 2016; Ullah et al., 2015).
Various risks associated with the existing microalgal products inhibiting their
commercialization have been explained in the following section.

8.4 TECHNICAL AND COMMERCIAL EXTERNALITIES OF TRADITIONAL UTILIZATION OF ALGAL BIOMASS

The real-time or large-scale application of microalgal technology for different
products is often hindered by the cost associated at the production stage.
Capital costs associated with algal biodiesel production occupy 50% of the
total production costs (Tredici et al., 2016). The cultivation and harvesting
steps have been reported to claim 20%–30% of total investments. Also, with
the current technology, algal biodiesel production seems to be 2.5 times
more energy-intensive compared to conventional diesel (Faried et al., 2017).
The US Department of Energy has projected the cost price of algal biodiesel
to be US $8 per gallon (Davis et al., 2016). In the case of developing coun-
tries like India, China, Taiwan, Myanmar, and Japan, the large-scale algal
biofuel industry is still in its infancy, with no large-scale production units
in operation (Behera et al., 2019a). Newer strategies like optimization of
culture media for improved biomass productivity and development of cost-
efficient harvesting technologies are increasingly being studied; however, no
real-time progress is visible. In a nutshell, even though algal biofuel has been
studied and reported in more than 1000 papers in literature, no commercial
facility actually sells algal biodiesel. Zhu et al. (2016) has summarized the

cost estimates for algal biofuel, considering different production units and routes, and none of the currently utilized strategies is economically viable. Inevitably, this apparently promising and sustainable third-generation biofuel opportunity is not feasible and applicable unless the economics associated with the luxurious fuel-only option is improved.

Further, with the development of interests in algal biofuel technology, the traditional applications of microalgal products in cosmetics and chemicals have faced potential recession over time. High process costs associated with the microalgal biofuels commercially make them uncompetitive with the petroleum fuel markets. Researchers also have a pessimistic opinion that the ever increasing interests in microalgal biofuels, which are not feasible for bulk application (due to high capital investment costs), will cause the disappearance of the existing market for fertilizers and chemicals due to the lack of focus (Zhu et al., 2016). Lifecycle analysis (LCA) and techno-economic analysis (TEA) have revealed that the plausible algal technology and related products could be commercially targeted only when the major drivers like biomass production costs, oil and by-product yields affecting the energy return on energy invested, and the rate of return on capital investment are addressed (Shurin et al., 2016). The main key in making algal production profitable and bringing the algal biofuel costs closer to those of petroleum fuels lies in exploring the potential by-products from the residual biomass, by developing synergistic relationships with water treatment systems and other industries.

8.5 BIOREFINERY APPROACH AS A PROMISING SOLUTION FOR SUSTAINABLE FUTURE

In order to improve the economics associated with the microalgal technology, algal biomass must be transformed into a range of materials, chemicals, and biochemicals that hold commercial potential. The following subsections explain in detail the algal biorefinery concept and elucidate the engineering aspects for process integration at the upstream and downstream levels to make the algal production process cheap and sustainable.

8.5.1 ALGAL BIOREFINERY CONCEPT AND SCOPE

Moving beyond the niche of the biodiesel market from the algal biomass, based on the strain-specific composition, other options are needed to be explored to make the entire process economically sustainable. The algal biorefinery concept integrates different types of biochemical and thermo-chemical conversion pathways to obtain the desirable products/by-products (Mohan et al., 2016). The major drivers behind identifying viable microalgal biorefinery concepts lie with the following aspects:

- Developing a compatible cultivation system to achieve the desired productivities;
- Identifying the composition of algal biomass and integrating it with the downstream conversion technologies to derive the targeted products and by-products; and
- Mapping in detail the possible markets for the bio-based products.

Unlike petroleum refineries, the algal biorefinery aims at obtaining higher volumes of low-value product (biofuels) and lower volumes of high-value products (food supplements, fertilizers, etc.) (De Bhowmick et al., 2018). The algal biorefinery with "zero-waste discharge" strategies integrates the process of utilizing wastewater resources and flue gas for cultivating microalgae for biofuels and using de-oiled biomass for deriving high-value products. The concept will not only supplement the energy demands but also make the entire process chain sustainable in terms of waste management. Economic benefits in terms of revenues can even be earned through carbon credits and through the sale of high-value, profitable products.

8.5.2   *ENGINEERING ASPECTS AND PROCESS INTEGRATION*

Microalgae are the futuristic sustainable feedstock for multiple products due to their attractive biochemical composition (Trivedi et al., 2015). Biorefining utilizes engineering strategies to integrate different chemical, thermochemical, and biochemical processes and related equipment to obtain a wide spectrum of marketable products.

Cultivation and harvesting are essential aspects for improving micro-algal biomass (Behera et al., 2019a). The range of products to be derived decides the cultivation regime and the harvesting conditions. The use of wastewater for nutrients and CO_2 derived from flue gas prohibits the use of the biomass for nutraceutical applications or use as protein supplement (De Bhowmick et al., 2018). The presence of metal ions or toxins could potentially have a negative impact on health during its use as food/feed supplement. Though a huge amount of literature (Behera et al., 2019a; Chu et al., 2014; Wen et al., 2016) exists regarding the cultivation procedure for microalgae, still detailed experiments are essential for the demonstration of major pathways and compositional shifts for the conversion of biomass to the desired by-products. The biochemical composition of microalgae varies dramatically with the cultivation conditions of the algae (Mohan et al., 2016). Different modes of cultivation have been discussed by Behera et al. (2019a), who emphasize on macromolecular accumulation. A sustainable microalgal biorefinery combines different bioprocess cultivation strategies like the nutrient depletion phase or stress conditions to direct the flux toward the accumulation of desired metabolites, followed by cultivation in the presence of excess nutrients to achieve higher biomass content (Mohan et al., 2016).

Open-pond reactors and closed photobioreactors (PBRs) are the two major cultivation systems for growing microalgae. Since microalgae are widely known for their diverse applications and production of food/feed and other bioactive compounds, challenges exist in selecting the appropriate cultivation system depending on the desirable end products. Open-pond reactors are widely used for large-scale culturing of microalgae, being cheap and easy to construct and maintain (Pawar, 2016). Certain species of microalgae, like *Chlorella.* sp., *Scenedesmus* sp., *Dunadiella* sp., and *Spirulina* sp., are easily acclimatized to the harsh environmental conditions and can be suitably cultivated in open algal ponds (Mata et al., 2010). These systems are only suitable for the production of higher volumes of low-value products such as biofuels, fertilizers, and animal feed, since these systems are prone to a certain level of contamination owing to poor control of cultivation conditions. High-value products like the bioactive compounds to be used as nutraceuticals or food applications can be best obtained from closed PBRs with an optimal and controlled cultivation system. Different types of PBRs have been discussed in detail by Behera et al. (2019a). Depending on the type of strain, the desired level of productivity, and the costs, different reactors can be designed and constructed. Each of these reactor systems has its own advantages and disadvantages. Closed PBRs have better mixing, light-utilization efficiency, temperature control, and efficient gas/liquid mass transfer, resulting in higher volumetric productivity compared to open ponds (Koller, 2015). A study by De Bhowmick et al. (2018) showed that moderate productivity comparable to that of closed PBRs can also be achieved with algal open ponds. For a system to be economically viable, the NER must be close to 1. While closed tubular PBRs have an NER of much less than 1, open-pond reactors and flat-panel PBRs with an NER close to 1 are considered economically viable and thus can be used commercially at the field scale. Studies have also reported a hybrid reactor where both the open ponds and closed PBR are suitably used at the desired stage and where one system overcomes the limitations of the other system (Moreno-Garcia et al., 2017). Further, the combined reactor system must be commercially efficient, with lower power consumption and efficient mass transfer, and must result in a high photosynthetic rate. Derakhshan et al. (2018) demonstrated significant removal of atrazine, total chemical oxygen demand (COD), nitrogen (N) and phosphorus (P) from secondary effluent of a wastewater treatment plant using a hybrid bacterial microalgal membrane photobioreactor with a biomass accumulation of 6 g L^{-1}. Table 8.2 describes the algal productivity obtained with hybrid reactor combinations that might be suitably explored in a microalgal biorefinery.

The next major challenge after cultivation is harvesting, as the microalgae-to-water ratio is very low and the lipid droplets are bound intracellularly. The harvesting of algal biomass usually involves 20%–30% of the total processing costs (Tredici et al., 2016). The choice of harvesting depends on the end use of the biomass targeted. Centrifugation alone, though energy intensive, is the preferred method of recovering high-quality biomass without contamination with salts or toxic metals. Similar is the case for filtration, where higher costs are involved with the need for replacement of the fouled membrane with time (Barros et al., 2015). The associated energy involved can be reduced by dividing the entire harvesting process into two steps, dewatering (sedimentation or flocculation) and thickening (centrifugation and filtration). The energy involved in centrifugation and filtration can be brought down from 13.8 to 1.83 MJ kg^{-1} dry cell weight, by supplementing the procedure with the initial step of flocculation (Brennan and Owende, 2010; De Bhowmick et al., 2018). The conventional flocculation procedure with the use of synthetic chemical flocculants would add toxic chemical/metal ions to the harvested biomass, thus inhibiting its use for food/feed supplements or as bioactive compounds. Natural flocculants (Behera and Balasubramanian, 2019) or bio-flocculation, with much less contamination effects, can be used to achieve the desired results (Rajendran and Hu, 2016). A continuous dewatering, followed by thickening with the suitable characteristics of water recycling, with lower energy and efficient biomass recovery (<95%), is essential to achieve the required biomass content to be used for different purposes. Other engineering strategies like multiparameter optimization using Monte Carlo simulations for robust optimal synthesis and design of integrated microalgal biorefinery systems to resolve the deterministic uncertainties have been utilized by Sy et al. (2018).

In the context of extraction of biological macromolecules in a biore-finery, the conventional approaches like utilizing mechanical or chemical methods are energy-intensive and costly (Ranjith Kumar et al., 2015). The

TABLE 8.2 Biomass Productivity of Microalgae in Hybrid Photobioreactors

S. No.	Microalgae	Reactor Configuration	Biomass Productivity	References
1.	*C. vulgaris*	Airlift tubular PBR with raceway ponds	1.8 g l^{-1} d^{-1}	Adesanya et al., 2014
2.	*H. pluvialis*	Continuous tubular reactors followed by open ponds	15.1 g m^{-2} d^{-1}	Huntly and Redalje. 2007
3.	*S. obliquus*	Tubular PBR integrated with thin layer cascades (modified open ponds)	8 g m^{-2} d^{-1}	Tramontin et al., 2018
4.	*Tetraselmis sp.* M8	Positive pressure driven airlift PBR combined with open raceway ponds	0.55 g m^{-2} d^{-1}	Narala et al., 2016
5.	*P. tricornutum*	Airlift driven external tubular loop photobioreactor submerged in a thermostatic pond with temperature of 20 ± 2 °C	20 g m^{-2} d^{-1}	Fernandez et al., 2001
6.	*Isochrysis sp.*	Microalgae grown on thin layer modules suspended in open cultivation system	0.6 g m^{-2} d^{-1}	Naumann et al., 2013
7.	*Nannochloropsis sp.*	Microalgae grown on thin layer modules suspended in open cultivation system	0.8 g m^{-2} d^{-1}	Naumann et al., 2013
8.	*Tetraselmis sp.*	Microalgae grown on thin layer modules suspended in open cultivation system	1.5 g m^{-2} d^{-1}	Naumann et al., 2013
9	*Phaeodactylum sp.*	Microalgae grown on thin layer modules suspended in open cultivation system	1.8 g m^{-2} d^{-1}	Naumann et al., 2013

integrated approach of recycling the residual solvents from extraction back into the process can reduce the process costs. Novel techniques of enzymatic disruption of microalgal cells to derive the desired products or the use of an immobilized enzyme system that can be reutilized several times have shown better macromolecular yields with little or no loss in the chemical/ functional properties (Mishra et al., 2017). Utilization of green solvents like subcritical water (Zakaria and Kamal, 2016) and free nitrous acid (Bai et al., 2014) as a cell disruption agent has also been shown to provide significant yields of lipids, carbohydrates, and proteins. Nevertheless, the conventional and novel techniques of bio-macromolecular extraction must be integrated in a technically wise manner to achieve better yields with lower investment of extraction energy. Last but not least, the conversion route also determines the technical feasibility of the process chain in a microalgal biorefinery. For instance, the bioprocess conversion of wet algal biomass, as in anaerobic digestion and hydrothermal liquefaction, is much more profitable due to less energy consumption compared to transesterification and pyrolysis, which require dried biomass (Kumar et al., 2017). The selection and integration of requisite upstream processing strategies are essential for establishing a sustainable algal biorefinery.

8.5.3   HARNESSING BIO-PRODUCTS IN A BIOREFINERY

Microalgae have the tremendous ability to transform the supplied nutrients into diverse biochemical constituents that can be harnessed and used as sources of energy or value-added products. Over years, researchers have focused only on the extraction of lipids that could be processed into biodiesel via transesterification. Owing to issues related to horrendous capital invest-ments and the differences in the amount of energy required to that of the output energy, the algal research has drifted toward exploring multiple products in a biorefinery approach derived from upstream and downstream methodical systems (De Bhowmick et al., 2018).

8.5.3.1   ENERGY-BASED PRODUCTS

Several researchers have identified microalgae as a feasible and reliable third-generation feedstock for meeting the ever-increasing global energy demands. This section discusses the energy-based options that can be obtained from microalgae via different conversion routes.

Microalgae are regarded as the most appropriate feedstock, having up to 60% of triacylglycerols, with other monoglycerides, diglycerides, and small amounts of phospholipids, tocopherols, etc. (Behera et al., 2019a). Transesterification is a biochemical process that converts the extracted cellular lipids into fatty acid alkyl esters (FAAEs). A typical transesterification reaction involves an alcohol-to-oil ratio varying from 3:1 to 6:1 and a suitable homogenous (acid/base) or heterogeneous catalyst (egg shell, nanocatalysts) (Galadima and Muraza, 2014). Park et al. (2015) and Skorupskaite et al. (2016) have reviewed the recent technical advances in biochemical transesterification of microalgal lipids into biodiesel. Depending on the optimal culture conditions, the phototrophic microalgae can yield lipids at an amount of 5000–100,000 L ha^{-1} day^{-1} with an energy content of 39–41 kJ g^{-1} (Chu et al., 2014; Mehrabadi et al., 2015). The transesterified algal oil contains oleic, palmitic, and stearic acids, and thus is considered suitable for use as biodiesel in engines. The low volatility and high viscosity of the algal biodiesel are usually 20 times more than that of the conventional diesel, which result in deposits over the engines during combustion ignition (De Bhowmick et al., 2018). The de-oiled remaining biomass after lipid extraction might also be subjected to other biochemical conversion, producing bioethanol, biogas, etc. in a biorefinery concept (Trivedi et al., 2015). It is noteworthy to mention that the conventional transesterification process often generates huge amounts of glycerol, and the use of heterogeneous catalyst often deactivates the nutrients in the residual biomass (Skorupskaite et al., 2016). A recent biorefinery approach proposed the recycling of crude glycerol produced via transesterification into the algal growth medium, as the carbon source to increase the growth rate and lipid productivity of microalgae (Leite et al., 2015). Further researchers have also reviewed the process of utilizing the enzyme based transesterification of triglycerides into FAAEs, to reduce the adverse effects of homogeneous and heterogeneous catalysts during in situ transesterification (Amini et al., 2017; Noraini et al., 2014).

Microalgae are very well known for the low-cost bioethanol production as their buoyant properties simplify the pretreatment steps (De Bhowmick et al., 2018). Under optimal environmental conditions, certain microalgae like *Chlorella* sp., *Dunaliella* sp., and *Scenedesmus* sp., due to high photon conversion efficiency, can accumulate carbohydrates as starch and cellulose up to 50% of their dry weight (Juneja et al., 2013). Mechanical or enzymatic methods are often used to break open the cell wall, to extract the intracellular starch. The extracted starch is thereby subjected to saccharification using cellulolytic hydrolysis enzymes to derive simple sugars, which are thereby fermented by a suitable yeast strain into bioethanol (Hernández et al., 2016).

The crude bioethanol obtained is subjected to a distillation process to remove the residual impurities and water, resulting in purified ethanol of 95% (De Bhowmick et al., 2018). Behera et al. (2019a) have recently reviewed and discussed the different parameters influencing algal bioethanol production like the use of multienzymatic mix and use of noble strains to achieve higher fermentation efficiency. The cellulose-extracted biomass after bioethanol fermentation can also be subjected to anaerobic digestion, thereby producing methane (biogas). Wieczorek et al. (2014) reported the production of biohydrogen by dark fermentation of *Chlorella vulgaris* (*C. vulgaris*), where the residual biomass was converted into biomethane. The biorefinery approach highlights the algal bioethanol production as potentially promising, since the CO_2 produced as by-product can be recycled and further used for biomass cultivation (Zhu et al., 2016).

Biogas is a combination of methane and carbon dioxide, mainly produced by anaerobic digestion, and can be used as fuel. John et al. (2011) reported that microalgal species with high organic loading, longer solid retention time, and under mesophilic and thermophilic conditions could produce methane at a rate of $0.2–0.41$ m^3 kg^{-1}. However, the cell wall resistance to anaerobic bacterial attack and ammonia-mediated inhibition often results in only 2% conversion into methane. Thus, microalgae-based methanogenesis, compared to other feedstocks, as the only energy option is not sustainable due to the low conversion rate and higher costs involved (De Bhowmick et al., 2018). Studies reported that the low C/N of 6–9, compared to 15–30 in the case of feedstock, is the major reason for the low productivity of biomethane by microalgae (Passos et al., 2014). Uggetti et al. (2017) reported that the low C/N can be increased by adding other carbon substrates, thereby increasing the yield. Producing biogas in a biorefinery model could be a viable option. Uggetti et al. (2017) and Behera et al. (2019a) have discussed the techniques for improving the methanization potential of microalgae, like strain modification, maintaining optimal operating conditions and hydraulic retention time to increase the biogas yields. Utilization of the biorefinery concept by coupling wet algal biomass and sludge produced by wastewater treatment in high-rate algal ponds (HRAPs) is also expected to boost the efficacy of the process (Singh et al., 2011; Uggetti et al., 2017). Gonzalez-Fernandez et al. (2011) also demonstrated a 123% increase in biogas production on utilization of the spent algal biomass after biohydrogen production.

Algal biohydrogen production has received considerable attention in recent years. Biohydrogen is produced from microalgae by one of the three processes, that is, biophotolysis of water, acidogenic phase of anaerobic digestion, or two-stage dark photo-fermentative production (He et al., 2016).

The hydrogenase activity of microalgae is often limited by the oxygen produced during photosynthesis; thus, only 15% of the theoretical yield can be reached practically, and the process seems to be costly (Behera et al., 2019a). Recent technical advancement and ongoing research in the field of zero-waste algal biorefinery are expected to improve the process economics of biohydrogen production (De Bhowmick et al., 2018). Sengmee et al. (2017) demonstrated that the recycling of crude glycerol produced during lipid transesterification as the carbon source for photo-fermentive generation of biohydrogen. Biohydrogen production through photolysis of water by microalgae is regarded as more economical and sustainable in terms of water utilization and pollutant CO_2 consumption compared to hydrogen production by simple hydrolysis of water (Mohan et al., 2016).

The thermochemical conversion of microalgal biomass into biofuel is regarded as a viable option due to the diverse range of products that can be obtained as output in a biorefinery strategy. Thermochemical conversion of dry biomass via pyrolysis, at a temperature ranging from 400 °C to 600 °C under anaerobic conditions, generates bio-oil (liquid product), biochar (solid residue), and other gaseous by-products (Tag et al., 2016). The bio-oil (combination of aromatics) is often upgraded via the utilization of different catalysts to reduce the oxygen content, making it more stable for engine application. The biochar obtained could be used either as a bio-adsorbent for wastewater treatment or as soil conditioner for enhancing the fertility, improving agricultural yields with simultaneous greenhouse gas (GHG) mitigation (Yu et al., 2017). The gaseous products obtained can be recirculated into the algal culture medium to supplement the nutrients (Behera et al., 2019a). The overall process of pyrolysis is, however, energy-intensive, owing to the drying of the harvested biomass (Thilakaratne et al., 2014). The associated disadvantages can be sorted by the process of hydrothermal liquefaction (HTL), which operates on a similar principle as pyrolysis, where the intracellular water acts as the catalyst, thereby thermo-chemically converting the biomass into biocrude oil (C17-C18 alkanes and polyaromatic hydrocarbons) that can be processed in a petroleum biorefinery producing a range of high-value products and biofuels (Biller et al., 2015). The hydrochar generated in HTL has better cation exchange property compared to pyrolytic biochar, and thus could serve as an improved soil ameliorator (Gollakota et al., 2017). The only disadvantage with biocrude oil as a potential fuel lies in the presence of high nitrogen and oxygen content along with high viscosity, causing environmental pollution during combustion, producing NO_x and particulate emissions. Hydrotreating the oil can cause 80% upgradation in oil quality, reducing the nitrogen and oxygen content (Elliot et al., 2015). The process is sustainable, since it utilizes the biorefinery concept, compared to other biofuel routes resulting in high net energy output. The cost economics of the process have been reportedly improved by recycling the aqueous waste generated back into the culture medium, improving the yield by 32.6% weight, thereby decreasing the water consumption (Hu et al., 2017).

There is a tremendous flexibility in designing an algal biorefinery system by integrating different bioprocess strategies—starting from cultivation to conversion—to achieve the desired products with significant cost reductions and economic benefits. Different biofuel production strategies, therefore, can be combined to achieve better energy yields and improve the process economics. Table 8.3 summarizes the combination of different conversion technologies for converting algal biomass into biofuels. Such promising strategies concurrently reduce the production costs and enhance the energy yield for real-time algal applications.

8.5.3.2 VALUE-ADDED PRODUCTS

Due to the absence of a suitable market and other economic hurdles associated with the microalgal fuel-only option, the futuristic template of microalgal biorefinery focuses on green phytochemicals and other bio-based products. The value-added chemicals produced by algal biomass have been explained previously. Recent studies by Khanra et al. (2018) and Haznedaroglu et al. (2017) have discussed the production pathways and practices for exploring the high-value platform chemicals from microalgae. More research is currently carried out in the area of sustainable plant-based proteins in order to reduce the dependency on animal-based proteins (Henchion et al., 2017). A recent review by Bleakley and Hayes (2017) which discussed the applications of algal protein also showed that the quality of microalgal protein is far better than that of the standard whey proteins currently in use as food supplements. Thus, protein from *Spirulina* sp. could be potentially utilized to replace the proteins in the health and food-processing industries. The only hurdle lies in separating the chlorophyll from the protein complexes (without affecting its functionality), which provides an intense color and bitter taste. Gattrel et al. (2014) used defatted algal biomass, having mostly proteins, as a replacement for corn and soybean meal in the diet of broiler chickens, egg-laying hens, and weanling pigs.

Kotrabacek et al. (2015) have reviewed the effects of supplementation of *Chlorella* sp. as animal feed and reported that the presence of organically bound selenium, iodine, pigments, antioxidants, vitamins, and growth factors

TABLE 8.3 Biorefinery Approach for Bioenergy Production Using Different Microalgal Strains

S. No.	Microalgae	Process	Biofuels Produced	References
1	*Laminaria digitata and Nannochloropsis oceanica*	Two stage batch fermentation for H_2 and CH_4	Biohydrogen (94.5 ml g^{-1} dry biomass) Biomethane (295.9 ml g^{-1} dry biomass)	Ding et al., 2016
2	*Chlamydomonas* sp. and *Scenedesmus* sp. with *Rhizobium* sp.	Photoheterotrophic cultivation of algal bacterial consortia for biohydrogen and anaerobic digestion of the residual biomass to methane	Biohydrogen (1.15 ml l^{-1} algal culture) Biomethane (253 ml g^{-1} dry biomass)	Wirth et al., 2015
3	*Scenedesmus bijugatus*	Sequential esterification and transesterification to biodiesel, and conversion of de-oiled biomass by fermentation to bioethanol	Biodiesel (0.21 g g^{-1} dry biomass) Bioethanol (0.158 g g^{-1} dry biomass)	Ashok Kumar et al., 2015
4	*Chlorella* sp. KR-1	Transesterification reaction followed by fermentation of residual biomass with *S. cerevisiae*	Biodiesel (0.29 g g^{-1} dry biomass) Bioethanol (0.16 g g^{-1} dry biomass)	Lee et al., 2015
5	Mixed algal biomass [*Scenedesmus sp.*, *Chlorella sp.*, *Ankistrosdemus sp.*, *Micromonas sp.*, and *Chlamydomonas* sp.]	ABE fermentation of microalgae pretreated with acid hydrolysis at 80–90 °C.	Acetone (1.33 g l^{-1} of pretreated algal slurry); Butanol (3.85 g l^{-1} of pretreated algal slurry); Ethanol (0.46 g l^{-1} of pretreated algal slurry)	Castro et al., 2015

enhanced their immunity and reproducibility. Enzing et al. (2014) reported that dried microalgal biomass (*Chlorella* sp. and *Spirulina* sp.) with rich proteins and carbohydrates can be directly used as dietary food supplements and also as a source of vitamins C, B12, and D. A huge amount of literature also exists on the extraction and utilization of value-added chemicals like astaxanthin (Begum et al., 2016; Rodrigues et al., 2015), antioxidants like β carotene (Zhao et al., 2017), and fatty acids like omega-3, DHA, and eicosapentaenoic acid (EPA) (Diaz et al., 2017). In order to provide better cost economics, and to improve the overall efficacy of algal technology, studies by Spolaore et al. (2006) and Tao et al. (2015) have proposed the use of defatted microalgal biomass (after lipid extraction) as an excellent source of the abovementioned value-added chemicals. Safafar et al. (2015), in a biorefinery context, studied the accumulation of tocopherols, β carotene, and phenolic compounds with industrial wastewater as the cultivation media. A recent study by Zhu and Hiltunen (2016) has discussed the cultivation of microalgal biomass using live feedstock wastes for biofuel and coproducts in a biorefinery framework.

Algal biomass is touted as a promising biomass for bioplastics having characteristics similar to petroleum-based plastics (Rahman and Miller, 2017). Zeller et al. (2015) showed that the mixed microalgal consortium of *Chlorella sp.* and *Spirulina* sp. could strive in municipal wastewater, and the protein-rich biomass could be thermo-mechanically polymerized into bio-based plastics. Rahman et al. (2015) reported a 31% polyhydroxybutyrate (PHB) production using a combined bacterial–microalgal biomass co-cultivated in wastewater. Khosravi-Darani et al. (2013) projected that PHB from microalgae is cost-efficient compared to bacterial bioplastics. *Synechococcus* sp. MA19 and *Spirulina subsalsa* could accumulate PHB up to 27% of their dry weight, utilizing GHG emissions from the industrial flue gas, and thereby would help in establishing a sustainable carbon-negative economy (Noreen et al., 2016). PHB accumulation is strain-specific and also depends on the culture conditions. A study by Kaewbai-ngam et al. (2016) reported that the filamentous *Calothrix scytonemicola* TISTR 8095 could sequester CO_2 at the rate of 729.2 ± 129.8 mg L^{-1}, resulting in the production of 356.5 ± 63.4 mg L^{-1} PHB under nitrogen-depleted conditions. Balaji et al. (2013) have reviewed the use of cheaper substrates for culturing a microalgal consortium, as well as genetically engineered algal strains with efficient PHB-accumulating properties, and further discussed novel noninvasive PHB extraction methods with reduced cost and energy efficiency. Das et al. (2018), in a biorefinery approach, used *Chlorella pyrenoidosa,* resulting in the production of 26.7 mL g^1 biodiesel and 27% PHB, further remediating the industrial wastewater, removing 33.89% nickel and 11.24% chromium. Recent studies by Hondo et al. (2015) and Carpine et al. (2017) have discussed the genetic engineering strategies for enhancing microalgal bioplastics' yield. Iwata (2015), Noreen et al. (2016), and Rahman and Miller (2017) have discussed the applications of microalgal bio-based plastics.

One of the potentially touted end-product uses of microalgal biomass is in the form of biochar, which not only acts as a soil conditioner to improve nutrient and water-retaining properties but also sequesters CO_2 (Bird et al., 2011). Biochar is a nutrient-rich by-product produced by thermal decomposition of algal biomass under oxygen-deficient conditions, during the process of pyrolysis, hydrothermal liquefaction, and carbonization (Chang et al., 2015). Algal biochar, with a high nutrient content and better pH properties, acts as a better cation exchange agent to improve soil pH and fertility, amending its properties and improving agricultural yields (Sun et al., 2017). Yu et al. (2017) have reviewed the properties and application of microalgal biochar with respect to environmental remediation and agricultural improvements. Several researchers like Johansson et al. (2016) and Zheng et al. (2017) have advocated that algal biochar could be used as a bio-adsorbent for remediating inorganic and organic pollutants from wastewaters due to the presence of many organic and inorganic functional groups. Wang et al. (2013) projected that the lipid-rich algal biomass could be utilized for biofuel production and the de-oiled biomass residue could be subjected to pyrolysis, thereby producing bio-oil and biochar in the biorefinery context. Sarkar et al. (2015) utilized mixed algal consortia grown heterotrophically to produce 320 mL g[1] biodiesel, and thereby utilized the de-oiled biomass to produce 100 mL g[-1] biomethane and 0.477 mg g[-1] biochar via pyrolysis. Pyrolysis of *C. vulgaris FSP-E* resulted in 29% bio-oil and 27% biochar production (Yu et al., 2018). Maurya et al. (2016b) and Santos and Pires (2018) have recently reviewed strategies for the production and application of biochar in the context of biorefinery. Nevertheless, exploration of high-value coproducts along with algal bioenergy options, as shown in Figure 8.2, will help in making algal production sustainable in the near future.

8.5.4 INDUSTRIAL SYMBIOSIS WITH THE CONVENTIONAL ALGAL BIOREFINERY TECHNOLOGY

With the increased environmental pollution and GHG emissions, there has been a switch from conventional fossil fuels to biofuels. The principal reasons for this are to have a negative GHG balance, and minimize land

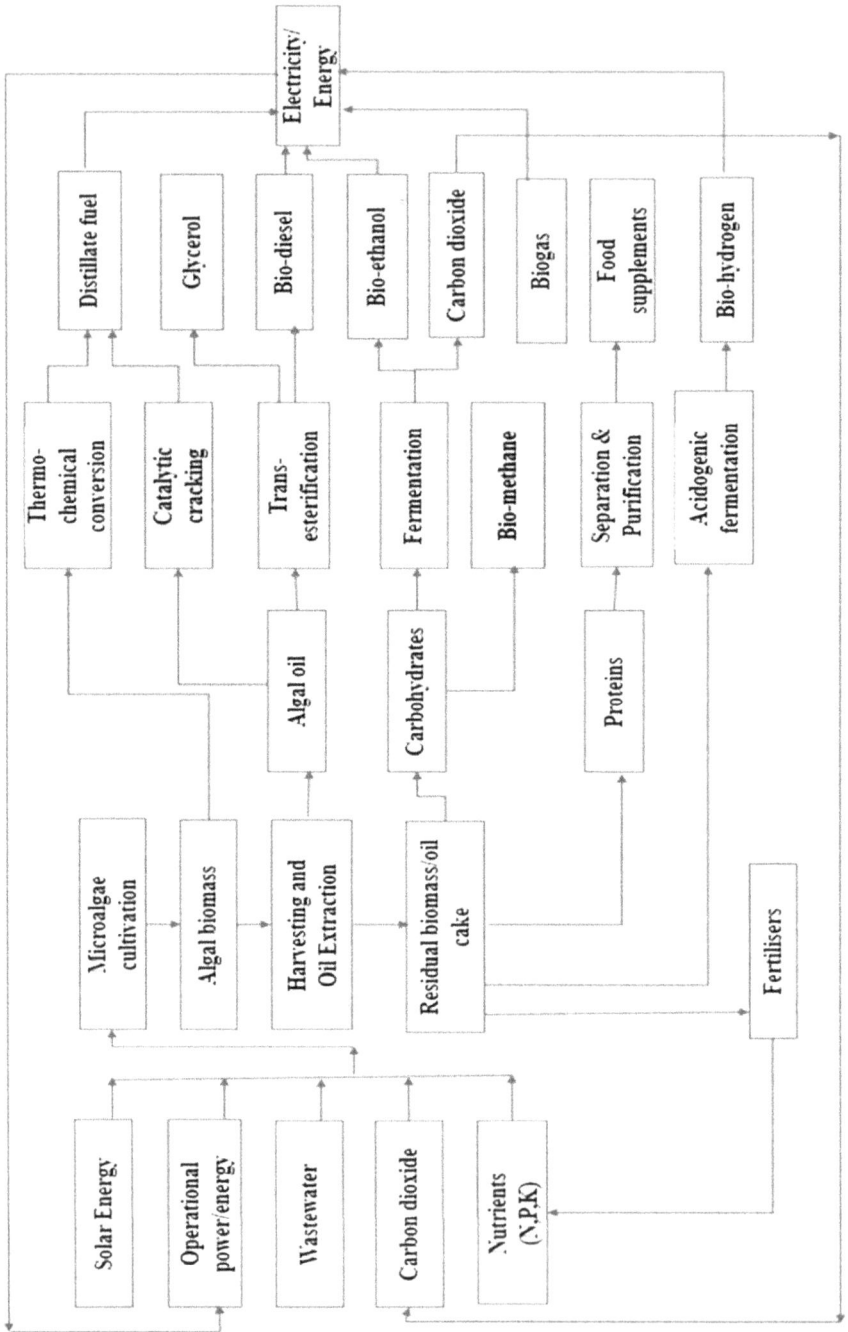

FIGURE 8.2 Algal biorefinery framework to harness bioenergy and high value products.

and water footprints, inclusive of co-product development. Microalgae are regarded as superior to other energy crops, with an overall energy yield of 793–4457 GJ ha^{-1} and the capability to grow in wastewater, utilizing flue gas from industries (Tandon and Jin, 2017). With technological advancements over the years, the upstream processing costs involved with the microalgal cultivation system have been minimized with the use of waste resources, thereby providing gains in terms of environmental sustainability and economic benefits. It is postulated that industrial processes must be operated in an integrated manner with the environmental research, resulting in a fruitful exchange of resources and a simultaneous platform for improved environmental advantages. Andersson et al. (2014) demonstrated that the industrial cluster could be successfully used as a material flow input to produce 36 MW of energy in the form of biodiesel and biogas as a part of the biorefinery concept. The following subsections provide an insight into the possible advantages that can be harnessed through the industrial integration of microalgal technology.

8.5.4.1   WASTEWATER REMEDIATION AND POLLUTION CONTROL

Conventional wastewater treatment plants installed at public places are often associated with intensive energy and cost requirements. Nutrients like carbon, nitrogen, and phosphorus are present in remarkable quantities in municipal digestate leachate, industrial and mine wastewaters, and runoffs from animal feedlots and shelters (Kumar and Pal, 2015). Microalgae can utilize the nutrients present in low-quality wastewater for growth and metabolism, assimilating bio-macromolecules that can be converted into biofuels and other bio-based products, with added benefits of nutrient recovery and pollution reduction, as illustrated in Figure 8.3.

Table 8.4 enlists the different microalgal strains cultured in various point wastewater sources, along with their nutrient removal efficiency, biochemical composition, and biomass accumulation. Microalgal biomass with diversified biochemical compositions can be transformed into a range of biofuels and value-added chemicals (Amulya et al., 2016). This integrated wastewater treatment and microalgal biorefinery results in a circular bio-economy, in which biomass conversion processes could be utilized to produce a range of biofuels and value-added products (Mohan et al., 2016). Caporgno et al. (2015) reported that *Chlorella kessleri* and *C. vulgaris* would remove 95% total nitrogen and 98% total phosphorus from urban wastewater, producing 2.7 g L^{-1} and 2.91 g L^{-1} algal biomass, respectively. The authors further reported that *C. kessleri* and *C. vulgaris* biomass could produce 0.074 g (biodiesel) g^{-1} and 346 mL (CH_4) g^{-1} and, 0.118 g (biodiesel) g^{-1} and 415 mL (CH_4) g^{-1}, respectively. The cultivation and remediation potential of *Spirulina platensis* in aquaculture wastewater was demonstrated by Wuang et al. (2016), with it showing 91.3% removal of ammonia and 70.2% chemical oxygen demand (COD). The accumulated algal biomass with 46%–58.3% carbohydrates (dcw) under mixotrophic conditions could be used as a potential biofertilizer. Thus, cultivation of microalgae in HRAPs utilizing wastewater from different sources not only will help in achieving cost-efficient cultivation but can also be used to produce a myriad of products.

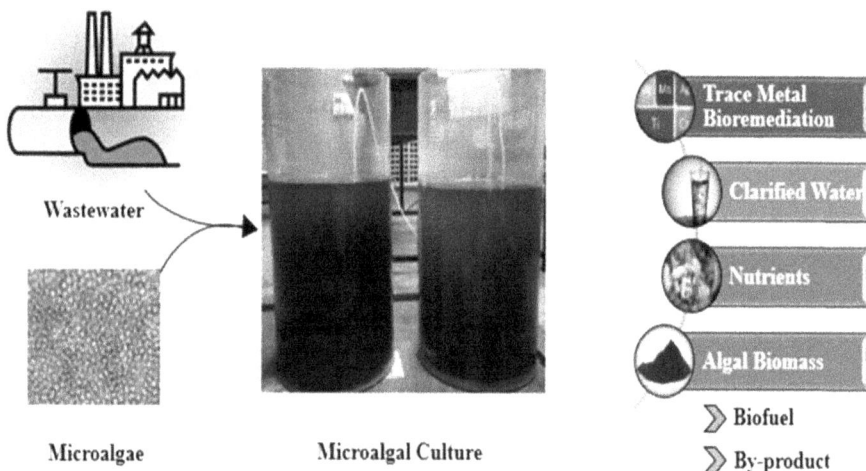

FIGURE 8.3   Co-integration of wastewater treatment with algal biomass cultivation.

8.5.4.2   *MICROALGAL CARBON DIOXIDE CAPTURE AND SEQUESTRATION*

The increased level of carbon dioxide emissions with the current global atmospheric value at 401.14 ppm, which is expected to soon reach an alarming value of 450 ppm with the rise of 2 ppm per year, is expected to have an adverse impact over the environment (Dineshbabu et al., 2017). Reports reveal that anthropogenic activities, especially indus-trial and vehicular emissions, are one of the major causes of increasing atmospheric CO_2 levels (contributing to about 50% of total emissions) leading to global warming (Cuellar-Bermudez et al., 2015). The melting of glaciers, increase in surface temperature, and rise in ocean levels due

TABLE 8.4 Wastewater Bioremediation Potential of Microalgae with the Biochemical Composition

Wastewater	Microalgae	Removal Efficiency	Biomass Characteristics	References
Urban wastewater	*Scenedesmus obliquus*	98% ammonium, 100% phosphorus 54% COD	25.2% lipids, 35.7% proteins, 42.6% sugars, 1% pigments	Batista et al., 2015
Urban wastewater	*Chlorella vulgaris*	96% ammonium, 77% phosphorus 36% COD	12.8% lipids, 34% proteins, 21.9% sugars, 0.5% pigments	Batista et al., 2015
Urban wastewater	Mixed algal consortium	99.5% ammonium, 100% phosphorus 3.8% COD	12.8% lipids, 45.3% proteins, 28.6% sugars, 0.8% pigments	Batista et al., 2015
Swine wastewater	*C. vulgaris* JSC 6	91.3% ammonium, 70.2% COD	58.3% sugars	Wang et al., 2015
Aquaculture wastewater	*Chlorella* sp.	77% ammonium 90% phosphorus 80% COD	23.7% lipids	Kuo et al., 2016
Piggery wastewater	*Chlorella vulgaris, Scenedesmus obliquus, Chlamydomonas reindhardtii*	99% ammonium, 82% phosphorus	34% carbohydrates; 35% lipids; 31% protein	Molinuevo-Salces et al., 2016
Piggery wastewater	*Chlorella vulgaris* (1.15 g l^{-1})	89.5% total nitrogen; 85.3% phosphorus	14% lipids; 28% proteins; 45% sugars	Wang et al., 2016
Pulp and paper industry wastewater	*Scenedesmus acuminatus* (8.2 g l^{-1})	99.9% ammonium 96.9% phosphorus	60.5% sugars	Tao et al., 2017
Municipal wastewater	*Chlorella zofingiensis* (2.5 g l^{-1})	93% total nitrogen; 90% phosphorus	26.9% carbohydrates; 25.46% lipids	Zhou et al., 2018
Domestic wastewater	*Chlorella vulgaris* (0.25 g l^{-1})	85% nitrogen; 35% phosphorus	32.7% lipids	Lam et al., 2017
Textile wastewater	*Chlorella* sp. G23 (0.008 g l^{-1})	84% ammonium; 78% COD	16.6% lipids	Wu et al., 2017
Daory wastewater	*Scenedesmus quadricauda* (0.47 g l^{-1})	86.21% nitrogen 89.83% phosphorus 64.47% COD	39.45% lipids	Daneshvar et al., 2018

to global warming have disturbed the food chain and led to the extinction of many species (Pires et al., 2012). Recently, techniques for CO_2 capture and storage (CCS) have grabbed the attention of researchers in order to trap GHG emissions from different sources. Different CCS strategies are currently being used to capture carbon from thermal power plants and cement industries, mostly relying on the principles of chemical looping combustion, adsorption and absorption by suitable solvents, membrane separation, and cryogenic distillations (Sreenivasulu et al., 2015; Wang et al., 2017). The real-time progress regarding the application of CO_2 sequestration strategies is quite slow, due to the excessive dependence on fossil fuels all over the world. Even when mitigation techniques are applied, the climate change effects would still continue, due to the resilient nature of the emitted CO_2 in the atmosphere (Moreira and Pires, 2016). Further, the conventional capture technologies are also associated with high maintenance and capital costs, with the problems of subsequent leakage over time (Kumar et al., 2011).

In lieu of the aforementioned challenges, to sort out the mitigation goals, negative emission technologies (NETs) are increasingly studied. NET implements methods to capture CO_2 from different locations at different point of times; further, the sequestration unit can be located near the diffused or the point sources of emissions. One of the most viable NETs is the biological sequestration of atmospheric CO_2 by microalgae. Microalgal carbon sequestration employs the process of photosynthetic carbon capture to assimilate biomass. The application of NET in the overall bioprocess cycle results in negative emissions. Further, it provides the flexibility of colocating an algal cultivation system inside the industrial premises, thereby reducing the land footprints (Moreira and Pires, 2016). Microalgal CO_2 sequestration has been widely applied in integration with industries to capture the flue gas generated from boilers, combined heat and power generation units, and lime kilns. The countries of Eastern and Western Europe, the United States, Canada, and Japan have deployed strategies of integrated industrial CO_2 sequestration (Ricci and Selosse, 2013). Table 8.5 enlists the carbon fixation efficiency of microalgal species that have been utilized to trap flue gas with different percentages of CO_2 at various industrial locations.

Gebreslassie et al. (2013) proposed the superstructure optimization strategy for a multiobjective microalgal hydrocarbon biorefinery with CO_2 sequestration from power plant flue gas. Gong and You (2014) demonstrated the design of an algal biorefinery framework for biological CO_2

TABLE 8.5 Carbon Sequestration Potential of Microalgae Using Flue Gas from Power Plants

Industrial Location	Strain	CO_2 fixation Efficiency (%)	References
Cogeneration Power Plant at MIT, Cambridge	*Dunaliella parva* (UTEXLB1983) *Dunaliella tertiolecta* (UTEXLB999)	82.3	Vunjak-Novakovic et al., 2005
Coke oven steel plant, China Corporation, Kaohsiung, Taiwan	*Chlorella* sp. MTF-7	60	Chiu et al., 2011
Oil-producing plant, West Bengal, India (15.6%)	*Chlorella sorokiniana*	4.1	Kumar et al., 2014
Coke oven plant (25% CO_2), China Steel Corporation, Southern Taiwan	*Chlorella* sp. MTF-15	25	Kao et al., 2014
Hot stove plant (26% CO_2), China Steel Corporation, Southern Taiwan	*Chlorella* sp. MTF-15	40	Kao et al., 2014
Power plant (24% CO_2), China Steel Corporation, Southern Taiwan	*Chlorella* sp. MTF-15	50	Kao et al., 2014
4 MW coal-fired boilers, Australian Country Choice Premises, Queensland, Australia (11.24% CO_2)	*Scenedesmus dimorphus* NT8c, *Chlorella vulgaris, Limnothrix planctonica*	12.3–14.9	Aslam et al., 2017
Coal burning unit at factory site of Pioneer Jellice Industries, Cuddalore, Tamil Nadu (15% CO_2)	*Phormidium valderianum*	56.4	Dineshbabu et al., 2017

sequestration, with processes encompassing purification of vent emissions, algal cultivation till harvesting, algal oil production, and upgrading and anaerobic digestion of remaining biomass to produce biogas. Microalgal CO_2 fixation not only will help mitigate GHG emissions but would also provide economic benefits to the society and the industry in a biorefinery concept, as shown in Figure 8.4.

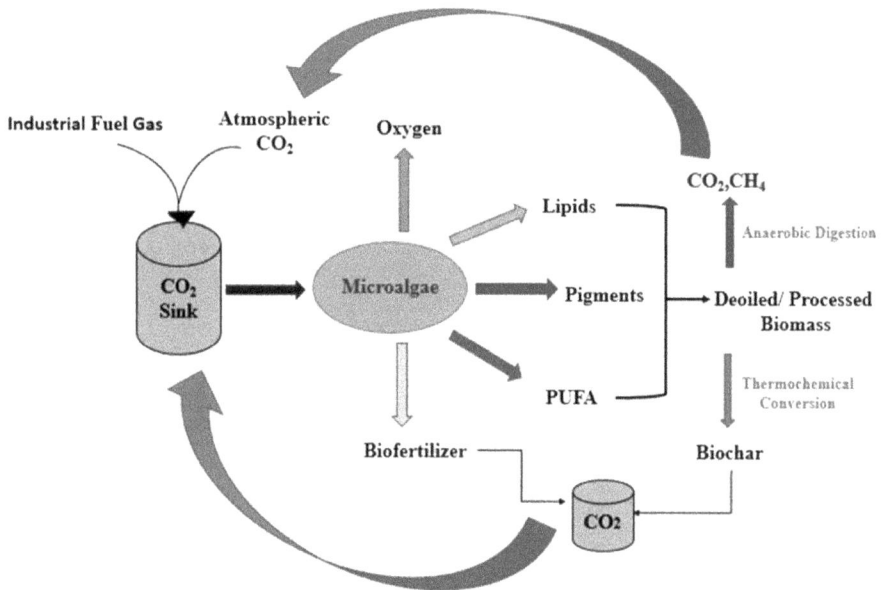

FIGURE 8.4   Algal biorefinery with biological carbon sequestration.

8.5.4.3   SOCIOECONOMIC BENEFITS

The major goal of microalgal biorefinery is reaping socioeconomic advantages via industrial symbiosis. An integrated microalgal industrial ecology will result in infrastructure development, as well as providing employment opportunities, especially to the rural population (Trivedi et al., 2015). An algal biorefinery produces energy in terms of heat or biofuels. These renewable biofuels generated will reduce the reliance on fuel imports, thus providing energy security, especially in developing countries (Gomiero, 2015; Milano et al., 2016). Algal biorefinery operated in a circular economy can also generate commercial benefits in terms of the production of high-value products like food supplements, fodder, cosmetics, vitamins, and bioplastics (Mohan et al., 2016). Microalgal biorefinery integrated with industry and operated as the multiproduct cascade system optimizes the material flows where the outputs of industries would be utilized in a productive manner, thus reducing the risks associated with the capital investments. Integration with power plants utilizes the heat for biomass drying and the power for mixing, thus reducing the overall energy requirements, leading to a higher NER. The decline in energy requirements under biorefinery approach will improve the cost economics of the process. Further, the integration of algal technology with the dairy industry would produce low-cost methanol (non-fossil based), which could be used as a substitute for synthetic methanol during microalgal lipid transesterification, reducing production costs (Trivedi et al., 2015). Process integration of algal technology with aquaculture industries could be used to produce proteins/ feed supplements to meet the growing protein demand in the market, thus earning benefits (Subhadra, 2011).

Economics of the microalgal technology can be further improved by using cheap resources during cultivation that occupies 20%–30% production costs. Integrating algal biorefineries with municipal or industrial water treatment plants would reduce the dependency on freshwater and synthetic chemicals as nutrients, thus bringing down the cultivation costs (Behera et al., 2019b). They will also provide a clarified effluent stream that could be utilized as a water resource in place of the water-treatment process of heat exchangers or washing equipment, thereby reducing the overall material requirements of the industry itself. The projected industrial CO_2 capture by microalgae can not only reduce GHG emissions but also be used to generate revenue via the international carbon credit exchange (Behera et al., 2019b). Carbon credit represents a commercial application-based approach of emission trading, which provides incentives for controlling the emissions of harmful gaseous pollutants to the industries (Kaufman et al., 2016). The United Nations runs the Clean Development Program (CDP) encouraging industrial development with simultaneous focus on meeting the demands of CO_2 reduction and carbon trade, as discussed in the Kyoto Protocol. Each credit equivalent to one metric ton CO_2 reduction can be sold in the global market (Singh et al., 2014).

Low-value algal biofuels, along with high-value by-products, can be coproduced through the microalgal biorefinery framework. While biofuel production can endure the high-value coproducts' extraction, these, on the other hand, increase the market value of biofuel. It is a well-known fact that the market demand for microalgal biofuels is large, but their economic value is low because of the high production and capital invest-ment costs. On the other hand, market value for the algal fine chemicals and bioactive compounds is high, despite the low market demands (Zhu and Hiltunen, 2016). Thus, cointegration of both these sectors could help to reduce the costs associated with land, water, and energy requirements, further earning revenues in terms of the sale of biofuels and high-value chemicals. Social and environmental protection achieved via waste reme-diation and reduced GHG emissions will improve the overall livelihood of people in the society.

8.6 CHALLENGES RELATED TO APPLICABILITY FOR MICROALGAL BIOREFINERY

Microalgal biorefinery approach promises products that could be harnessed in a vertically integrated closed-loop production pathway to provide a wide range of bio-based products (Trivedi et al., 2015). Even though the zero-discharge biorefinery concept promises attractive commercial benefits along with waste bioremediation, certain risk factors and uncertainties associated with the microalgal biorefinery must be rightly assessed (Mohan et al., 2016). There is a great insecurity associated with the economic and environmental aspects of algal biorefinery technology with respect to field application (De Bhowmick et al., 2018). Further, no proper market exists for the sale of microalgal biofuel or by-products (Zhu and Hiltunen, 2016). To promote the real-time integration and application of microalgal biorefinery, several pre-investigations are needed to identify and sort out the critical loopholes.

Resource assessment and feasibility analysis via the use of mathematical models must be performed to identify the geographical sites with suitable meteorological conditions for year-round cultivation of microalgae with significant biomass yields (Aly and Balasubramanian, 2017; Aly et al., 2017; Behera et al., 2019e). The theoretical mathematical model was simulated by Behera et al. (2018) to study the influence of climatologic variables and reactor-operating mode to pre-estimate the site-specific algal productivity and CO_2 sequestration potential. LCA and TEA are identified as essential strategies to evaluate the environmental and economic feasibility of microalgal biorefinery processes (Behera et al., 2019b; Juneja and Murthy, 2017). Gong and You (2016) have summarized the related challenges and futuristic solutions associated with LCA of microalgal biorefinery. These methods have been designed to identify the critical drivers/barriers and do further research to negate their effects. The research and development (R&D) activities must be carried out to identify the appropriate cultivation and harvesting strategy based on the desired end products. The residual fraction must be evaluated to identify the desirable characteristics for further applications. Review of different assessment considerations clearly shows that the algal biorefinery technique can be successful by robust selection of the appropriate engineering processes, starting from the cultivation, harvesting, and thereafter biomass conversion (De Bhowmick et al., 2018). There is a need to develop modified protocols and multiobjective optimization strategies to reduce the influence of pH, temperature, solvents, salts, and other contaminants that decrease functionality over time in the case of real-time applications. Suitable market and related subsidies and legislations must be identified for efficient policymaking. It is often assumed that the market for high-value products from microalgae would reach a saturation point, and recession might occur due to increased focus on biofuels that alone are not feasible (Zhu and Hiltunen, 2016).

The traditional microalgal industry for fodder, food supplements, and cosmetics must be integrated with the algal biofuel markets. Additional high-value products like bioplastics and biochar must be explored to improve the process economics, as one of the major barriers in making the algal biorefinery cost-inefficient lies in the presence of techniques for coproduct identification. The market legislations must be framed in such a manner that the algal industry produces products depending on the market size. The existing industrial power plants must be integrated with algal technology to expand the domain for achieving intangible environmental benefits in terms of wastewater treatment (WWT) and sequestration of CO_2 streams. Additional income can also be obtained in terms of reclaimed water and carbon credits. Policies toward GHG emissions and taxes can affect the microalgal biofuel practices. Government policies toward cleaner practices and GHG reduction via algal technology can generate funding, which would help to commercialize new technologies for biofuels and value-added products, thus reducing dependency on fuel imports. Further, it is expected to create employment opportunities for people, which, along with energy security, will help in establishing a long-term, sustainable and stable economy. Integration of these key strategies (Figure 8.5) will help in making the algal biorefinery concept a commercial reality.

- Site specific
  feasibility analysis
- Lifecycle assessment
- Techno-economic
  analysis

- Integrated facility
  development

- Optimization of
  operational
  conditions

Resource and Feasibility Assessment

Research and Development (R&D)

Policies and Legislations

Marketing Strategies

- Dissemination of
  knowledge
- Demonstration of
  projects

- Identification of
  market demands
- Market subsidies
- Financial support

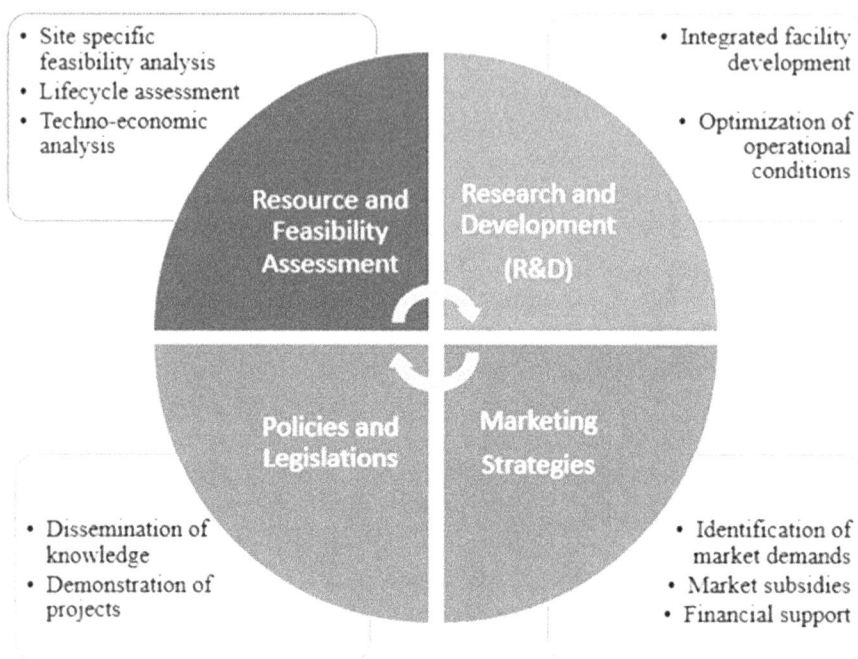

FIGURE 8.5   Strategies for promoting the microalgal biorefinery approach.

8.7   CONCLUDING REMARKS

Microalgal biorefinery has gained tremendous attention over recent years due to its enormous potential to act as a promising sustainable feedstock for a wide range of bioenergy sources and by-products, to be utilized in food, fodder, cosmetics, pharmaceuticals, and biomaterials. Commercial hurdles related to the high market costs of algal biofuel can be sorted out only through the implementation of the integrated zero-waste microalgal biorefinery concept. Engineering aspects and process integration strategies for cultivation, harvesting, and extraction of products in a cheap and energy-efficient framework is essential for innovative algal refinery. The methodical multivariate exploration of bioenergy with simultaneous utilization of high-value products will bring down the overall production costs. Industrial symbiosis will aid in ecological application via WWT and GHG reduction, thereby reducing environmental pollution. Integration of algal technology with the industries will provide a win-win situation, where the algal production cost can be brought down via the utilization of the energy, power, and waste resources; on the other hand, industry could earn profits in terms of revenues from the sale of biofuels and value-added chemicals. Additional economic returns would also be earned by the industries in terms of carbon tax. As a whole, the overall process would become sustainable and profitable. Societal benefits could be obtained by establishing a cleaner and greener environment, reducing reliance on energy imports, and generating more employment opportunities. Challenges related to the real-time application of the algal biorefinery model like resource feasibility, LCA, and TEA of the systematic pathway to be operated at a particular site must be resolved, and the demands and market-driving forces for algal products must be identified. Successful integration of the R&D activities in academia and industries, with the right implementation of government and corporate policies, would aid in establishing a sustainable bio-economy.

ACKNOWLEDGMENTS

The authors acknowledge the Ministry of Human Resources Development (MHRD), Government of India, for sponsoring the PhD programme of the first author and the Department of Biotechnology and Medical Engineering of National Institute of Technology (NIT), Rourkela, for providing the necessary research facilities.

KEYWORDS

- **biorefinery**
- **biofuels**
- **microalgae**
- **industrial symbiosis**
- **waste remediation**
- **sustainable development**

REFERENCES

Adesanya, V.O.; Cadena, E.; Scott, S.A.; Smith, A.G. Life cycle assessment on microalgal biodiesel production using a hybrid cultivation system. *Bioresource Technology* 2014, 163, 343–355.

Aly, N.; Balasubramanian, P. Effect of geographical coordinates on carbon dioxide sequestration potential by microalgae. *International Journal of Environmental Science and Development* 2017, 8, 147–152.

Aly, N.; Tarai, R.K.; Kale, P.G.; Paramasivan, B. Modelling the effect of photoinhibition on microalgal production potential in fixed and trackable photobioreactors in Odisha, India. *Current Science* 2017, 113, 272–283.

Amini, Z.; Ilham, Z.; Ong, H.C.; Mazaheri, H.; Chen, W.H. State of the art and prospective of lipase-catalyzed transesterification reaction for biodiesel production. *Energy Conversion and Management* 2017, 141, 339–353.

Amulya, K.; Dahiya, S.; Mohan, S.V. Building a bio-based economy through waste remediation: Innovation towards sustainable future. In: *Bioremediation and Bioeconomy*; Vara Prasad M.N., Ed.; Elsevier, the Netherlands, 2016; pp. 497.

Andersson, V.; Viklund, S.B.; Hackl, R.; Karlsson, M.; Berntsson, T. Algae-based biofuel production as part of an industrial cluster. *Biomass Bioenergy* 2014, 71, 113–124.

Ashok Kumar, V.; Salam, Z.; Tiwari, O.N.; Chinnasamy, S.; Mohammed, S.; Ani, F.N. An integrated approach for biodiesel and bioethanol production from *Scenedesmus bijugatus* cultivated in a vertical tubular photobioreactor. *Energy Conversion and Management* 2015, 101, 778–786.

Aslam, A.; Thomas-Hall, S.R.; Mughal, T.A.; Schenk, P.M. Selection and adaptation of microalgae to growth in 100% unfiltered coal-fired flue gas. *Bioresource Technology* 2017, 233, 271–283.

Bai, X.; Naghdi, F.G.; Ye, L.; Lant, P.; Pratt, S. Enhanced lipid extraction from algae using free nitrous acid pretreatment. *Bioresource Technology* 2014, 159, 36–40.

Balaji, S.; Gopi, K.; Muthuvelan, B. A review on production of poly β hydroxybutyrates from cyanobacteria for the production of bio plastics. *Algal Research* 2013, 2, 278–285.

Banskota, A.H.; Sperker, S.; Stefanova, R.; McGinn, P.J.; O'Leary, S.J. Antioxidant properties and lipid composition of selected microalgae. *Journal of Applied Phycology* 2018, 31, 309–318.

Barros, A.I.; Gonçalves, A.L.; Simões, M.; Pires, J.C. Harvesting techniques applied to microalgae: A review. *Renewable & Sustainable Energy Reviews* 2015, 41, 1489–1500.

Batista, A.P.; Moura, P.; Marques, P.A.; Ortigueira, J.; Alves, L.; Gouveia, L. *Scenedesmus obliquus* as feedstock for biohydrogen production by *Enterobacter aerogenes* and *Clostridium butyricum. Fuel* 2014, 117, 537–543.

Batista, A. P.; Ambrosano, L.; Graça, S.; Sousa, C.; Marques, P. A.; Ribeiro, B.; Botrel, E.P.; Neto, P.C.; Gouveia, L. (2015). Combining urban wastewater treatment with biohydrogen production–an integrated microalgae-based approach. *Bioresource Technology*, 184, 230–235.

Becker, E.W. Micro-algae as a source of protein. *Biotechnology Advances* 2007, 25, 207–210.

Begum, H.; Yusoff, F.M.; Banerjee, S.; Khatoon, H.; Shariff, M. Availability and utilization of pigments from microalgae. *Critical Reviews in Food Science and Nutrition* 2016, 56, 2209–2222.

Behera, B.; Aly, N.; Balasubramanian, P. Biophysical modeling of microalgal cultivation in open ponds. *Ecological Modelling* 2018, 388, 61–71.

Behera, B.; Acharya, A.; Gargey, I.A.; Aly, N.; Balasubramanian, P. Bioprocess engineering principles of microalgal cultivation for sustainable biofuel production. *Bioresource Technology Reports* 2019a, 5, 297–316.

Behera, B.; Aly, N.; Balasubramanian, P. Biophysical model and techno-economic assessment of carbon sequestration by microalgal ponds in Indian coal based power plants. *Journal of Cleaner Production* 2019b, 221, 587–597.

Behera, B.; Selvanayaki, S.; Jayabalan, R.; Balasubramanian, P. An in-silico approach for enhancing the lipid productivity in microalgae by manipulating the fatty acid biosynthesis. In: *Soft Computing for Problem Solving*; Bansal J., Das K., Nagar A., Deep K., Ojha A., Eds.; Springer, Singapore, 2019c, Vol. 816, p. 877.

Behera, B.; Balasubramanian, P. Natural plant extracts as an economical and ecofriendly alternative for harvesting microalgae. *Bioresource Technology* 2019d, 283, 45–52.

Behera, B.; Aly, N.; Rajkumar, M.A.; Balasubramanian, P. Theoretical estimation of the microalgal potential for biofuel production and carbon dioxide sequestration in India. In *Soft Computing for Problem Solving*; Bansal J., Das K., Nagar A., Deep K., Ojha A., Eds.; Springer, Singapore, 2019e, Vol. 816, p. 775.

Benavides, A.M.S.; Torzillo, G.; Kopecký, J.; Masojídek, J. Productivity and biochemical composition of *Phaeodactylum tricornutum* (Bacillariophyceae) cultures grown outdoors in tubular photobioreactors and open ponds. *Biomass Bioenergy* 2013, 54, 115–122.

Biller, P.; Sharma, B.K.; Kunwar, B.; Ross, A.B. Hydroprocessing of bio-crude from continuous hydrothermal liquefaction of microalgae. *Fuel* 2015, 159, 197–205.

Bird, M.I.; Wurster, C.M.; de Paula Silva, P.H.; Bass, A.M.; De Nys, R. Algal biochar–production and properties. *Bioresource Technology* 2011, 102, 1886–1891.

Bleakley, S.; Hayes, M. Algal proteins: Extraction, application, and challenges concerning production. *Foods* 2017, 6, 33–67.

Blifernez-Klassen, O.; Chaudhari, S.; Klassen, V.; Wördenweber, R.; Steffens, T.; Cholewa, D.; Niehaus, K.; Kalinowski, J; Kruse, O. Metabolic survey of *Botryococcus braunii*: Impact of the physiological state on product formation. *PLoS One* 2018, 13, 1–23.

Brennan, L.; Owende, P. Biofuels from microalgae—A review of technologies for production, processing, and extractions of biofuels and co-products. *Renewable & Sustainable Energy Reviews* 2010, 14, 557–577.

Burgard, A.P.; Pharkya, P.; Maranas, C.D. Optknock: A bilevel programming framework for identifying gene knockout strategies for microbial strain optimization. *Biotechnol Bioeng* 2003, 84(6), 647–657.

Caporgno, M.P.; Taleb, A.; Olkiewicz, M.; Font, J.; Pruvost, J.; Legrand, J.; Bengoa, C. Microalgae cultivation in urban wastewater: Nutrient removal and biomass production for biodiesel and methane. *Algal Research* 2015, 10, 232–239.

Carpine, R.; Du, W.; Olivieri, G.; Pollio, A.; Hellingwerf, K.J.; Marzocchella, A.; dos Santos, F.B. Genetic engineering of *Synechocystis* sp. PCC6803 for poly-β-hydroxybutyrate overproduction. *Algal Research* 2017, 25, 117–127.

Castro, Y.A.; Ellis, J.T.; Miller, C.D.; Sims, R.C. Optimization of wastewater microalgae saccharification using dilute acid hydrolysis for acetone, butanol, and ethanol fermentation. *Applied Energy* 2015, 140, 14–19.

Chang, Y.M.; Tsai, W.T.; Li, M.H. Chemical characterization of char derived from slow pyrolysis of microalgal residue. *Journal of Analytical and Applied Pyrolysis* 2015, 111, 88–93.

Chew, K.W.; Yap, J.Y.; Show, P.L.; Suan, N.H.; Juan, J.C.; Ling, T.C.; Lee, D.J.; Chang, J.S. Microalgae biorefinery: High value products perspectives. *Bioresource Technology* 2017, 229, 53–62.

Chiu, S.Y.; Kao, C.Y.; Huang, T.T.; Lin, C.J.; Ong, S.C.; Chen, C.D.; Chang J.S.; Lin, C.S. Microalgal biomass production and on-site bioremediation of carbon dioxide, nitrogen oxide and sulfur dioxide from flue gas using *Chlorella* sp. cultures. *Bioresource and Technology* 2011, 102, 9135–9142.

Chu, F.F.; Chu, P.N.; Shen, X.F.; Lam, P. K.; Zeng, R.J. Effect of phosphorus on biodiesel production from *Scenedesmus obliquus* under nitrogen-deficiency stress. *Bioresource Technology* 2014, 152, 241–246.

Cuellar-Bermudez, S.P.; Garcia-Perez, J.S.; Rittmann, B.E.; Parra-Saldivar, R. Photosynthetic bioenergy utilizing CO_2: An approach on flue gases utilization for third generation biofuels. *Journal of Cleaner Production* 2015, 98, 53–65.

Daneshvar, E.; Zarrinmehr, M.J.; Hashtjin, A.M.; Farhadian, O.; Bhatnagar, A. Versatile applications of freshwater and marine water microalgae in dairy wastewater treatment, lipid extraction and tetracycline biosorption. *Bioresource Technology* 2018, 268, 523–530.

Das, S.K.; Sathish, A.; Stanley, J. Production of biofuel and bioplastic from *Chlorella Pyrenoidosa*. *Materials Today: Proceedings* 2018, 5, 16774–16781.

Davis, R.; Markham, J.; Kinchin, C.; Grundl, N.; Tan, E.C.; Humbird, D. *Process design and economics for the production of algal biomass: Algal biomass production in open pond.* National Renewable Energy Laboratory; 2016.

De Bhowmick, G.; Sarmah, A.K.; Sen, R. Zero-waste algal biorefinery for bioenergy and biochar: A green leap towards achieving energy and environmental sustainability. *Science of the Total Environment* 2018, 650, 2467–2482.

Derakhshan, Z.; Ehrampoush, M. H.; Mahvi, A. H.; Dehghani, M.; Faramarzian, M.; Eslami, H. A comparative study of hybrid membrane photobioreactor and membrane photobioreactor for simultaneous biological removal of atrazine and CNP from wastewater: A performance analysis and modeling. *Chemical Engineering Journal* 2018, 355, 428–438.

Díaz, M.T.; Pérez, C.; Sánchez, C.I.; Lauzurica, S.; Cañeque, V.; González, C.; De La Fuente, J. Feeding microalgae increases omega 3 fatty acids of fat deposits and muscles in light lambs. *Journal of Food Composition and Analysis* 2017, 56, 115–123.

Dineshbabu, G.; Uma, V.S.; Mathimani, T.; Deviram, G.; Ananth, D.A.; Prabaharan, D.; Uma, L. On-site concurrent carbon dioxide sequestration from flue gas and calcite formation in ossein effluent by a marine cyanobacterium *Phormidium valderianum* BDU 20041. *Energy Conversion and Management* 2017, 141, 315–324.

Ding, L.; Cheng, J.; Xia, A.; Jacob, A.; Voelklein, M.; Murphy, J.D. Co-generation of biohydrogen and biomethane through two-stage batch co-fermentation of macro-and micro-algal biomass. *Bioresource Technology* 2016, 218, 224–231.

Elliott, D.C.; Biller, P.; Ross, A.B.; Schmidt, A.J; Jones, S.B. Hydrothermal liquefaction of biomass: Developments from batch to continuous process. *Bioresource Technology* 2015, 178, 147–156.

Enzing, C.; Ploeg, M.; Barbosa, M.; Sijtsma, L. Microalgae-based products for the food and feed sector: An outlook for Europe. *JRC Scientific and Policy Reports* 2014, 19–37.

Faried, M.; Samer, M.; Abdelsalam, E.; Yousef, R.S.; Attia, Y.A.; Ali, A.S. Biodiesel production from microalgae: Processes, technologies and recent advancements. *Renewable & Sustainable Energy Reviews* 2017, 79, 893–913.

Fernandez, F.A.; Sevilla, J.F.; Pérez, J.S.; Grima, E.M.; Chisti, Y. Airlift-driven external-loop tubular photobioreactors for outdoor production of microalgae: Assessment of design and performance. *Chemical Engineering Science* 2001, 56, 2721–2732.

Fu, W.; Chaiboonchoe, A.; Khraiwesh, B.; Nelson, D.; Al-Khairy, D.; Mystikou, A.; Alzahmi, A.; Salehi-Ashtiani, K. Algal cell factories: Approaches, applications, and potentials. *Marine Drugs* 2016, 14, 225–244.

Galadima, A.; Muraza, O. Biodiesel production from algae by using heterogeneous catalysts: A critical review. *Energy* 2014, 78, 72–83.

Gattrel, S.; Lum, K.; Kim, J.; Lei, X.G. Potential of defatted microalgae from biofuel industry as an ingredient to replace corn and soybean meal in swine and poultry diets. *Journal of Animal Science* 2014, 92, 1306–1314.

Gebreslassie, B.H.; Waymire, R.; You, F. Sustainable design and synthesis of algae-based biorefinery for simultaneous hydrocarbon biofuel production and carbon sequestration. *AIChE Journal* 2013, 59, 1599–1621.

Gerber, L.N.; Tester, J.W.; Beal, C.M.; Huntley, M.E.; Sills, D.L. Target cultivation and financing parameters for sustainable production of fuel and feed from microalgae. *Environmental Science and Technology* 2016, 50, 3333–3341.

Gollakota, A.R.K.; Kishore, N.; Gu, S.A. review on hydrothermal liquefaction of biomass. *Renewable and Sustainable Energy Reviews* 2017, 81, 1378–1392.

Gomiero, T. Are biofuels an effective and viable energy strategy for industrialized societies? A reasoned overview of potentials and limits. *Sustainability* 2015, 7, 8491–8521.

Gong, J.; You, F. Optimal design and synthesis of algal biorefinery processes for biological carbon sequestration and utilization with zero direct greenhouse gas emissions: MINLP model and global optimization algorithm. *Industrial & Engineering Chemistry Research* 2014, 53, 1563–1579.

Gong, J.; You, F. Life cycle algal biorefinery design. In: *Alternative Energy Sources and Technologies*; Martín, M., Ed.; Springer, Cham, 2016; p. 363.

González-Fernández, C.; Molinuevo-Salces, B.; García-González, M.C. Evaluation of anaerobic codigestion of microalgal biomass and swine manure via response surface methodology. *Applied Energy* 2011, 88, 3448–3453.

Hathwaik, L.T.; Cushman, J.C.; Love, J.; Bryant, J.A. Strain selection strategies for improvement of algal biofuel feedstocks. In: *Biofuels Bioenergy*; John Wiley & Sons; 2017, pp. 173–189.

Haznedaroglu, B.Z.; Rismani-Yazdi, H.; Allnutt, F.C.T.; Reeves, D.; Peccia, J. Algal biorefinery for high-value platform chemicals. In: *Platform Chemical Biorefinery*; Brar S.K., Sharma S.J., Pakshirajan, K., Eds.; Elsevier, the Netherlands; 2017, p. 333.

He, Z.Z.; Qi, H.; He, M.J.; Ruan, L.M. Experimental research on the photobiological hydrogen production kinetics of *Chlamydomonas reinhardtii* GY-D55. *International Journal of Hydrogen Energy* 2016, 41, 15651–15660.

Henchion, M.; Hayes, M.; Mullen, A.M.; Fenelon, M.; Tiwari, B. Future protein supply and demand: Strategies and factors influencing a sustainable equilibrium. *Foods* 2017, 6, 53–74.

Hernández, D.; Riaño, B.; Coca, M.; Solana, M.; Bertucco, A.; García-González, M.C. Microalgae cultivation in high rate algal ponds using slaughterhouse wastewater for biofuel applications. *Chemical Engineering Journal* 2016, 285, 449–458.

Ho, S.H.; Huang, S.W.; Chen, C.Y.; Hasunuma, T.; Kondo, A.; Chang, J.S. Bioethanol production using carbohydrate-rich microalgae biomass as feedstock. *Bioresource Technology* 2013, 135, 191–198.

Hondo, S.; Takahashi, M.; Osanai, T.; Matsuda, M.; Hasunuma, T.; Tazuke, A.; Nakahira, Y.; Chohnan, S.; Hasegawa, M.; Asayama, M. Genetic engineering and metabolite profiling for overproduction of polyhydroxybutyrate in cyanobacteria. *Journal of Bioscience and Bioengineering* 2015, 120, 510–517.

Hu, Y.; Feng, S.; Yuan, Z.; Xu, C.C.; Bassi, A. Investigation of aqueous phase recycling for improving bio-crude oil yield in hydrothermal liquefaction of algae. *Bioresource Technology* 2017, 239, 151–159.

Hudek, K.; Davis, L.C.; Ibbini, J.; Erickson, L. Commercial products from algae. In *Algal Biorefineries*; Bajpai, R., Prokop, A., Zappi, M., Eds; Springer, Dordrecht; 2014, p. 275.

Huntley, M.E.; Redalje, D.G. CO_2 mitigation and renewable oil from photosynthetic microbes: A new appraisal. *Mitigation and Adaptation Strategies for Global Change* 2007, 12, 573–608.

Hwang, J.H.; Church, J.; Lee, S.J.; Park, J.; Lee, W.H. Use of microalgae for advanced wastewater treatment and sustainable bioenergy generation. *Environmental Engineering Science* 2016, 33, 882–897.

Iwata, T. Biodegradable and bio-based polymers: Future prospects of eco-friendly plastics. In: *Angewandte Chemie International*; Ed; Wiley, 2015, Vol. 54, pp. 3210–3215.

Johansson, C.L.; Paul, N.A.; de Nys, R.; Roberts, D.A. Simultaneous biosorption of selenium, arsenic and molybdenum with modified algal-based biochars. *Journal of Environmental Management* 2016, 165, 117–123.

John, R.P.; Anisha, G.S.; Nampoothiri, K.M.; Pandey, A. Micro and macroalgal biomass: A renewable source for bioethanol. *Bioresource Technology* 2011, 102, 186–193.

Juneja, A.; Murthy, G.S. Evaluating the potential of renewable diesel production from algae cultured on wastewater: Techno-economic analysis and life cycle assessment. *AIMS Energy* 2017, 5, 239–257.

Juneja, A.; Ceballos, R.M.; Murthy, G.S. Effects of environmental factors and nutrient availability on the biochemical composition of algae for biofuels production: A review. *Energies* 2013, 6, 4607–4638.

Kaewbai-ngam, A.; Incharoensakdi, A.; Monshupanee, T. Increased accumulation of polyhydroxybutyrate in divergent cyanobacteria under nutrient-deprived photoautotrophy: An efficient conversion of solar energy and carbon dioxide to polyhydroxybutyrate by *Calothrix scytonemicola* TISTR 8095. *Bioresource Technology* 2016, 212, 342–347.

Kao, C.Y.; Chen, T.Y.; Chang, Y.B.; Chiu, T.W.; Lin, H.Y.; Chen, C.D.; Chang J.S.; Lin, C. S. Utilization of carbon dioxide in industrial flue gases for the cultivation of microalga *Chlorella* sp. *Bioresource Technology* 2014, 166, 485–493.

Karemore, A.; Pal, R.; Sen, R. Strategic enhancement of algal biomass and lipid in *Chlorococcum infusionum* as bioenergy feedstock. *Algal Research* 2013, 2, 113–121.

Kaspar, H.F.; Keys, E.F.; King, N.; Smith, K.F.; Kesarcodi-Watson, A.; Miller, M.R. Continuous production of *Chaetoceros calcitrans* in a system suitable for commercial hatcheries. *Aquaculture* 2014, 420, 1–9.

Kaufman, N.; Obeiter, M.; Krause, E. Putting a price on carbon: Reducing emissions. *World Resources Institute Issue Brief*, January 2016, pp. 1–36.

Khanra, S.; Mondal, M.; Halder, G.; Tiwari, O.N.; Gayen, K.; Bhowmick, T.K. Downstream processing of microalgae for pigments, protein and carbohydrate in industrial application: A review. *Food and Bioproducts Processing* 2018, 110, 60–84.

Khosravi-Darani, K.; Mokhtari, Z.B.; Amai, T.; Tanaka, K. Microbial production of poly (hydroxybutyrate) from C1 carbon sources. *Applied Microbiology and Biotechnology* 2013, 97, 1407–1424.

Kim, J.; Reed, J. L. OptORF: Optimal metabolic and regulatory perturbations for metabolic engineering of microbial strains. *BMC Systems Biology*, 2010, 4(1), 53–72.

Koller, M.; Design of closed photobioreactors for algal cultivation. In: *Algal Biorefineries*; Bajpai, R.; Prokop, A.; Zappi, M., Eds; Springer, Dordrecht; 2015, Vol. 1, p. 133.

Kotrbáček, V.; Doubek, J.; Doucha, J. The chlorococcalean alga *Chlorella* in animal nutrition: A review. *Journal of Applied Phycology* 2015, 27, 2173–2180.

Kumar, R.; Pal, P. Assessing the feasibility of N and P recovery by struvite precipitation from nutrient-rich wastewater: A review. *Environmental Science and Pollution Research* 2015, 22, 17453–17464.

Kumar, K.; Banerjee, D.; Das, D. Carbon dioxide sequestration from industrial flue gas by *Chlorella sorokiniana*. *Bioresource Technology* 2014, 152, 225–233.

Kumar, K.; Dasgupta, C.N.; Nayak, B.; Lindblad, P.; Das, D. Development of suitable photobioreactors for CO_2 sequestration addressing global warming using green algae and cyanobacteria. *Bioresource Technology* 2011, 102, 4945–4953.

Kumar, G.; Shobana, S.; Chen, W.H.; Bach, Q.V.; Kim, S.H.; Atabani, A.E.; Chang, J.S. A review of thermochemical conversion of microalgal biomass for biofuels: Chemistry and processes. *Green Chemistry* 2017, 19, 44–67.

Kuo, C.M.; Jian, J.F.; Lin, T.H.; Chang, Y.B.; Wan, X.H.; Lai, J.T.; Chang, J.S.; Lin, C.S. Simultaneous microalgal biomass production and CO_2 fixation by cultivating *Chlorella* sp. GD with aquaculture wastewater and boiler flue gas. *Bioresource Technology* 2016, 221, 241–250.

Lam, M.K.; Yusoff, M.I.; Uemura, Y.; Lim, J.W.; Khoo, C.G.; Lee, K.T.; Ong, H.C. Cultivation of *Chlorella vulgaris* using nutrients source from domestic wastewater for biodiesel production: Growth condition and kinetic studies. *Renewable Energy* 2017, 103, 197–207.

Lee, O.K.; Oh, Y.K.; Lee, E.Y. Bioethanol production from carbohydrate-enriched residual biomass obtained after lipid extraction of *Chlorella* sp. KR-1. *Bioresource Technology* 2015, 196, 22–27.

Leite, G.B.; Paranjape, K.; Abdelaziz, A.E.; Hallenbeck, P.C. Utilization of biodiesel-derived glycerol or xylose for increased growth and lipid production by indigenous microalgae. *Bioresource Technology* 2015, 184, 123–130.

Lerat, Y.; Cornish, M.L.; Critchley, A.T. Applications of algal biomass in global food and feed markets: From traditional usage to the potential for functional products. In: *Blue Biotechnology: Production and Use of Marine Molecules*; La Barre, S., Bates, S.S., Eds.; John Wiley & Sons, 2018, Vol. 1, p. 143.

Markou, G.; Angelidaki, I.; Georgakakis, D. Microalgal carbohydrates: An overview of the factors influencing carbohydrates production, and of main bioconversion technologies for production of biofuels. *Applied Microbiology and Biotechnology* 2012, 96, 631–645.

Mata, T.M.; Martins, A.A.; Caetano, N.S. Microalgae for biodiesel production and other applications: A review. *Renewable and Sustainable Energy Reviews* 2010, 14, 217–232.

Maurya, R.; Chokshi, K.; Ghosh, T.; Trivedi, K.; Pancha, I.; Kubavat, D.; Mishra, S.; Ghosh, A. Lipid extracted microalgal biomass residue as a fertilizer substitute for *Zea mays* L. *Frontiers in Plant Science* 2016a, 6, 1–10.

Maurya, R.; Paliwal, C.; Ghosh, T.; Pancha, I.; Chokshi, K.; Mitra, M.; Ghosh, A.; Mishra, S. Applications of de-oiled microalgal biomass towards development of sustainable biorefinery. *Bioresource Technology* 2016b, 214, 787–796.

Mehrabadi, A.; Craggs, R.; Farid, M.M. Wastewater treatment high rate algal ponds (WWT HRAP) for low-cost biofuel production. *Bioresource Technology* 2015, 184, 202–214.

Milano, J.; Ong, H.C.; Masjuki, H.H.; Chong, W.T.; Lam, M.K.; Loh, P.K.; Vellayan, V. Microalgae biofuels as an alternative to fossil fuel for power generation. *Renewable and Sustainable Energy Reviews* 2016, 58, 180–197.

Mishra, V.; Dubey, A.; Prajapti, S.K. Algal biomass pretreatment for improved biofuel production. In: *Algal Biofuels*; Gupta, S.K., Malik, A., Bux, F., Ed.; Springer, Cham, 2017, p. 259.

Mohan, S.V.; Nikhil, G.N.; Chiranjeevi, P.; Reddy, C.N.; Rohit, M.V.; Kumar, A.N.; Sarkar, O. Waste biorefinery models towards sustainable circular bioeconomy: Critical review and future perspectives. *Bioresource Technology* 2016, 215, 2–12.

Mohr, A.; Raman, S. Lessons from first generation biofuels and implications for the sustainability appraisal of second generation biofuels. *Energy Policy* 2013, 63, 114–122.

Molinuevo-Salces, B.; Mahdy, A.; Ballesteros, M.; González-Fernández, C. From piggery wastewater nutrients to biogas: Microalgae biomass revalorization through anaerobic digestion. *Renewable Energy* 2016, 96, 1103–1110.

Moreira, D.; Pires, J.C. Atmospheric CO_2 capture by algae: Negative carbon dioxide emission path. *Bioresource Technology* 2016, 215, 371–379.

Moreno-Garcia, L.; Adjallé, K.; Barnabé, S.; Raghavan, G.S.V. Microalgae biomass production for a biorefinery system: Recent advances and the way towards sustainability. *Renewable and Sustainable Energy Reviews* 2017, 76, 493–506.

Mourelle, M.L.; Gómez, C.P.; Legido, J.L. The potential use of marine microalgae and cyanobacteria in cosmetics and thalassotherapy. *Cosmetics* 2017, 4, 46–60.

Narala, R.R.; Garg, S.; Sharma, K.K.; Thomas-Hall, S R.; Deme, M.; Li, Y.; Schenk, P.M. Comparison of microalgae cultivation in photobioreactor, open raceway pond, and a two-stage hybrid system. *Frontiers in Energy Research* 2016, 4, 29–39.

Naumann, T.; Çebi, Z.; Podola, B.; Melkonian, M. Growing microalgae as aquaculture feeds on twin-layers: A novel solid-state photobioreactor. *Journal of Applied Phycology* 2013, 25, 1413–1420.

Noraini, M.Y.; Ong, H.C.; Badrul, M.J.; Chong, W.T. A review on potential enzymatic reaction for biofuel production from algae. *Renewable and Sustainable Energy Review* 2014, 39, 24–34.

Noreen, A.; Zia, K.M.; Zuber, M.; Ali, M.; Mujahid, M. A critical review of algal biomass: A versatile platform of bio-based polyesters from renewable resources. *International Journal of Biological Macromolecules* 2016, 86, 937–949.

Park, J.Y.; Park, M.S.; Lee, Y.C.; Yang, J.W. Advances in direct transesterification of algal oils from wet biomass. *Bioresource Technology* 2015, 184, 267–275.

Passos, F.; Uggetti, E.; Carrère, H.; Ferrer, I. Pretreatment of microalgae to improve biogas production: A review. *Bioresource Technology* 2014, 172, 403–412.

Pawar, S. Effectiveness mapping of open raceway pond and tubular photobioreactors for sustainable production of microalgae biofuel. *Renewable and Sustainable Energy Review* 2016, 62, 640–653.

Pignolet, O.; Jubeau, S.; Vaca-Garcia, C.; Michaud, P. Highly valuable microalgae: Biochemical and topological aspects. *Journal of Industrial Microbiology & Biotechnology* 2013, 40, 781–796.

Pires, J.C.M.; Alvim-Ferraz, M.C.M.; Martins, F.G.; Simoes, M. Carbon dioxide capture from flue gases using microalgae: Engineering aspects and biorefinery concept. *Renewable and Sustainable Energy Review* 2012, 16, 3043–3053.

Rahman, A.; Miller, C.D. Microalgae as a source of bioplastics. In: *Algal Green Chemistry*; Rastogi, R., Madamwar, D., Pandey, A., Eds.; Elsevier, the Netherlands; 2017, p. 121.

Rahman, A.; Putman, R.J.; Inan, K.; Sal, F.A.; Sathish, A.; Smith, T.T.; Nielsen, C.; Sims, R.C.; Miller, C.D. Polyhydroxybutyrate production using a wastewater microalgae based media. *Algal Research* 2015, 8, 95–98.

Rajendran, A.; Hu, B. Mycoalgae biofilm: Development of a novel platform technology using algae and fungal cultures. *Biotechnology for Biofuels* 2016, 9, 112–125.

Rangabhashiyam, S.; Behera, B.; Aly, N.; Balasubramanian, P. Biodiesel from microalgae as a promising strategy for renewable bioenergy production-A review. *Journal of Environment and Biotechnology Research* 2017, 6, 260–269.

Ranjith Kumar, R.; Hanumantha Rao, P.; Arumugam, M. Lipid extraction methods from microalgae: A comprehensive review. *Frontiers in Energy Research* 2015, 2, 1–9.

Rashidi, B.; Trindade, L.M. Detailed biochemical and morphologic characteristics of the green microalga *Neochloris oleoabundans* cell wall. *Algal Research* 2018, 35, 152–159.

Ricci, O.; Selosse, S. Global and regional potential for bioelectricity with carbon capture and storage. *Energy Policy* 2013, 52, 689–698.

Rodrigues, D.B.; Menezes, C.R.; Mercadante, A.Z.; Jacob-Lopes, E.; Zepka, L. Q. Bioactive pigments from microalgae *Phormidium autumnale*. *Food Research International* 2015, 77, 273–279.

Safafar, H.; Van Wagenen, J.; Møller, P.; Jacobsen, C. Carotenoids, phenolic compounds and tocopherols contribute to the antioxidative properties of some microalgae species grown on industrial wastewater. *Marine Drugs* 2015, 13, 7339–7356.

Santos, F.M.; Pires, J.C. Nutrient recovery from wastewaters by microalgae and its potential application as bio-char. *Bioresource Technology* 2018, 267, 725–731.

Sarkar, O.; Agarwal, M.; Kumar, A.N.; Mohan, S.V. Retrofitting hetrotrophically cultivated algae biomass as pyrolytic feedstock for biogas, bio-char and bio-oil production encompassing biorefinery. *Bioresource Technology* 2015, 178, 132–138.

Sengmee, D.; Cheirsilp, B.; Suksaroge, T.T.; Prasertsan, P. Biophotolysis-based hydrogen and lipid production by oleaginous microalgae using crude glycerol as exogenous carbon source. *International Journal of Hydrogen Energy* 2017, 42, 1970–1976.

Sharma, Y.C.; Singh, V. Microalgal biodiesel: A possible solution for India's energy security. *Renewable and Sustainable Energy Review* 2017, 67, 72–88.

Shurin, J.B.; Burkart, M.D.; Mayfield, S.P.; Smith, V.H. Recent progress and future challenges in algal biofuel production. *F1000Research* 2016, 5, 1–7.

Singh, S.K.; Dixit, K.; Sundaram, S. Algal-based CO_2 sequestration technology and global scenario of carbon credit market: A review. *American Journal of Engineering Research* 2014, 3, 35–37.

Singh, A.; Nigam, P.S.; Murphy, J.D. Renewable fuels from algae: An answer to debatable land based fuels. *Bioresource Technology* 2011, 102, 10–16.

Skorupskaite, V.; Makareviciene, V.; Gumbyte, M. Opportunities for simultaneous oil extraction and transesterification during biodiesel fuel production from microalgae: A review. *Fuel Processing Technology* 2016, 150, 78–87.

Spolaore, P.; Joannis-Cassan, C.; Duran, E.; Isambert, A. Commercial applications of microalgae. *Journal of Bioscience and Bioengineering* 2006, 101, 87–96.

Sreenivasulu, B.; Gayatri, D.V.; Sreedhar, I.; Raghavan, K.V. A journey into the process and engineering aspects of carbon capture technologies. *Renewable and Sustainable Energy Review* 2015, 41, 1324–1350.

Subhadra, B. Algal biorefinery-based industry: An approach to address fuel and food insecurity for a carbon-smart world. *Journal of Science of Food and Agriculture* 2011, 91, 2–13.

Sun, H.; Lu, H.; Chu, L.; Shao, H.; Shi, W. Biochar applied with appropriate rates can reduce N leaching, keep N retention and not increase NH_3 volatilization in a coastal saline soil. *Science of the Total Environment* 2017, 575, 820–825.

Sy, C.L.; Ubando, A.T.; Aviso, K.B.; Tan, R.R. Multiobjective target oriented robust optimization for the design of an integrated biorefinery. *Journal of Cleaner Production* 2018, 170, 496–509.

Tag, A.T.; Duman, G.; Ucar, S.; Yanik, J. Effects of feedstock type and pyrolysis temperature on potential applications of biochar. *Journal of Analytical and Applied Pyrolysis* 2016, 120, 200–206.

Tandon, P.; Jin, Q. Microalgae culture enhancement through key microbial approaches. *Renewable and Sustainable Energy Review* 2017, 80, 1089–1099.

Tao, L.; Barcus, M.; Lei, X.G. *Can Feeding Defatted Microalgae Produce Healthier Animal Foods?* Cornell University; 2015, pp. 1–4.

Tao, R.; Kinnunen, V.; Praveenkumar, R.; Lakaniemi, A.M.; Rintala, J.A. Comparison of *Scenedesmus acuminatus* and *Chlorella vulgaris* cultivation in liquid digestates from anaerobic digestion of pulp and paper industry and municipal wastewater treatment sludge. *Journal of Applied Phycology* 2017, 29, 2845–2856.

Thilakaratne, R.; Wright, M.M.; Brown, R.C. A techno-economic analysis of microalgae remnant catalytic pyrolysis and upgrading to fuels. *Fuel* 2014, 128, 104–112.

Tramontin, D. P.; Gressler, P. D.; Rörig, L.R.; Derner, R.B.; Pereira-Filho, J.; Radetski, C.M.; Quadri, M.B. Growth modeling of the green microalga *Scenedesmus obliquus* in a hybrid photobioreactor as a practical tool to understand both physical and biochemical phenomena in play during algae cultivation. *Biotechnology and Bioengineering* 2018, 115, 965–977.

Tredici, M.R.; Rodolfi, L.; Biondi, N.; Bassi, N.; Sampietro, G. Techno-economic analysis of microalgal biomass production in a 1-ha Green Wall Panel (GWP®) plant. *Algal Research* 2016, 19, 253–263.

Trivedi, J.; Aila, M.; Bangwal, D.P.; Kaul, S.; Garg, M.O. Algae based biorefinery—How to make sense? *Renewable and Sustainable Energy Review* 2015, 47, 295–307.

Uggetti, E.; Passos, F.; Solé, M.; Garfí, M.; Ferrer, I. Recent achievements in the production of biogas from microalgae. *Waste and Biomass Valorization* 2017, 8, 129–139.

Ullah, K.; Ahmad, M.; Sharma, V.K.; Lu, P.; Harvey, A.; Zafar, M.; Sultana, S. Assessing the potential of algal biomass opportunities for bioenergy industry: A review. *Fuel* 2015, 143, 414–423.

Vardon, D.R.; Sharma, B.K.; Blazina, G.V.; Rajagopalan, K.; Strathmann, T.J. Thermochemical conversion of raw and defatted algal biomass via hydrothermal liquefaction and slow pyrolysis. *Bioresource Technology* 2012, 109, 178–187.

Vizcaíno, A.J.; López, G.; Sáez, M.I.; Jiménez, J.A.; Barros, A.; Hidalgo, L.; Camacho-Rodríguez, J.; Martínez, T.F.; Cerón-García, M.C; Alarcón, F.J. Effects of the microalga *Scenedesmus almeriensis* as fishmeal alternative in diets for gilthead sea bream, *Sparus aurata*, juveniles. *Aquaculture* 2014, 431, 34–43.

Vunjak-Novakovic, G.; Kim, Y.; Wu, X.; Berzin, I.; Merchuk, J.C. Air-lift bioreactors for algal growth on flue gas: Mathematical modeling and pilot-plant studies. *Industrial and Engineering Chemistry Research* 2005, 44, 6154–6163.

Wagner, A.; Zarecki, R.; Reshef, L.; Gochev, C.; Sorek, R.; Gophna, U.; Ruppin, E. Computational evaluation of cellular metabolic costs successfully predicts genes whose expression is deleterious. *Proc Natl Acad Sci* 2016, 110(47), 19166-19171.

Wang, K.; Brown, R.C.; Homsy, S.; Martinez, L.; Sidhu, S.S. Fast pyrolysis of microalgae remnants in a fluidized bed reactor for bio-oil and biochar production. *Bioresource Technology* 2013, 127, 494–499

Wang, Y.; Zhao, L.; Otto, A.; Robinius, M.; Stolten, D. A review of post-combustion CO_2 capture technologies from coal-fired power plants. *Energy Procedia* 2017, 114, 650–665.

Wang, M.; Yang, Y.; Chen, Z.; Chen, Y.; Wen, Y.; Chen, B. Removal of nutrients from undiluted anaerobically treated piggery wastewater by improved microalgae. *Bioresource Technology* 2016, 222, 130–138.

Wang, Y.; Guo, W.; Yen, H.W.; Ho, S.H.; Lo, Y.C.; Cheng, C.L.; Ren, N.; Chang, J.S. Cultivation of *Chlorella vulgaris* JSC-6 with swine wastewater for simultaneous nutrient/COD removal and carbohydrate production. *Bioresource Technology* 2015, 198, 619–625.

Wen, X.; Du, K.; Wang, Z.; Peng, X.; Luo, L.; Tao, H.; Zhu, Y.; Zhang, D.; Geng, Y.; Li, Y. Effective cultivation of microalgae for biofuel production: A pilot-scale evaluation of a novel oleaginous microalga *Graesiella* sp. WBG-1. *Biotechnology for Biofuels* 2016, 9, 123–135.

Wieczorek, N.; Kucuker, M.A.; Kuchta, K. Fermentative hydrogen and methane production from microalgal biomass *(Chlorella vulgaris)* in a two-stage combined process. *Applied Energy* 2014, 132, 108–117.

Wirth, R.; Lakatos, G.; Maróti, G.; Bagi, Z.; Minárovics, J.; Nagy, K.; Kondorosi, E.; Rakhely, G.; Kovács, K.L. Exploitation of algal-bacterial associations in a two-stage biohydrogen and biogas generation process. *Biotechnology for Biofuels* 2015, 8, 59–73.

Wu, J.Y.; Lay, C.H.; Chen, C.C.; Wu, S.Y. Lipid accumulating microalgae cultivation in textile wastewater: Environmental parameters optimization. *Journal of the Taiwan Institute of Chemical Engineers* 2017, 79, 1–6.

Wuang, S.C.; Khin, M.C.; Chua, P.Q.D.; Luo, Y.D. Use of *Spirulina* biomass produced from treatment of aquaculture wastewater as agricultural fertilizers. *Algal Research* 2016, 15, 59–64.

Yan, L.; Lim, S.U.; Kim, I.H. Effect of fermented chlorella supplementation on growth performance, nutrient digestibility, blood characteristics, fecal microbial and fecal noxious gas content in growing pigs. *Asian-Australasian Journal of Animal Science* 2012, 25, 1742–1747.

Yen, H.W.; Hu, I.C.; Chen, C.Y.; Ho, S.H.; Lee, D.J.; Chang, J.S. Microalgae-based biorefinery–from biofuels to natural products. *Bioresource Technology* 2013, 135, 166–174.

Yu, K.L.; Lau, B.F.; Show, P.L.; Ong, H.C.; Ling, T.C.; Chen, W.H.; Ng, E.P.; Chang, J.S. Recent developments on algal biochar production and characterization. *Bioresource Technology* 2017, 246, 2–11.

Yu, K.L.; Show, P.L.; Ong, H.C.; Ling, T.C.; Chen, W.H.; Salleh, M.A.M. Biochar production from microalgae cultivation through pyrolysis as a sustainable carbon sequestration and biorefinery approach. *Clean Technologies and Environmental Policy* 2018, 20, 1–9.

Zakaria, S.M.; Kamal, S.M.M. Subcritical water extraction of bioactive compounds from plants and algae: Applications in pharmaceutical and food ingredients. *Food Engineering Reviews* 2016, 8, 23–34.

Zeller, M.A.; Hunt, R.; Jones, A.; Sharma, S. Bioplastics and their thermoplastic blends from *Spirulina* and *Chlorella microalgae*. *Journal of Applied Polymer Science* 2015, 130, 3263–3275.

Zhao, Y.; Wang, Y.; Li, Y.; Santschi, P.H.; Quigg, A. Response of photosynthesis and the antioxidant defense system of two microalgal species (*Alexandrium minutum* and *Dunaliella salina*) to the toxicity of BDE-47. *Marine Pollution Bulletin* 2017, 124, 459–469.

Zheng, H.; Guo, W.; Li, S.; Chen, Y.; Wu, Q.; Feng, X; Yin, R.; Ho, S.H.; Ren, N.; Chang, J. S. Adsorption of p-nitrophenols (PNP) on microalgal biochar: Analysis of high adsorption capacity and mechanism. *Bioresource Technology* 2017, 244, 1456–1464.

Zhou, W.; Wang, Z.; Xu, J.; Ma, L. Cultivation of microalgae *Chlorella zofingiensis* on municipal wastewater and biogas slurry towards bioenergy. *Journal of Bioscience and Bioengineering* 2018, 126, 644–648.

Zhu, L. Biorefinery as a promising approach to promote microalgae industry: An innovative framework. *Renewable and Sustainable Energy Review* 2015, 41, 1376–1384.

Zhu, L.D.; Hiltunen, E. Application of livestock waste compost to cultivate microalgae for bioproducts production: A feasible framework. *Renewable and Sustainable Energy Review* 2016, 54, 1285–1290.

CHAPTER 9

Fruit Peel Waste "Biorefinery" and Sustainability Issues

PRANAV D. PATHAK[1,2], SAURABH N. JOGLEKAR[1],
SACHIN A. MANDAVGANE[1*], and BHASKAR D. KULKARNI[3]

[1]*Department of Chemical Engineering, Visvesvaraya National Institute of Technology, Nagpur, Maharashtra 440010, India*

[2]*MIT School of Bioengineering Sciences & Research, Pune, Maharashtra 440010, India*

[3]*CSIR-National Chemical Laboratory, Pune, Maharashtra 411008, India*

*Corresponding author. E-mail: sam@che.vnit.ac.in.

ABSTRACT

Different climatic conditions in India favor the growth of different plants and crops. Fruits such as banana, pomegranate, orange, mango, pineapple, apple, and guava are grown in various parts of the country. These fruits are mostly consumed afresh and are commonly used in food-processing industries for the making of juices, desserts, jams, jellies, salads, fruit cocktail, pickles, etc. After the utilization of their edible part, the "fruit peel" is generally considered as an environmental burden and waste. Fruit peel, in fact, contains a high amount of cellulose, hemicellulose, and phenolic compounds. Thus, it can prove to be a sustainable, renewable, and low-cost alternative feedstock for the production of numerous high-value products that include biofertilizers, dietary fiber for consumption, industrial enzymes, animal feed and clean energy (biogas, methane, or biohydrogen), and substrate for the production of several bioactive compounds, and the synthesis of nanomaterials. This chapter reviews the utilization of seven fruit peels and proposes a generalized valorization and biorefinery approach for the treatment of fruit peel wastes (FPWs). The chapter highlights the importance of sustainability issues in the design phase of a biorefinery and tools to evaluate its environmental, economic, and social impacts. Sustainability assessment tools like life-cycle assessment and multicriteria decision analysis methods are discussed for developing an integrated FPW biorefinery.

9.1   INTRODUCTION

Food-processing industries play an important role in connecting the agricultural and industrial sectors of the economy. Food processing industries generate significant quantities of fruit peel wastes (FPWs). Typically, a peel accounts for 15%–50% of the total fruit weight that is discarded. Dumping of this waste, especially by food-processing industries, has contributed to the growing problem of environmental pollution, as landfills where these are dumped occupy large volumes of land. The phenolic compounds in these wastes are primary land pollutants. In the past few years, many studies have identified numerous applications of FPWs because of their physicochemical characteristics. FPWs mainly consist of cellulose and lignin as major constituents, and may also contain other functional groups, such as aldehydes, ketones, alcohols, carboxylic groups, hydroxides, phenolic groups, and ether groups, which can be easily converted into high-value products. Furthermore, FPW possesses a unique chemical composition, and considering its abundance and low cost, it is a viable option for valorization (Bhatnagar et al., 2015; Pathak et al., 2015, 2016a).

The food-processing sector is a highly fragmented industry, and includes the following subsegments: (a) fruits and vegetables, (b) milk and milk products, (c) beer and alcoholic beverages, (d) meat and poultry, (e) seafood, (f) grain processing, and (g) packaged or convenience food and packaged drinks.

Figure 9.1 presents a general idea of India's contribution to the global waste generation of fruit and vegetables (Wadhwa and Bakshi, 2013). The waste generated mainly includes fruit and vegetable peels and their seeds. In some cases, the amount of waste generated is larger than the product itself (Ayala-Zavala et al., 2011; García et al., 2015). Fruits such as banana, citrus fruits (e.g., orange, lemon, and grapefruit), mango, pineapple, pomegranate, and grapes are mostly used in their natural forms or as preserved pulps or derivatives (e.g., juices) and have a very high demand (Ayala-Zavala et al., 2011; Schieber et al., 2001). In most instances, the wastes from these fruits (e.g., seeds and peels) are discarded, which eventually fill up land spaces and cause significant environmental problems due to their foul smell and high biodegradable capacity. Various studies have evaluated the utilization of this waste (biomass) in different applications. In fact, the peel of these fruits is rich in high-value compounds, including antioxidants, lignins, dietary fibers, and bioactive compounds, among others, and their recovery is economically beneficial. In addition, recycling these wastes to obtain other valuable by-products helps in their safe disposal from the environment, which also reduces the problem of solid-waste handling (Pathak et al., 2015).

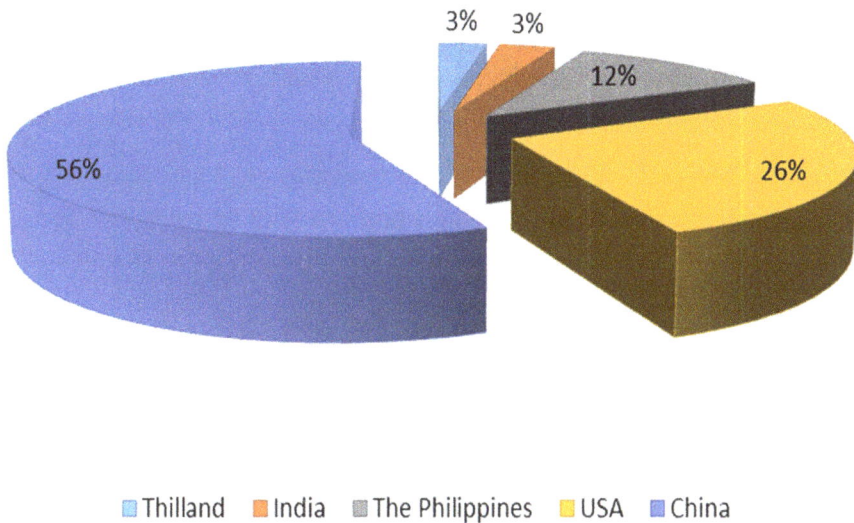

FIGURE 9.1 Generation of fruit peel waste.

This chapter discusses various ways for the utilization of major fruit peels, such as those from banana, citrus fruits, mango, pomegranate, pineapple, apple, and guava. In addition, the chapter also proposes the biorefinery approach for efficient utilization of these peels. Finally, the environmental impact of fruit peel biorefinery, through a life-cycle assessment (LCA), is also discussed.

9.2 VALORIZATION OF FRUIT PEEL WASTE

Solid-waste management and its safe disposal are problems caused by the generation of huge amounts of FPWs by food-processing industries. Their proper handling is compounded by the high cost required for their transportation and limited land spaces available for their disposal. Improper landfill disposal causes serious environmental problems due to their fast degeneration, and thus FPWs eventually become a source of fungus, bacteria, and insect multiplication. In addition, high biological oxygen demand and chemical oxygen demand of FPWs can create further problems in their disposal. Another method currently used to dispose FPWs is burning them on-site. However, burning in open fields pollutes the air, thereby making this method unsuitable for FPW disposal. From an environmental perspective, it is important to identify disposal methods that are both eco-friendly and economical (Pathak et al., 2016b; Puligundla et al., 2014).

FPW contains many functional groups and elements which make it a better feedstock for the production of value-added by-products. Therefore, it is important to explore the valorization of FPW through different techniques to reduce the ecological burden and contribute to the global economy. The valorization of several FPWs (banana, orange, pomegranate, pineapple, mango, apple, and guava) is presented here. Finally, this section concludes with the overall valorization of FPW in the form of a diagrammatic representation.

9.2.1 BANANA PEEL

Banana (*Musa* sp.) is a tropical fruit which is cultured in more than 130 countries across the globe (Mohapatra et al., 2010). Banana peel (BP) contributes nearly 30%–40% to the total weight of this fruit and is considered a waste. In 2014, approximately 102 MMT of bananas (mainly *Musa* sp.—*Musa* AA and *Musa* AAA) were cultivated worldwide (García et al., 2015). BP has an extensive range of applications. It is primarily used as an animal feed for the pectin production (Emaga et al., 2008), biogas and methane (Tumutegyereize et al., 2011), and ethanol (Itelima et al., 2013). BP is a rich source of phenolic acids, flavonoids, and several other antioxidants. The phenolic compounds in BP exhibit good antioxidant properties. According to Fatemeh et al. (2012), the total phenolic compounds extracted from BP varied from 75.01 to 685.57 mg GAE/100 g of dry peel, and the total flavonoid compounds extracted varied in the range of 39.01–389.33 mg CEQ/100 g of dry matter (Fatemeh et al., 2012). BP also contains norepinephrine, serotonin, and dopamine in minor quantities, and these compounds are shown to have both antifungal and antibiotic properties (Kumar et al., 2012b). BP is also a rich source of single-cell protein (Yousufi, 2012). Other bioactive compounds present in BP include sterols, triterpenes (Akihisa et al., 1998), esters and vitamin E (Waghmare and Kurhade, 2014), active amines (Udenfriend et al., 1959), carotenoid pigments (Knappa and Nicholasa, 1969), and water-soluble and water-insoluble sugars (Chandraju et al., 2011). BP contains dietary fiber, which includes both soluble (gums, pectins, etc.) and insoluble (cellulose, hemicelluloses, and lignin) fractions (Pathak et al., 2016b).

BP is rich in polymers such as cellulose, hemicellulose, lignin, and pectins, and has been used in the synthesis of nanoparticles such as silver (Bankar et al., 2010b), gold (Bankar et al., 2010c), palladium (Bankar et al., 2010a), Mn_3O_4 (Yan et al., 2014), cadmium sulfide (Zhou et al., 2014), and hydroxyapatite. BP has been studied for the removal of synthetic pollutants such as dyes, organic pollutants, heavy metals, and pesticides from wastewater (Pathak et al., 2015). Membranes produced using BP fibers are effective and comparable with the commercial spiral-wound membranes used in reverse osmosis (Datta et al., 2012). BP has also been used as a refining agent to adsorb peroxides from a carcinogenic substance in waste cooking oil, which increases the brightness of the oil (Taqiuddin and Aliah, 2014).

Solid-state fermentation (SSF) of BP gives value-added chemicals such as citric acid (Karthikeyan and Sivakumar, 2010), biosurfactant (Chooklin et al., 2014), thermostable levansucrase (Vaidya and Prasad, 2012), xylitol (birch sugar) (Rehman et al., 2013), and enzymes α-amylase (Kirankumar et al., 2011), β-amylase (Adeniran et al., 2010), cellulase (Omojasola and Jilani, 2009), polygalacturonase, xylanase (Mohamed et al., 2013), oxidase (Jadhav et al., 2013), laccase, manganese peroxidase (Elisashvili et al., 2008), etc. BP contains lipofuscin fluorescent pigments (Maguire and Haard, 1976) and anthocyanin, both of which are used as natural food colorants (Schieber et al., 2001). BP flour contains digestible starch, resistant starch, total starch, and soluble and insoluble dietary fibers. Because of its high phenolic content, BP flour has high antioxidant activity (Ramli et al., 2009, 2010; Rebello et al., 2014). High-density Na–Li ion battery anode is synthesized using low-surface-area carbon prepared from BP (Lotfabad et al., 2014). BP carbon modified with HNO_3 has also been used as an electrode for Li-ion capacitors (Arie and Lee, 2012). Grounded BP can be utilized for making ply boards, which can be a partial replacement for the costly prefabricated wall materials used currently (Ganiron, 2013). An antimicrobial edible film was prepared using BP in combination with clove oil (antimicrobial) and glycerol (plasticizer), which can be used for food packaging (Astuti and Erprihana, 2014). Thus, BP proves itself as an economical and encouraging bioresource that has an extensive variety of applications in different areas.

The valorization of BP was explained by Pathak et al. (2016b), who revealed that BP can be directly used as animal feedstock and, after drying, for energy generation by gasification (pyrolysis) or combustion. It is possible to extract value-added products such as enzymes, protein, ethanol, high-value chemicals, biogas/methane, antioxidants, medically useful compounds, phenolic compounds, and even color pigments from BP through a combination of several physical and chemical pretreatment methods and other unit operations and processes.

9.2.2 POMEGRANATE PEEL

Pomegranate (*Punica granatum* L.), also known as "granular apple" or "seeded apple," is considered as a symbol of abundance, fertility, and good luck. The pleasant taste, high nutritional value, and medicinal properties make the fruit one of the significantly consumed fruits all over the world. The fruit is consumed as a whole and is used for making jams, juice, jellies, wine, etc. (Ay et al., 2012; Bhatnagar et al., 2015; El-Ashtoukhy et al., 2008; Miguel et al., 2010; Moghadam et al., 2013). Being a rich source of dyes, alkaloids, tannins (Mirdehghan and Rahemi, 2007), and antioxidants (Miguel et al., 2010), different portions of the fruit plant are used for treating numerous diseases.

The global pomegranate production in 2014 approximately exceeded 1.5 MT (Farag et al., 2014). Seeds, arils, and peels constitute about 10%, 40%, and 50%, respectively, of the total weight of the pomegranate fruit (Christaki et al., 2011). Thus, pomegranate peel (PP) is one major waste generated in pomegranate processing. PP contains significant quantities of phenolic compounds that are responsible for 92% of its antioxidant activity (Ismail et al., 2012). These phenolic compounds include flavonoids (anthocyanins, catechins, and other complex flavonoids) and hydrolyzable tannins (pedunculagin, punicalagin, punicalin, ellagic acid, and gallic acid). Thus, the extract from PP has found application in cosmetics, food recipes, and tincture and healing formulations (Yasoubi et al., 2007).

The primary polyphenols observed in PP are tabulated in Table 9.1 (Farag et al., 2014; Sreekumar et al., 2014). Traditionally, PP is used as a medicine due to its high astringency. PP is used for the treatment of malaria, mouth ulcers, diarrhea, and parasitic infections (Bhattacharya, 2004; Dell'Agli et al., 2009; Farag et al., 2014). The medicinal properties of PP are attributed to the presence of steroids, glycosides, vitamin C, triterpenoids, ellagitannins, anthocyanins, and carbohydrates (Bhandary et al., 2012; Christaki et al., 2011). These compounds are also responsible for its antimitotic, antimicrobial, anticancer, antioxidant, anti-inflammatory, and immune-modulatory effects

(Abarghuei et al., 2013; Dikmen et al., 2011; Lansky and Newman, 2007). PP also shows good results in the prevention and treatment of cardiovascular disease, diabetes, cancer, microbial infections, erectile dysfunction, dental conditions, and skin repair, and in protection from ultraviolet (UV) radiation. Its potential utilization for the treatment of Alzheimer disease, arthritis, male infertility, infant brain ischemia, obesity, and dermal wounds is being examined extensively (Bhandary et al., 2012; Middha et al., 2013).

TABLE 9.1 Primary Polyphenols Observed in PP

Sr. No	Polyphenol	Concentration
1	Protocatechuic acid	14.512
2	Coumarin	3.534
3	Gallic acid	14.147
4	Caffeine	6.42
5	Chlorogenic acid	2.355
6	Vanillic acid	3.851
7	Quercetin	0.949
8	Caffeic acid	2.748
9	Ferulic acid	1.857
10	Oleuropein	0.590

PP is rich in several bioactive compounds. These bioactive compounds have high economic value and an extensive application range. Thus, extraction of these compounds can be economically beneficial. Bioactive compounds extracted from PP include ellagitannins (punicalagin, corilagin, tellimagrandin, punicalin, pedunculagin, gallagyldilacton, casuarinin, granatin A, and granatin B); flavonoids (catechin, flavan-3-ol, kaempferol, kaempferol 3-*O*-rhamnoglucoside, epicatechin, quercetin kaempferol, naringin, rutin, epigallocatechin 3-gallate, 3-*O*-glycoside, luteolin, and luteolin 7-*O*-glycoside), chlorogenic acid, pelletierine alkaloids (pelletierine), quinic acid, caffeic acid, p-coumaric acid, polyphenols (saponins, ellagic tannins, ellagic acid, and gallic acid), and anthocyanidins (Abarghuei et al., 2014; Aslam et al., 2006; Viladomiu et al., 2013); moreover, PP contains ascorbic acid (Ayala-Zavala et al., 2011), glycosides, triterpenoids, vitamin C steroids, carbohydrates (Bhandary et al., 2012), and tannins (Clifford and Scalbert, 2000; Seeram et al., 2005; Viladomiu et al., 2013). PP powder is rich in fiber ingredients and polyphenols, which exhibit antioxidant properties. The dietary fibers can be utilized to produce baked products such as cookies, bread, and other bakery products (Altunkaya et al., 2013; Hossin, 2009; Ismail et al., 2014; Sayed-Ahmed, 2014).

Because of the presence of higher polyphenols and micronutrients, PP and its extracts are found to be suitable for use in animal feed. Addition of supplements such as antibiotics, defaunating agents, ionophores, and methane inhibitors to ruminants helps in the fermentation process, which subsequently improves meat quality and milk efficiency and helps to keep diseases at bay (Abarghuei et al., 2013, 2014; Kushwaha et al., 2013).

PP has a higher heating value of 15.173 ± 35 J/g (Garcia et al., 2012), and thus it can be used for production of energy by either direct burning or preparation of briquettes. The biogas produced from PP by anaerobic digestion contains 61%–74% of methane, which has a potential for energy generation (Kirtane et al., 2009). PP is also a good substrate for the production of enzymes such as invertase (Uma et al., 2010) and tannase (Srivastava and Kar, 2009), and single-cell proteins in SSF (Khan et al., 2010).

PP is used as an adsorbent in wastewater treatment (Pathak et al., 2015, 2016a) and in the synthesis of various nanoparticles, such as gold, silver (Abdelmonem and Amin, 2014), cobalt oxide, and nickel (Ullah et al., 2014), along with anchored Fe_3O_4 magnetic nanorods (Venkateswarlu et al., 2014).

The yellow dye content in PP can also be used to color silk, wool, and cotton with acceptable fastness properties (Saxena and Raja, 2014). After the extraction of antioxidants, a toothpaste is prepared from the PP residue remaining (Abbas, 2014). A PP extract-loaded nanostructure lipid carrier shows strong antityrosinase activity, making it suitable for cosmetic applications (Tokton et al., 2014).

In addition, PP extract can be used as a safe preservative and has the potential to increase the shelf life of fish products (Zarei et al., 2015), chicken products (Kanatt et al., 2010), and ready-to-eat meats. In addition, PP is also reported to be a strong antioxidant used for sunflower oil stabilization (Iqbal et al., 2008).

9.2.3 ORANGE PEEL

Orange (*Citrus sinensis*) belongs to the Rutaceae family. It is cultivated worldwide and includes a wide range of varieties. India is the leading producer of oranges. In 2013, 6.5 MT of oranges were produced in India. Orange peels (OPs) are discarded as a waste (about 40%–50%) (Knappa and Nicholasa, 1969). OP is mainly composed of pectin, hemicellulose, cellulose, lignin, chlorophyll pigments, and other low–molecular weight hydrocarbons (Bhatnagar et al., 2015). OP also contains essential oils that are used as flavoring agents in the food industry. It is possible to extract more than 5 kg oil from about 1000 oranges. D-Limonene is the primary biochemical in orange oil (~90%), which is used as a flavoring agent in the manufacture of medicines and food (Braddock et al., 1986; Hull et al., 1953). Traditionally, orange wastes are used to enhance microbial growth and lactation of ruminants (Bampidis and Robinson, 2006). Hence, OP can be a good food source for ruminants for their weight gain and high yield of milk.

OP contains a substantial amount of pectin (42.50%), which is an important source for the high-value food industry (Rivas et al., 2008). OP is also a good source of biofertilizer, as it contains minerals such as phosphorus, potassium, calcium, magnesium, chloride, boron, and sodium, along with substantial quantities of carbon and nitrogen (carbon-to-nitrogen ratio [C/N] = 7), which aid in plant growth (Van Heerden et al., 2002).

Because of the presence of high amounts of cellulose and hemicellulose, OP can be easily fermented into many fermentable products, such as ethanol and biogas. Pectinase (Siessere and Said, 1989), pectate lyase, pectic poly-galacturonase, β-xylosidase, xylanase, and invertase (Rombouts and Pilnik, 1986) can be produced from OP by SSF. In submerged cultures, extracellular hydrolytic enzymes can be produced as well (Mahmood et al., 1998). Single-cell protein is another product obtained by the fermentation of OP, which has a very high commercial value (Locurto et al., 1992).

In addition, OP exhibits antioxidant, germicidal, and anticarcinogenic properties, which are utilized for treating colon and breast cancers, stomach upset, muscle pain, skin inflammation, and ringworm infections (Foo and Hameed, 2012; López et al., 2010). The other applications of OP are its utilization in wastewater treatment as an adsorbent (Pathak et al., 2015) and in the extraction of organic acids (López et al., 2010).

9.2.4   *PINEAPPLE PEEL*

Pineapple (*Ananas comosus*) belongs to the Bromeliaceae family. It is an herbaceous perennial which originated in the southern parts of Paraguay and Brazil. The fruit is mainly consumed afresh and is used in food-processing industries for preparing fruit cocktails, juices, salads, jams, and desserts. According to Food and Agriculture Organization estimate, India produced about 1.6 MT of pineapples in 2013. The fruits are harvested throughout the year, and only 52% of the total fruit is utilized for different industrial processes, with the leaves (13% of dry weight) and pineapple peel (PAP;

35% of dry weight) being discarded; however, these leaves and PAP are a rich source of valuable compounds (Bardiya et al., 1996; Foo et al., 2011; Krishni et al., 2014). Bromelain is perhaps the most valuable and the most studied compound from PAP. Bromelain has anti-inflammatory, anti-edematous, antithrombotic, and fibrinolytic properties, and also has potential as an anticancer agent (Bhui et al., 2009; Chobotava et al., 2009; Ketnawa et al., 2011). It is also used as a dietary supplement and meat tenderizer in food industries (Maurer, 2001).

PAP has fermentable sugars such as glucose, sucrose, and fructose in low quantities, and about 86%–92% of alcohol (ethanol) is produced from PAP (Nigam, 1999, 2000; Tanaka et al., 1999). The peel also produces biogas (yield range 0.41–0.67 mg/kg), with methane content of 41%–65% (Upadhyay et al., 2010). PAP is a rich source of phenolic antioxidants (2.01 mmol FRAP/100 g wet weight) (Changjiang et al., 2003). Because of the presence of phenolic compounds, PAP also exhibits antioxidant and antimicrobial activities, and is also used as a fodder.

PAP has been utilized for the production of various organic acids, such as lactic acid (Idris and Suzana, 2006), citric acid, and ferulic acid (Tilay et al., 2008), by fermentation. The PAP, the leaves, and the core obtained from pineapple-canning industries, called "bran," are used as a feed for ruminants (Tran, 2006). In addition, PAP is used as a substrate for the production of polygalacturonase and multienzymes (Anuradha et al., 2010), and as an adsorbent for heavy metals and dyes (as an anti-dyeing agent) (Pathak et al., 2015).

9.2.5 MANGO PEEL

Mango (*Mangifera indica* L.) is native to India and is widely grown worldwide, because of its lovable taste and high nutritive content. Mango peel (MP) accounts for 35%–55% of the fruit's total weight. MP is a good source of phytonutrients like pectin, cellulose, carotenoids, polyphenols, hemicelluloses, lipids, proteins, and vitamins E and C. The polyphenolic contents in MP are higher than those in the mango pulp. Generally, the contents of reducing and nonreducing sugars, cellulose, and proteins depend on the variety (Ajila et al., 2007; Imran et al., 2013). MPs are used as a substrate for the production of organic acids (lactic acid), ethanol, biogas, and single-cell protein through fermentation (Puligundla et al., 2014).

MP has been used for the production of bioethanol, yielding about 9.68% ethanol (Arumugam and Manikandan, 2011). Mango wine fermentation using yeast and MP as a biocatalyst has been reported. Using this process, the overall quality of wine improved significantly (Varakumar and Reddy, 2012). Biogas produced from MP at a level of 0.33 m^3 biogas/kg has methane content of 53% at 15 days of a hydraulic retention time (Deepak et al., 2001).

MP has the ability to produce enzymes such as pectinases (Kumar et al., 2013), carboxymethyl cellulose (Kumar et al., 2012a), and cellulose (Saravanan et al., 2012). MP is also a good source of pectin. The maximum yield of pectin (21%) was obtained by soaking finely ground, defatted MP in H_2SO_4 solution at pH 2.5 and 80°C for 120 min (Rehman et al., 2004). The pectin obtained from MP can be used as a pharmaceutical excipient to formulate solid oral dosage forms (Malviya and Kulkarni, 2012).

MP can be utilized as a livestock feed due to its higher energy value than maize silage. Studies have shown that the addition of MP in ruminants' diets increases their digestibility and has the potential to attenuate rumen methanogenesis (AzevêdoI et al., 2011; Geerkens et al., 2013).

The presence of phenolic compounds (mainly quercetin, mangiferin pentoside, syringic acid, and ellagic acid) (Ajila et al., 2010) gives MP some beneficial properties. It is also a good source of mangiferin (C-glucosyl xanthone), a heat-stable and pharmacologically active phytochemical. Mangiferin present in MP possesses several bioactivities, including antidiabetic, anti-inflammation, antitumor, immunomodulatory, antioxidant (Luo et al., 2012), and antibacterial properties (Berardini et al., 2005; Zgórka and Kawka, 2001). The high content of carotenoids in MP is shown to possess antioxidative capacity and high vitamin A activity, due to its high β-carotene content (Mercadante and Rodriguez-Amaya, 1998). Ethanol and ester extracts of MP showed significant antifungal activity on pathogenic fungi *Rhizoctonia solani* Kühn and *Rhizoctonia cerealis* van der Hoven (Qin et al., 2007).

Because of the presence of carotenoids, polyphenols, and vitamins, owing to its varied health-promoting properties, MP holds a good potential for the development of healthy functional food alternatives, such as noodles, sponge cakes, bread, biscuits, and other bakery products (Aziz et al., 2012). The properties of MP powder, such as solubility and oil and water absorption, play an important role in its utilization in food products (Sogi et al., 2013). MP fiber has high hydration capacities and is thus used in the preparation of dietary fiber-rich foods. The antioxidant capacity of MP dietary fiber is greater than that of DL-α-tocopherol (Koubala et al., 2012; Larrauri et al., 1997). In addition, MP powder is used as a suitable adsorbent of heavy metals from wastewater (Pathak et al., 2015).

9.2.6   APPLE PEEL

Apples are one of the most popular and commonly consumed fruits world-wide and are among the key sources of antioxidants and phytochemicals in the human diet. In 2015, approximately 70 MT of apples were produced globally. Apple peel (AP) constitutes approximately 5%–15% of the whole fruit weight. The polyphenolic content of AP was found to be six times greater than that of the flesh (Massias et al., 2015; Nileeka Balasuriya and Rupasinghe, 2012). Intake of AP polyphenols was shown to have beneficial actions on inflammation and oxidative stress (Massias et al., 2015). AP is rich in minerals such as calcium, potassium, magnesium, sodium, iron, and zinc. Besides, it is a good source of dietary fiber and bioactive compounds, which provide more nutritive and medicinal benefits (Leontowicz et al., 2007; Manzoor et al., 2012). AP powder is rich in phenolic compounds, flavonoids, anthocyanins, and triterpenoids, and hence may be used in several food products to improve the intake of phytochemicals and promote good health. AP powder is also reported to have a strong antiproliferative effect against human HepG2 liver cancer cells, Caco-2 colon cancer cells, and MCF-7 breast cancer cells (He and Liu, 2007; Vieiraa et al., 2011; Wolfe and Liu, 2003).

AP contains 1.21% pectin (Virk and Sogi, 2007). Flavonoids and triterpenes are the two major bioactive compounds in AP. The extract obtained from flavonoids and triterpenes has potential as a dietary supplement to lower blood cholesterol levels (Thilakarathnaa et al., 2012). High levels of triterpenoids and anthocyanins present in AP increase life span (Palermo et al., 2012). In addition, the polyphenols isolated from AP are potential natural antioxidants for stabilizing omega-3 polyunsaturated fatty acid-enriched fish oil (Sekhon-Loodu et al., 2013). AP has quercetin and quercetin glycosides as physiologically active flavonol molecules, which have numerous health benefits (Rupasinghe et al., 2011).

AP may be a low-cost raw material for cosmetic and food industries with a possibility to reduce glycation stress (Parengkuan et al., 2013). Ingredients developed from AP powder may be used in the preparation of functional foods and beverages. They have nutrients and a series of bioactive phytochemicals with putative health-beneficial effects (Henríquez et al., 2013). Edible films made from AP polyphenols can be used commercially to protect food from contamination by pathogenic bacteria (Du et al., 2011). Extracts of AP, lemon peel, OP, BP, and clove oil were used to formulate a polyherbal toothpaste (Abhay et al., 2014). Treated and untreated AP can be successfully used as adsorbent for the removal of dyes and heavy metals from wastewater (Krishnaa and Sree 2013; Pathak et al., 2015; Sartape et al., 2013).

Apple pomace is the solid residue containing a complex mixture of peel, seed, calyx, core, stem, and soft tissue. This residual material is a poor animal feed supplemented with very low protein content and high sugar (Vendruscolo et al., 2008). Apple pomace can be used for the production of several products by SSF or submerged fermentation (SMF), including β-glucosidase (Hang and Woodams, 1998), lignocellulolytic enzymes (Villas-Boas et al., 2002), pectin methylesterase (Joshi et al., 2006), pectinases (Pericin et al., 1999), pectolytic enzymes (Berovic and Ostroversnik, 1997), polygalacturonase (Hang and Woodams, 1994), phenolic compounds (Zheng and Shetty, 2000), aroma compounds (Christen et al., 2000), fruity aroma compounds (Bramorski et al., 1998), and single-cell protein (Albuquerque et al., 2006).

Apple pomace can be used as an animal feed after SSF (Joshi and Sandhu, 1996). Organic acids such as citric acid (Shojaosadati and Babaeipour, 2002) and γ-linolenic acid (Stredansky et al., 2000) were produced through SSF. A high concentration of ethyl alcohol was produced from apple pomace through SSF, while propyl, butyl, and amyl alcohols were obtained in very low concentrations. Apple pomace extract can be used as a carbon source in an aerobic-fed batch culture in the manufacture of baker's yeast. Powdered apple pomace was also used as a low-cost material for pigment production (Vendruscolo et al., 2008) and in the cultivation of mushrooms in SSF (Poppe, 2000; Worrall and Yang, 1992).

9.2.7 GUAVA PEEL

Guava (Family: Myrtaceae; *Psidium guajava* Linn.) is a tropical berry which has a fleshy pulp and many small seeds. Different parts of the fruit, like pulp, peel, and seed, are potentially nutritious (Rai et al., 2009; Rejal, 2010). In India, it is the fifth most significant fruit crop, covering about 3.38% of the total fruit cultivation area (Shiva et al., 2017). From guava fruit, after processing guava peel (GP), leaves, seeds, and bark are generated as waste.

Due to the presence of minerals such as Ca (17.31 ppm), Mg (206.65 ppm), Na (2.04 ppm), ascorbic acid (Packer et al., 2015), and phenolic compounds (596.67 mg/L) in GP (Rejal, 2010), it shows good antioxidant properties (Guo et al., 2003). Also, GP aqueous extracts show hypoglycemic and antidiabetic effects on the blood glucose level (Rai et al., 2009) and have the potential to decrease oxidative stress in the pancreas of diabetic rats (Budin et al., 2013). The gallic acid and ferulic acid present in the GP aqueous extracts show antimicrobial activity against bacteria such as *Escherichia coli*, *Pseudomonas aeruginosa*, and *Listeria monocytogenes* (Abdelmalek et al., 2016). Polyphenols such as cinnamoyl-*O*-hexoside, benzophenones, guavin B isomer, and abscisic acid are present in pink guava. Also, flavonoids like quercetin-*O*-hexoside, quercetin-*O*-pentoside, and adimethoxycinnamoyl-*O*-hexoside are available in GP. Rojas-Garbanzo et al. (2017) found about 42 more valuable compounds in GP (Rojas-Garbanzo et al., 2017).

GP has some applications in the food sector also. GP flour up to 30% can be used with wheat flour for the preparation of cookies, thereby enhancing the aroma, flavor, and texture, and it also gives good nutritional advantages such as increased amounts of protein and fiber and also decreased levels of fat and carbohydrates. Thus, GP flour can significantly increase the nutritional value of cookies without affecting the quality of the product (Bertagnolli et al., 2014). In addition to this, jellies prepared from GP show good properties in color, taste, aroma, texture, appearance, and global impression (Pereira et al., 2015).

In terms of extraction of chemicals, various chemicals can be extracted from GP. Bhat and Singh (2014) extracted pectin from GP using HCl (Pectin yield: 3.87% to 16.8%) and citric acid (Pectin yield: 2.65% to 11.12%) at different conditions (Bhat and Singh, 2014). Some enzymes like proteases can be extracted from GP, which has the maximum enzymatic activity with the pH of 5.0 (Ismail and Faizal, 2016). Pectinase is the other enzyme that can be extracted using response surface methodology (RSM) using a four-factor central composite design (CCD) from GP, using ultrasound. Amid et al. (2016) obtained the optimized conditions for the extraction of pectinase as a temperature of 40°C, sonication time of 20 minutes, and pH of 5.0, using a sample-to-solvent ratio of 1:4 mL/g for achieving a maximum pectinase yield of 96.2% (Amid et al., 2016).

Based on previous discussion, a general valorization scheme is presented in Figure 9.2. FPW can be used in the production of animal feed, energy production, medicines, and nutritional food.

9.3 BIOREFINERY APPROACH

Global concerns regarding environmental issues have necessitated the evaluation of fruit and vegetable discards as potential bioremediation agents. This huge amount of waste generated has given rise to a new problem of solid-waste management and its safe disposal. Landfill has been the most common method of FPW disposal, but in some cases, burning in open fields is preferred. However, these methods of FPW disposal cause serious

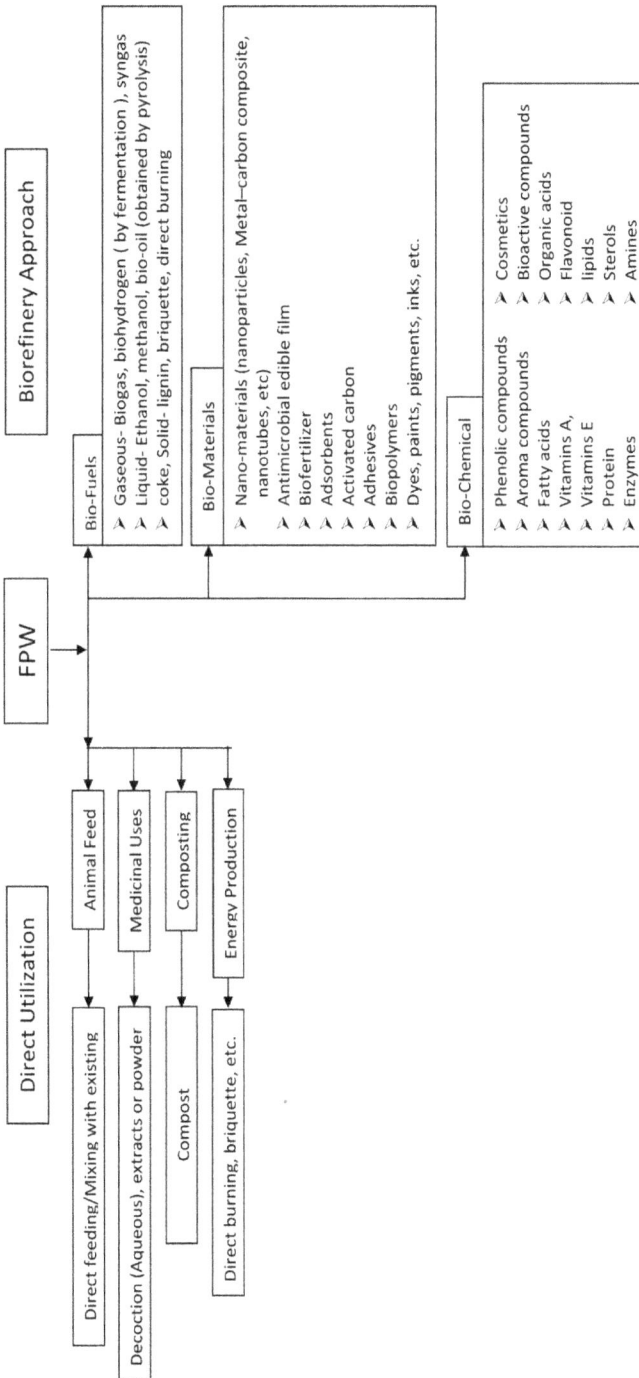

FIGURE 9.2 Overall valorization of FPW.

environmental problems (Anwar et al., 2010; Emaga et al., 2008; Mohapatra et al., 2010; Padam et al., 2012). From an environmental perspective, it is important to reuse FPW and explore it as a potential raw material for producing an array of value-added products. This would help reduce the burden on ecology and contribute to the world economy.

Thus, a methodology/process should be developed to reuse FPW to produce useful and profitable products (i.e., generating "wealth from waste"). The valorization expressed in the aforementioned sections is very comprehensive yet scrambled, due to which effective and economical utilization of FPW is not guaranteed. Thus, an integrated approach for the extraction of multiple products from FPW (termed as "Biorefinery") is to be developed. A biorefinery (Figure 9.3) is a facility in which biomass can be converted into value-added products through chemical, thermal, thermochemical, or biochemical means. This idea of a biorefinery is equivalent to that of the petroleum refinery in which an array of products are produced from raw materials. This concept can be applied to the crop industry wherein an array of value-added products can be produced from various agriculture-waste materials. By producing multiple products, a biorefinery can take advantage of the various biomass compounds and intermediates by increasing the value of derived products and leaving behind a very small amount of waste or, sometimes, zero waste. Importantly, farmers also benefit from adopting this approach, as the waste they dump or burn in open fields gets an additional value, maximizing their profits. In a nutshell, the idea is to generate "wealth from the waste." Furthermore, the biorefinery approach not only reduces the environmental burden by reducing the release of harmful gases, but also prevents the problems related to the dumping of these wastes.

9.3.1   PROPOSED BIOREFINERY OF FRUIT PEEL WASTE

With the increasing knowledge on OP and BP, it is now possible to enhance the value of these important agricultural wastes using a biorefinery approach.

Based on the valorization studies and previous studies on biorefineries, the following generalized biorefinery of FPW is proposed (Figure 9.4). Various paths for the production of different by-products are defined. Biorefinery of FPW mainly has three routes of utilization. These routes may vary based on the physicochemical properties of the raw material. The first route is the production of fermentable products such as enzymes, organic acids, and protein via SSF. The remaining edible solid residue can be used as animal feed. The second route is associated with energy production.

By taking this route, energy may be produced by fermentation, pyrolysis, methylation, or direct burning in the form of powder or briquets. In this route, the FPW can be either used for the production of ethanol and biogas/ methane through fermentation or used as a feed to the gasifier to produce gases and, eventually, energy. The third route concerns the production of a wide range of chemicals, with different unit operations and unit processes such as liquid extraction, leaching, or fermentation.

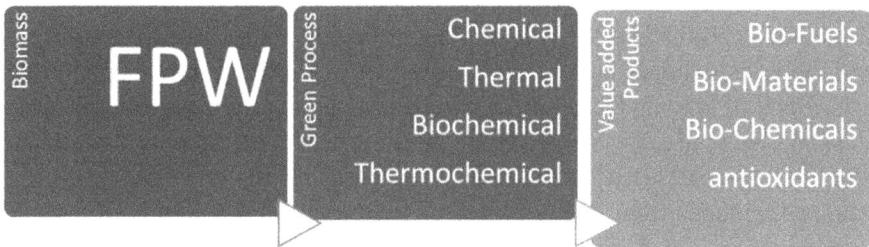

FIGURE 9.3 Biorefinery concept.

In this case, various permutations and combinations can be applied to identify the optimal route that can maximize profitability.

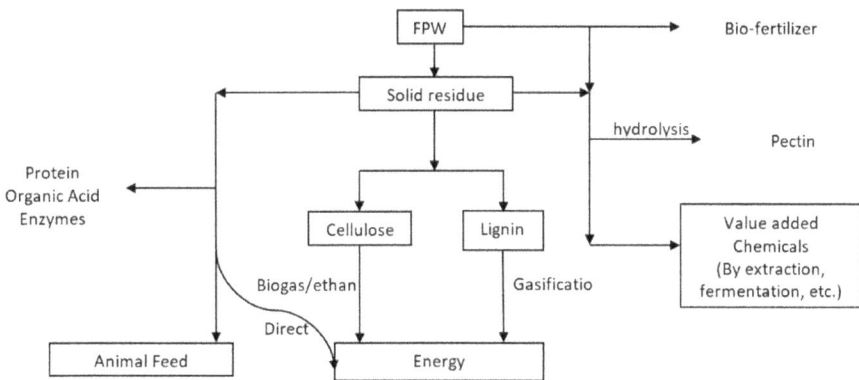

FIGURE 9.4 Generalized biorefinery scheme of FPW.

9.4 CURRENT PROBLEMS IN UTILIZATION AND FUTURE WORK

Although fruit wastes have the potential to be used in a variety of applications, their utilization on an industrial scale is not an easy task, as several factors restrict their use. The problems start with the collection of peels from their source of generation, that is, food-processing industry or households. Fruit peels are mainly composed of lignocellulosic materials, which are responsible for their rapid degradation. Therefore, their long-term storage is an issue to be considered. In addition, the moisture content in peels is very high, which helps them to degrade rapidly and reduces their heating value when utilized for generation of energy through direct combustion or gasification. Therefore, efficient drying of peels is necessary, which also significantly improves their value. More importantly, for the utilization of fruit peels in the production of value-added products at the industrial level, their supply should be continuous and their costs should be lower. However, this is difficult to manage, owing to the lack of knowledge on processing and transporting these raw materials (Katongolea et al., 2011; Tock et al., 2010). Thus, there is an urgent need to identify low-cost, highly advanced technologies for the conversion of high-value compounds from FPWs.

9.5 SUSTAINABILITY AND LIFE CYCLE ASSESSMENT

Brundtland Commission's report on the global environment and development in 1987 gave impetus to the need of sustainable development as a part of policy decision (Redclift, 2005). The integrated FPW biorefinery engages in extracting various value-added chemicals from a waste biomass, but has many disadvantages like excess land use, more eutrophication, and higher reaction time. All these disadvantages lead to higher environmental impacts and reduced economic viability, thereby hampering the overall sustainability of the process. Figure 9.5 shows the various factors to be considered during the sustainability assessment of an integrated FPW biorefinery.

It is imperative to evaluate the environmental impacts due to different chemical-processing activities to achieve the sustainability goals of a biorefinery. The IPAT equation was presented to evaluate the environmental aspect of sustainability of a process/technology:

$$I = P \times T \times A,$$

where

I = calculated environmental impact;
P = number of habitants the process/ technology is affecting;
T = technology factor; and
A = average per capita consumption of the population.

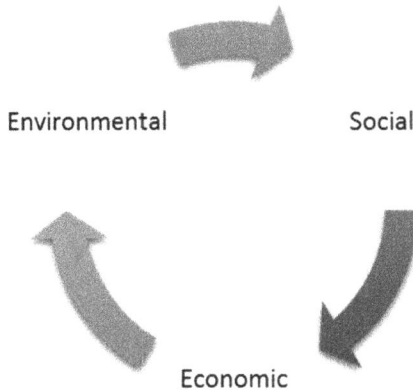

FIGURE 9.5   Factors affecting for sustainability assessment.

LCA precisely addresses the need for evaluation of the environmental impact of technology, which enables decision-makers to regulate the per-unit environmental impact (T of the IPAT equation) (Moltesen and Bjørn, 2018).

LCA provides a potent tool for analyzing any product or process from an environmental viewpoint. It is a compendious tool that quantitatively estimates the environmental impacts through various midpoint and end-point indicator values over the entire life cycle of a product (i.e., raw material acquisition, processing, manufacturing, use, and disposal) (Joglekar et al., 2018b). ISO 14040 has prescribed a framework to perform LCA. Figure 9.6 shows the framework for LCA.

The framework includes the following four steps:

1. *Definition of goal and scope:* The goal and scope play an important role in LCA. This section describes the system boundaries set for the analysis and the functional unit. The functional unit is generally used to compare the environmental impacts of the different alternatives under consideration (ISO14041:1998).

2. *Life cycle inventory analysis:* This section describes all the input and output categories of the process. Compilation of heat and material balance of the process is done in line with the functional unit set during the goal and scope definition (ISO14040:2006).

3. *Life cycle impact assessment (LCIA) method:* Various methods are used to calculate the environmental impacts. The environmental impacts are classified into two general sets of indicators—end-point indicators and midpoint indicators. The general framework for LCIA is drafted as per ISO 14042 (ISO14040:2006).

4. *Interpretation*: The primary objective of life cycle interpretations is to analyze results and to reach conclusions, explaining the limitations of and recommendations for the process. The ISO 14043 standard provides the framework for life cycle interpretation.

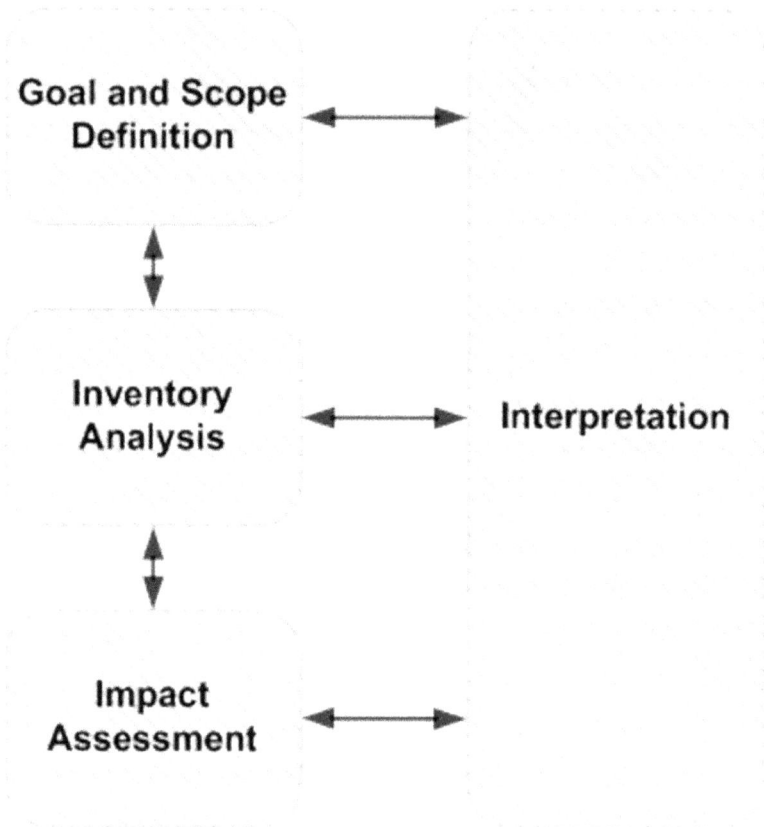

FIGURE 9.6 Framework for LCA.

The outcomes of performing LCA on FPW biorefinery can be broadly identified as follows.

1. To evaluate the overall environmental burden of the FPW biorefinery process;
2. To locate the potential hot spots (unit operations and processes that contribute most to the overall environmental burden). Such identification is important in finding alternative methods/processes. Alternative

methods can be developed using techniques like process intensification and green-engineering principles. The alternatives proposed can then be evaluated for their environmental impacts. There are several studies available that use sophisticated methods to replace the conventional setups (Boukroufa et al., 2015; Mohan et al., 2016). LCA can help in the evaluation of the overall sustainability of such methods; and

3. To make an informed decision as regards prioritization of the processing setup and evaluation of the processing scheme and design improvement based on the environmental indicators.

Several case studies can be cited wherein researchers have used LCA as a tool to select a processing route based on its corresponding environmental impacts (Cherubini and Jungmeier, 2010; Gnansounou et al., 2015; Joglekar et al., 2018b; Uihlein and Schebek, 2009).

Joglekar et al. (2018a) have used the LCA tool to evaluate the environmental impacts of various bagasse valorization routes. Based on the detailed inventory analysis, four different routes were evaluated for their environmental impacts and a decision was made based on the results (Joglekar et al., 2018b). The LCA methodology can be extended to select the FPW biorefinery design. We are also engaged in evaluating the environmental impacts of sophisticated processes like microwave- and ultrasound-assisted technologies and comparing them with conventional systems. It is interesting to note that although with the use of modern technologies the environmental impacts might seemingly increase, the environmental impact per unit rate of production is considerably lowered. Such analysis is specific to decision-making for an FPW biorefinery system.

Social LCA (S-LCA) addresses the issue of incorporation of societal implications or perceptions of the FPW biorefinery design. Very limited work has been published on S-LCA, yet a framework for the S-LCA methodology has been prescribed. The tool can be used to analyze the negative impacts of the FPW biorefinery design on the society and can help in achieving the overall sustainability goal. The impacts of S-LCA are classified into two types—obligatory and optional. Such types of impact categories are prescribed to include the evaluation of different design systems. S-LCA holds promise to promote the economic and societal parameters during the design phase of an FPW biorefinery (Dreyer et al., 2006).

Addition of societal and monetary viewpoints to the conventional LCA is proposed to make a life cycle sustainability assessment (LCSA) (Moltesen and Bjørn, 2018). Multi-criteria decision analysis (MCDA) has been increasingly gaining popularity in decision-making, because of its ability to integrate different criteria and its aggregation (Wang et al., 2009). The common intent of MCDA methods is to evaluate and select the most suitable alternative based on a systematic analysis, overcoming the limitations of unstructured individual and group opinions. One major advantage of MCDA approaches is the ability to highlight the identical and potential areas of conflict between different views, resulting in a more complete understanding of the process/ product (Kiker et al., 2009). FPW biorefinery should be analyzed in terms of the environmental, social, and economic aspects of sustainability, using MCDA methods like MIVES (Alarcon et al., 2011), triple bottom line (Elkington, 1997), analytic hierarchy process (Schaidle et al., 2011), etc.

Joglekar et al. (2018a) have used the MIVES methodology to calculate the sustainability index of bricks used for brickwork in a low-cost housing scheme based on four different criteria—economic, environmental, social, and technical (Joglekar et al., 2018a). A sustainability index is assigned to the alternatives based on the different indicators. A sustainability index can be assigned to the FPW biorefinery design based on the economic, environmental, and societal parameters.

9.6   CONCLUSION

Apart from being used as animal feed, fruit peels can be utilized as a feedstock for several value-added products' production, including biochemicals, biofertilizers, clean energy (biogas or methane), industrial enzymes, and functional ingredients and dietary fibers in bakery products. In this manner, the environmentally polluting FPWs can be converted into value-added products that have a higher economic value. However, to achieve this feat, better and advanced technologies should be developed, which should be not only efficient but also economically viable. Tools like LCA can be used for identifying hot spots in the process and improving the same by using more efficient and state-of-the-art equipment. MCDA can be used to decide the most suitable route for sequential utilization of FPW.

ACKNOWLEDGMENT

The authors thank the Department of Science and Technology, Government of India, for financial support, and BDK acknowledges support as an SERB distinguished fellow.

KEYWORDS

- **fruit peel waste**
- **valorization**
- **biorefinery**
- **sustainability**
- **LCA**

REFERENCES

Abarghuei, M.J.; Rouzbehan, Y.; Salem, A.-F. The influence of pomegranate-peel extracts on in vitro gas production kinetics of rumen inoculum of sheep. *Turkish Journal of Veterinary and Animal Sciences* 2014, 38, 212–219.

Abarghuei, M.J.; Rouzbehan, Y.; Salem, A.Z.M.; Zamiri, M.J. Nutrient digestion, ruminal fermentation and performance of dairy cows fed pomegranate peel extract. *Livestock Science* 2013, 157, 452–461.

Abdelmalek, S.; Mohsen, E.; Awwad, A.; Issa, R. Peels of Psidium guajava fruit possess antimicrobial properties. *The International Arabic Journal of Antimicrobial Agents* 2016, 6, 1–9.

Abdelmonem, A.M.; Amin R.M. Rapid green synthesis of metal nanoparticles using pomegranate polyphenols. *International Journal of Sciences: Basic and Applied Research* 2014, 15, 57–65.

Abhay, S.; Dinnimath, B.M.; Hullatti, K.K. Formulation and spectral analysis of new poly herbal toothpaste. *Journal of Drug Delivery and Therapeutics* 2014, 4, 68–74.

Adeniran, H.A.; Abiose, S.H.; Ogunsua, A.O. Production of fungal β-amylase and amyloglu-cosidase on some Nigerian agricultural residues. *Food and Bioprocess Technology* 2010, 3, 693–698.

Ajila, C.M.; Bhat, S.G.; Prasada Rao, U.J.S. Valuable components of raw and ripe peels from two Indian mango varieties. *Food Chemmistry* 2007, 102, 1006–1011.

Ajila, C.M.; Rao, L.J.; Rao, U.J.S.P. Characterization of bioactive compounds from raw and ripe Mangifera indica L. peel extracts. *Food and Chemical Toxicology* 2010, 48, 3406–3411.

Akihisa, T.; Kimura, Y.; Tamurab, T. Cycloartane triterpenes from the fruit peel of Musa sapientum. *Phytochemistry* 1998, 47, 1107–1110.

Alarcon, B.; Aguado, A.; Manga, R.; Josa, A. A value function for assessing sustainability: Application to industrial buildings. *Sustainability* 2011, 3, 35–50.

Albuquerque, P.M.; Koch, F.T.G.T.; Esposito, E.; Ninow, J.L. Production of Rhizopus oligosporus protein by solid state fermentation of apple pomace. *Brazilian Archives of Biology and Technology* 2006, 49, 91–100.

Altunkaya, A.; Hedegaard, R.V.; Brimer, L.; Gokmen, V.; Skibsted, L.H. Antioxidant capacity versus chemical safety of wheat bread enriched with pomegranate peel powder. *Food Function* 2013, 4, 722–727.

Amid, M.; Murshid, F.S.; Manap, M.Y.; Sarker, Z. Optimisation of ultrasound-assisted extraction of pectinase enzyme from Guava (Psidium guajava) peel enzyme recovery, specific activity, temperature and storage stability. *Preparative Biochemistry & Biotechnology* 2016, 46, 91–99.

Anuradha, K.; Padma, P.N.; Venkateshwar, S.; Reddy, G. Fungal isolates from natural pectic substrates for polygalacturonase and multienzyme production. *Indian Journal of Microbiology* 2010, 50, 339–344.

Anwar, J.; Shafique, U.; Waheed-uz-Zaman; Salman, M.; Dar, A.; Anwar, S. Removal of Pb(II) and Cd(II) from water by adsorption on peels of banana. *Bioresource Technology* 2010, 101, 1752–1755.

Arie, A.A.; Lee, J.K. *Activated carbon synthesized from banana peel as electrodes in Li-Ion capacitors.* 224th ECS Meeting; 2012.

Arumugam, R.; Manikandan, M. Fermentation of pretreated hydrolyzates of banana and mango fruit wastes for ethanol production. *Asian Journal of Experimental Biological Sciences* 2011, 2, 246–256.

Aslam, M.N.; Lansky, E.P.; Varani, J. Pomegranate as a cosmeceutical source: Pomegranate fractions promote proliferation and procollagen synthesis and inhibit matrix metallo-proteinase-1 production in human skin cells. *Journal of Ethnopharmacology* 2006, 103, 311–318.

Astuti, P.; Erprihana, A.A. Antimicrobial edible film from banana peels as food packaging. *American Journal of Oil and Chemical Technologies* 2014, 2, 65–70.

Ay, C.Ö.; Özcan, A.S.; Erdogan, Y.; Özcan, A. Characterization of Punica granatum L. peels and quantitatively determination of its biosorption behavior towards lead(II) ions and Acid Blue 40. *Colloids Surf B Biointerfaces* 2012, 100, 197–204.

Ayala-Zavala, J.F.; Vega-Vega, V.; Rosas-Domínguez, C.; Palafox-Carlos, H.; Villa-Rodriguez, J.A.; Siddiqui, M.W.; Dávila-Aviña, J.E.; González-Aguilar, G.A. Agro-industrial potential of exotic fruit byproducts as a source of food additives. *Food Research International* 2011, 44, 1866–1874.

AzevêdoI, J.A.G.; Filho, S.d.C.V.; Pina, D.d.S.; Chizzotti, M.L.; Valadares, R.F.D. Intake, total digestibility, microbial protein production and the nitrogen balance in diets with fruit by-products for ruminants. *Revista Brasileira de Zootecnia* 2011, 40, 1052–1060.

Aziz, N.A.A.; Wong, L.M.; Bhat, R.; Cheng, L.H. Evaluation of processed green and ripe mango peel and pulp flours (Mangifera indica var. Chokanan) in term of chemical composition, antioxidant compounds and functional properties. *Journal of Science of Food and Agriculture* 2012, 92, 557–563.

Bampidis, V.A.; Robinson, P.H. Citrus by-products as ruminant feeds: A review. *Animal Feed Science and Technology* 2006, 128, 175–217.

Bankar, A.; Joshi, B.; Kumar, A.R.; Zinjarde, S. Banana peel extract mediated novel route for the synthesis of palladium nanoparticles. *Materials Letters* 2010a, 64, 1951–1953.

Bankar, A.; Joshi, B.; Kumar, A.R.; Zinjarde, S. Banana peel extract mediated novel route for the synthesis of silver nano-particles. *Colloids and Surfaces A: Physicochemical and Engineering Aspects* 2010b, 368, 58–63.

Bankar, A.; Joshi, B.; Kumar, A.R.; Zinjarde, S. Banana peel extract mediated synthesis of gold nanoparticles. *Colloids and Surfaces B: Biointerfaces* 2010c, 80, 45–50.

Bardiya, N.; Somayaji, D.; Khanna, S. Biomethanation of banana peel and pineapple waste. *Bioresource Technology* 1996, 58, 73–76.

Berardini, N.; Knodler, M.; Schieber, A.; Carle, R. Utilization of mango peels as a source of pectin and polyphenolics. *Innovative Food Science and Emerging Technologies* 2005, 6, 442–452.

Berovic, M.; Ostroversnik, H. Production of Aspergillus niger pectolytic enzymes by solid state bioprocessing of apple pomace. *Journal of Biotechnology* 1997, 53, 47–53.

Bertagnolli, S.M.M.; Silveira, M.L.R.; Fogaça, A.D.O.; Umann, L.; Penna, N.G. Bioactive compounds and acceptance of cookies made with Guava peel flour. *Food Science and Technology, Campinas* 2014, 34, 303–308.

Bhandary, S.K.; Kumari, S.N.; Bhat, V.S.; Sharmila, K.P.; Bekal, M.P. Preliminary phytochemical screening of various extracts of punica granatum peel, whole fruita and seeds. *Nitte University Journal of Health Science* 2012, 2, 34–38.

Bhat, S.A.; Singh, E.R. Extraction and characterization of pectin from extraction and characterization of pectin from guava fruit peel. *International Journal of Research in Engineering & Advanced Technology* 2014, 2, 1–7.

Bhatnagar, A.; Sillanpää, M.; Witek-Krowiak, A. Agricultural "waste peels" as versatile biomass for water purification—A review. *Chemical Engineering Journal* 2015, 270, 244–271.

Bhattacharya, D. Punica granatum as a human use, wide-spectrum prophylactic against malaria and viral diseases in India. *The American Society of Tropical Medicine and Hygiene* 2004, 71, 288.

Bhui, K.; Prasad, S.; George, J.; Shukla, Y. Bromelain inhibits COX-2 expression by blocking the activation of MAPK regulated NF-kappa B against skin tumorinitiating triggering mitochondrial death pathway. *Cancer Letters* 2009, 28, 167–176.

Boukroufa, M.; Boutekedjiret, C.; Petigny, L.; Rakotomanomana, N.; Chemat, F. Bio-refinery of orange peels waste: A new concept based on integrated green and solvent free extraction processes using ultrasound and microwave techniques to obtain essential oil, polyphenols and pectin. *Ultrasonics Sonochemistry* 2015, 24, 72–79.

Braddock, R.J.; Temelli, F.; Cadwallader, K.R. Citrus essential oils—A dossier for material safety data sheets. *Food Technology* 1986, 40, 114–116.

Bramorski, A.; Soccol, C.R.; Christen, P.; Revah, S. Fruit aroma production by Ceratocystis fimbriata in solid cultures from agroindustrial wastes. *Revista de Microbiologia (online)* 1998, 29, 208–212.

Budin, S.B.; Ismail, H.; Chong, P.L. Psidium guajava fruit peel extract reduces oxidative stress of pancreas in streptozotocin-induced diabetic rats. *Sains Malaysiana* 2013, 42, 707–713.

Chandraju, S.; Thejovathi, C.; Kumar, C.S.C. Extraction, isolation and identification of sugars from banana peels (Musa Sapientum) by HPLC coupled LC/MS instrument and TLC analysis. *Journal of Chemical and Pharmaceutical Research* 2011, 3, 312–321.

Changjiang, G.; Yang, J.; Wei, J.; Li, Y.; Xu, J.; Jiang, Y. Antioxidant activities of peel, pulp, and seed fractions of common fruits as determined by FRAP assay. *Nutrition Research* 2003, 23, 1719–1726.

Cherubini, F.; Jungmeier, G. LCA of a biorefinery concept producing bioethanol, bioenergy, and chemicals from switchgrass. *The International Journal of Life Cycle Assessment* 2010, 15, 53–66.

Chobotava, K.; Vernallis, A.B.; Majid, F.A.A. Bromelain's activity and potential as an anti-cancer agent: Current evidence and perspectives. *Cancer Letters* 2009, 290, 148–156.

Chooklin, C.S.; Maneerat, S.; Saimmai, A. Utilization of banana peel as a novel substrate for biosurfactant production by halobacteriaceae archaeon AS65. *Applied Biochemistry and Biotechnology* 2014, 173, 624–645.

Christaki, E.V.; Bonos, E.M.; Florou-Paneri, P.C. Dietary benefits of pomegranates in humans and animals. *Journal of Food, Agriculture & Environment* 2011, 9, 142–144.

Christen, P.; Bramorski, A.; Revah, S.; Soccol, C.R. Characterization of volatile compounds produced by Rhizopus strains grown on agroindustrial solid wastes. *Bioresource Technology* 2000, 71, 211–215.

Clifford, M.N.; Scalbert, A. Review: Ellagitannins—nature, occurrence and dietary burden. *Journal of the Science of Food and Agriculture* 2000, 80, 1118–1125.

Datta, S.; Karmoker, S.; Sowgath, M.T. *Membrane development from banana peel fibers for waste water treatment at low cost.* 12 AIChE Annual Meeting; 2012.

Deepak, S.; Padshetty, N.S.; Nand, K. Recycling of mango-peel waste for biogas production. *Asian Journal of Microbiology, Biotechnology & Environmental Science* 2001, 3, 339–341.

Dell'Agli, M.; Galli, G.V.; Corbett, Y.; Taramelli, D.; Lucantoni, L.; Habluetzel, A.; Maschi, O.; Caruso, D.; Giavarini, F.; Romeo, S.; Bhattacharya, D.; Bosisio, E. Antiplasmodial activity of Punica granatum L. fruit rind. *Journal of Ethnopharmacology* 2009, 125, 279–285.

Dikmen, M.; Ozturk, N.; Ozturk, Y. The antioxidant potency of Punica granatum L. Fruit peel reduces cell proliferation and induces apoptosis on breast cancer. *Journal of Medicinal Food* 2011, 14, 1638–1646.

Dreyer, L.C.; Hauschild, M.Z.; Schierbeck, J. A framework for social life cycle impact assessment. *International Journal of LCA* 2006, 11, 88–97.

Du, W.-X.; Olsen, C.W.; Avena-Bustillos, R.J.; Friedman, M.; McHugh, T.H. Physical and antibacterial properties of edible films formulated with apple skin polyphenols. *Journal of Food Science* 2011, 76, M149–M155.

El-Ashtoukhy, E.-S.Z.; Amina, N.K.; Abdelwahab, O. Removal of lead (II) and copper (II) from aqueous solution using pomegranate peel as a new adsorbent. *Desalination* 2008, 223, 162–173.

Elisashvili, V.; Kachlishvili, E.; Penninckx, M. EVect of growth substrate, method of fermentation, and nitrogen source on lignocellulose-degrading enzymes production by white-rot basidiomycetes. *Journal of Industrial Microbiology & Biotechnology* 2008, 35, 1531–1538.

Elkington, J. *Cannibals with forks: The triple bottom line of twentieth century business.* Capstone, Oxford; 1997.

Emaga, T.H.; Robert, C.; Ronkart, S.N.; Wathelet, B.; Paquo, M. Dietary fibre components and pectin chemical features of peels during ripening in banana and plantain varieties. *Bioresources Technology* 2008, 99, 4346–4354.

Farag, R.S.; Abdel-Latif, M.S.; Emam, S.S.; Tawfeek, L.S. Phytochemical screening and polyphenol constituents of pomegranate peels and leave juices. *Agriculture and Soil Sciences* 2014, 1, 86–93.

Fatemeh, R.; Saifullah, S.; Abbas, R.; Azhar, F.M.A. Total phenolics, flavonoids and antioxidant activity of banana pulp and peel flours influence of variety and stage of ripeness. *International Food Research Journal* 2012, 19, 1041–1046.

Foo, K.Y.; Hameed, B.H. Preparation, characterization and evaluation of adsorptive properties of orange peel based activated carbon via microwave induced K_2CO_3 activation. *Bioresources Technology* 2012, 104, 679–686.

Foo, L.P.Y.; Tee, C.Z.; Raimy, N.R.; Hassell, D.G.; Lee, L.Y. Potential Malaysia agricultural waste materials for the biosorption of cadmium(II) from aqueous solution. *Clean Technologies and Environmental Policy* 2011, 14, 273–280.

Ganiron, T.U. An investigation of moisture performance of sawdust and banana peels ply board as non-veneer panel. *International Journal of u and e Service* 2013, 6, 43.

García, M.Á.A.; Vargas, J.H.L.; Molina, D.A.R. Agro-industrial fruit co-products in Colombia, their sources and potential uses in processed food industries: A review. *Revista Facultad Nacional de Agronomía* 2015, 68, 7729–7742.

Garcia, R.; Pizarro, C.; Lavin, A.G.; Bueno, J.L. Characterization of Spanish biomass wastes for energy use. *Bioresource Technology* 2012, 103, 249–258.

Geerkens, C.H.; Schweiggert, R.M.; Steingass, H.; Boguhn, J.; Rodehutscord, M.; Carle, R. Influence of apple and citrus pectins, processed mango peels, a phenolic mango peel extract, and gallic acid as potential feed supplements on in vitro total gas production and rumen methanogenesis. *Journal of Agricultural and Food Chemistry* 2013, 61, 5727–5737.

Gnansounou, E.; Vaskan, P.; Pachón, E.R. Comparative techno-economic assessment and LCA of selected integrated sugarcane-based biorefineries. *Bioresource Technology* 2015, 196, 364–375.

Guo, C.; Yang, J.; Wei, J.; Li, Y.; Xu, J.; Jiang, Y. Antioxidant activities of peel, pulp and seed fractions of common fruits as determined by FRAP assay. *Nutrition Research* 2003, 23, 1719–1726.

Hang, Y.D.; Woodams, E.E. Production of fungal polygalacturonase from apple pomace. *LWT-Food Science and Technolnology* 1994, 27, 194–196.

Hang, Y.D.; Woodams, E.E. Apple pomace: A potential substrate for production of β-glucosidase by Aspergillus foetidus. *LWT-Food Science and Technology* 1998, 27, 587–589.

He, X.; Liu, R.H. Triterpenoids isolated from apple peels have potent antiproliferative activity and may be partially responsible for apple's anticancer activity. *Journal of Agricultural and Food Chemistry* 2007, 55, 4366–4370.

Henríquez, M.; Almonacid, S.; Lutz, M.; Simpson, R.; Valdenegro M. Comparison of three drying processes to obtain an apple peel food ingredient. *CyTA—Journal of Food* 2013, 11, 127–135.

Hossin, F.L.A. Effect of pomegranate (Punica granatum) peels and it's extract on obese hypercholesterolemic rats. *Pakistan Journal of Nutrition* 2009, 8, 1251–1257.

Hull, W.Q.; Lindsay, C.W.; Baier, W.E. Chemicals from oranges. *Industrial and Engineering Chemistry* 1953, 45, 876–890.

Idris, A.; Suzana, W. Effect of sodium alginate concentration, bead diameter, initial pH and temperature on lactic acid production from pineapple waste using immobilized Lactobacillus delbrueckii. *Process Biochemistry* 2006, 41, 1117–1123.

Imran, M.; Butt, M.S.; Anjum, F.M.; Sultan, J.I. Chemical profiling of different mango peel varieties. *Pakistan Journal of Nutrition* 2013, 12, 934–942.

Iqbal, S.; Haleem, S.; Akhtar, M.; Zia-ul-Haq, M.; Akbara. J. Efficiency of pomegranate peel extracts in stabilization of sunflower oil under accelerated conditions. *Food Research International* 2008, 41, 194–200.

Ismail, N.; Faizal, M. Characterization and purification of protease extracted from Guava (Psidium guajava) peel. *Science Letters* 2016, 10, 4–7.

Ismail, T.; Sestili, P.; Akhtar, S. Pomegranate peel and fruit extracts: A review of potential anti-inflammatory and anti-infective effects. *Journal of Ethnopharmacology* 2012, 143, 397–405.

Ismail, I.; Akhtar, S.; Riaz, M.; Ismail, A. Effect of pomegranate peel supplementation on nutritional, organoleptic and stability properties of cookies. *International Journal of Food Sciences and Nutrition* 2014, 65, 661–666.

ISO14041. *Environmental management-life cycle assessment-goal and scope definition and inventory analysis*. International Organization for Standardization; 1998.

ISO14040:2006. Indian standard "environmental management—Life cycle assessment—Principles and framework," First Revision; 2009. New Delhi, India.

Itelima, J.; Onwuliri, F.; Onwuliri, E.; Onyimba, I.; Oforji, S. Bio-ethanol production from banana, plantain and pineapple peels by simultaneous saccharification and fermentation process. *International Journal of Environmental Science and Development* 2013, 4, 213–216.

Jadhav, U.; Salve, S.; Dhawale, R.; Padul, M.; Dawkar, V.; Chougale, A.; Waghmode, T.; Salve A.; Patil, M. Use of partially purified banana peel polyphenol oxidase in the degradation of various phenolic compounds and textile dye blue 2RNL. *Textiles and Light Industrial Science and Technology* 2013, 2, 27–35.

Joglekar, S.N.; Kharkar, R.A.; Mandavgane, S.A.; Kulkarni, B.D. Sustainability assessment of brick work for low-cost housing: A comparison between waste based bricks and burnt clay bricks. *Sustainable Cities and Society* 2018a, 37, 396–406.

Joglekar, S.N.; Tandulje, A.P.; Mandavgane, S.A.; Kulkarni, B.D. Environmental impact study of Bagasse valorization routes. *Waste and Biomass Valorization* 2018b, 10, 2067–2078.

Joshi, V.K.; Sandhu, D.K. Preparation and evaluation of an animal feed byproduct produced by solid state fermentation of apple pomace. *Bioresource Technology* 1996, 56, 251–255.

Joshi, V.K.; Parmar, M.; Rana, N.S. Pectin esterase production from apple pomace in solid-state and submerged fermentations. *Food Technology and Biotechnology* 2006, 44, 253–256.

Kanatt, S.R.; Chander, R.; Sharma, A. Antioxidant and antimicrobial activity of pomegranate peel extract improves the shelf life of chicken products. *International Journal of Food Science & Technology* 2010, 45, 216–222.

Karthikeyan, A.; Sivakumar, N. Citric acid production by Koji fermentation using banana peel as a novel substrate. *Bioresource Technology* 2010, 101, 5552–5556.

Katongolea, C.B.; Sabiitib, E.; Bareebaa, F.; Ledinc, I. Utilization of market crop wastes as animal feed in urban and peri-urban livestock production in Uganda. *Journal of Sustainable Agriculture* 2011, 35, 329–342.

Ketnawa, S.; Chaiwut, P.; Rawdkuen, S. Aqueous two-phase extraction of bromelain from pineapple peels ('Phu Lae' cultv.) and its biochemical properties. *Food Science and Biotechnology* 2011, 2, 1219–1226.

Khan, M.; Khan, S.S.; Ahmed, Z.; Tanveer A. Production of single cell protein from Saccharomyces cerevisiae by utilizing fruit wastes. *Nanobiotechnica Universale* 2010, 1, 127–132.

Kiker, G.A.; Bridges, T.S.; Varghese, A.; Seager, T.P.; Linkov I. Application of multicriteria decision analysis in environmental decision making. *Integrated Environmental Assessment and Management* 2009, 1, 95–108.

Kirankumar, V.; Sankar, N.R.; Shailaja, R.; Saritha, K.; Siddhartha, E.; Ramya, S.; Giridhar, D.; Sahaja, R.V. Purification and characterization of α-Amylase produced by Aspergillus Niger using banana peels. *Journal of Cell and Tissue Research* 2011, 11, 2775–2780.

Kirtane, R.D.; Suryawanshi, P.C.; Patil, M.R.; Chaudhari, A.B.; Kothari, R.M. Optimization of organic loading rate for diffrent biomethanization. *Journal of Scientific & Industrial Research* 2009, 68, 252–255.

Knappa, F.F.; Nicholasa, H.J. Cycloartenyl palmitate: A naturally occurring ester that forms a Cholesteric Mesophase. *Molecular Crystals* 1969, 6, 319–328.

Koubala, B.B.; Kansci, G.; Garnier, C.; Ralet, M.C.; Thibault, J.F. Mango (Mangifera indica) and Ambarella (Spondias cytherea) peel extracted pectins improve viscoelastic properties of derived jams. *African Journal of Food, Agriculture, Nutrition and Development* 2012, 12, 6200–6212.

Krishnaa, D.; Sree, R.P. Removal of chromium from aqueous solution by custard apple (Annona Squamosa) peel powder as adsorbent. *International Journal of Applied Science and Engineering* 2013, 11, 171–194.

Krishni, R.R.; Foo, K.Y.; Hameed, B.H. Food cannery effluent, pineapple peel as an effective low-cost biosorbent for removing cationic dye from aqueous solutions. *Desalination and Water Treatment* 2014, 52, 6096–6103.

Kumar, D.; Ashfaque, M.; Muthukumar, M.; Singh, M.; Garg, N. Production and characterization of carboxymethyl cellulase from Paenibacillus polymyxa using mango peel as substrate. *Journal of Environmental Biology* 2012a, 33, 81–84.

Kumar, K.P.S.; Bhowmik, D.; Duraivel, S.; Umadevi, M. Traditional and medicinal uses of banana. *Journal of Pharmacognosy and Phytochemistry* 2012b, 1, 57–70.

Kumar, Y.S.; Kumar, P.V.; Reddy, O.V.S. Pectinase production from mango peel using Aspergillus foetidus and its application in processing of mango juice. *Food Biotechnology* 2013, 26, 107–123.

Kushwaha, S.C.; Bera, M.B.; Kumar, P. Nutritional composition of detanninated and fresh pomegranate peel powder. *IOSR Journal Of Environmental Science, Toxicology And Food Technology* 2013, 7, 38–42.

Lansky, E.P.; Newman, R.A. Punica granatum (pomegranate) and its potential for prevention and treatment of inflammation and cancer. *Journal of Ethnopharmacology* 2007, 109, 177–206.

Larrauri, J.A.; Rupérez, P.; Saura-Calixto, F. Mango peel fibres with antioxidant activity. *Zeitschrift für Lebensmitteluntersuchung und -Forschung A* 1997, 205, 39–42.

Leontowicz, H.; Leontowicz, M.; Gorinstein, S.; Belloso, M.; Trakhtenberg, S. Apple peels and pulp as a source of bioactive compounds and their influence on digestibility and lipid profile in normal and atherogenic rats. *Medycyna Weterynaryjna* 2007, 63, 1434–1436.

Locurto, R.; Tripodo, M.M.; Leuzzi, U.; Giuffre, D.; Vaccarino, C. Flavonoids recovery and SCP production from orange peel. *Bioresource Technology* 1992, 42, 83–87.

López, J.Á.S.; Li, Q.; Thompson, I.P. Biorefinery of waste orange peel. *Critical Reviews in Biotechnology* 2010, 30, 63–69.

Lotfabad, E.M.; Ding, J.; Cui, K.; Kohandehghan, A.; Kalisvaart, W.P.; Hazelton, M.; Mitlin D. High-density sodium and lithium ion battery anodes from banana peels. *ACS Nano* 2014, 8, 7115–7129.

Luo, F.; Lv, Q.; Zhao, Y.; Hu, G.; Huang, G.; Zhang, J.; Sun, C.; Li, X.; Chen, K. Quantification and purification of mangiferin from Chinese mango (*Mangifera indica* L.) cultivars and its protective effect on human umbilical vein endothelial cells under H_2O_2-induced stress. *International Journal of Molecular Sciences* 2012, 13, 11260–11274.

Maguire, Y.P.; Haard, N.F. Isolation of lipofuscin-like fluorescent products from ripening banana fruit. *Journal of Food Science* 1976, 41, 551–554.

Mahmood, A.U.; Greenman, J.; Scragg, A.H. Orange and potato peel extracts: Analysis and use as Bacillus substrates for the production of extracellular enzymes in continuous culture. *Enzyme and Microbial Technology* 1998, 22, 130–137.

Malviya, R.; Kulkarni, G.T. Extraction and characterization of mango peel pectin as pharmaceutical excipient. *Polymers in Medicine* 2012, 42, 185–190.

Manzoor, M.; Anwar, F.; Saari, N.; Ashraf, M. Variations of antioxidant characteristics and mineral contents in pulp and peel of different apple (Malus domestica Borkh.) cultivars from Pakistan. *Molecules* 2012, 17, 390–407.

Massias, A.; Boisard, S.; Baccaunaud, M.; Leal Calderon, F.; Subra-Paternault, P. Recovery of phenolics from apple peels using CO2+ethanol extraction: Kinetics and antioxidant activity of extracts. *The Journal of Supercritical Fluids* 2015, 98, 172–182.

Maurer, H.R. Bromelain: Biochemistry, pharmacology and medical use. *Cell Molecular Life Science* 2001, 58, 1234–1245.

Mercadante, A.Z.; Rodriguez-Amaya, D.B. Effects of ripening, cultivar differences and processing on the carotenoid composition of mango. *Journal of Agricultural and Food Chemistry* 1998, 35, 262–265.

Middha, S.K.; Usha, T.; Pande, V. A review on antihyperglycemic and antihepatoprotective activity of eco-friendly punica granatum peel waste. *Evidence-based Complementary and Alternative Medicine* 2013, 2013, 656172.

Miguel, M.G.; Neves, M.A.; Antunes, M.D. Pomegranate (*Punica granatum* L.) A medicinal plant with myriad biological properties—A short review. *Journal of Medicinal Plants Research* 2010, 4, 2836–2847. [Vol. 4(25), 29 December Special Review]

Mirdehghan, S.H.; Rahemi, M. Seasonal changes of mineral nutrients and phenolics in pomegranate (*Punica granatum* L.) fruit. *Scientia Horticulturae* 2007, 111, 120–127.

Moghadam, M.R.; Nasirizadeh, N.; Dashti, Z.; Babanezhad, E. Removal of Fe(II) from aqueous solution using pomegranate peel carbon: equilibrium and kinetic studies. *International Journal of Industrial Chemistry* 2013, 4, 1–6.

Mohamed, S.A.; Al-Malki, A.L.; Khan, J.A.; Kabli, S.A.; Al-Garni, S.M. Solid state production of polygalacturonase and xylanase by Trichoderma species using cantaloupe and watermelon rinds. *Journal of Microbiology* 2013, 51, 605–611.

Mohan, S.V.; Modestra, J.A.; Amulya, K.; Butti, S.K.; Velvizhi, G. A circular bioeconomy with biobased products from CO$_2$ sequestration. *Trends in Biotechnology* 2016, 34, 506–519.

Mohapatra, D.; Mishra, S.; Sutar, N. Banana and it's by-product utilization: An overview. *Journal of Scientific & Industrial Research* 2010, 69, 232–329.

Moltesen, A.; Bjørn, A. LCA and sustainability. In *Life Cycle Assessment*; Hauschild, M., Rosenbaum, R., Olsen, S., Eds.; Springer, Cham; 2018.

Nigam, J.N. Continuous ethanol production from pineapple cannery waste. *Journal of Biotechnology* 1999, 78, 197–202.

Nigam, J.N. Continuous ethanol production from pineapple cannery waste using immobilized yeast cells. *Journal of Biotechnology* 2000, 80, 189–193.

Nileeka Balasuriya; Rupasinghe, H.P.V. Antihypertensive properties of flavonoid-rich apple peel extract. *Food Chemistry* 2012, 135, 2320–2325.

Omojasola, P.F.; Jilani, O.P. Cellulase production by Trichoderma longi, Aspergillus niger and Saccharomyces cerevisae cultured on plantain peel. *Research Journal of Microbiology* 2009, 4, 67–74.

Packer, V.G.; Melo, P. S.; Bergamaschi, K.B.; Selani, M.M.; Villanueva, N.D.M.; Alencar, S.M.d.; Contreras-Castillo, C.J. Chemical characterization, antioxidant activity and application of beetroot and guava residue extracts on the preservation of cooked chicken meat. *Journal of Food Science and Technology* 2015, 52, 7409–7416.

Padam, B.S.; Tin, H.S.; Chye, F.Y.; Abdullah, M.I. Banana by-products: An under-utilized renewable food biomass with great potential. *Journal of Food Science and Technology* 2012, 51, 3527–3545.

Palermo, V.; Mattivi, F.; Silvestri, R.; Regina, G.L.; Falcone, C.; Mazzoni, C. Apple can act as anti-aging on yeast cells. *Oxidative Medicine and Cellular Longevity* 2012, 2012, 491759.

Parengkuan, L.; Yagi, M.; Matsushima, M.; Ogura, M.; Hamada, U.; Yonei, Y. Anti-glycation activity of various fruits. *Anti-Aging Medicine* 2013, 10, 70–76.

Pathak, P.D.; Mandavgane, S.A.; Kulkarni, B.D. Fruit peel waste as a novel low-cost bio adsorbent. *Reviews in Chemical Engineering* 2015, 31, 361–381.

Pathak, P.D.; Mandavgane, S.A.; Kulkarni, B.D. Characterizing fruit and vegetable peels as bioadsorbents. *Current Science* 2016a, 110, 2114–2123.

Pathak, P.D.; Mandavgane, S.A.; Kulkarni, B.D. Valorization of banana peel: A biorefinery approach. *Reviews in Chemical Engineering* 2016b. [Accepted, In press]

Pericin, D.M.; Antov, M.G.; Popov, S.D. Simultaneous production of biomass and pectinases by Polyporus squamosus. *Acta Periodica Technologica* 1999, 29, 183–189.

Poppe, J. Use of agricultural waste materials in the cultivation of mushrooms. *Mushroom Science* 2000, 15, 3–23.

Puligundla, P.; Obulam, V.S.R.; Oh, S.E.; Mok, C. Biotechnological potentialities and valorization of mango peel waste: A review. *Sains Malaysiana* 2014, 43, 1901–1906.

Qin, L.J.; Wang, Q.; Wu, L.-Y. Stability of antimicrobial activities of mango (*Mangifera indica* L.) peel extracts. *Guangxi Agricultural Sciences* 2007, 4, 423–426.

Rai, P.K.; Jaiswal, D.; Mehta, S.; Watal, G. Anti-hyperglycaemic potential of Psidium guajava raw fruit peel. *Indian Journal of Medical Research* 2009, 129, 561–565.

Ramli, S.; Alkarkhi, A.F.M.; Yong, Y.S.; Easa, A.M. Utilization of banana peel as a functional ingredient in yellow noodle. *Asian Journal of Food and Agro-Industry* 2009, 2, 321–329.

Ramli, S.; Ismail, N.; Alkarkhi, A.F.M.; Easa, A.M. The use of principal component and cluster analysis to differentiate banana peel flours based on their starch and dietary fibre components. *Tropical Life Sciences Research* 2010, 21, 91–100.

Rebello, L.P.G.; Ramos, A.M.; Pertuzatti, P.B.; Barcia, M.T.; Castillo-Muñoz, N.; Hermosín-Gutiérrez, I. Flour of banana (Musa AAA) peel as a source of antioxidant phenolic compounds. *Food Research International* 2014, 55, 397–403.

Redclift, M. Sustainable development (1987–2005): An oxymoron comes of age. *Sustainable Development* 2005, 13, 212–227.

Rehman, S.; Nadeem, M.; Ahmad, F.; Mushtaq, Z. Biotechnological production of Xylitol from banana peel and its impact on physicochemical properties of rusks. *Journal of Agricultural Science and Technology* 2013, 15, 747–756.

Rehman, Z.U.; Salariya, A.M.; Habib, F.; Shah, W.H. Utilization of mango peels as a source of pectin. *Journal of the Chemical Society of Pakistan* 2004, 26, 73–76.

Rejal, S.Z.B. *Extraction of antioxidant acivity, phenolic content and mineral content from guava peel*. Bachelor of Engineering, Faculty of Chemical & Natural Resources Engineering Universiti Malaysia Pahang; 2010.

Rivas, B.; Torrado, A.; Torre, P.; Converti, A.; Domínguez, J.M. Submerged citric acid fermentation on orange peel autohydrolysate. *Journal of Agricultural and Food Chemistry* 2008, 56, 2380–2387.

Rojas-Garbanzo, C.; Zimmermann, B.F.; Schulze-Kaysers, N.; Schieber, A. Characterization of phenolic and other polar compounds in peel and flesh of pink guava (*Psidiumguajava* L. cv. 'Criolla') by ultra-high performance liquid chromatography with diode array andmass spectrometric detection. *Food Research International* 2017, 100, 445–453.

Rombouts, F.M.; Pilnik, W. Pectinases and other cell-wall degrading enzymes of industrial importance. *Symbiosis* 1986, 2, 79–90.

Rupasinghe, H.P.V.; Kathirvel, P.; Huber, G.M. Ultra-sonication-assisted solvent extraction of quercetin glycosides from 'Idared' apple peels. *Molecules* 2011, 16, 9783–9791.

Saravanan, P.; Muthuvelayudham, R.; Viruthagiri, T. Application of statistical design for the production of cellulase by Trichoderma reesei using mango peel. *Enzyme Research* 2012, Article ID 157643, 7.

Sartape, A.S.; Mandhare, A.M.; Jadhav, V.V.; Raut, P.D.; Anuse, M.A.; Kolekar, S.S. Removal of malachite green dye from aqueous solution with adsorption technique using *Limonia acidissima* (wood apple) shell as low cost adsorbent. *Arabian Journal of Chemistry* 2013, 69(S2), 1–10.

Saxena, S.; Raja, A.S.M. Natural dyes: Sources, chemistry, application and sustainability issues. In: *Roadmap to Sustainable Textiles and Clothing*; Subramanian Senthilkannan Muthu, Ed.; Springer; 2014, pp. 37–80.

Sayed-Ahmed, E.F. Evaluation of pomegranate peel fortified pan bread on body weight loss. *International Journal of Nutrition and Food Sciences* 2014, 3, 411–420.

Schaidle, J.A.; Moline, C.J.; Savage, P.E. Biorefinery sustainability assessment. *Environmental Progress & Sustainable Energy* 2011, 30, 743–753.

Schieber, A.; Stintzing, F.C.; Carle, R. By-products of plant food processing as a source of functional compounds—recent developments. *Trends in Food Science & Technology* 2001, 12, 401–413.

Seeram, N.; Lee, R.; Hardy, M.; Heber, D. Rapid large-scale purification of ellagitannins from pomegranate husk, a by-product of the commercial juice industry. *Separation Purification Technology* 2005, 41, 49–55.

Sekhon-Loodu, S.; Warnakulasuriya, S.N.; Rupasinghe, H.P.V.; Shahidi, F. Antioxidant ability of fractionated apple peel phenolics to inhibit fish oil oxidation. *Food Chemistry* 2013, 140, 189–196.

Shiva, B.; Nagaraja, A.; Srivastav, M.; Kumari, S.; Goswami, A.K.; Singh, R.; Arun, M.B. Characterization of guava (Psidium guajava) germplasm based on leaf and fruit parameters. *Indian Journal of Agricultural Science* 2017, 87, 634–638.

Shojaosadati, S.A.; Babaeipour, V. Citric acid production from apple pomace in multi-layer packed bed solid-state bioreactor. *Process Biochemistry* 2002, 37, 909–914.

Siessere, V.; Said S. Pectic enzymes production in solid-state fermentation using citrus pulp pellets by *Talaromyces flavus*, *Tubercularia vulgaris* and *Penicillium charlessi*. *Biotechnology Letters* 1989, 11, 343–344.

Sogi, D.S.; Siddiq, M.; Greiby, I.; Dolan, K.D. Total phenolics, antioxidant activity, and functional properties of 'Tommy Atkins' mango peel and kernel as affected by drying methods. *Food Chemistry* 2013, 141, 2649–2655.

Sreekumar, S.; Sithul, H.; Muraleedharan, P.; Azeez, J.M.; Sreeharshan, S. Pomegranate fruit as a rich source of biologically active compounds. *Biomed Research International* 2014, 2014, 686921.

Srivastava, A.; Kar, R. Characterization and application of tannase produced by Aspergillus Niger Itcc 6514.07 on pomegranate rind. *Brazilian Journal of Microbiology* 2009, 40, 782–789.

Stredansky, M.; Conti, E.; Stredanska, S.; Zanetti, S. γ-Linolenic acid production with Thamnidium elegans by solid state fermentation on apple pomace. *Bioresource Technology* 2000, 73, 41–45.

Tanaka, K.; Hilary, Z.D.; Ishizaki, A. Investigation of the utility of pineapple juice and pineapple waste material as low cost substrate for ethanol fermentation by *Zymomonas mobilis*. *Journal of Bioscience and Bioengineering* 1999, 87, 642–646.

Taqiuddin, R.; Aliah, N.Y. Banana peels: An economical refining agent for carcinogenic substance in waste cooking oil. *APEC Youth Scientist Journal* 2014, 4, 62–73.

Thilakarathnaa, S.H.; Wangb, Y.; Rupasinghea, H.P.V.; Ghanam, K. Apple peel flavonoid- and triterpene-enriched extracts differentially affect cholesterol homeostasis in hamsters. *Journal of Functional Foods* 2012, 4, 963–971.

Tilay, A.; Bule, M.; Kishenkumar, J.; Annapure, U. Preparation of ferulic acid from agricultural wastes: its improved extraction and purification. *Journal of Agricultural and Food Chemistry* 2008, 56, 7664–7648.

Tock, J.Y.; Lai, C.L.; Lee, K.T.; Tan, K.T.; Bhatia, S. Banana biomass as potential renewable energy resource A Malaysian case study. *Renewable Sustainable Energy Review* 2010, 14, 798–805.

Tokton, N.; Ounaroon, A.; Panichayupakaranant, P.; Tiyaboonchai, W. Development of ellagic acid rich pomegranate peel extract loaded nanostructured lipid carriers (Nlcs). *International Journal of Pharmacy and Pharmaceutical Sciences* 2014, 6, 259–265.

Tran, A.V. Chemical analysis and pulping study of pineapple crown leaves. *Industrial Crops and Products* 2006, 24, 66–74.

Tumutegyereize, P.; Muranga, F.I.; Kawongolo, J.; Nabugoomu, F. Optimization of biogas production from banana peels Effect of particle size on methane yield. *African Journal of Biotechnology* 2011, 10, 18243–18251.

Udenfriend, S.; Lovenberg, W.; Sjoerdsma, A. Physiologically active amines in common fruits and vegetables. *Archives of Biochemistry and Biophysics* 1959, 85, 487–490.

Uihlein, A.; Schebek, L. Environmental impacts of a lignocellulose feedstock biorefinery system: an assessment. *Biomass and Bioenergy* 2009, 33, 793–802.

Ullah, M.; Naz, A.; Mahmood, T.; Siddiq, M.; Bano, A. Biochemical synthesis of nickel & cobalt oxide nano-particles by using biomass waste. *International Journal of Enhanced Research in Science Technology & Engineering* 2014, 3, 415–422.

Uma, C.; Gomathi, D.; Muthulakshmi, C.; Gopalakrishnan, V.K. Production, purification and characterization of invertase by aspergillus flavus using fruit peel waste as substrate. *Advances in Biological Research* 2010, 4, 31–36.

Upadhyay, A.; Lama, J.P.; Tawata, S. Utilization of pineapple waste: A review. *Journal of Food Science and Technology Nepal* 2010, 6, 10–18.

Vaidya, V.; Prasad, D.T. Thermostable levansucrase from Bacillus subtilis BB04, an isolate of banana peel. *Journal of Biochemical Technology* 2012, 3, 322–327.

Van Heerden, I.; Cronjé, C.; Swart, S.H.; Kotzé J.M. Microbial, chemical and physical aspects of citrus waste composting. *Bioresource Technology* 2002, 81, 71–76.

Varakumar, S.; Reddy, K.N.O.V.S. Preparation of mango (*Mangifera indica* L.) wine using a new yeast-mango peel immobilized biocatalyst system. *Czech Journal of Food Sciences* 2012, 30, 557–566.

Vendruscolo, F.; Albuquerque, P.M.; Streit, F.; Esposito, E.; Ninow, J. Apple pomace: A versatile substrate for biotechnological applications. *Critical Reviews in Biotechnology* 2008, 28, 1–12.

Venkateswarlu, S.; Kumar, B.N.; Prathima, B.; SubbaRao, Y.; Jyothi, N.V.V. A novel green synthesis of Fe3O4 magnetic nanorods using Punica Granatum rind extract and its application for removal of Pb(II) from aqueous environment. *Arabian Journal of Chemistry* 2014, 12(4), 588–596.

Vieiraa, F.G.K.; Borgesa, G.D.S.C.; Copettia, C.; Pietrob, P.F.D.; Nunesc, E.d.C.; Fetta, R. Phenolic compounds and antioxidant activity of the apple flesh and peel of eleven cultivars grown in Brazil. *Scientia Horticulturae* 2011, 128, 261–266.

Viladomiu, M.; Hontecillas, R.; Lu, P.; Bassaganya-Riera, J. Preventive and prophylactic mechanisms of action of pomegranate bioactive constituents. *Evidence-based Complementary and Alternative Medicine* 2013, 2013, 789764.

Villas-Boas, S.G.; Esposito, E.; Mendonca, M.M. Novel lignocellulolytic ability of Candida utilis during solid state cultivation on apple pomace. *World Journal of Microbiology and Biotechnology* 2002, 18, 541–545.

Virk, B.S.; Sogi, D.D.S. Extraction and characterization of pectin from apple (Malus Pumila. Cv Amri) peel waste. *International Journal of Food Properties* 2007, 7, 693–703.

Wadhwa, M.; Bakshi, M.P.S. *Utilization of fruit and vegetable wastes as livestock feed and as substrates for generation of other value-added products*; Makkar, H.P.S.; FAO; 2013, pp. 1–67.

Waghmare, J.S.; Kurhade, A.H. GC-MS analysis of bioactive components from banana peel (Musa sapientum peel). *European Journal of Experimental Biology* 2014, 4, 10–15.

Wang, J.-J.; Jing, Y.-Y.; Zhang, C.-F.; Zhao, J.-H. Review on multi-criteria decision analysis aid in sustainable energy decision-making. *Renewable and Sustainable Energy Reviews* 2009, 13, 2263–2278.

Wolfe, K.L.; Liu, R.H. Apple peels as a value-added food ingredient. *Journal of Agricultural and Food Chemistry* 2003, 51, 1676–1683.

Worrall, J.J.; Yang, C.S. Shiitake and oyster mushroom production on apple pomace and sawdust. *HortScience* 1992, 27, 1131–1133.

Yan, D.; Zhang, H.; Chen, L.; Zhu, G.; Wang, Z.; Xu, H.; Yu, A. Supercapacitive properties of Mn3O4 nanoparticles bio-synthesized from banana peel extract *RSC Advances* 2014, 4, 23649–23652.

Yasoubi, P.; Barzegar, M.; Sahari, M.A.; Azizi, M.H. Total phenolic contents and antioxidant activity of pomegranate (*Punica granatum* L.) peel extracts. *Journal of Agricultural Science and Technology* 2007, 9, 35–42.

Yousufi, M.K. Utilization of various fruit wastes as substrate for the production of single cell protein using *Aspergillus oryzae* and *Rhizopus oligosporus*. *International Journal of Advanced Scientific and Technical Research* 2012, 2, 92–95.

Zarei, M.; Ramezani, Z.; Ein-Tavasoly, S.; Chadorbaf, M. Coating effects of orange and pomegranate peel extracts combined with chitosan nanoparticles on the quality of refrigerated silver carp fillets. *Journal of Food Processing and Preservation* 2015, 39, 2180–2187.

Zgórka, G.; Kawka, S. Application of conventional UV, photodiode array (PDA) and fluorescence (FL) detection to analysis of phenolic acids in plant material and pharmaceutical preparation. *Journal of Pharmaceutical and Biomedical Analysis* 2001, 24, 1065–1072.

Zheng, Z.; Shetty, K. Enhancement of pea (*Pisum sativum*) seedling vigour and associated phenolic content by extracts of apple pomace fermented with *Trichoderma* spp. *Process Biochemistry* 2000, 36, 79–84.

Zhou, G.J.; Li, S.H.L.; Zhang, Y.C.; Fu, Y.Z. Biosynthesis of CdS nanoparticles in banana peel extract. *Journal of Nanoscience and Nanotechnology* 2014, 14, 4437–4442.

CHAPTER 10

Utilization of Bioresources: Towards Biomass Valorization for Biofuels

SAMUEL ESHORAME SANNI[1*], OLURANTI AGBOOLA[1],
EMETERE MOSES[2,3], EMEKA OKORO[4], SAM SUNDAY ADEFILA[1],
ROTIMI SADIKU[5], and PETER ALABA[1]

[1]*Department of Chemical Engineering, Covenant University, Ota,
Ogun State, Nigeria*

[2]*Department of Physics, Covenant University, Ota, Ogun State, Nigeria*

[3]*Department of Petroleum Engineering, Covenant University, Ota,
Ogun State, Nigeria*

[4]*Department of Mechanical Engineering, University of Johannesburg,
Auckland Park, South Africa*

[5]*Department of Chemical, Metallurgical and Materials Engineering,
Tshwane University of Technology, Pretoria, South Africa*

Corresponding author. E-mail: adexz3000@yahoo.com

ABSTRACT

Biomass valorization has recently gained wide attention owing to the vast availability of wastes from which essential oils can be extracted, processed, and subsequently converted to energy utilities and value-added products by taking advantage of the free fatty acid contents of the parent bio-oils. The waste materials undergo a conversion process to give the prospective fuels. The two major final products of the conversion process are bioethanol and biodiesel. Recent findings have also indicated the usefulness of these products as intermediate products/raw materials for end products such as simple sugars or bio-lubes; this goes further to emphasize the huge potential in these substrates (agricultural wastes) when used as raw materials. In recent times, hydrogen from biomass is being sought as a means of energy to power cars, and this also brings to bear the need to underscore how far efforts have been made to introduce bioethanol-driven cars. Fossil fuels, besides being nonbiodegradable relative to bio-oils, are gradually being depleted, with no measures put in place for their replacement. This then suggests the need to look beyond the current situation for viable alternatives. This chapter provides background information on biofuels, their production processes, and their physicochemical characterization, application, and sustainability, since the world is like a vicious circle that currently tilts in the direction of "engineering for a sustainable world". Furthermore, the chapter unveils some gray areas of research that still remain untapped, thus opening up doors for revenue generation.

10.1   INTRODUCTION

The continuous search for environment-friendly replacement fuels for those derived from fossils has geared renewed interest in the persistent exploitation of biomass systems for biofuel production. Several biomasses have been used as parent materials for biofuel synthesis. First-generation biofuels are those sourced from food crops, including sugarcane and corn (Alexander et al., 2009). For second-generation biofuels, non-food feedstocks and agricultural wastes have proven to be the parent materials (Andersen and Stenby, 2007), while third-generation biofuels are sourced from synthetic fibers (Badger, 2002) or microorganisms such as microalgae (Maity et al., 2014). Studies have also shown that the environmental interactions of microalgae at wastewater treatment sites help to foster microbial growth, which subsequently translates to high volumes of bacterial mass ready for use as raw material for biofuel production (Chisti, 2007; Solimeno and Garcia, 2017; Solimeno et al., 2017). Several prior studies have revealed that, despite the prohibitive costs involved in the production of biofuels from microalgae, efforts can be channeled toward developing wastewater treatment plants with very low-energy footprints from which marketable products can be generated, besides water, which has been the usual focus aforetime (Craggs et al., 2011; Dalrymple et al., 2013; Park et al., 2011). Furthermore, Suganyaa et al. (2016) mentioned that high rate of algal ponds attracted the generation of electrical power from biofuels sourced from biomass, and the ponds were operated at reduced design/operating costs. Reports from Abdoulmooumine et al. (2015) bother on the biomass gasification as well as on the application of cold and hot gas clean-up techniques to remove basic significant but undesirable contaminants including CO, CO_2, H_2, tar, nitrogen-based compounds, mercaptans, hydrogen sulfide, hydrogen halides (i.e., HCl, HF, etc.), CH_4 and trace elements such as Na, K, etc., to appreciable/tolerable limits in order to render the gas suitable for downstream applications. According to literature, biomasses remain the only earth-sourced materials that have the inborn potential to store solar energy within their chemical bonds at their developmental stages. Considering the energy storage capacity of a biomass material, the stored energy can then be converted to combustible gases, such as producer gas (Figure 10.1), which can be converted to liquid fuels/heating fuel for heat energy generation in furnaces or internal combustion engines and for power generation in fuel cells using thermochemical techniques (Fischer–Tropsch method) to purify the gases obtained from the biomass (Asadullah, 2014).

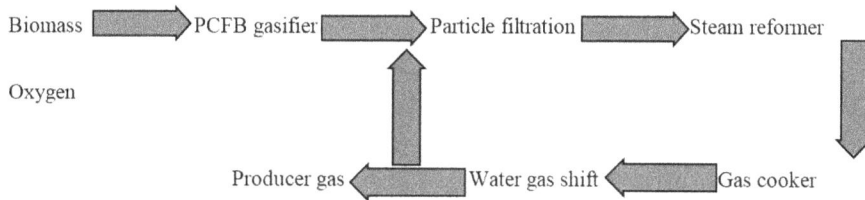

FIGURE 10.1 A typical process flow scheme for producer gas synthesis from biomass.

Alternately, the already highlighted advantages inherent in the process, biomass gasification technology still poses challenges owing to the presence of particulate matter and impurities which must be removed in order to tap the enormous advantages the produced biogas offers (Chiang et al., 2013; Paethanom et al., 2012; Shen and Yoshikawa, 2013; Sulc et al., 2012; Xu et al., 2010).

Figures 10.2 and 10.3 are illustrations for the production of bioethanol, biogas, biodiesel, and bioethanol, respectively.

10.2 BIOMASS SOURCES FOR BIORESOURCES

Biofuels/bio-oils can be sourced from several sources ranging from biomass to waste cooking oils, agricultural, industrial, and municipal wastes, as well as animal fats, via thermochemical/biochemical methods (Bhatia et al., 2017; Corral et al., 2017; Demirbas, 2017; Palmer and Brigham, 2016). In the application of the thermochemical method, the biomass undergoes

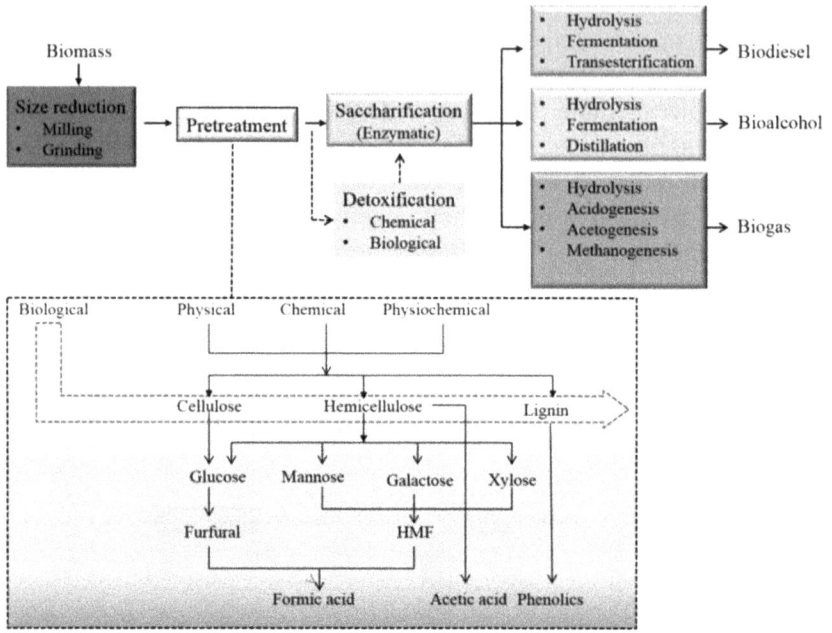

FIGURE 10.2   Schematic illustration of second-generation biofuel (biodiesel, bioalcohol, and biogas) synthesis via microbial fermentation.

Source: Reprinted with permission from Bhatia, et al., 2017. © Elsevier.

pyrolysis or gasification, whereas during biochemical treatment, the biomass is initially transformed to sugars, which are further hydrolyzed by enzymes/microbes or acids to obtain biofuel/bio-alcohol (Huang et al., 2016; Le et al., 2017; Nhien et al., 2017). Classification of biofuels may be according to their source. First-generation biofuels are those sourced from sugars, starchy components, and edible vegetable oils. Second-generation biofuels have their origin in nonedible biomass. Third-generation biofuels are sourced from algae, while biofuels of the fourth generation are those produced by advanced technologies such as trapping CO_2 (Aro, 2016; Naik et al., 2010; Sanni et al., 2018). A large portion of the currently produced biofuels belong to the class of first-generation biofuels. In addition, countries such as the United States and Brazil obtain their biofuels from corn/sugarcane wastes, while in Europe, wheat and barley are the major sources of the fuel (Lopes et al., 2016; Singh et al., 2016). One of the recent advances made in the conversion of biofuels to bio-oils is in the field of fast pyrolysis; however, the resulting oils have distinct properties from those of petroleum fuels (Oasmaa and Czernik, 1999;

Oasmaa et al., 2005; Zhang et al., 2007). The process involves the anaerobic breakdown of complex biomass molecules to simpler ones at 500 °C while supplying heat to the system at 103–105 Coloumbs/s in less than 2 s, in order to form liquid products, liquefied gases, noncondensable gases, and char, alongside rapidly quenched vapor from pyrolysis (Qiang et al., 2009). Types of pyrolytic reactors include the bubble-oriented fluidized bed, moving bed, the rotating-mode fluid bed, rotating conic vacuum, and ablative- and screw-type pyrolytic reactors (Bridgwater and Peacocke, 2000; Bridgwater et al., 2001; Bridgwater and Maniatis, 2004).

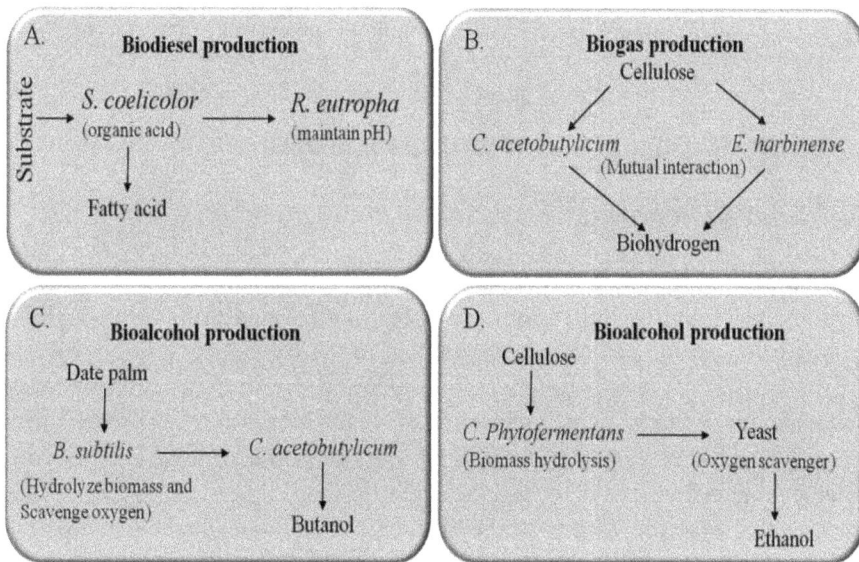

FIGURE 10.3 Consotia of microbes for biofuel synthesis. (A) Interactive behaviour of microbes result in a rise in fatty acid production (i.e. substrate fermentation and hydrolysis by *S. coelicolor* provides for fatty acids storage and organic acid production, while *Rhizophaeutropha* exploits the released acids, and regulates the medium-pH to support the development/growth of *S. coelicolor*) (Bhatia et al., 2015). (B) *C. acetobutylicum and E. harbinense* convert cellulose to biohydrogen while, a co-culturing of the microbes results in higher biohydrogen yield (Wang et al., 2009). (C) *B. subtilis* aids biomass hydrolyses via aerobic digestion, supplies nutrients and makes available a suitable environment/favourable conditions for *C. acetobutylicum* development/replication and butanol synthesis (Abd-Alla and El-Enany, 2012). (D) This is the Yeast-strain oxygen scavenger which creates a favourable condition (devoid of O_2) for *C. phytofermentans* development and biomass hydrolysis. *C. phytofermentans* supplies the required nutrients for the growth of yeast and ethanol production (Zuroff et al., 2013).

Source: Reprinted with permission from Bhatia, et al., 2017. © Elsevier.

10.3 BIOFUEL CLASSIFICATION

Biofuels can be categorized based on source, state, year of discovery/ exploitation, and application (Institute of Intellectual Property Research and Development, 2018). In terms of source, the classes are (1) edible crops, for example, corn and cassava; (2) nonedible crops, for example, agricultural wastes; and (3) microbes, such as algae and fungi. Classes in terms of physical state include solid (wood), liquid (bioethanol), and gaseous (biohydrogen and biomethane) biofuels. In terms of year of discovery/exploitation, the classes are first-fourth-generation biofuels; in terms of application, the groups are transport fuel, such as diesel for tractors, and industrial fuel, for firing heaters and boilers.

10.4 CHARACTERISTICS OF BIO-OILS/BIOFUELS

10.4.1 HOMOGENEITY

Most bio-oils/biofuels are considered homogenous, though a number of biomasses belonging to the group of forestry and agricultural products differ in terms of percent extractives, as well as in their products from pyrolysis, owing to their relative differences in solubility, density, and polarity when compared to normal oils. Literature has it that bio-oils from feedstocks that are rich in extractives usually separate out into binary phases, with a top layer rich in extractives (25–50 wt%), pyroligneous lignin (20–40 wt%), and compounds miscible with water, while the bottom layer has the semblance of normal oil (Oasmaa et al., 2003; Perez et al., 2007). However, several cellulosic-waste materials only harbor a few extractives, and hence cannot undergo this kind of phase separation, which is the reason they appear some-what distributed within the bio-oil matrix.

10.4.2 MULTIPHASE MICROSCOPIC STRUCTURE

Biofuels/bio-oils may contain some highly polar substances such as H_2O, acids, and alcohols/less polar components, including esters, aldehydes/ ketones, and phenolic alcohols, and nonpolar substances, for example, hexane, all of which are insoluble. A study by Radlein (2002) shows that bio-oils possess properties similar to those of microemulsions in the aqueous phase, whereas other immiscible components are suspended as micelles in the oil. In this oil are multipolar components/emulsifiers which give stability to their structural matrices. The multiphase nature of bio-oils synthesized from softwood and hardwood via vacuum pyrolysis was investigated, and based on their findings, the complex structural forms of bio-oils are traceable to the inherent char, wax, condensate, and high-micelle compounds in the biomass. At 25 °C, large, crystal-containing droplets in the range of 20–80 μm, as well as aqueous droplets of 5–10 μm, were formed from the softwood; however, it was also reported that phase separation of the aforementioned macromolecular emulsions is possible at 60 °C. In addition, only a few crystals and chars were reported to be suspended in the oil recovered from the hardwood. Based on further investigations, the results obtained from measurements from a differential scanning calorimeter showed that despite the release of wax as one of the products, which may also pose hindrance to the quantity of extract obtained, the top, extract-rich layer easily separates out as a result of the absorbed heat, which subsequently melts the waxy crystals. Based on the study conducted by Fratini et al. (2006), bio-oils were reported to be nanostructured fluids which constitute a complex continuous phase and an association of monomer units of pyrolytic lignin.

10.4.3 OXYGEN CONTENT

After processing biomass feedstock, on wet basis, most of the oxygen (i.e., 35–60 wt%) inherent in the biomass is largely retained in the resulting bio-oil. Also, there is ample evidence that it constitutes one of the reasons behind the vast difference in the nature of bio-oils/biofuels and that of mineral oils/ petroleum fuels recovered from refining processes (Kalam and Masijuki, 2002). The presence of oxygen makes for the immiscibility of the polar bio-oils/biofuels with the nonpolar petroleum oils or fuels; however, this poses some limitations to the use of bio-oils/biofuels. Furthermore, the low calorific values, corrosive potential, and unstable nature of some bio-oils can be attributed to their high oxygen contents, and vice versa.

10.4.4 WATER CONTENT

Biofuels are known to be associated with water, as they form azeotropes (e.g., ethanol) even on attempting to have total separation of the fuels from water. Bio-oils form an emulsion with water that can be separated out on allowing the mixture to settle out into clear phases. Water is produced from biomass via the application of heat, and the process is known as dehydration; water in bio-oils usually ranges from 15 to 30 wt% which depends on the initial moisture of the biomass precursor and the temperature of dehydration. Water is present in biofuels as hydrated aldehyde or is hydrogen-bonded to organic molecules, which can be determined using Karl Fischer titration. The limited solubility of bio-oils in water is caused by an increase in the water content of the oils. When this happens, the initial microemulsion structure is deformed, causing the separation of the mix into light aqueous and heavy organic phases. This type of separation poses some difficulty in the use of bio-oils; hence, in order to restore/maintain uniformity in the distribution of components in the oils, water in the mix is usually not allowed to exceed 30–35 wt% (Radlein, 2002), whereas most biofuels are highly soluble in water. This then informs the need to keep the wt% of water in the feedstock at <10 wt% in order to keep the water level in the synthesized oil within appreciable limits. However, water has negative (low heating value, increased ignition delay in automobile engines, premature evaporation, injection difficulties, reduction in burn rates and adiabatic flame temperature during such operations) consequences in the storage and use of bio-oils. Its advantages in bio-oils include reduction in oil viscosity and an increase in atomization of oil molecules when used as fuel, for example, as fuel oil in industrial heaters which releases less bio-oil emissions. Furthermore, water helps to control the formation of NO_x gases by lowering and regulating the temperature profile of the heating chamber. Micro-explosion of bio-oil droplets is also made possible by the presence of water which aids sufficient combustion; nonetheless, the hydroxyl group in water can prevent soot formation, which may in turn accelerate its oxidative potential.

10.4.5   CALORIFIC VALUE

Due to the low percentage of oxygen in biofuels, they are seen to have lower caloric values, which range from 14–18 MJ/kg, as compared to those of their petroleum counterparts, whose calorific values are in the range of 41–43 MJ/kg, though their relative densities are 1.2 g/mL and 0.8–1.0 g/mL, respectively, which is informative of the higher volumetric density of bio-oils; the calorific value of bio-oils can be calculated using Equation (10.1), as proposed by Oasmaa et al. (1997):

$$LHV = HHV - 218:3\ H\%\ (wt\%)\ kJ/kg \qquad (10.1)$$

10.4.6 ASH/SOLID MATTER

The quantification of particulate matter in bio-oils is usually done in terms of ash content using optical instruments (Agblevor and Besler, 1996; Oasmaa et al., 1997; Oasmaa et al., 2005), and oftentimes, ethanol, a biofuel, is used to keep bio-oil extracts free, whereas neutral solvents such as methanol and dichloromethane, or a mixture of both, perform the reverse and in turn improve the solubility of extractives in the oils (Oasmaa and Meier, 2005). During thermal degradation of lignocellulosic biomass, ash particulates are 3–8 times more than those of the feedstocks; the ash contents of biomass are in the range of 15–50 wt%. The ash content of bio-oil simply refers to the amount of ash obtained from bio-oil when it is heated to 775 °C (Oasmaa et al., 1997). However, directly heating bio-oils may cause unwanted foam formation and splashing of the oils due to the presence of H_2O; hence, it is necessary to first control water evaporation at 105 °C prior to heating to 775 °C (Oasmaa et al., 1997). According to Bakhshi and Adjaye (1994), the presence of metals such as compounds of group 1 metals (i.e., compounds of Li, Na, and K) in ashes has been confirmed to be responsible for the low melting points of ashes; hence, they can stick to turbine blades and thus act as stimulants of high-temperature corrosion of the blades. Despite the zero tolerance for the presence of solids and ashes in bio-oils, attempts to remove them via filters can result in bio-oil char/sludge (Agblevor and Besler, 1996; Bakhshi and Adjaye, 1994; Elliot, 1994; Morris, 2001), though National Renewable Energy Laboratory (NREL) and Technical Research Centre of Finland filtration procedures have been validated to be appropriate for obtaining bio-oils with <0.01 wt% ash and <10 ppm group 1 metals (Oasmaa and Czernik, 1999; Scahill et al., 1997). According to Oasmaa and Czernik (1999), the presence of particulates in bio-oils has negative consequences/ influences on the storage and heating capacities of bio-oils, because in a filtration process, after the removal of large particles, the char particulates slowly agglomerate/cluster to form larger particles and, thereafter, rest at the vessel bottom, which in turn increases the bio-oil viscosity with negative consequences such as difficulty in pumping and atomization of the bio-oils. Due to an increase in the viscosity of the particles, the particles may erode and clog the fuel injection system or act as catalysts that speed up the aging process of bio-oils; therefore, there is a resultant phase separation of the bio-oil from particulates that are absorbed with lignin to produce sticky tars. In addition, the char particulates produce slow-heating carbon-like ceno-spheres and unburnt particulates in the waste gas (Qiang et al., 2009).

10.4.7 VISCOSITY

The resistance a fluid offers to shear or angular deformation is termed viscosity. Viscosity of bio-oils can be measured using capillary viscometers, rotaviscotester (Oasmaa and Meier, 2005), and falling ball viscometer. The two forms of viscosity are kinematic and dynamic viscosity. It is difficult to take viscosity measurements of fluids at high temperatures, and thus, it is recommended to use the falling ball viscometer as a means of tackling the problem; viscosities of bio-oils at 40 °C have been reported to be in the range of 10–100 cP. Generally, bio-oils have relatively lower viscosities than biodiesel but higher than those of bioethanol. At moderate temperatures, bio-oils become easier to pump and atomize, either directly or with the aid of atomizing steam; however, Boucher et al. (2000) carried out an investigation in which they found that properties of bio-oils become totally altered at temperatures above 80 °C.

10.4.8 RHEOLOGY

The behavioral property of a bio-fluid is represented by Equation (10.2)

$$\tau = \tau_0 + \mu D^n, \tag{10.2}$$

where τ is the shear stress, τ_0 the yield stress, D is shear rate, and n is an index. For Newtonian fluids, yield stress = 0 and $n = 1$, and hence they do not have apparent viscosities, because their viscosities are unchanged under applied forces. The relationship between viscosity and kinematic viscosity is density, that is,

$$\mu/\rho = v, \tag{10.3}$$

where μ is the oil viscosity, ρ the oil density, and v the oil kinematic viscosity. Most bio-oils are known to have constant viscosities at <80 °C, though extractive-rich bio-oils may possess non-Newtonian behaviors at such temperatures. For non-Newtonian oils, their viscosities are characterized by apparent viscosities, which are functions of the changes in shear rates of the fluids, and examples include Bingham plastics, thixotropic fluids, and dilatants. An investigation on the steady and dynamic rheological properties of softwood biomass residue (SWBR)-derived bio-oils was carried out by Ba et al. (2004a, b), in which they found that there existed a phase change when the oil attained a temperature of 46 °C. Based on the study, the sample bio-oil was described as one exhibiting loss of modulus behavior (non-Newtonian), which is evident in its three-dimensional (3D) structure and the presence of waxy substances, lignin, and particles in the oil at <46 °C, whereas the oil retained its Newtonian behavior at temperatures above 46 °C; these findings were also validated by the works of Pérez et al. (2006) and Boucher et al. (2000).

10.4.9   STABILITY

Bio-oils are not thermodynamically stable, owing to the various interactions their components undergo during storage. According to Oasmaa and Kuopalla (2003), the unstable nature of bio-oils can be traced to their altered viscosities during storage, a drastic increase in viscosity upon heating, evaporation of volatiles, and oxidation in air, whereas Diebold (1997) asserted that the reactions may be caused by the process of aging in the bio-oils. Aldehydes in bio-oils are the most highly unstable fractions of bio-oils because of their relatively high potentials to become hydrated. With alcohols they form hemiacetals, acetals, and H_2O, with phenol-based compounds they produce resinous substances and H_2O, with proteins, dimers are formed, while with other aldehydes oligomers and resins are formed. Their findings further revealed that the physiochemical changes in oil property occur within the first 6 months of storage, with a resultant increase in molar mass of polymeric lignin as well as low yield in aldehydes and ketones/monomeric lignin.

10.4.10   FLASH, POUR, AND CLOUD POINTS

The point at which a liquid fuel flashes is the lowest temperature at which the vapor above the fuel/oil ignites when in contact with the flame. Based on the percentage of organic volatiles, this point is usually determined using a flash-cup tester, with an operating temperature in the range of 40–100 °C or >100 °C. Flash-point determination at temperatures in the range of 70–100 °C should not be encouraged, because vaporization of moisture prevents the fuel's vapor from igniting. The lowest temperature at which a biofuel/oil can flow without resistance to its movement is regarded as its pour point. Resistance to flow of liquid fuels results from increased viscosity and wax crystallization in the fuel. Pour point helps to characterize the fluidity of a liquid fuel at low temperatures. The pour point of bio-oils sourced from wood ranges from 12–33 °C (Oasmaa et al., 1997), whereas that of other materials may be above this range, for example, that of bio-oil from straw is about 36 °C (Sipila et al.,

1988). Liquid fuel cloud point is the highest temperature at which a cloud of waxy crystals is first detected; the cloud point of bio-oil obtained from SWBR has been reported to have a 3D structure below 40–50 °C (Ba et al., 2004a, 2004b; Perez et al., 2007). However, their investigation was criticized as it does not follow the standard procedures based on the results obtained by Oasmaa et al. (1997) in which they observed a bio-oil without any waxy crystals. In order to explore the possibility of the formation of wax by other components in the oil, the oil was cooled to 21 °C and no crystals of wax was formed.

10.4.11 SPECIFIC HEAT CAPACITY, SURFACE TENSION, AND THERMAL CONDUCTIVITY

The surface tension of a liquid fuel is the force within which the surface of a liquid fuel still acts as a skin because of the attractive force/force of cohesion between the fluid molecules. It is a vital parameter for atomizing liquid fuels. Bio-oils have been reported to have a surface tension of 28–40 mN/m at room temperature, and this is quite higher than that of mineral oils, though there are speculations that the higher surface tension of bio-oils relative to those of mineral oils is a result of the presence of moisture, whose surface tension at 25–27 °C is 72 mN/m. Generally, the surface tension of most bio-oils/liquid fuels decreases slightly at increased temperatures, for example, Perez et al. (2001) observed that bio-oil from SWBR showed a sharp decline in its surface tension at 45 °C owing to the melting of its 3D structure at that temperature. Specific heat capacity of a liquid fuel is the heat energy requirement to raise its temperature by one degree, while the thermal conductivity of a liquid fuel is the heat containment and transport potential of the fuel. Both parameters are very relevant for adequate design of heat transfer and atomization equipment such as heat exchangers or furnaces. Based on available information, bio-oils have thermal conductivities and specific heat capacities in the range of 0.35–4 W/m K and 2.5–3.5 kJ/kg K at temperatures of 20–60 °C (Qiang et al., 2009).

10.4.12 ATOMIZATION POTENTIAL

The breaking of liquid droplets aided by steam or some pressure, such that they spray as small pockets during combustion, is known as atomization; this is largely influenced by the high relative velocity of liquid fuel to the surrounding air. The spray quality of the liquid fuel is characterized by its droplet size and velocity/spray angle. The droplet size of bio-oils increases with surface tension and viscosity. For this reason, bio-oils have higher viscosities and surface tensions than light mineral oils, and hence, precautionary measures ought to be taken for successful atomization of bio-oils. According to Chiaramonti et al. (2005), a bio-oil sample can be atomized by preheating the oil at 80 °C while maintaining the Sauter Mean Diameter (SMD) of the oil droplet size at 50 µm at pressures >0.6 MPa. A correlation for estimating the SMD of bio-oils is given by Equation (10.4)

$$\text{SMD} = 599.2 \text{ FN}^{-0.393} \Delta P^{-0.418} \gamma^{0.251} \sigma^{0.277}, \tag{10.4}$$

where FN is flow number, ΔP is the differential pressure, and γ is the surface tension.

10.4.13 IGNITION POTENTIAL AND CETANE AND OCTANE NUMBERS

Because of the viscosities, high latent heats of vaporization and chemical compositions of bio-oils, they are hardly ignited, whereas biodiesel and bioethanol are easily ignited because of their thin and volatile natures, respectively; hence, the fuels are characterized by high octane numbers, whereas the oils are characterized by cetane numbers. However, by preheating the air/oil–air mix using pilot flames, and with the inclusion of ignition improvers and pilot flames, they are easily ignited. Cetane number defines the ignition potential of a liquid fuel. It also measures the tendency of a fuel to automatically ignite when enclosed in a diesel engine. High cetane numbers inform low ignition delays prior to combustion, and vice versa. Low cetane numbers can result in engine malfunction and knocked pistons. Mineral oils have cetane numbers of about 48, whereas those of bio-oils are difficult to measure, because pure bio-oils hardly ignite when in conventional diesel engines or when in contact with flammable sources. The method proposed by Ikura et al. (2003) is useful for determining the cetane number of bio-oils. NREL and Ensyn procedures were adopted to synthesize two bio-oils, as well as measure their ignition delays, in a high-speed diesel engine (HSDE), and the oils were found to exhibit far higher ignition delays/activation potentials than a typical diesel oil (Shihadeh and Hochgreb, 2000). They also investigated the relationship between the pyrolytic conditions and thermal characteristics/properties of the oils and found that the bio-oil experimented by using NREL procedure was more accurate relative to that experimented results obtained using Ensyn procedure. However, they concluded that the results obtained may have been as a result of the oil from NREL having lower moisture and mean molar-weight relative to that obtained by Ensyn method (Shihadeh and Hochgreb, 2002).

10.4.14   LUBRICITY

The ability of a liquid fuel to reduce wear and friction is termed lubricity, and it is a function of the fuel's properties, for example, its viscosity and composition. Fuels characterized by low lubricities will primarily cause engine/machine wear. Tribological characteristics of a bio-oil were determined using a high-frequency friction tester (open system) and a four-ball wear machine (Oasmaa et al., 1997). The high-frequency friction tester failed in taking the appropriate measurement owing to the evaporation of oil during the test whereas the four-ball tester was confirmed suitable for taking reliable friction measurements at a loading speed of 392 N, spindle speed of 1430 rpm at a total time of 1 h and obtained Wear Sour Diameter in the range of 0.5–0.7 mm for the bio-oil as compared to 1.7 mm, which was obtained for a typical diesel fuel; hence, confirming the greater lubricity of the bio-oil. The work of Lu et al. (2008) investigated the tribological properties of oil obtained from pyrolyzed rice husk using a four-ball tester within a maximum nonseizure loading of 470 N. The wear scar diameter and mean friction coefficient were obtained as 0.57 mm and 0.082, respectively, for a loading speed, spindle speed, and test duration of 196 N, 1450 rpm, and 30 min, respectively.

10.4.15   ACIDITY, CORROSION, AND ANTICORROSION PROPERTY

Acid content, pH, and acid number of bio-oils usually range from 7–12 wt%, 2–4, and 50–100 mg KOH/g oil, respectively. Raw bio-oils are known to be corrosive to metals like aluminum, mild steel, and nickel-based alloys; however, when these oils are processed/transesterified to form esters/polyols, as well as when mixed with additives to form biolubes, they become anticorrosive (Sanni et al., 2017a, 2017b). The corrosion rates of raw bio-oils are accelerated by high moisture or increased temperatures (Aubin and Roy, 1990). Metals such as cobalt and stainless steel, and polymers such as polyethylene, polyester resins, and polypropylene, are highly resistant to bio-oil corrosion or other forms of corrosion. The study conducted by Fuleki et al. (1999) confirmed that bio-oil produced using Ensyn procedure was not corrosive on brass and austenitic steel but had some effect on mild steel and aluminum. However the bio-oil from soft wood biomass residue will form oxides and hydroxide films on the surfaces of aluminum, copper, and austenitic steel at 80 °C which will not impede further oxidation of the metals, whereas, for the steel, the film (Cr_2O_3) formed was found to be passive.

10.4.16 COMBUSTIBILITY

Combustion of a bio-oil is the ability of the oil to burn in an appreciable quantity of oxygen in air. Recent advancements in technology allow for the in situ imaging of the behavioral patterns of bio-oils in combustion chambers. One method employs the fiber suspension of a single-oil droplet which takes advantage of the simultaneous measurements of temperature/ pressure and changes in oil behavior in the engine, though the droplet size is determined by the fiber diameter, with a significant influence of the fiber on oil combustion, while the other adopts the concept of a free-falling monodispersed droplet of the oil; the second approach is directly an opposite of the first method. Laminar streams of monodispersed droplets of bio-oil directed into a reactor were investigated using an oil droplet diameter of 300 μm, temperature of 1600 K, and oxygen concentrations of 14%–33%. Several identified stages of the bio-oil combustion included droplet ignition, quiet/slow combustion, micro-explosion, and disruptive soot combustion of fragmented droplets, alongside production and burnout of cenosphere particulates, but the mineral oils only underwent slow sooty burning from ignition to burnout within similar conditions (Allessio et al., 1998; Calabria et al., 2000; Calabria et al., 2007; Perez et al., 2006; Shaddix and Hardesty, 1999; Shaddix and Huey, 1997; Wornat et al., 1994).

10.4.17 TOXICITY

Gratson (1994) conducted toxicological tests of two bio-oils produced in a vortex pyrolytic reactor. Confirmatory toxicological effects of the oils were ascertained for their eye irritations, acute breathing/lung damage, and Ames mutagenicity resulting from bacterial infection when direct contacts were made with the oils or when air contaminated with the oils was inhaled. Also, a study by Diebold (1997) reported that the harmful effects of bio-oils are caused by their inherent aldehydes, unsaturated oxygenates, and furans; although the toxicities of these compounds were found to decline at longer times, there are projections indicating the toxicity of bio-oils in the region of 700 mg/kg body weight. Also, the presence of polycyclic aromatic hydrocarbons (PAHs) in bio-oils has been reported to be cancerous in humans; the presence of the cyclic aromatics was judged to be caused by the pyrolysis temperature (i.e., 500 °C) of formation of the bio-oils from their parent biomasses (Diebold, 1997; Williams and Horne, 1995). Tsai et al. (2007) analyzed the PAHs formed from agricultural wastes via inductive heating at 500 °C. The resulting PAHs (i.e., 1.10–2.45 mg/L for naphthene and 0.72–7.61 mg/L for acenaphthene) in the bio-oils were primarily of low molecular weights. Also, there are speculations that a lot of PAHs would be formed when catalytically upgrading bio-oils using zeolite catalysts (Williams and Horne, 1994; Vitolo et al., 2001). According to a study by Girard and Blin (2003), bio-oils seem to exhibit no toxic effects but exhibit slight mutagenic effects; thus, overall, they are less harmful compared to conventional diesel fuels.

10.4.18   BIODEGRADABILITY

At the production, storage, and transportation stages of bio-oils, leaks from ruptured or corroded/aged pipes or vessels might ensue, which might cause bio-oil spills, but owing to their biodegradable potential, they can be transformed into harmless substances. In biodegradation studies by Piskorz and Radlein (1999), aquatic and soil environments were considered where the test bio-oils were inoculated using treated sludge. The biological oxygen demand was recorded over a period of 5 days, and the calculated biodegradation rates/patterns, though similar, were higher than those of diesel fuels. The study also highlighted the enhancement of biodegradability of the oils by reacting them with a base such as lime. The work of Biln et al. (2007) involves the examination of the aerobic biodegradable properties of several bio-oils in freshwater using the Modified Sturm (OECD 301B) Test method. Inoculation of the bio-oils was done using freshly obtained, activated sludge from a sewage treatment plant. After 8 days, fast degradation of the bio-oils was observed. At the end of 28 days, there was similarity in the degradation curves observed with the 41–50 wt% bio-oils as compared with an EN590 diesel fuel.

10.5   ECONOMICS AND SUSTAINABILITY

Generally, biomass refers to materials comprising cellulose, hemicellulose, lignin, lipids, proteins, simple sugars, and starches; however, the three major constituents of any biomass include cellulose, hemicellulose, and lignin. Water, carbon (51 wt%), and oxygen (42 wt%) constitute above 90% of the dry weight of a sample biomass. Other trace elements include hydrogen (5 wt%), nitrogen (0.9 wt%), and chlorine (0.01–2 wt%) (Bolyos et al., 2009). Renewable energy crops, such as wood and agricultural/forest residues, are major/predominant sources of renewable energy. Wastes recovered from processed food, sewers, municipal solid wastes, and wastes from wood pulping can also be considered as biomass (Demirbas, 2000; International Energy Agency, 2006, 2007). Biomasses can be used as heating sources, though they have lower heating values, that is, 15–19 GJ/t, on a similar weight basis, whereas those of agricultural residues and woody materials are in the ranges of 15–17 GJ/t and 18–19 GJ/t, respectively, which are comparative to that of coal, that is, 20–30 GJ/t. For biomass, the bulk density is 10%–40% of the density of fossil fuels (Scurlock, 2009). The relative ratio of volatile matter in fossil fuel to that of biomass is 20%:80% (Bolyos et al., 2009). The high ignition stability of biomass makes for easy processing of the raw material thermochemically to other higher-value fuels, including methanol (C_2H_5OH) and hydrogen (H_2).

In Figure 10.4, several biomass sources are presented alongside their prospective energy applications. Several waste biomass materials, including food residues, agricultural crops, and animal wastes or municipal solid wastes, have been exploited as potential sources of energy and bio-products, which find applications in power generation and transportation, as well as in the production of biomaterials for biomedical applications. The sustainability of the use of biomass as a renewable energy source is justified by its contribution to 14% of the world's energy requirement, which makes it the fourth in the hierarchy of the world's energy sources (Veringa, 2009). Canada, despite being one of the richest in fossil fuels, still has approximately over 4.7% of its national primary energy being sourced from renewable biomass/ waste from 2006 till date. Furthermore, there are projections that this value will rise by 6%–9% in the next 20 years (Natural Resources Canada, 2009).

Figure 10.5 gives an overview of the four major processes that a biomass will undergo in the course of its utilization for the production of solid fuel (charcoal), liquid fuel (bio-oil), and gas, for example, synthetic gas, while Figure 10.6 is an illustration of the power generation in European Union nations in the years 1990, 1995, and 2000, with a steady increase in net energy generation from municipal waste and biomass. Also, the International Energy Agency data for power generation from biomass has been on a mean annual rise of 2.5% over the last decade (International Energy Agency, 2003).

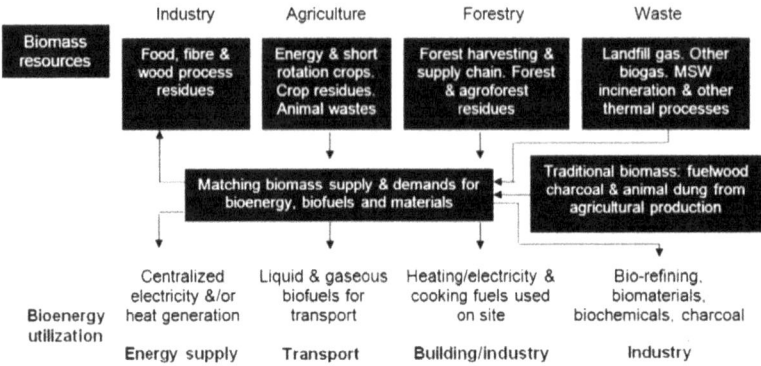

FIGURE 10.4   Different biomass resources and their utilization for bioenergy (Reprinted from International Energy Agency, 2007, with reprint permission obtained from Elsevier (Zhang et al., 2010)).

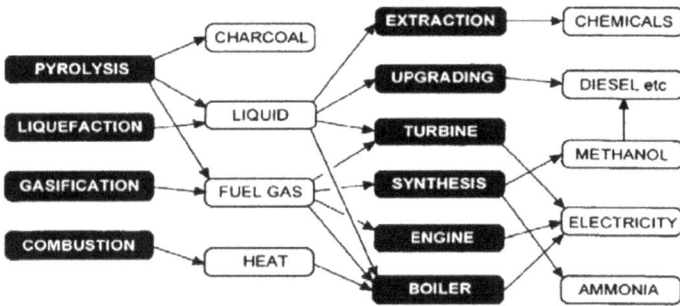

FIGURE 10.5   Thermochemical processes for energy production and other value-added products.

Source: Reprinted with permission from Bridgewater and Peacocke, 2000. © Elsevier.

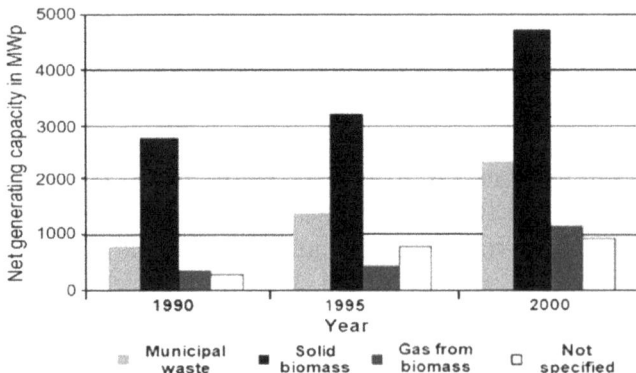

FIGURE 10.6   Net energy from municipal waste, solid biomass, biogas and other sources for three different years (adopted from International Energy Agency, 2000).

Two main processes by which biomass can be converted to bioenergy are the thermochemical and biochemical/biological processes (Ni et al., 2006). According to Bridgewater (2001), higher efficiencies are usually achieved for thermochemical processes than for biochemical/biological processes, especially when estimated in terms of the time requirement, that is, days, weeks, or months (Bridgewater, 2001), as well as the potential to render most of the organic constituents unreactive. For instance, lignocellulosic materials are deemed nonfermentable, which makes it difficult for them to easily decompose through biological means; however, their decomposition is made easy by thermochemical means, that is, by direct combustion, pyrolysis, gasification, and liquefaction (Bridgewater, 2000, 2001; Jenkins et al., 1998; Williams et al., 2003). The energy stored in biomass is released as heat of combustion when co-firing to form charcoal, bio-oils, and syngas through pyrolysis, liquefaction, and gasification processes, respectively. Also, another major way to sustain the use of biomass for energy generation is to encourage local energy generation, as well as implement several land-use policies while creating awareness on the bioresource potential of the untapped resources lying around us as waste (Figure 10.7). According to Karmarkar-Deshmukh and Pray (2009), multinational companies, such as Royal Dutch/Shell, Conoco-Philips, and British Petroleum, are intensely studying the effects of several bioethanol fractions/blends on automobile engine performance, which has made manufacturers like Ford, Chevrolet, and tractor manufacturers shift their engineering budgets to cater to the development of modified engine structures in order to suit the properties of the fuel or its blends. This further implies that the new/future Flexible Fuel Vehicles are such that they can run on ethanol-blended gasoline or pure ethanol. Petrobas, a company in Brazil, produces ethanol from sugarcane. Table 10.1 shows the current trends in the exploitation of biomass for biofuels across the globe, which is an indication of good sustainability measures put in place by the companies highlighted; also, there is a high tendency for those nations to expand/triple their GDPs if the market/growing demand for biofuels is sustained in the next 10 years.

Among all alternative sources for biofuel synthesis, the production of biofuels from natural renewable bacteria such as microalgae is seen to be the most sustainable option for biodiesel and bioethanol production, which in turn contributes less to greenhouse gas emissions (Adenle et al., 2013; Najafi et al., 2011; Ndimba et al., 2013). Bacteria are very useful for biomass production, because they comprise 77% dry cell mass, generate lipids via photosynthesis, as well as improve biofuel yield (Amaro et al., 2012; Gurung et al., 2012; Haik et al., 2011; Hu et al., 2013; Lin et al., 2014;

Bioresource Utilization and Management

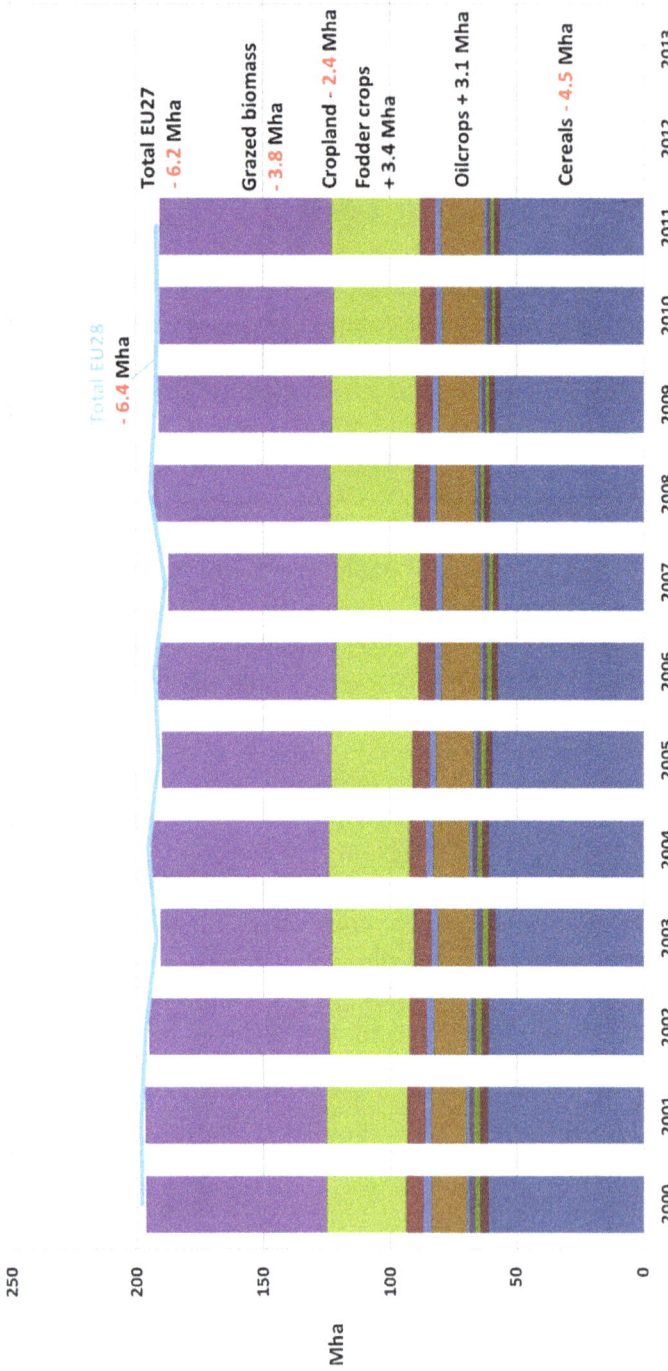

FIGURE 10.7   Agricultural land area statistics in EU-27 from 2000 to 2011.

Source: Schütz (2003), WI, based on FAOSTAT.

TABLE 10.1 Recent Data on Biofuel Trends Across the Globe

Country	Company	Product	Feedstock	Capacity	Status
Brazil	Petrobras	Ethanol	Sugarcane	2 billion L	Running plants for 30 years, and has production plants in18 countries
		Biodiesel	Plant oil/animal fat	326,000 cm^3	–
	Raizen	Ethanol	Sugarcane	2.1 billion L	Brazil's fifth largest company in terms of revenue
	GranBio	Ethanol	bagasse	82 million L	Operating since 2014
India	Praj Ind. Ltd	Ethanol	Sugarcane	30,000 L	Demonstration level
	Indian oil company with Praj Industries	Ethanol	Sugarcane	$(100–150) \times 10^3$ L	Under plan
Japan	Blue Fire				
	Renewables	Ethanol	Rice and wheat straw	44.5 million L	First plant to use acid hydrolyzed cellulose
U.S.	Coskata	Ethanol	Wood	208 million L	Able to utilize agriculture and municipal waste
	Renewable Energy Group Inc. (Nasdaq: REGI)	Biodiesel	Inedible corn oil	972 million L	Largest biodiesel producer in the United States, and running 10 plants
	POET-DSM	Ethanol	Corn crop residue	3.8 billion L	POET developing ethanol production technology since 2001
	Verenium	Ethanol	Softwood and grass	5.3 million L	Running demonstration level plant
	DuPont Industrial Biosciences Business unit	Ethanol	Cellulose	104 million L	Running plant demonstration level since 2008

TABLE 10.1 *(Continued)*

Country	Company	Product	Feedstock	Capacity	Status
	Thailand PTT	Ethanol	Cassava and Molasses	4.8 million L	Under plan
China	TMO renewables	Ethanol	Cassava stock	13.6 million L	Demonstration scale
Malaysia	Pioneer Bio Industries Corp.	Ethanol	Nipah palm	530 million L	Project under plan
Canada	Enerkem	Ethanol	Municipal solid waste	40 million L	Under construction
Norway	Weyland	Ethanol	Cellulosic material	200,000 L	Demonstration level
Italy	Beta Renewables	Ethanol	Wheat straw	75 million L	First commercial scale cellulosic ethanol plant in Crescentino, Italy, and able to process 270,000 tons of biomass

Source: Adapted with permission from Bhatia, et al., 2017. © Elsevier.

Najafi et al., 2011; Parmar et al., 2011; Singh and Gu, 2010; Shuping et al., 2010). The high photosynthetic ability of microalgae suffices not only for effective lipid storage/generation but also for efficient production of oxygen, sequestration of carbon, and balancing of the nitrogen cycle (Kumar et al., 2011; Tsai et al., 2012).

10.6 INTELLECTUAL PROPERTY DOCUMENTATION ON BIOFUELS

This section reviews the role of increased economic/technological trends and their transformations and influence on the ownership of intellectual property rights in biofuel research. Countries leading in productive biofuel research are, in the following order: United States, Germany, Japan, France, and the United Kingdom (Figure 10.8). Here, a review of patent publications on lignocellulosic biofuels is considered in order to buttress the large spectrum of economic and regulatory factors that influence the advancement of technologies in biofuel research. In recent times, patent applications have gained an eightfold jump in the United States, and are in the range of 130–150 annually; this is suggestive of a growing pace of biofuel research and a potential market for biofuels. Also, industry-technology interdependence is likely to rise as the likelihood for broad, exclusive patent regimes diminishes, hence increasing the tendency for seeking collaborative technological solutions in R&D as well as investing in newer production equipment in the future.

1. Developing processes for the conversion of carbohydrates, such as sugars, sugar alcohols, celluloses, lignocelluloses, hemicelluloses, and lignocellulosic biomass, to higher hydrocarbons, such as synthetic gasoline, diesel fuel, chemicals, and jet fuels, for use as transportation fuels is usually associated with some challenges, which include process cost and complexity, thus making it a difficult task to accomplish with traditional resources (mineral fuels), hence the need to search for other viable alternatives. A patent describing the production of bioethanol from the consumption of a variety of biomass-derived substrates via anaerobic digestion was filed by the Trustees of Dartmouth College (2008). Inactivation of hydrogenase genes from *Thermoanaerobacterium saccharolyticum* was done in order to produce mutant strains with less enzymic activities; the straining of a mutant gene indicated a significant rise in ethanol production. By manipulating hydrogenase activities, a new approach was developed for improving substrate utilization and ethanol yield via fermentation.

FIGURE 10.8 Biofuel (bioethanol, biodiesel, biojet kerosene, and biogas) producing countries and their filed patent statistics.

Source: Reprinted from IEA data (2015). Open access.

Triacylglycerols are generated from the sequential acylation of glycerol-3-phosphate transferases (GPAT) backbone with 3-acyl-CoAs-catalyzed acyltransferases (Figure 10.9). The acylation results in lysophosphoric acid which is further acylated by lysophosphatidic acid transferase to generate phosphatidic acid which hastens the removal of phosphate groups to produce diacylglycerol (DAG). The oil production is then catalyzed by diacylglycerol acyl transferase from DAG to TAG (stored fat/lipids) in the algal cell (Cagliari et al., 2011; Fan et al., 2011).

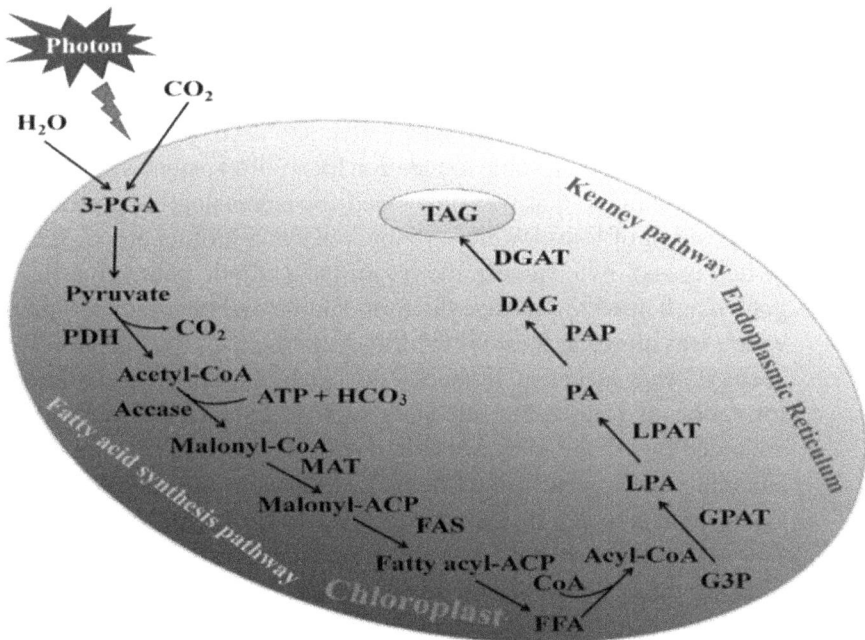

FIGURE 10.9 Synthetic pathway for tri-acyl-glycerols (TAGs) as storage systems for chemical energy in lipids found in microalgae.

Source: Reprinted with permission from Maity, et al., 2014. © Elsevier.

2. The market for cellulosic ethanol from wheat straw was successfully created by Iogen Inc. with an industrial process plant that produces 10 million gallons yearly, which is the first ever in Canada. The annual production of ethanol has grown in the United States to a net value of 41% increase in the number of ethanol plants between 2000 and 2005. New emerging companies in the bioethanol market are enzyme- and biotech-based firms, and among the enzyme-based

research companies, Genencor and Novozymes are in possession of 60% of the patents on biofuels produced from enzymatic reactions (Clark et al., 2008).

3. A patent was filed by Syngenta Participations AG (1997) on transgenic plants expressing cellulolytic enzymes. The patent describes new methods of regulating gene expression in plastids with the aid of an inducible transactivator-mediated system. The invention also includes the description of the production of cellulose-degrading enzymes in plants through genetic engineering, which facilitates the conversion of lignocellulose to biofuel.

4. Transgenic plants describing one or multiple cell wall-degrading enzymes that can disintegrate lignocellulosic biomasses to form fermentable sugars, which serve as intermediates for ethanol production, was filed by Michigan State University (2007).

5. The Samuel Roberts Noble foundation Inc. (2009) patent was filed for a novel technique that was developed for increasing the percentage composition of fermentable carbohydrates in a biofuel crop (alfalfa/switch grass), by changing the lignin biosynthetic pathway using a gene that helped to increase the fermentable carbohydrate polymers which was aimed at improving biofuel yield.

6. Methods that establish fermentation reactions which result in the formation of four-carbon alcohols such as butanol and butan-1-ol are aided by recombinant bacterium cells having one or more genes encoding an enzyme, which helps to catalyze biochemical conversions, were patented by Butamax Advanced Biofuels (2006, 2009).

7. The use of a fuel blend comprising a conventional fuel (i.e., diesel fuel, jet fuel, kerosene, or gasoline) additive and at least a C5 isoprenoid compound obtained from sugar fermentation was patented by Amyris, Inc. (2007).

10.7 CONCLUSION

The world's continuous growing demand for energy is informed by the rapid increase in the average global population. Since there is no guarantee that the world can solely rely on fossil fuels to cater to her geometrically projecting energy demands, it is evident in the provisions of this chapter that considering an alternative source of energy such as biomass/bioresource which has huge potential for biofuel production can be sustained. However, going by the advancements in technology and the evidential production statuses of some biofuel plants across the globe, as highlighted in this chapter, biofuels have competing properties with their mineral-sourced counterparts and can directly replace them or be modified to suit the energy requirement. Three major advantages of biofuels buttress these facts, which include their biodegradability, cost-effectiveness, and sustainability. This is supported by the fact that wastes from biomass can be easily sourced, cultivated, and processed to generate energy and valuable products such as lube oils, biodiesel, aromatics for petrochemicals, etc., as highlighted in this chapter. Although processing them might be somewhat energy-intensive, they will attract lower costs when compared to mineral-based fuels. Based on the data gathered in this study, the market for biofuels is gradually being established, as advances in technology are still being unveiled to reduce costs of production significantly. The properties of biofuels were also discussed in order to understand their chemistry, underscore their potential for sustainability, as well as advocate for the need to take advantage of the oils as an alternative source of fuel, owing to their properties and underscored advantages. Bio-algae and other materials were also discussed as parent materials for biofuels, and this goes further to imply that the future is being watched and protected in order to sustain the technology involved without jeopardizing the environment. Based on the discussions in this chapter, biofuels are indeed the future fuels to complement gasoline/automotive gas oil, which is obtained from petroleum to be used as fuel in automobile engines; this goes further to imply that the market for biofuels can be fully established, as done in some countries, such that manufacturers of gasoline/diesel engines may eventually consider remodifying the internal walls of some conventional gasoline engines so as to accommodate bioethanol/biodiesel as means of complementing or ensuring a diversified system that does not solely rely on fossils as energy sources.

KEYWORDS

- **biomass**
- **biofuels**
- **biomass valorization**
- **bio-oils**
- **value-added products**

REFERENCES

Abdoulmoumine, N.; Adhikari, S.; Kulkarni, A.; Chattanathan, S. A review on biomass gasification syngas clean-up. *Applied Energy* 2015, 155, 294–307.

Adenle, A.A.; Haslam, G.E.; Lee, L. Global assessment of research and development for algae biofuel production and its potential role for sustainable development in developing countries. *Energy Policy* 2013, 61, 182–195.

Agblevor, F.A.; Besler, S. Inorganic compounds in biomass feed stocks. Effect on the quality of fast pyrolysis oils. *Energy Fuel* 1996, 10, 293–298.

Alexander, D.; Richard, D.; Scott, G.; Ryan, P. *Enzymatic hydrolysis of cellulosic biomass for the production of second generation biofuels*. Worcester Polytechnic Institute, Worcestershire, UK; 2009.

Allessio, J.D.; Lazzaro, M.; Massoli, P.; Moccia, V. *Thermo-optical investigation of burning biomass pyrolysis oil droplets*. Im Twenty-seventh Symposium (International) on combustion. The Combustion Institute, Pittsburgh; 1998, pp. 1915–1922.

Amaro, H.M.; Macedo, Â.C. F.; Malcata, X. Microalgae: An alternative as sustainable source of biofuels? *Energy* 2012, 44(1), 158–166.

Amyris Biotechnologies Inc. *Fuel components, fuel compositions and methods of making and using same*. CA 2652732; 2007.

Andersen, N.; Stenby, E.H. *Enzymatic hydrolysis of cellulose: Experimental and modelling studies*. Technical University of Denmark, Denmark, Germany; 2007.

Asadullah, M. Biomass gasification gas cleaning for downstream applications: A comparative critical review. *Renewable and Sustainable Energy Reviews* 2014, 40, 118–132.

Aro, E.-M. From first generation biofuels to advanced solar biofuels. *Ambio* 2016, 45, 24–31.

Aubin, H.; Roy, C. Study on the corrosiveness of wood pyrolysis oils. *Fuel Science Technology International* 1990, 8, 77–86.

Ba, T.; Chaala, A.; Pérez, M.G.; Rodrigue, D.; Roy, C. Colloidal properties of bio-oils obtained by vacuum pyrolysis of softwood bark: Characterization of water-soluble and water-insoluble fractions. *Energy and Fuel* 2004a, 18, 704–712.

Ba, T.; Chaala, A.; Pérez, M.G.; Roy, C. Colloidal properties of bio-oils obtained by vacuum pyrolysis of softwood bark, storage stability. *Energy and Fuel*, 2004b, 18, 188–201.

Badger, P.C. *Ethanol from cellulose: A general review*. ASHS Press, Alexandria, VI, USA, 2002.

Bakhshi, N.N.; Adjaye, J.D. Properties and characteristics of ENSYN bio-oil. In: *Biomass Pyrolysis Oil Properties and Combustion Meeting*; National Renewable Energy Laboratory, Colorado, 1994, pp. 54–66.

Bhatia, S.K.; Kim, S.-H.; Yoon, J.-J.; Yang, Y.-H. Current status and strategies for second generation biofuel production using microbial systems. *Energy Conversion and Management* 2017, 148, 1142–1156.

Biln, J.; Volle, G.; Girard, P.; Bridgwater, T.; Meier, D. Biodegradability of biomass pyrolysis oils: Comparison to conventional petroleum fuels and alternatives fuels in current use. *Fuel* 2007, 86, 2679–2686.

Bolyos, E.; Lawrence, D.; Nordin, A. *Biomass as an energy source: The challenges and the path forward*. 2009. http://www.ep.liu.se/ecp/009/003/ecp030903.pdf (accessed September, 2018).

Boucher, M.E.; Chaala, A.; Roy, C. Bio-oils obtained by vacuum pyrolysis of softwood bard as a liquid fuel for gas turbines. Part I: Properties of bio-oil and its blends with methanol and a pyrolytic aqueous phase. *Biomass Bioenergy* 2000, 19, 337–350.

Boucher, M.E.; Chaala, A.; Pakdel, H.; Roy, C. Bio-oils obtained by vacuum pyrolysis of softwood bark as a liquid fuel for gas turbines. Part II: Stability and ageing of bio-oil and its blends with methanol and a pyrolytic aqueous phase. *Biomass Bioenergy* 2000, 19, 351–361.

Bridgewater, A.V. *Thermal conversion of biomass and waste: The status.* Bio-Energy Research Group, Aston University, Birmingham, UK, 2001.

Bridgwater, A.V.; Maniatis, K. The production of biofuels by the thermochemical processing of biomass. In: *Molecular to Global Photosynthesis*; Archer, Md., Barber, J., Eds.; IC Press, London, 2004, pp. 521–612.

Bridgwater, A.V.; Peacocke, G.V.C. Fast pyrolysis processes for biomass. *Renewable and Sustainable Energy Review* 2000, 4, 1–73.

Bridgwater, A.V.; Czernik, S.; Piskorz, J. An overview of fast pyrolysis technology. In: *Progress in Thermochemical Biomass Conversion*; Bridgwater, A.V., Ed.; Oxford, Blackwell, 2001, pp. 977–997.

Butamax Advanced Biofuels LLC. *Fermentative production of four carbon alcohols.* CA 2619458, 2006.

Butamax Advanced Biofuels LLC. *Carbon pathway optimized production hosts for the production of Isobutanol.* CA 2737112, 2009.

Calabria, R.; Chiariello, F.; Massoli, P. Combustion fundamentals of pyrolysis oil based fuels. *Experimental Thermal and Fluid Science* 2007, 31, 413–420.

Calabria, R.; Allessio, J.D.; Lazzaro, M.; Massoli, P.; Moccia, V. *Bio-fuel oil. Upgrading by hot filtration and novel physical methods, Task 5: Fundamental behavior of BFO in combustion.* Final Report, JOR3-CT97–0253, 2000.

Cagliari, A.; Margis, R.; Maraschin, F.S.; Turchetto-Zolet, A.C.; Loss, G.; Margis-Pinheiro, M. Biosynthesis of triacylglycerols (TAGs) in plants and algae. *International Journal of Plant Biology* 2011, 2(1), 40–52.

Chiang, K.-Y.; Lu, C.-H.; Lin, M.-H.; Chien, K.-L. Reducing tar yield in gasification of paper-reject sludge by using a hot-gas cleaning system. *Energy* 2013, 50, 47–53.

Chiaramonti, D.; Riccio, G.; Baglioni, P.; Bonini, M.; Milani, S.; Soldaini, I., et al. Sprays of biomass pyrolysis oil emulsions: Modeling and experimental investigation of preliminary results and modelling. In: Proceeding of the 14th European Biomass Conference & Exhibition, Paris, 2005.

Chisti, Y. Biodiesel from microalgae. *Biotechnology Advances* 2007, 25, 294–306.

Clark, K.; Patel, R.; Jensen, K.; Bennett, A. Lignocellulosic derived bioethanol patent landscape. In: *The Public Intellectual Property Resources for Agriculture (PIPRA)*; Davis, C.A., Ed. University of California, Davis, 2008.

Corral, B.M.; Lostado, L.R.; Escribano, G.R.; Somovilla, G.F.; Vergara G.E. An improvement in biodiesel production from waste cooking oil by applying thought multi-response surface methodology using desirability functions. *Energies* 2017, 10, 130. http://dx.doi.org/10.3390/ en10010130.

Craggs, R.J.; Heubeck, S.; Lundquist, T.J.; Benemann, J.R. Algae biofuel from wastewater treatment high rate algal ponds. *Water Science and Technology* 2011, 63(4), 660–665.

Dalrymple, O.K.; Halfhide, T.; Udom, I.; Gilles, B.; Wolan, J.; Zhang, Q.; Ergas S. Wastewater use in algae production for generation of renewable resources: A review and preliminary results. *Aquatic Biosystems* 2013, 9(1), 2.

Demirbas, A. Mechanisms of liquefaction and pyrolysis reactions of biomass. *Energy Conversion and Management* 2000, 41, 633–646.

Demirbas, A. Biodiesel from municipal sewage sludge (MSS): Challenges and cost analysis. *Energy Sources Part B* 2017, 12, 351–357.

Diebold, J.P. *A review of the toxicity of biomass pyrolysis liquids formed at low temperatures.* National Renewable Energy Laboratory, NREL/TP-430–22739; 1997.

Elliott, D.C. Water, alkali and char in flash pyrolysis oils. *Biomass Bioenergy* 1994, 7, 179–185.

Fan, J.; Andre, C.; Xu, C. A chloroplast pathway for the De Novo biosynthesis of triacylglycerol in *Chlamydomonas reinhardtii*. *FEBS Letters* 2011, 585, 1985–1991.

Fratini, E.; Bonini, M.; Oasmaa, A.; Solantausta, Y.; Teiseira, J.; Baglioni, P. SANA analysis of the microstructural evolution during the ageing of pyrolysis oils from biomass. *Langmuir* 2006, 22, 306–312.

Fuleki, D. Bio-fuel system materials testing. *PyNe Newsletter* 1999, 7, 5–6.

Girard, P.; Blin, J. Environmental, health and safety Pyne working group: Assessment of bio-oil toxicity for safe handling and transportation. In: *Pyrolysis and Gasification of Biomass and Waste*; Bridgwater, A.V., Ed.; Newbury, CPL Press; 2003, pp. 155–60.

Gratson, D. A. Results of toxicological testing of whole wood oils derived from the fast pyrolysis biomass. In: *Biomass Pyrolysis Oil Properties and Combustion Meeting*; Colorado, 1994, 203–211.

Gurung, A.; Van-Ginkel, S.W.; Kang, W.C.; Qambrani, N.A.; Oh, S.E. Evaluation of marine biomass as a source of methane in batch tests: A lab-scale study. *Energy* 2012, 43, 396–401.

Haik, Y.; Selim, M.Y.E.; Abdulrehman, T. Combustion of algae oil methyl ester in an indirect injection diesel engine. *Energy* 2011, 36, 1827–1835.

Hu, Z.; Zheng, Y.; Yan, F.; Xiao, B.; Liu, S. Bio-oil production through pyrolysis of blue-green algae blooms (BGAB): Product distribution and bio-oil characterization. *Energy* 2013, 52, 119–125.

Huang, Y.F.; Chiueh, P.T.; Lo, S.L. A review on microwave pyrolysis of lignocellulosic biomass. *Sustainable. Environment Resources* 2016, 26, 103–109.

IEA (International Energy Agency). *Renewables for power generation, status & prospects.* http://www.iea.org/textbase/nppdf/free/2000/renewpower_2003.pdf (accessed September, 2018).

IEA (International Energy Agency). *IEA bioenergy annual report.* 2006. http://www.energytech.at/pdf/iea_bereport06.pdf (accessed September, 2018).

IEA (International Energy Agency). *Bioenergy project development & biomass supply.* http://www.iea.org/textbase/nppdf/free/2007/biomass.pdf (accessed September, 2018).

Ikura, M.; Stanciulescu, M.; Hogan, E. Emulsification of pyrolysis derived bio-oil in diesel fuel. *Biomass Bioenergy* 2003, 24, 221–31.

Institute of Intellectual Property Research and Development (IIPRD). *Landscape analysis of biofuels in India.* Landscape Report, IIPRD, 2018.

Jenkins, B.M.; Baxter, L.L.; Miles Jr. T.R.; Miles, T.R. Combustion properties of biomass. *Fuel Processing Technology* 1998, 54, 17–46.

Kalam, M.A.; Masjuki, H.H. Biodiesel from palm oil—Analysis of its properties and potential. *Biomass Bioenergy* 2002, 23(6), 471–479.

Karmarkar-Deshmukh, R.; Pray, C.E. Private sector innovation in biofuels in the United States: Induced by prices or policies? *Ag-Bio Forum—Journal of Agrobiotechnology Management and Economics* 2009, 12, 1, article 13.

Kumar, K.; Dasgupta, C.N.; Nayak, B.; Lindblad, P; Das, D. Development of suitable photobioreactors for CO_2 sequestration addressing global warming using green algae and cyanobacteria. *Bioresource Technology* 2011, 102(8), 4945–4953.

Le, R.K.; Wells Jr. T.; Das, P.; Meng, X.; Stoklosa, R.J.; Bhalla, A.; et al. Conversion of corn stover alkaline pre-treatment waste streams into biodiesel via *Rhodococci*. *Royal Society of Chemistry Advances* 2017, 7, 4108–4115. http://dx.doi.org/10.1039/ C6RA28033A.

Lin, K.C.; Lin, Y.C.; Hsiao, Y.H. Microwave plasma studies of spirulina algae pyrolysis with relevance to hydrogen production. *Energy* 2014, 64, 567–574.

Lopes, M.L.; Paulillo, S.C.; Godoy, A.; Cherubin, R.A.; Lorenzi, M.S.; Giometti, F.H.C.; et al. Ethanol production in Brazil: A bridge between science and industry. *Brazilian Journal of Microbiology*, 2016, 47(Supplement 1), 64–76. http://dx.doi.org/10.1016/j.bjm.2016.10.003.

Lu, Q.; Yang, X.L.; Zhu, X.F. Analysis on chemical and physical properties of bio-oil pyrolyzed from rice husk. *Journal of Analytical and Applied Pyrolysis* 2008, 82, 191–198.

Maity, J. P.; Bundschuh, J.; Chen, C-Y.; Bhattacharya, P. Microalgae for third generation biofuel production, mitigation of greenhouse gas emissions and wastewater treatment: Present and future perspectives—A mini review. *Energy* 2014, 78, 104–113.

Michigan State University. *Production of β-glucosidase, Hemicellulose and Ligninase in E1 and Flc-Cellulase-transgenic plants*. CA 2589657, 2007.

Morris, K.W. Fast pyrolysis of bagasse to produce bio-oil fuel for power generation. *International Sugar Journal* 2001, 103, 259–263.

Naik, S. N.; Goud, V. V.; Rout, P. K.; Dalai, A. K. Production of first and second generation biofuels: A comprehensive review. *Renewable and Sustainable Energy Review* 2010, 14, 578–597.

Najafi, G.; Ghobadian, B.; Yusaf, T.F. Algae as a sustainable energy source for biofuel production in Iran: A case study. *Renewable and Sustainable Energy Review* 2011, 15(8), 3870–3876.

Natural Resources Canada. *Canada report on bioenergy*. http://www.canbio.ca/documents/publications/canadacountryreport2009.pdf (accessed September, 2018).

Ndimba, B.K.; Ndimba, R.J.; Johnson, T.S.; Waditee-Sirisattha, R; Baba, M.; Sirisattha, S.; et al. Biofuels as sustainable energy source: an update of the applications of proteomics in bioenergy crops and algae. *Journal of Proteomics* 2013, 93, 234–244.

Nhien, L.C.; Long, N.V.D.; Lee, M. Novel heat–integrated and intensified biorefinery process for cellulosic ethanol production from lignocellulosic biomass. *Energy Conversion and Management* 2017, 141, 367–377.

Ni, M.; Leung, D.Y.C.; Leung, M.K.H.; Sumathy, K. An overview of hydrogen production from biomass. *Fuel Processing Technology* 2006, 87, 461–472.

Oasmaa, A.; Czernik, S. Fuel oil quality of biomass pyrolysis oils—State of the art for the end users. *Energy Fuel* 1999, 13, 914–921.

Oasmaa, A.; Kuoppala, E. Fast pyrolysis of 3 forestry residue, 3. Storage stability of liquid fuels. *Energy Fuel* 2003, 17, 1075–1084.

Oasmaa, A.; Kuoppala, E.; Gust, S.; Solantausta, Y. Fast pyrolysis of forestry residue: Effect of extractive on phase separation of pyrolysis liquid. *Energy Fuel* 2003, 17, 1–12.

Oasmaa, A.; Leppamaki, E.; Koponen, P.; Levander, J.; Tapola, E. *Physical characterization of biomass-based pyrolysis liquids: Application of standard fuel oil analyses*. Espoo. Technical Research Centre of Finland; 1997.

Oasmaa, A.; Peacocke, C.; Gust, S.; Meier, D.; McLellan, R. Norms and standards for pyrolysis liquids, ender-user requirements and specifications. *Energy Fuel* 2005, 19, 2155–2163.

Paethanom, A.; Nakahara, S.; Kobayashi, M.; Prawisudha, P.; Yoshikawa, K. Performance of tar removal by absorption and adsorption for biomass gasification. *Fuel Processing and Technology* 2012, 104, 144–154.

Palmer, J.D.; Brigham, C.J. Feasibility of triacylglycerol production for biodiesel, utilizing RhodococcusOpacus as a biocatalyst and fishery waste as feedstock. *Renewable and Sustainable Energy Review* 2016, 56, 922–928.

Park, J.B.K.; Craggs, R.J.; Shilton, A.N. Wastewater treatment high rate algal ponds for biofuel production. *Bioresource Technology* 2011, 102(1), 35–42.

Parmar, A.; Singh, N.K.; Pandey, A.; Gnansounou, E.; Madamwar, D. Cyanobacteria and microalgae: A positive prospect for biofuels. *Bioresource Technolgy* 2011, 102(22), 10163–10172.

Pérez, M.G.; Lappas, P.; Hughes, P.; Dell, L.; Chaala, A.; Kretschmer, D.; et al. Evaporation and combustion characteristics of biomass vacuum pyrolysis oils. *IFRF Combustion Journal* 2006, 1, 1–28.

Pérez, M.G.; Chaala, A.; Pakdel, H.; Kretschmer, D.; Rodrigue, D.; Roy, C. Multiphase structure of bio-oils. *Energy Fuel* 2006, 20, 364–375.

Pérez, M.G.; Chaala, A.; Padel, H.; Kretschmer, D.; Roy, C. Vacuum pyrolysis of softwood and hardwood biomass: Comparison between product yields and bio-oil properties. *Journal of Analytical and Applied Pyrolysis* 2007, 78, 104–116.

Piskorz, J.; Radlein, D. Determination of biodegradation rates of bio-oil by respirometry. In: *Fast Pyrolysis of Biomass: A Handbook*; Newbury, CPL Press, 1999.

Qiang, L.; Wen-Zhi, L.; Xi-Feng, Z. Overview of fuel properties of biomass fast pyrolysis oils. *Energy Conversion and Management* 2009, 50, 1376–1383.

Radlein, D. Study of levoglucosan production—A review. In *Fast Pyrolysis of Biomass: a Handbook*; Vol. 2, Newbury, CPL Press, 2002.

Sanni, S.E.; Emetere, M.E.; Efeovbokhan, V.E.; Udonne. J.D. Process optimization of the transesterification processes of palm kernel and soybean oils for lube oil synthesis. *International Journal of Applied Engineering Research* 2017a, 12(14), 4113–4129.

Sanni, S.E.; Odigure, J.O.; Efeovbokhan, V.; Emetere, M.E. Comparative study of lube oils synthesized from chemically modified castor and soybean oils using additive. *Science and Engineering Applications* 2017b, 2, 134–141.

Sanni, E.S.; Olasubomi, A.; Ojima, E.Y.; Fagbiele, O.O.; Oluranti, A. Chemical kinetics of alkaline pretreatment of Napier grass (*Pennisetum purpureum*) prior enzymatic hydrolysis. *Open Chemical Engineering Journal* 2018, 12, 36–56.

Scahill, J.W.; Diebold, J.P.; Feik CJ. Removal of residual char fines from pyrolysis vapors by hot gas filtration. In: *Developments in Thermochemical Biomass Conversion*; Bridgwater, A.V. Ed.; Blackie Academic & Professional, London, 1997, pp. 253–266.

Schütz, H. Economy-wide material flow accounts, land use accounts andderived indicators for Germany "MFA Germany". Final report to the Commission of the European Communities. 2003

Scurlock, J. *Bioenergy feedstock characteristics*, http://bioenergy.ornl.gov/ papers/misc/ biochar_factsheet.html; 2009 (accessed September, 2018).

Shaddix, C.R.; Hardesty, D.R. *Combustion properties of biomass flash pyrolysis oils: Final project report*. SAND99–8238, Sandia National Laboratory, 1999.

Shaddix, C.R.; Huey, S.P. Combustion characteristics of fast pyrolysis oils derived from hybrid poplar. In: *Developments in Thermochemical Biomass Conversion*; Bridgwater, A.V., Ed.; Blackie Academic and Professional, London; 1997, pp. 465–480.

Shen, Y.; Yoshikawa, K. Recent progresses in catalytic tar elimination during biomass gasification or pyrolysis—A review. *Renewable and Sustainable Energy Review* 2013, 21, 371–392.

Shihadeh, A.; Hochgreb, S. Diesel engine combustion of biomass pyrolysis oils. *Energy Fuel* 2000, 14, 260–274.

Shihadeh, A.; Hochgreb, S. Impact of biomass pyrolysis oil process conditions on ignition delay in compression ignition engines. *Energy and Fuel* 2002, 16, 552–561.

Shuping, Z.; Yulong, W.; Mingde, Y.; Kaleem, I.; Chun, L.; Tong, J. Production and characterization of bio-oil from hydrothermal liquefaction of microalgae *Dunaliella tertiolecta* cake. *Energy* 2010, 35, 5406–5411.

Singh, J.; Gu, S. Commercialization potential of microalgae for biofuels production. *Renewable and Sustainable Energy Review* 2010, 14(9), 2596–2610.

Singh, A.K.; Garg, N.; Tyagi, A.K. Viable feedstock options and technological challenges for ethanol production in India. *Current Science* 2016, 111, 815–822.

Sipilä, K.; Keoppala, E.; Fagernäs, L.; Oasmaa, A. Characterization of biomass-based flash pyrolysis oils. *Biomass Bioenergy* 1988, 14, 103–113.

Solimeno, A.; García, J. Microalgae-bacteria models evolution: From microalgae steady-state to integrated microalgae-bacteria wastewater treatment models—A comparative review. *Science of the Total Environment* 2017b, 607–608, 1136–1150.

Solimeno, A.; Parker, L.; Lundquist, T.; García, J. Integral microalgae-bacteria model (BIO_ALGAE): Application to wastewater high rate algal ponds. *Science of the Total Environment* 2017, 601–602, 646–657.

Suganyaa, T.; Varmana, M.; Masjukia, H.H.; Renganathanb, S. Macroalgae and microalgae as a potential source for commercial applications along with biofuels production: A biorefinery approach. *Renewable and Sustainable Energy Review* 2016, 55, 909–941.

Šulc, J.; Štojdl, J.; Richter, M.; Popelka, J.; Svoboda, K.; Smetana, J.; et al. Biomass waste gasification—can be the two stage process suitable for tar reduction and power generation? *Waste Management* 2012, 32, 692–700.

The Samuel Roberts Noble Foundation, Inc. *Biofuel production methods and compositions*. CA 2670096, 2009.

The Trustees of Dartmouth College. *Modification of hydrogenase activities in thermophilic bacteria to enhance ethanol production*. CA 2708818, 2008.

Tsai, D.D.; Ramaraj, R.; Chen, P.H. Growth condition study of algae function in ecosystem for CO_2 bio-fixation. *Journal of Photochemistry and Photobiology* B 2012, 107, 27–34.

Tsai, W.T.; Mi, H.H.; Chang, Y.M.; Yang, S.Y.; Chang, J.H. Polycyclic aromatic hydrocarbons (PAHs) in bio-crudes from induction-heating pyrolysis of biomass wastes. *Bioresource Technology* 2007, 98, 1133–1137.

U.S. Non-Provisional Application Ser. No. 12/972,141 Filed Dec. 17, 2010, which claims the benefit of U.S. Provisional Application Ser. No. 61/291,567, Filed Dec. 31, 2009.

Veringa, H.J. *Advanced techniques for generation of energy from biomass and waste*. 2009. http://www.ecn.nl/fileadmin/ecn/units/bio/Overig/pdf/Biomassa_voordelen.pdf (accessed September, 2018).

Vitolo, S.; Bresci, B.; Seggiani, M.; Gallo, M.G. Catalytic upgrading of pyrolytic oils over HZSM-5 zeolite: Behavior of the catalyst when used in repeated upgrading-regeneration cycles. *Fuel* 2001, 80, 17–26.

Williams, P.T.; Horne, P.A. Characterisation of oils from the fluidised bed pyrolysis of biomass with zeolite catalyst upgrading. *Biomass Bioenergy* 1994, 7, 223–236.

Williams, P.T.; Horne, P.A. Analysis of aromatic hydrocarbons in pyrolytic oil derived from biomass. *Journal of Analytical and Applied Pyrolysis* 1995, 31, 15–37.

Williams, R.B.; Jenkins, B.M.; Nguyen, D. *Solid waste conversion: a review and database of current and emerging technologies—final report*. California Integrated Waste Management Board, 2003.

Wornat, M.J.; Porter, B.G.; Yang, N.Y.C. Single droplet combustion of biomass pyrolysis oils. *Energy Fuel* 1994, 8, 1131–1142.

Xu, C.; Donald, J.; Byambajav, E.; Ohtsuka, Y. Recent advances in catalysts for hot gas removal of tar and NH_3 from biomass gasification. *Fuel* 2010, 89, 1784–1795.

Zhang, L.; Xu, C.C.; Champagne, P. Overview of Recent Advances in Thermo-Chemical Conversion of Biomass. Energy Conversion and Management 2010, 51, 969–982.

Zhang, Q.; Chang, J.; Wang, T. J.; Xu, Y. Review of biomass pyrolysis oil properties and upgrading research. *Energy Conversion and Management* 2007, 48, 87–92.

CHAPTER 11

Lignocellulosic Waste-Derived Bioethanol as a Potential Biofuel for the Transportation Sector: Sources and Technological Advancement

SHITARASHMI SAHU[1], KRISHNA PRAMANIK[1*], and NEELAM MEHER[2]

[1]Department of Biotechnology and Medical Engineering, National Institute of Technology, Rourkela, Odisha 769008, India

[2]Academy of Management and Information Technology, Khordha, Odisha 752057, India

*Corresponding author. E-mail: kpr@nitrkl.ac.in

ABSTRACT

The various negative consequences of greenhouse gases, particulate emissions, fluctuations in price, and unbalanced demand–supply relationship using conventional transportation fuel derived from fossil fuel have drawn attention worldwide toward the production of biofuel as an alternative resource of transportation fuel. Among biofuels, bioethanol is considered as the most promising alternative to the petroleum-based transportation fuel, and hence, a lot of research work has been directed in the last two decades for exploring various sources and viable technology for its production because it is derived from renewable resources and clean fuel, and it offers reduced environmental pollution. Among the alternative sources, bioethanol produced from a variety of lignocellulosic waste such as agricultural, wood, forest, and industrial wastes is the most attractive. This chapter enlightens the potential sources of lignocellulosic waste worldwide and technological advances in its conversion to bioethanol as a potential fuel substitute for the transportation sector.

11.1 INTRODUCTION

The depletion of resources, high cost but still uncertainty in availability and major environmental pollution concern of petroleum-based fossil fuel has offered a big headache to the transportation sector and ultimately led to the search of alternative sources based on renewable resources for producing transportation fuel (Lee et al., 1983; Pramanik, 2004; Rogner et al., 2000; Williums et al., 1995). In this context, bioethanol can provide a key solution to replace fossil fuel for the transportation sector. Bioethanol is economically competitive, environmentally adaptable, and can be produced from widely available raw material including more attractive and abundant lignocellulosic waste biomass (Hamelinck et al., 2005; Kumari and Pramanik, 2012a; Sánchez and Cardona, 2008; Solange et al., 2010). Bioethanol can be used as an octane enhancer when mixed with gasoline and is adaptable to the presently used vehicles, designed with an internal combustion engine. The use of bioethanol/gasoline mixture as fuel also reduces the release of gases like CO_2 responsible for global warming, and volatile organic compounds and particulate matter. Therefore, efforts have been put in place to produce bioethanol from lignocellulosic waste biomass instead of energy crops that are associated with the high demand of consumption of water and land for their growth (Balat, 2011). Furthermore, the use of sugarcane and corn to produce biofuel is highly discouraged due to the current scenario in food price (Rajendran et al., 2015). To avoid food-feed-fuel conflicts, it is important to incorporate all kinds of waste biomass into a promising biomass economy (Mahro and Timm, 2007). Although the technology for the conversion of lignocellulosic waste has long been considered to be expensive, however, the recent increase in grain prices and advancements in its conversion technology strongly diverted the attention toward the use of lignocellulosic waste biomass for the production of transportation biofuels. Biofuels offer a reduction in competition with grain for food and feed and allow the utilization of a variety of materials which would otherwise go to waste. Various sources of lignocellulosic waste such as agricultural waste, municipality waste, weeds, wood, grasses, agricultural residues, and industrial waste are considered as a potential feedstock for bioethanol production (Kim and Dale, 2004; Kumari and Pramanik, 2012b). Due to stringent environmental regulations, the disposal of lignocellulosic-based waste biomass is one of the biggest problems because of the negative environmental consequences of the waste biomass that are faced by various industries all over the world (Balat, 2011). The huge quantity of this waste biomass can be used as potential raw material for the production of bioethanol if an effective and feasible conversion process is developed (Balat, 2011).

Lignocellulosic waste is comprised of cellulose, hemicellulose, and lignin as the main components. The three major steps involved in the conversion of any lignocellulosic waste to bioethanol are as follows: (1) pretreatment for the release of cellulose and hemicellulose components by removing lignin, (2) hydrolysis for converting released sugar components to fermentable sugars, and (3) fermentation of sugars to bioethanol. Among these, the major technological hurdle lies in the pretreatment process for the removal of lignin. Though the pretreatment of lignocellulosic waste using dilute sulfuric acid has been the most efficient and widely used method so far, besides being hazardous, this method produces toxic by-products which affect the fermentation of hydrolysate, resulting in the low yield of bioethanol (Han et al., 2009; Kumari and Pramanik, 2012b). Another important challenge lies in the development of efficient and stable microbial strains that can co-ferment pentose (C5) and hexose (C6) sugars released by hydrolysis to bioethanol.

11.2   WHY BIOETHANOL FOR TRANSPORTATION SECTOR?

Due to continuous depletion of petroleum oil reserve, uncertainty in its availability, rising prices, and environmental consequences of its use have drawn attention toward the production of bioethanol as an alternative energy resource for the transportation sector. The current and future economic development critically depends on the availability of energy sources that are easy to get, reasonably priced, and environmentally friendly (Rajendran et al., 2015). Bioethanol is a clear colorless liquid, flammable, biodegradable, and relatively harmless to the environment. Bioethanol is an "oxygenated" fuel due to its higher oxygen content and possesses high octane number. Therefore, it is suitable as a blending element of gasoline or as a raw material to make high octane fuel (Silverstein, 2005). The combustion of fuel gasoline offers gaseous pollutants such as carbon monoxide (CO), hydrocarbons, and particulates. The main cause of global warming is the release of CO_2 from the burning of coal and oil. It is reported that cellulosic ethanol has the potential for removing 86% of greenhouse gas emissions (Kumar et al., 2009). Therefore, the addition of bioethanol or other oxygenated fuels to gasoline can reduce CO production by providing more oxygen and promoting complete combustion (Jeoh, 1998). Bioethanol emits 35% less CO, 79% less CO_2, 42% less nitrogen oxides, 39% less particulate matter, and 43% less hydrocarbons than the fossil fuel-based petroleum oil (Prasad et al., 2007).

The combustion of oxygenated fuels produces CO_2 as the end product rather than CO. It helps not only the reduction of CO concentration and to minimize health risks In view of the environmental benefits and the depletion of crude oil, the industry has been moving toward potential bioethanol production for use as fuel for the transportation sector (Jeoh, 1998). Therefore, ligno-cellulosic bioethanol is represented to be a promising alternative for total energy gain and emission of contaminants (Talebnia, 2008). Bioethanol fuel blends are effectively used in some countries. The most common blends are E5 (5% bioethanol and 95% petrol) and E85 (85% volume bioethanol and petrol) (Rajendran et al., 2015; Talebnia, 2008). Well-organized research and advanced technological development for the conversion of lignocellulosic to bioethanol are still in progress. So, an efficient combination of approachable systems analysis and design of economical techniques should emerge for potential second-generation (lignocellulosic biomass) biofuel production (Limayem and Ricke, 2012). Thus, up to 491 GL per year of bioethanol can be generated from lignocellulosic biomass, which is about 16 times higher than the current world bioethanol production, and 32% of the global gasoline consumption can be replaced using bioethanol in E85 fuel (Balat, 2011; Han et al., 2009). India is probably the first among the developing countries to use biofuel for flying planes, which used 75% aviation turbine fuel and 25% of bio-fuel in aircraft engine as reported on August 27, 2018 (web page).

11.3 ALTERNATIVE SOURCE: LIGNOCELLULOSIC BIOMASS

Lignocellulosic biomass constitutes the world's largest renewable resource and is abundantly available on the Earth. It mainly consists of cellulose, hemicellulose (complex carbohydrates), and lignin. Any biomass containing sugars or converted sugars can further be used for bioethanol production. First-generation bioethanol is usually generated from sugarcane in Brazil or corn in the USA (Kim and Dale, 2004). However, to enable a more substantial increase in bioethanol production capacity worldwide, lignocellulosic substrates need to be exploited. There are various types of lignocellulosic raw materials that are differentiated by their composition, origin, and structure. Lignocellulosic feedstocks can be categorized into five main groups: forest wood, agricultural residues (hardwood and soft wood), energy crops, industrial waste, and municipal waste. Energy crops such as sugarcane, corn, maize, soya beans, and sugar beet are not preferable because of their cost and availability, as these are primarily used as food. Therefore, the main groups of raw materials that are considered for bioethanol production are agricultural waste biomass (waste wood, straws, corn, and stover), some herbs, municipal solid waste, weeds (*Eicchornia crassipes, Ipomoea carnea, Lantana camara, Saccharum spontaneum,* and *Prosopis juliflora*), industrial waste (sugar cane bagasses, wood residues, cotton gin waste, paper sludge), etc. (Mtui, 2009; Sarkar et al., 2012). These cellulosic wastes do not need extra economic input for their generation. These feedstocks can produce substantial bioethanol as a value-added product, which could solve the problem of their disposal as well as environmental pollution. Generally, most of the lignocellulosic biomass is not directly fermentable because sugar components are in polymeric form. Furthermore, lignocellulosic biomass is a carbon-neutral resource of energy, as lignocellulosic bioethanol produces no net CO_2 into the environment. Conversion of these lignocellulosic residues to bioethanol is a potential and attractive way to increase fossil fuels.

The composition of any lignocellulosic biomass varies according to their season of harvesting, origin, and land quality. Table 11.1 shows various types of lignocellulosic wastes and their compositional variation (Howard et al., 2003; John et al., 2006; Nigam et al., 2009; Sahu and Pramanik, 2015; Solange et al., 2010; Spence et al., 2010).

11.4 CONVERSION TECHNIQUES: LIGNOCELLULOSIC BIOMASS TO BIOETHANOL

Generally, bioethanol produced from lignocellulosic biowaste involves three important conversion processes such as pretreatment, enzymatic hydrolysis, and fermentation, as depicted in Figure 11.1 (Carere et al., 2008). Pretreatment is a process that is used for removing or modifying lignin, decrystallizing cellulose, extracting hemicellulose, removing acetyle group from hemicellulose, reducing polymerization of cellulose, and expanding the structure to increase pore value and internal surface area, so that saccharification of carbohydrate portion to monomeric sugars can be achieved quickly with maximum yields (Aita and Kim, 2010; Kaparaju et al., 2009; Zhao et al., 2011). Cellulose and hemicelluloses polymers are broken down into sugar monomers, which are then fermented into bioethanol. The main factors leading the lignocelluloses to fermentable monosaccharides are the reduction of cellulose crystallinity and lignin removal (Kumar et al., 2009). Several advanced technologies and methodologies have been developed for the pretreatment of lignocellulosic waste to achieve efficient accessibility of the cellulosic components for enzymatic hydrolysis. The methods are classified into pulsed electric field treatment; physicochemical, CO_2 explosion,

TABLE 11.1 Types and Variation in the Composition of Lignocellulosic Wastes and Residues

Lignocellulose Waste	Hemicellulose (wt%)	Cellulose (wt%)	Lignin (wt%)
Wheat straw	24–30	32.9–50	8.9–15
Sunflower stalks	29.7	42.1	13.4
Barley straw	21. 9	33.8	13.8
Corn cobs	20–30	33.7–45	6.1
Oat straw	27.1	39.4	17.5
Rye straw	30.5	37.6	19.0
Rice straw	19.0	36.2	9.9
Cotton gin waste	15.0	40–55	19.8
Cotton seed hair	5–20	80–95	0
Soya stalks	24.8	34.5	19.8
Sugarcane bagasses	27.0	40.0	10.0
Wood pulp	29.2	65–79	31.2
Hardwood stem	24–40	40–55	18–25
Softwood stem	25–35	45–50	25–35
Grasses	35–50	25–40	10–30
Waste paper	10–20	60–70	5–10
Switch grass	31.4	45	12
Paper and pulp waste	60–70	10–20	5–10
Banana waste	13.2	14.8	14
swine waste	28	6.0	Na
Leaves	80–85	15–20	0

SOURCE: Adapted from Kumar et al. (2009), Solange et al. (2010), John et al. (2006), Nigam et al. (2009), Spence et al. (2010), Sahu and Pramanik (2015), Howard et al. (2003), and Rezania et al. (2017).

physical, chemical (organic, inorganic, alkaline, ammonia explosion, salt and conventional acids) and biological pretreatment (fungal, bacterial, mutated, and genetically modified microbes) (Agbor et al., 2011; Kuhad et al., 2010; Kumar et al., 2009; Sahu and Pramanik, 2017a). It is reported that a different pretreatment method affects biomass in different ways (Mosier et al., 2005; Wyman et al., 2005). In the near future, the use of lignocellulolytic enzymes or microorganisms may lead to new, environment-friendly technologies. Hydrolysis process is used to extract various monosaccharides present in cellulose (glucose) and hemicellulose (xylose, arabinose, galactose, and mannose) by acid or enzymatic hydrolysis (Alvira et al., 2010; Sahu and Pramanik, 2015). Hydrolysis methods include an enzymatic method (fungal or commercial enzyme), steam explosion, dilute and concentrated acid methods for lignocellulosic biomass have been reported. Lignocellulolytic enzymes specifically α-amylase, cellulase, xylanase, and protease have been produced by solid state of fermentation (SSF) on various lignocellulosic biomass (Kumar et al., 2009). The cellulase enzymes employed for the hydrolysis of cellulose to glucose are mainly categorized into three groups: endoglucanases, exoglucanases, and beta-glucosidases. Enzymatic hydrolysis is beneficial over acid hydrolysis, as it offers mild working conditions, higher yields, low toxic by-product formation, and low energy requirements. The third and final step is to improve the yield of bioethanol production from monomeric C_5 and C_6 sugars (Sahu and Pramanik, 2017b). Once the carbohydrate polymers are hydrolyzed into free monomeric sugar, they can be fermented into bioethanol using different ethanologenic microorganisms (Kumari, 2011). Fermentation is the final stage of the conversion of lignocellulosic waste to produce bioethanol. Evaluation of bioethanol production by fermentation is of utmost importance to quantify the performance of the final process. The efficient conversion of lignocellulosic biomass into bioethanol depends upon the utilization of C5 and C6 sugars. In this context, the selection of yeast strain for the conversion of these sugars is the most important. Another important challenge lies in the development of efficient and stable microbial strains that can co-ferment C5 and C6 sugar components released by hydrolysis to bioethanol. Yeast is the

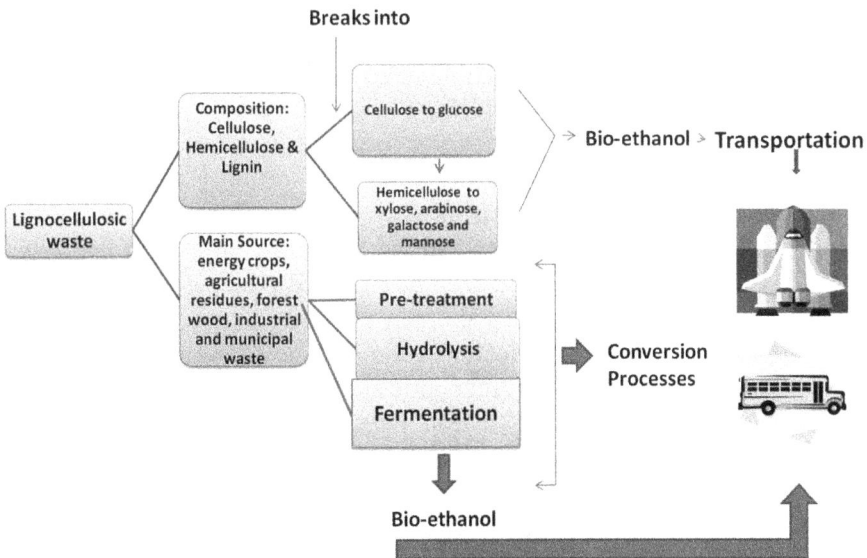

FIGURE 11.1 Conversion process of lignocellulosic waste biomass to bioethanol.

most commonly used microorganism for fermentation processes. However, few species of bacteria like *Zymomonas mobilis* and *Escherichia coli* are also used for fermentation (Aita and Kim, 2010; Sahu and Pramanik, 2015). The evaluation of bioethanol production by fermentation is of utmost importance to quantify the performance of the final process.

11.5   RECENT TECHNOLOGICAL ADVANCES IN CONVERSION PROCESSES

Although some recent research with promising results has been achieved, still there are several hurdles in the utilization of lignocellulosic waste for bioethanol production. Besides the removal of lignin component present in this waste that resists the release of sugar components and also acts as the major inhibitor for their hydrolysis, high cost involved in pretreatment stage and efficient conversion of C5 and C6 sugars by fermentation process are the critical factors that need to be solved. The challenge lies in the pretreatment process which include the development of a more efficient, environment-friendly and cost-effective pretreatment technology. Current efforts for developing different techniques are not economically feasible and also very complex in nature. (Kumari, 2011; Safartalab et al., 2014). Inefficient pretreatment of waste leads to the carbohydrate polymer to be easily hydro-lyzable by cellulase or xylase enzyme, resulting in the low release of ferment-able sugar and produces toxic by-products that cause growth inhibition of fermentative microbial strains and thus eventually, low bioethanol yield is obtained (Agbor et al., 2011; Kodali and Pogaku, 2006). The development of an efficient pretreatment agent and process of its use for delignification of biomass play a critical role. The use of organic acids (lactic acid, oxalic acid, citric acid, and maleic acid) in place of hazardous acid pretreatment of an industrially generated cotton gin waste for the removal of lignin was reported (Sahu and Pramanik, 2018). Among these, maleic acid was found to be the most efficient, producing maximum xylose sugar at optimum pretreat-ment condition. A single pretreatment method, which can offer complete delignification of biomass in an economic and environment friendly manner, has not been established so far. Instead, integrated processes combining two or more pretreatment techniques seem to be beneficial. Though combined pretreatment methods have been successful to an extent, still a lot of research needs to be done in developing combined pretreatment methods to their full potential (Kumar and Sharma, 2017).

The major challenges in enzymatic hydrolysis lie in the reduction in cellulose crystallinity that hinders cellulose to be hydrolyzed to release monomers sugar (Kumari and Pramanik, 2013). Most organisms are efficient to improve biofuel yield of naturally existing cellulose-degrading microorganisms or genetically engineered microbes (Pérez et al., 2002; Rastogi and Shrivastava, 2018). It was reported that PvKN1 broadly controlled the lignin biosynthesis pathway, decreased the lignin percentage, and altered the expression of hemicellulose and cellulose biosynthetic genes, thereby increasing the yield of sugar release (Wuddineh et al., 2016). A similar work reported that the overexpression of GA20-OXIDASE1 in maize increased the hydrolysis yield of biomass using NaOH pretreatment (Voorend et al., 2016).

The efficient conversion of both C5 and C6 sugar hydrolysates by fermentation is another main factor for bioethanol to be produced efficiently and economically. The major issues such as nonflammability, less toxicity, and easy recovery of fermentable sugars after extraction should be established. Furthermore, to increase sugar yields, efficient release of C5 and C6 sugars and their subsequent conversion by fermentation are important, and this will also reduce the cost of ethanol production (Agbor et al., 2011). Efforts can be made to improve bioethanol production by developing genetically modified microorganism or hybrid yeast strains combining pentose and hexose fermenting microbial strains (Kang et al., 2014; Pramanik, 2004; Pramanik and Rao, 2005). Although submerge and solid state fermentation system can be used for bioethanol production, SSF system is more efficient because of providing some advantages such as no need of prefractionation of raw materials, higher yield of products and environmental friendly methods (Kang et al., 2014). The control of the key variables is another important aspect for fermentation to be successful, because the fermentation process being nonlinear, complex and time varying in nature requires a reliable online estimate through the development of a suitable simulation and prediction method (Pramanik, 2004). Detail cost analysis is done to access the feasibility of different conversion processes by conducting the study at pilot scale for the large-scale production of bioethanol.

Recent advances on different conversions like simultaneous saccharification, separate hydrolysis, consolidated hydrolysis for efficient saccharification, co-fermentation, and fermentation of lignocellulosic biomass were reported (Rastogi and Shrivastava, 2018). Ultrafiltration, nanofiltration, reverse osmosis, membrane filtration, and membrane separation technologies have been examined in different stages of the bioethanol production (Kang et al., 2014).

11.6 CONCLUSION

Lignocellulosic biomass-derived bioethanol is considered to be a promising and sustainable transportation fuel alternative to conventional fossil fuel. However, there are several technological challenges are prevailed in utilizing this vast amount of waste as a potential raw material source for bioethanol production. Considering the recent technological progress in various conversion processes, use of combined pretreatment process, genetically modified yeast strains or hybrid strains for converting both C5 and C6 sugar components and development of solid-state fermentation system with online control facility may offer a feasible solution for bioethanol production from lignocellulosic waste biomass in future in an efficient and cost-effective manner.

ACKNOWLEDGMENT

We are grateful to the Ministry of Environment and Forest, Government of India for their financial support vide section no (19/14/27-RE) to carry out this research work.

KEYWORDS

- **bioethanol**
- **lignocellulosic biomass**
- **biofuel**
- **transportation**
- **fermentation**

REFERENCES

Agbor, V.B.; Cicek, N.; Sparling, R.; Berlin, A.; Levin, D.B. Biomass pretreatment: fundamentals toward application. *Biotechnology Advances* 2011, 29, 675–685.
Aita, G.M.; Kim, M. Pretreatment technologies for the conversion of lignocellulosic materials to bioethanol. *ACS Symposium Series* 2010, 117–145.
Alvira, P.; Tomás-Pejó, E.; Ballesteros, M.; Negro, M.J. Pretreatment technologies for an efficient bioethanol production process based on enzymatic hydrolysis: a review. *Bioresource Technology* 2010, 101, 4851–4861.

Balat, M. Production of bioethanol from lignocellulosic materials via the biochemical pathway: a review. *Energy Conversion and Management* 2011, 52, 858–875.

Carere, C.R.; Sparling, R.; Cicek, N.; Levin, D.B. Third generation biofuels via direct cellulose fermentation. *International Journal of Molecular Science* 2008, 9, 1342–1360.

Hamelinck, C.N.; Van Hooijdonk, G.; Faaij, A.P. Ethanol from lignocellulosic biomass: techno-economic performance in short-, middle-and long-term. *Biomass and Bioenergy* 2005, 28, 384–410.

Han, M.; Moon, S.K.; Kim, Y.; Kim, Y.; Chung, B.; Choi, G.W. Bioethanol production from ammonia percolated wheat straw. *Biotechnology and Bioprocess Engineering* 2009, 14, 606–611.

Howard, R.L.; Abotsi, E.; Jansen, E.L.; Rensburg, V.; Howard, S. Lignocellulosic biotechnology: issues of bioconversion and enzyme production. *African Journal of Biotechnology* 2003, 2, 602–619.

Jeoh, T. *Steam explosion pretreatment of cotton gin waste for fuel ethanol production* (master's thesis). Virginia Tech. University, Blacksburg, VA, 1998, p. 153.

John, F.; Monsalvem, G.P.; Medinam, I.V.; Ruiz, C.A.A. Ethanol production of banana shell and cassava starch. *Dyna University National Colombia* 2006, 73, 21–27.

Kang, Q.; Appels, L.; Tan, T.; Dewil, R. Bioethanol from lignocellulosic biomass: current findings determine research priorities. *Scientific World Journal* 2014, 2014, 1–13.

Kaparaju, P.; Serrano, M.; Thomsen, A. B.; Kongjan, P.; Angelidaki, I. Bioethanol, biohydrogen and biogas production from wheat straw in a biorefinery concept. *Bioresource Technology* 2009, 100, 2562–2568.

Kim, S.; Dale, B.E. Global potential bioethanol production from wasted crops and crop residues. *Biomass and Bioenergy* 2004, 26, 361–375.

Kodali; Pogaku, B.R. Pretreatment studies of rice bran for the effective production of cellulase. *Electronic Journal of Environmental, Agricultural and Food Chemistry* 2006, 5, 1253–1264.

Kuhad, R.C.; Gupta, R.; Khasa; Singh, Y.P.A. Bioethanol production from Lantana camara (red sage): pretreatment, saccharification and fermentation. *Bioresource Technology* 2010, 101, 8348–8354.

Kumar, A.K.; Sharma, S. Recent updates on different methods of pretreatment of lignocellulosic feedstocks: a review. *Bioresources and Bioprocessing* 2017, 4, 7.

Kumar, P.; Barrett, D.M.; Delwiche, M.J.; Stroeve, P. Methods for pretreatment of lignocellulosic biomass for efficient hydrolysis and biofuel production. *Industrial and Engineering Chemistry Research* 2009, 48, 3713–3729.

Kumari, R. *Development of hybrid yeast strains for the production of bioethanol from lignocellulosic biomass* (doctoral dissertation). NIT, Rourkela, 2011.

Kumari, R.; Pramanik, K. Improved bioethanol production using fusants of *Saccharomyces cerevisiae* and xylose-fermenting yeasts. *Applied Biochemistry and Biotechnology* 2012a, 167, 873–884.

Kumari, R.; Pramanik, K. Improvement of multiple stress tolerance in yeast strain by sequential mutagenesis for enhanced bioethanol production. *Journal of Bioscience Bioengineering* 2012b, 114, 622–629.

Kumari, R.; Pramanik, K. Bioethanol production from *Ipomoea carnea* biomass using a potential hybrid yeast strain. *Applied Biochemistry and Biotechnology* 2013, 171, 771–785.

Lee, J.M.; Pollard, J.F.; Coulman, G.A. Ethanol fermentation with cell recycling: computer simulation. *Biotechnology and Bioengineering* 1983, 25, 497–511.

Limayem, A.; Ricke, S.C. Lignocellulosic biomass for bioethanol production: current perspectives, potential issues and future prospects. *Progress in Energy and Combustion Science* 2012, 38, 449–467.

Mahro, B.; Timm, M. Potential of biowaste from the food industry as a biomass resource. *Engineering in Life Sciences* 2007, 7, 457–468.

Mosier, N.; Wyman, C.; Dale, B.; Elander, R.; Lee, Y.Y.; Holtzapple, M.; Ladisch, M. Features of promising technologies for pretreatment of lignocellulosic biomass. *Bioresource Technology* 2005, 96, 673–686.

Mtui, G.Y. Recent advances in pretreatment of lignocellulosic wastes and production of value added products. *African Journal of Biotechnology* 2009, 8, 1398–1415.

Nigam, P.S.; Gupta, N.; Anthwal, A. Pre-treatment of agro-industrial residues. In: *Biotechnology for Agro-industrial Residues Utilization*; Nigam, P.S., Pandey, A., Eds.; 1st Ed., Springer, The Netherlands, 2009, pp. 13–33.

Pérez, J.; Munoz-Dorado, J. Dela.; Rubia, TD.; Martinez J. Biodegradation and biological treatments of cellulose, hemicellulose and lignin: An overview. *International Microbiology* 2002, 5, 53–63.

Pramanik, K. Use of artificial neural networks for prediction of cell mass and ethanol concentration in batch fermentation using *Saccharomyces cerevisiae* yeast. *Journal of Institute of Engineers, India* 2004, 85, 31–35.

Pramanik, K.; Rao, D.E. Kinetic study on ethanol fermentation of grape waste using *Saccharomyces cerevisiae* yeast isolated from toddy. *The Institution of Engineers (India)* 2005, 85, 53–58.

Prasad, S.; Singh, A.; Joshi, H.C. Ethanol as an alternative fuel from agricultural, industrial and urban residues. *Resources Conservation & Recycling* 2007, 50, 1–39.

Rajendran, R.; Radhai, R.; Sundaram, K.S.; Rajalakshmi. V. Utilization of cellulosic biomass as a substrate for the production of bioethanol. *International Journal of Environmental Science and Technology* 2015, 5, 743–753.

Rastogi, M.; Shrivastava, S. Current methodologies and advances in bio-ethanol production. *Journal of Biotechnology and Bioresearch* 2018, 1, 1–8.

Rogner, H.H. Energy resources. In: *World Energy Assessment; Energy and the Challenge of Sustainability*; Goldberg, J.W., Baker, H., Khatib, S., Daw, Ba-N'., Popescu, A., Viray, F.L. Eds.; United National Development Programme (UNPD), New York, NY, 2000, pp. 135–171.

Safartalab, K.; Dadashian, F.; Vahabzadeh, F. Fed batch enzymatic hydrolysis of cotton and viscose waste fibers to produce ethanol. *Universal Journal of Chemistry* 2014, 2, 11–15.

Sahu, S.; Krishna, P. Evaluation and optimization of organic acid pretreatment of cotton gin waste for enzymatic hydrolysis and bioethanol production. *Applied Biochemistry and Biotechnology* 2018, 186, 1047–1060.

Sahu, S.; Pramanik, K. Bioconversion of cotton gin waste to bioethanol. In: *Environmental Microbial Biotechnology*; Sukla, L.B.; Pradhan, N.; Panda, S.; Mishra, B.K., Eds.; Chapter. 14, Springer International Publishing, Switzerland, 2015, vol. 45, pp. 267–288.

Sahu, S.; Pramanik, K. Biological treatment of lignocellulosic biomass to bioethanol. *Advances in Biotechnology and Microbiology* 2017a, 5, 555–674.

Sahu, S.; Pramanik, K. Evaluating fungal mixed culture for pretreatment of cotton gin waste to bioethanol by enzymatic hydrolysis and fermentation using co-culture. *Polish Journal of Environmental Studies* 2017b, 26, 1215–1223.

Sánchez, O.J.; Cardona, C.A. Trends in biotechnological production of fuel ethanol from different feedstocks. *Bioresource Technology* 2008, 99, 5270–5295.

Sarkar, N.; Ghosh, S.K.; Bannerjee, S.; Aikat, K. Bioethanol production from agricultural wastes: an overview. *Renewable Energy* 2012, 37(1), 19–27.

Silverstein, R.A. *A comparison of chemical pretreatment methods for converting cotton stalks to ethanol.* MS Dissertation, Raleigh, NC: North Carolina State University, 2005.

Solange, I.; Mussatto, J.; Teixeira, A. Lignocellulose as raw material in fermentation processes. In: *Current research, technology and education topics in applied microbiology and microbial biotechnology*; Mendez-Vilas, A. MS, Ing. 2010, 4710–4057.

Spence, K.L.; Venditti, R.A.; Habibi, Y.; Rojas, O.J.; Pawlak, J.J. The effect of chemical composition on microfibrillar cellulose films from wood pulps: mechanical processing and physical properties. *Bioresource Technology* 2010, 101, 5961–5968.

Talebnia, F. *Ethanol production from cellulosic biomass by encapsulated Saccharomyces cerevisiae.* Chalmers University of Technology, Department of Chemical and Biological Engineering, 2008.

Voorend, W.; Nelissen, H.; Vanholme, R.; De Vliegher, A.; Van Breusegem, F.; Boerjan, W.; Roldán-Ruiz, I.; Muylle, H.; Inzé, D. Overexpression of GA 20-OXIDASE 1 impacts plant height, biomass allocation and saccharification efficiency in maize. *Plant Biotechnology Journal* 2016, 14, 997–1007.

Williams, R.H.; Larson, E.D.; Katofsky R.E.; Chen, J. Methanol and hydrogen from biomass for transportation, with comparisons to methanol and hydrogen from natural gas and coal. *PU/CEES Report 292.* Princeton University/Center for Energy and Environmental Studies, Princeton, NJ, 1995, p. 47.

Wuddineh, W.A.; Mazarei, M.; Zhang, J.Y.; Turner, G.B.; Sykes, R.W.; Decker, S.R.; Davis, M.F.; Udvardi, M.K.; Stewart Jr, C.N. Identification and overexpression of a Knotted1-like transcription factor in switchgrass (*Panicum virgatum* L.) for lignocellulosic feedstock improvement. *Frontiers in Plant Science* 2016, 7, 520.

Wyman, C.E.; Dale, B.E.; Elander, R.T.; Holtzapple, M.; Ladisch, M.R.; Lee, Y.Y. Coordinated development of leading biomass pretreatment technologies. *Bioresource Technology* 2005, 96, 1959–1966.

Zhao, X.; Wu, R.; Liu, D. Production of pulp, ethanol and lignin from sugarcane bagasse by alkali-peracetic acid delignification. *Biomass and Bioenergy* 2011, 35, 2874–2882.

## CHAPTER 12

Cost-Effective Production of Biofuels from Lignocellulosic Biomass: Prospects and Challenges

K. C. SAMAL* and LAXMIPREEYA BEHERA

Department of Agricultural Biotechnology, College of Agriculture, Odisha University of Agriculture and Technology, Bhubaneswar, Odisha 751003, India

*Corresponding author. E-mail: samalkcouat@gmail.com

ABSTRACT

The present scenario of biofuels is found to be most effective and is alternative source to fossil-based fuels and natural gases. Biofuels have become an innovative alternative that offers a wide variety of exceptional benefits. Biofuels are making resurgence due to increasing oil prices and dwindling fossil fuel reserves, as these are eco-friendly, renewable, and a reliable source of energy. There are four generations of biofuels: first-, second-, third-, and fourth-generation biofuels. They are categorized by their sources of biomass, their limitations as a renewable source of energy, and their technological requirement. The main limitation of first-generation biofuels is that they are produced from biomasses which are also used for food purpose. In future, its production will impose problems during shortage of food production to the growing population. Second-generation biofuels are produced from nonfood biomass, but the land utilized for food production is diverted for this purpose. Third-generation biofuels are a good alternative source because algal biomass is used as feedstock for biofuel production. In fourth-generation biofuels, the feedstock is tailored not only to improve the processing efficiency, but it is also designed to capture more carbon dioxide, as the crop grows in cultivation. Fourth-generation biofuels are found to be carbon neutral or even carbon negative compared to the other generation biofuels and result in reducing greenhouse gas emissions. Biofuels such as bioethanol and biohydrogen can be produced by fermentation process of sugars obtained from cellulosic agricultural and industrial wastes. In the recent years, significant improvement has been made with the help of biotechnological and molecular approaches so as to improve microbial activity and enzymes. The use of genetically modified organisms is considered to be the most efficient and quick method for efficient conversion of lignocellulosic biomass to biofuels. Lignocellulosic biomasses contain cellulose (30%–50%), hemicellulose (20%–40%), and lignin (20%–30%). Hence, in the present study, major attention is focused on improvement of cellulase enzyme production for biofuel industry via genetic modification and its utilization for biofuel production by employing thermostable cellulase enzymes via genetic modification and characterization and optimization of thermostable cellulase enzymes for biofuel production.

12.1 INTRODUCTION

Increase in energy demand on one hand and depletion of fossil fuel availability in the coming years forces the scientists to think alternate source of energy. It has been found that biofuel can be a potential alternative to fossil fuel. Biofuels are produced through existing biological processes, such as agriculture and anaerobic digestion, rather than fossil fuels, such as coal and petroleum, from prehistoric biological matter (Ojeda et al., 2011). But the high cost of production of biofuel has impeded its commercialization in most parts of the world. Fossil fuels are actually ancient biofuels as they are made from decomposed plants and animals that have been buried in the ground for millions of years. Biofuels are similar, except that they are made from plants grown today. Biofuels are grouped into two main classes: primary biofuels and secondary biofuels (Gnansounou, 2010; Sun and Cheng, 2002). Primary biofuels are also known as natural biofuels, which are produced from firewood, plants, wood chips, forest, animal waste, landfill gas, crop residues, etc. (Werther et al., 2000). Secondary biofuels are again subdivided into four major groups (first generation, second generation, third generation, and fourth generation) on the basis of their creation. First-generation biofuels include bioethanol production from wheat, barley, corn, potato, sugarcane, beet, oil seeds (soybeans, coconut, sunflower, and rapeseed), animal fat, and used cooking oil (Baeyens et al., 2015). Second-generation biofuels include bioethanol and biodiesel derived from cassava, jatropha, straw, grass, and wood (Boluda-Aguilar and López-Gómez, 2013). Third-generation biofuel production is made from microalgae and microbes (Lu, 2014). The current and advanced era of biofuel production aims to reduce the destruction of biomass and this includes electrofuels and photobiological solar fuels (Sawin et al., 2014).

Biofuels are made from locally available renewable sources. The fuel is nontoxic, biodegradable, eco-friendly, and has increased safety in storage and transport. It reduces vehicle emission and increases engine performance because it has a higher octane number (51.0) as compared to petrodiesel (47.8). It reduces the release of greenhouse gases at least by 3.3 kg CO_2 equivalent per kg of biodiesel (Hill et al., 2006). Production of biodiesel in India will reduce the reliance on foreign suppliers, thus being helpful in price stability. A comparative property of biofuel with petroleum/diesel is described in Table 12.1.

TABLE 12.1 Basic Comparison Between Biofuel and Petroleum-Based Fuel

Property	Biofuel (Biodiesel)	Petroleum/Diesel
Viscosity (cp) (30 °C)	5.51	3.60
Specific gravity (15 °C/4 °C)	0.917	0.841
Solidifying point (°C)	2.0	0.14
Cetane value	51.0	47.8
Flash point (°C)	110	80
Carbon residue (%)	0.64	0.05
Distillation (°C)	284–295	350
Sulfur (%)	0.13–0.16	1.0
Acid value	1.0–38.2	–
Saponification value	188–198	–
Iodine value	90.8–112.5	–
Refractive index (30 °C)	1.47	–

12.2 GENERATION OF BIOFUELS

12.2.1 FIRST-GENERATION BIOFUELS

First-generation biofuels are derived from food-based feedstock (like maize, wheat, sugar cane, or vegetable oil) involving process technologies like fermentation (for ethanol) and transesterification (for biodiesel). First-generation biofuels are produced, involving processes, like fermentation, distillation, and transesterification. In this case, sugars and starches are fermented to produce ethanol, butanol, propanol, etc. Ethanol has one-third of the energy density of gasoline, and it burns cleaner than gasoline and therefore produces less greenhouse gases. Biodiesel is produced through transesterification of vegetable oils. Biodiesel can be used in place of petroleum diesel in many diesel engines or in a mixture of the two. Biodiesel is an ecofriendly, alternative diesel fuel prepared from domestic renewable resources, that is, vegetable oils (edible or nonedible oil) and animal fats. These natural oils and fats are primarily made up of triglycerides. These triglycerides, when reacted chemically with lower alcohols in the presence of a catalyst, result in fatty acid esters. These esters show a striking similarity to petroleum-derived diesel and are called "biodiesel." Many organizations have developed methodologies and established commercial units for the production of biodiesel from plant or vegetable oils (Fangrui and Hanna, 1999; Grapen, 2005; Ingole and Kakde, 2012).

Biotechnology can help to accelerate the selection of varieties that are more suited to biofuel production—with increased biomass per hectare, increased content of oils (biodiesel crops) or fermentable sugars (ethanol crops), or improved processing characteristics that facilitate their conversion to biofuels (Sun and Cheng, 2002). Fermentation of sugars is central to the production of ethanol from biomass. However, the most commonly used industrially fermentative microorganism, yeast (*Saccharomyces cerevisiae*) cannot directly ferment starchy material, such as maize starch. The biomass must first be broken down (hydrolyzed) to fermentable sugars using enzymes called amylases. Many of the current commercially available enzymes, including amylases, are produced using genetically modified microorganisms. Research continues on developing efficient genetic yeast strains that can produce the amylases themselves, so that the hydrolysis and fermentation steps can be combined.

First-generation biofuels improve agricultural industries and rural communities through increased demand for crops. The advantages of first generation biofuel threat to food reduce stability of the prices. Corn and sugar beet provide a small benefit over fossil fuels in regard to greenhouse gases since they still require high amount of energy to grow, collect, and process and have the potential to have a negative impact on biodiversity and competition for water and land use.

12.2.2   *SECOND-GENERATION BIOFUELS*

Second-generation biofuels tackle many issues associated with first-generation biofuels. They are often referred to as "cellulosic ethanol" and "cellulosic biobutanol," as they are derived from lignocellulosic biomass which contains mainly cellulose (30%–50%), hemicellulose (20%–40%), and lignin (20%–30%). Compared with the production of ethanol from first-generation feedstocks, the use of lignocellulosic biomass is more complicated because the polysaccharides are more stable and the pentose sugars are not readily fermentable by *S. cereviseliae* (Velmurugan et al., 2012). For biofuel production, the polysaccharides must first be hydrolyzed or broken down into simple sugars using either acid or enzymes. Several biotechnology-based approaches have been utilized for the development of strains of that can ferment pentose sugars. The use of alternative yeast species that naturally ferment pentose sugars are able to break down cellulose and hemicellulose into simple sugars. Apart from agricultural, forestry, and other by-products, the main source of lignocellulosic biomass for second-generation biofuels is likely to be from "dedicated biomass feedstocks," such as certain perennial grass and forest tree species (Sun and Cheng, 2002). Genomics, genetic modification, and other biotechnologies are all being investigated as tools to produce plants with desirable characteristics for second-generation biofuel production, for example, plants that produce less lignin (a compound that cannot be fermented into liquid biofuel), that produce enzymes themselves for cellulose and/or lignin degradation, or that produce increased cellulose or overall biomass yields.

Second-generation biofuels are gaining acceptance. They do not compete between fuels and food crops, since they come from the specified biomass. Second-generation biofuels also generate higher energy yields per acre than first-generation fuels. They are grown in low fertility and degraded lands with limited water where food crops may not be able to grow. The technology is fairly immature, so it still has a potential of cost reduction and increased production efficiency as scientific advances occur. However, some biomasses for second-generation biofuels still compete with land use, since some of the biomass grows in the same climate as food crops. This leaves farmers and policymakers with the hard decision of which crop to grow. Cellulosic sources that grow alongside food crops could be used for biomass, such as corn stover (leaves, stalk, and stem of corn) (Maitan-Alfenas et al., 2015). However, this would take away too many nutrients from the soil and would need to be replenished through fertilizers. In addition to this, the processing procedure of second-generation fuels is more complicated than first-generation biofuels because it requires pretreating the biomass to release the trapped sugars. This requires more energy and materials.

12.2.3 *THIRD-GENERATION BIOFUELS*

Third-generation biofuels is considered a good alternative source because they do not compete with food but faces challenges in making them economically feasible. The third-generation biofuels utilize algae as feedstock. Nowadays, the production of biofuel from microalgae is an option that has attracted strong interest of the scientific community and should be evaluated to determine the technical, technological, economic, and environmental sustainability of the process (Sun and Cheng, 2002). The use of algae for biofuel production has received a significant interest because of their fast growth rate and increased photosynthetic efficiency. But fermentative ethanol production from algal feedstocks and transesterification of algal oils require extensive downstream processing, like dewatering, which is regarded as one of the major hurdles in algal biofuel commercialization. In case of fourth-generation biofuels, the feedstock is tailored not only to improve the processing efficiency, but it is also designed to capture more CO_2.

12.2.4 *FOURTH-GENERATION BIOFUELS*

Fourth-generation biofuels contribute more in reducing greenhouse gas emissions by being more carbon neutral or even carbon negative than other generation biofuels. Industrial biotechnology, with its competitive, clean, and clever use of bio-based technologies, can play a key role in sustainable production and utilization of biofuels. The journey of production and utilization of biofuels from first generation to fourth generation is gaining attention worldwide because of several benefits such as reducing global warming and global energy crises (Pandey and Kumar, 2017).

12.3 LIGNOCELLULOSIC-BASED BIOFUEL PRODUCTION

Various types of biological residues generated from agricultural and industrial activity have been suggested as prospective nutritional sources for microbial cultures. Lignocellulosic biomass is found to be the most abundant residue from agricultural crops (Abu Yousuf, 2012); this by-product has been given top-priority consideration as a source of biomass for producing biodiesel. Lignocellulosic biomass can be transformed into bioethanol via different approaches, of which two basic approaches are most important, such as enzymatic hydrolysis and chemical hydrolysis. Both routes involve disruption of the recalcitrant cell wall structure of lignocelluloses break into lignin, hemicellulose, and cellulose (Alya and Steven, 2012). Each polysaccharide is hydrolyzed into sugars that are converted into bioethanol followed by a purification process. Enzymatic approach is found promising, as it yields high sugar and eliminates the need for large quantities of chemicals and the formation of inhibitory by-products during dilute acid hydrolysis (Taherzadeh and Karimi, 2008). In the production of ethanol, first the release of cellulose and hemicelluloses from lignocellulosic material is required. Based on the specific lignocellulosic structure, the bioconversions of lignocellulosic materials to bioethanol normally require multistep processes. Four stages, namely pretreatment, hydrolysis, fermentation, and distillation, are involved in the production of lignocellulosic-based ethanol (Zhu et al., 2010). During the past decades, substantial developments have been made in genetic and enzymatic technologies that have helped to improve different steps of ethanol production and expand the capability of *S. cerevisiae* for fermenting different sugars. During cellulose hydrolysis, lignocellulosic material is broken down or converted to sugar molecules (Zhao et al., 2009). In the next "separation phase," the sugar solution is separated from the residual materials, notably lignin. During "microbial fermentation," sugar solution is converted into ethanol and then "distillation" is done to produce alcohol (Figure 12.1).

12.3.1 PRETREATMENT OF LIGNOCELLULOSIC MATERIAL

Hemicellulose and lignin provide a protective sheath around cellulose, which must be modified or removed before efficient hydrolysis. Pretreatment involves delignification of the feedstock in order to make cellulose more accessible in the hydrolysis step, using physical, physicochemical, chemical, and biological treatment. Different types of physical processes such as milling and irradiation can be used to improve the enzymatic hydrolysis of lignocellulosic waste materials. Milling can improve enzymatic degradation of lignocellulosic materials toward ethanol by reducing the size of the materials and increasing their surface area. Sulphuric acid is often added in order to reduce the production of inhibitors and improve the solubilization of hemicelluloses. Treatment of lignocellulosic materials with acid at a high temperature can efficiently improve the enzymatic hydrolysis. Sulfuric acid, hydrochloric acid, and nitric acid are commonly used (Taherzadeh and Karimi, 2008). Acid pretreatment either can operate with dilute acid under a high temperature (e.g., 180 °C for 5 min) or can be done under a low temperature (e.g., 120 °C for 30–90 min) and high acid concentration (concentrated acid pretreatment). Dilute acid hydrolysis is the suitable method among the chemical pretreatment methods. It can be used either as a pretreatment of lignocelluloses for enzymatic hydrolysis or as the actual method of hydrolyzing to fermentable sugars. The main advantage of dilute acid pretreatment is the possibility to recover a high portion (e.g., 90%) of hemicellulose sugars. Dilute acid pretreatment is not effective in dissolving lignin, but it can disrupt lignin and increases the cellulose's susceptibility to enzymatic hydrolysis (Yang and Wyman, 2004).

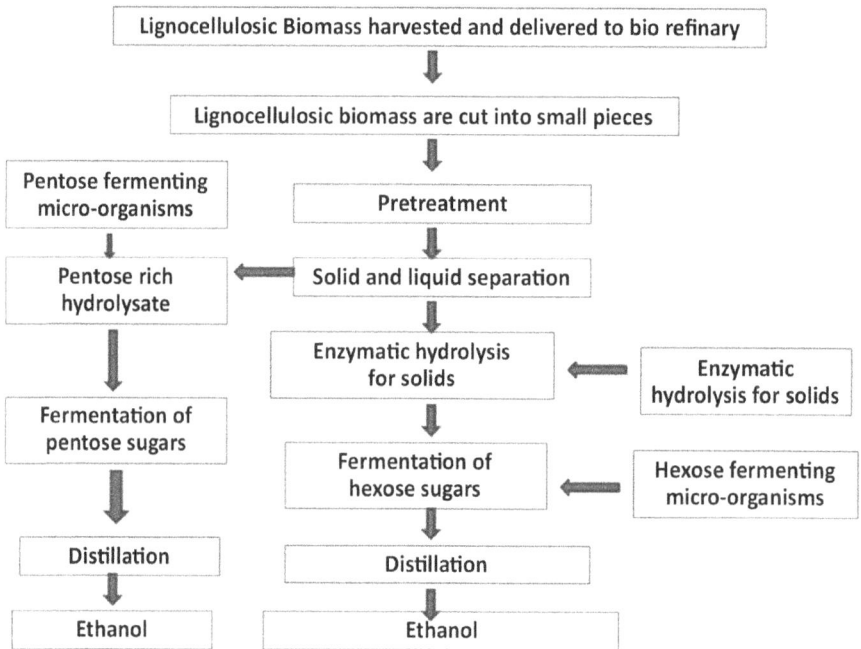

FIGURE 12.1 Flowchart of bioethanol production from lignocellulosic biomass.

Alkaline pretreatment utilizes NaOH; $Ca(OH)_2$ (lime) or ammonia is generally applied to remove lignin and a part of the hemicelluloses (Khuong et al., 2014). It is performed at low temperatures under high alkaline concentration for long duration. It results in pores in the material due to the extensive swelling facilitated by removal of the crosslinks. Pretreatment of lignocellulosic materials can also be performed by treatment with ozone, which effectively degrades lignin and part of hemicellulose.

Ozonolysis and pretreatment of lignocellulosic are performed by treatment with ozone, and this technique can effectively degrade lignin and part of hemicellulose. But this method is expensive, since a large amount of ozone is required (Sun and Cheng, 2002). The main parameters in ozonolysis pretreatment are moisture content of the sample, particle size, and ozone concentration in the gas flow.

12.3.2 HYDROLYSIS OF PRETREATED LIGNOCELLULOSIC MATERIAL

In this stage, cellulose is converted into glucose. Two different types of hydrolysis processes such as acidic (sulfuric acid) or enzymatic reactions are employed. Acidic reaction can be divided into dilute or concentrated acid hydrolysis. Dilute hydrolysis (3%) requires a high temperature of 200°C–240 °C to disrupt cellulose crystals. Dilute sulfuric acid (H_2SO_4) flows continuously to the biomass at a high temperature of 200 °C–240 °C in a short period of time, allowing for a greater sugar recovery. Concentrated acid hydrolysis is the more prevalent method. But during enzymatic hydrolysis, the enzyme activity was inhibited by the increased degradation products from pretreatment and the increased sugar concentration. Similarly, the increased solid loading resulted in water constraint and high viscosity, which led to the poor efficiency of mass transfer concomitantly (Zhang et al., 2016).

12.3.3 FERMENTATION

Fermentation is the biological process that converts the hexose and pentose sugars into ethanol using bacteria, yeast, or fungi, alone or in combination. These microorganisms are employed to ferment sugar into alcohol, lactic acid, or other end products. The yeast *S. cerevisiae* is commonly used for first-generation ethanol production, but it cannot metabolize xylose. Other yeasts and bacteria are under investigation to ferment xylose and other pentoses into ethanol (Ho et al., 1998). Genetically engineered fungi that produce large volumes of cellulase, xylanase, and hemicellulase enzymes are under investigation. These could convert agricultural residues (e.g., corn stover, straw, and sugar cane bagasse) and trees into fermentable sugars. *S. cerevisiae* ferments hexose sugars, mainly glucose, into ethanol in a large tank bioreactor under anaerobic conditions and controlled temperature. Yeast-based fermentation is always accompanied by the formation of CO_2

by-products and supplemented by nitrogen to enhance the reaction. This conventional strain is optimal at a temperature of approximately 30 °C and resists a high osmotic pressure in addition to its tolerance to low pH levels of 4.0 as well as inhibitory products. *S. cerevisiae* can generate a high yield of ethanol from hexose sugars. Conversion of agricultural biomass and residual materials into sugars and ethanol was achieved by employing different fermentation methods such as separate hydrolysis and fermentation (SHF), simultaneous saccharification and fermentation (SSF), simultaneous saccharification and co-fermentation, and consolidated bioprocessing (Chen and Liu, 2017). Traditionally, SHF sequential steps are used in bioethanol production (Singh and Bishnoi, 2014).

12.4 BIOTECHNOLOGICAL TOOLS AND TECHNIQUES FOR PRODUCTION OF LIGNOCELLULOSIC-BASED BIOFUELS

In recent years, as the need to develop alternative, non-petroleum-based transportation fuels has become more projected, there has been a growing interest in using advanced biotechnological processes to improve biofuel production. Specifically, numerous advanced biotechnological techniques are being used to create improved biofuel products or processes, involving the creation of engineered or synthetic microorganisms for use in production of ethanol, biodiesel, or other fuels, or genetically engineered ("transgenic") plants as improved fuel feedstocks (Sawin, 2014). Engineered microorganisms with enhanced capability have been developed for the fermentation of ethanol, butanol, and other fuels. Different biotechnological approaches are employed to engineer microorganisms or plants to manufacture enzymes used in fuel production, to improve algal strains for biofuel production, and to develop selected or engineered plant species with favorable traits for use as improved biofuel feedstocks (Weber et al., 2010).

Plant biomass contains huge calorific value, but most of it makes up robust cell walls, an unappetizing evolutionary advantage that helped grasses to survive foragers and prosper for more than 60 million years. The dilemma is that this robustness makes them less digestible in the rumen of cows and sheep and difficult to process in bioenergy refineries for ethanol fuel. But recently, a multinational team of researchers, from the UK, Brazil, and the USA, has pinpointed a gene involved in the stiffening of cell walls whose suppression increased the release of sugars by up to 60%. In the team's genetically modified plants, a transgene suppresses the endogenous gene responsible for feruloylation to around 20% of its normal activity. In this way, the biomass produced is less feruloylated than it would otherwise be in an unmodified plant. Studies at the US Department of Energy's Brookhaven National Laboratory create a new era of how a sugar-signaling molecule helps regulate oil production in plant cells.

However, in most cases, the technology strategies will utilize genetic engineering methods based on recombinant DNA technology. Genetic engineering methods enable the insertion of genes from any source into a target "host" organism, thereby conferring on the host organism a genetic trait or a biochemical capability not naturally found in that organism. Through genetic transcription and followed by translation, the newly introduced genes synthesize specific enzymes in host cells, which catalyze specific biochemical reactions.

There are also a number of companies and academic research laboratories using more advanced technologies for improvement of microbial or plant performance. Many of these methodologies utilize recombinant DNA for the development of organisms with improved and or specific desired function. The different biotechnology techniques for improvement of organism are directed toward evolution or DNA shuffling. This also use a combination of genetic tricks and enhanced selective pressure to greatly enhance the activity of a targeted enzyme or its pathway, while other techniques like synthetic biology allow creation of novel or enhanced metabolic pathways in organisms.

12.5 FUTURE PROSPECTS OF BIOFUEL

India is the world's fourth largest consumer of crude and petroleum products after the USA, China, and Japan. The oil import by India resulted in a huge burden on government exchequer. Therefore, India's energy security would remain vulnerable until alternative fuels based on indigenously produced renewable feedstock are developed to substitute or supplement petro-based fuels. Eco-friendly sustainable biofuel production faces several challenges like biofuel versus food competition, recalcitrance of biomass for biofuel production, and low physical properties of biofuels. The "first-generation" biofuels are not socially and economically sustainable because of their impact on food prices and the environment. The controversy surrounding first-generation biofuels has helped to articulate issues and challenges that need to be considered in implementing second-generation biofuels. Given the current state of technology, and the uncertainty remaining about the future breakthroughs that would potentially make some second-generation biofuels cost-competitive, policymakers need to carefully consider what goals are to be pursued in providing support to different biofuels. The future of biofuels relies on environmental performance and sustainability of biofuels, as well as to encourage provision of a more abundant and geographically extensive feedstock supply.

Recent biotechnological development made to produce transgenic biofuel feedstocks, particularly cellulosic ethanol, and the regulatory process in the United States oversees the development and commercialization of new transgenic plants (Lee et al., 2008).

We hope to illustrate that the level of regulation for transgenic organisms is not proportional to their potential risk to human health or the environment, and that revisions to the regulatory system in the United States currently under consideration are necessary to streamline the process.

The latest emerging genome-editing tools (e.g., CRISPR-Cas9, TALEN, and ZFN) used in editing the genomes of nuclei, mitochondria, and chloroplasts of microalgae are thoroughly surveyed. Although all the techniques mentioned above demonstrate their abilities to perform gene editing and desired phenotype screening, there still need to overcome higher production cost and lower biomass productivity, to achieve efficient production of the desired products in micro algal biorefineries (Ng et al., 2017).

12.6 CONCLUSION

Due to the ever-increasing demand of petroleum and depletion of their stocks, the demand for liquid transportation fuels is constantly increasing, and biodiesel might be one of the most important solutions for this problem. Although lignocellulosic biomass may be cheap, the costs of processing are high. Technologies for converting lignocellulosic biomass to biodiesel have been developed, but suffer from numerous bottlenecks. The technologies such as pretreatment of biomass, enzymatic saccharification of the pretreated biomass, and fermentation of the hexose and pentose sugars released by hydrolysis, and saccharification should be refined for efficient conversion. However, various strategies such as synergistic system of multiple enzymes and the fed-batch mode should be combined together to overcome the limitation factors and improve production efficiency.

KEYWORDS

- **biofuel**
- **lignocellulosic biomass**
- **pretreatment**
- **thermostable cellulase enzyme**

REFERENCES

Abu Yousuf. Biodiesel from lignocellulosic biomass—Prospects and challenges. *Waste Management* 2012, 32, 2061–2067.

Alya, L.; Steven, R.C. Lignocellulosic biomass for bioethanol production: current perspectives, potential issues and future prospects. *Progress in Energy and Combustion Science* 2012, 38, 449–467.

Baeyens, J.; Kang Q.; L. Appels, R.; Dewil, Y.; Tan, T. Challenges and opportunities in improving the production of bio-ethanol. *Progress in Energy and Combustion Science* 2015, 47, 60–88.

Boluda-Aguilar, M.; López-Gómez, A. Production of bioethanol by fermentation of lemon (*Citrus limon* L.) peel wastes pretreated with steam explosion. *Industrial Crops and Products* 2013, 41, 188–197.

Chen, H.; Liu, Z.H. Enzymatic hydrolysis of lignocellulosic biomass from low to high solids loading. *Engineering in Life Sciences* 2017, 5, 489–499.

Fangrui, M.A.; Milford, A.H. Biodiesel production: a review. *Bioresource Technology* 1999, 70, 1–5.

Gnansounou, E. Production and use of lignocellulosic bioethanol in Europe: current situation and perspectives. *Bioresource Technology* 2010, 101, 4842–4850.

Grapen, J.V. Biodiesel processing and production. *Fuel Processing Technology* 2005, 86, 1097–1107.

Hill, J.; Nelson, E.; Tilman, D.; Polasky, S.; Tiffany, D. Environmental, economic and energetic costs and benefits of biodiesel and ethanol biofuels. *PNAS* 2006, 103(30), 11206–112010.

Ho, N.W.Y.Z.; Chen, D.; Brainard, A.P. Genetically engineered Saccharomyces yeast capable of effective co-fermentation of glucose and xylose. *Applied and Environmental Microbiology* 1998, 64(5), 1852–1859.

Khuong, L.D.; Kondo, R.; Leon, R.D.; Anh, T.K.; Shimizu, K.; Kamei, I. Bioethanol production from alkaline-pretreated sugarcane bagasse by consolidated bio-processing using *Phlebia* sp. MG-60. *International Biodeterioration and Biodegradation* 2014, 88, 62–68.

Lee, D.1.; Chen, A.; Nair, R. Recent developments on genetic engineering of microalgae for biofuels and biobased chemicals. *Biotechnology and Genetic Engineering Reviews* 2008, 25, 331–336.

Lu, X. *Biofuels: From microbes to molecules*. Caister Academic Press, Norfolk, VA, 2014.

Maitan-Alfenas, G.P.; Visser, E.M.; Guimarães, V.M. Enzymatic hydrolysis of lignocellulosic biomass: converting food waste in valuable products. *Current Opinion in Food Science* 2015, 1, 44–49.

Ng, I.S.; Tan S.I.; Kao, P.H.; Chang, Y.K.; Chang, J.S. Recent developments on genetic engineering of microalgae for biofuels and bio-based chemicals. *Biotechnology Journal* 2017, 12 (10), 1–13.

Ojeda, K.; Sánchez, E.; Kafarov, V. Sustainable ethanol production from lignocellulosic biomass-application of energy analysis. *Energy* 2011, 36(4), 2119–2128.

Pandey, R.; Kumar, G. A comprehensive review on generations of biofuels: current trends, development and scope. *International Journal of Emerging Technologies* 2017, 8(1), 561–565.

Sangita Pradeep Ingole; Kakde, A.U. Evaluation of various plant species for biodiesel production. *Current Botany* 2012, 3(3), 22–25.

Sawin, J.L.; Sverrisson F. *Renewable global status report*. REN21 Secretariat REN21, Paris, 2014.

Singh, B.S.; Bishnoi, N.R. Enzymatic hydrolysis of microwave alkali pretreated rice husk for ethanol production by *Saccharomyces cerevisiae*, *Scheffersomyces stipitis* and their co-culture. *Fuel* 2014, 116, 699–702.

Sun, Y.; Cheng, J. Hydrolysis of lignocellulosic materials for ethanol production: a review. *Bioresource Technology* 2002, 83, 1–11.

Taherzadeh M.J.; Karimi K. Pretreatment of lignocellulosic wastes to improve ethanol and biogas production. *International Journal of Molecular Sciences* 2008, 9(9), 1621–1651.

Velmurugan, R.; Muthukumar, K. Sono-assisted enzymatic saccharifcation of sugarcane bagasse for bioethanol production. *Biochemical Engineering Journal* 2012, 63, 1–9.

Weber, C.; Farwick, A.; Benisch, F. Trends and challenges in the microbial production of lignocellulosic bioalcohol fuels. *Applied Microbiology and Biotechnology* 2010, 87(4), 1303–1315.

Werther, J.; Saenger, M.; Hartge, E.U.; Ogada, T.; Siagi, Z. Combustion of agricultural residues. *Progress in Energy and Combustion Science* 2000, 26(1), 1–27.

Yang, B.; Wyman, C.E. Effect of xylan and lignin removal by batch and flow through pretreatment on the enzymatic digestibility of corn stover with water. *Biotechnology and Bioengineering* 2004, 86, 88–95.

Zhao X., Cheng K., Liu D. Organosolv pretreatment of lignocellulosic biomass for enzymatic hydrolysis. *Applied Microbiology and Biotechnology* 2009, 82(5), 825–827.

Zhu, J.Y; Xuejun, P.; Ronald, S.; Zalesny J.R. Pretreatment of woody biomass for biofuel production: energy efficiency, technologies, and recalcitrance. *Applied Microbiology and Biotechnology* 2010, 87, 847–857.

CHAPTER 13

Monitoring Innovations in Biofuel Production from Lignocellulosic Biomass: A Patent-based Approach

SURUCHEE SAMPARNA MISHRA[1], SONALI MOHAPATRA[2*], and
HRUDAYANATH THATOI[3]

[1]*Department of Biological Sciences, Indian Institute of Science
Education and Research, Berhampur, Odisha 760010, India*

[2]*Department of Biotechnology, College of Engineering and Technology,
Biju Patnaik University of Technology, Bhubaneswar, Odisha 751003,
India*

[3]*Department of Biotechnology, North Orissa University,
Sriram Chandra Vihar, Takatpur, Baripada, Odisha 757003, India*

Corresponding author. E-mail: sonalimohapatra85@gmail.com

ABSTRACT

The technology and inventions that have been implemented for the past two decades have brought remarkable transformations in biofuel industries. These technological inventions greatly affect the nature of the intellectual property rights (IPR) in the biofuel industries. Thus, this chapter highlights the up-to-date progress of the patent applications that have been granted for the different processes involved in production of biofuels from lignocellulosic biomass. More specifically, we emphasize either on the patents that have been granted for selection of raw material and microorganisms that are involved for production of bioethanol or on the different steps involved in bioethanol production such as pretreatment, enzymatic hydrolysis, and ethanol production. We also discuss on the patents that encompass the advancement in downstream strategies of bioethanol production. It is expected that this chapter will expand our knowledge on the current trends in technologies that are applied in industrial sectors and thus will be helpful for exploring technological solutions from an industrial prespective.

13.1   INTRODUCTION

Bioenergy has been the center of attraction for attaining current energy demand concurrently controlling the detrimental impact of the greenhouse gases on human health. Bioenergy, explicitly bioethanol, has been proved to be an efficient alternative for petroleum and thus studied by various research communities for its beneficial effects for a long time. With the aim of comprehending the current status of biofuel, precisely, bioethanol, respective patents were looked into in this chapter using various databases, that is, Indian Patent, European Patent Office database, United States Patent, World Intellectual Property Organization, and Trademark Office.

Information derived from the aforementioned databases was categorized into three strategies: upstream, mainstream, and downstream strategies. The upstream strategies included patents on selection of fermentative strains and genetic engineering approaches adopted for them, while the mainstream strategies cover the patents focusing on different pretreatment strategies, enzymatic hydrolysis, and novel innovations in the field of fermentation, bioreactor design, media composition, nutrients, and CO_2 supply. Furthermore, downstream strategies section covered the inventions in ethanol extraction processes, reprocessing microbial cultures, postdistillation processes, and processing.

Because of the high energy demand, biofuel production has become the new trend in recent times which leads to the development of a number of novel inventions on various aspects of its production and processing. Hence, a thorough review on the recent patent publications in this field could be of great support to the inventors and patent examiners.

Furthermore, this chapter is not being inquisitorial regarding the patentability of any inventions related to biofuel production, rather it is more into endeavoring to highlight the up-to-date progress in this field in the form of patent applications.

13.2   OVERVIEW OF PATENT

Patenting a novel invention is a type of grant of right to a property of inventor from the government that reflects on the guarantee for a limited period of time, the exclusive privilege of making selling, and using the invention for which a patent has been granted. The patent is to ensure the commercial returns to the inventors for the time and money spent in generating a new product.

According to the United Stated Patent and Trademark Office (USPTO), a patentable invention is any "new and useful" process, machine, manufacture, or composition of matter. USPTO has set five elements for patent eligibility, which include (1) it must be novel or new, (2) it must have an "utility," (3) it must be a patentable subject matter, (4) it must be "nonobvious," and (5) it must not be "disclosed" to the public prior to the application for the patent.

Patents can be categorized into two types, such as product patent and process patent. Product patent passes the inventor an exclusive right on a product that is, there will not be any other competitor for the producer. This patent system provides a higher level of protection to the inventor as there will not be any other patent holder. On the other hand, process patent involves the grant for a particular manufacturing process, and not for the product itself. This patent system does not provide a high-level security; thus, anyone can produce the same product using a different or improved process, modifying the various parameters. Based on the criteria defined for patentability, USPTO has categorized the patents into three broader categories such as (1) utility patents, which depict the process of using a new machine, process, or system, (2) design patents, which protect the wholesome design of a useful item, and (3) plant patents, which cover the patents for plants, focusing more on conventional horticulture. Filling a patent application involves several steps starting from filling of application and examination of the application to either acceptance or rejection of the patent. The complete process is well depicted in Figure 13.1.

13.3   PATENTS ON BIOFUEL

Depending on the processing and production strategies, patents granted on biofuel production can be classified into three broader groups. The first group involves the patents on the selection of raw material and microorganisms used for production of bioethanol. The second group includes patents covering different processing steps involved in bioethanol production, such as pretreatment, enzymatic hydrolysis, and ethanol production. The third group involves advancement in downstream strategies such as ethanol separation process, recycling the fermentative microbe, as well as the proper utilization of fermentation waste.

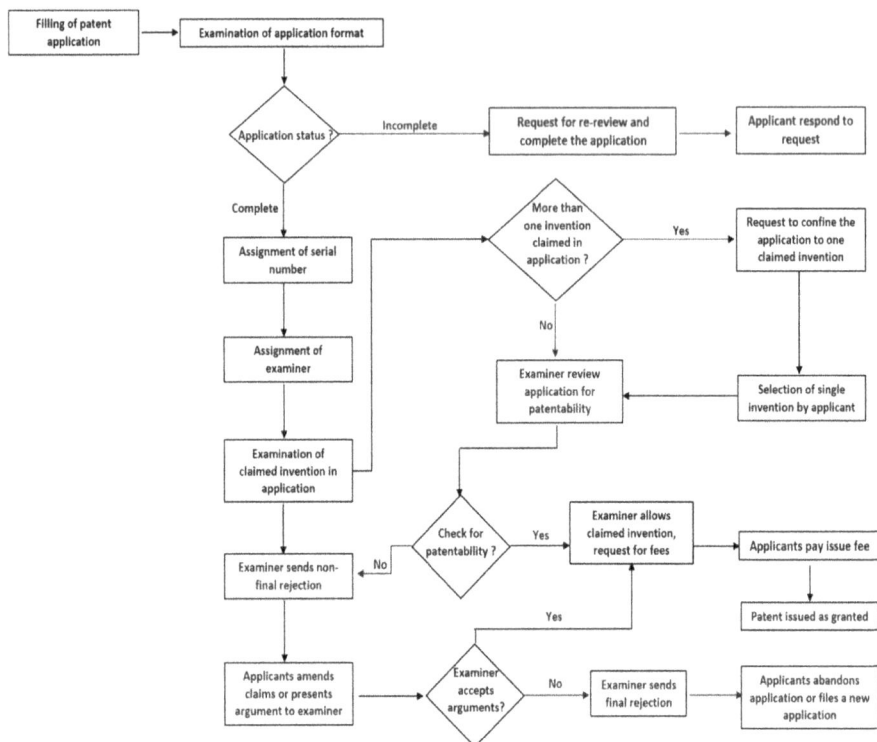

FIGURE 13.1   Process of filling a patent application form.

13.3.1   UPSTREAM STRATEGIES

Upstream strategies are considered to be the building blocks of biochemical processes such as bioethanol production that involves the selection of cellu-lolytic and ethanologenic strains for hydrolysis and fermentation, respec-tively. It is generally easier to aim for pre-established strains to enhance their capability and productivity rather than searching for newly emerging strains owing to which, not only the limitations encountered by those strains can be overcome, but also their productivity can be improved. This section will focus on selection of different types of strains to be employed in bioethanol production. Furthermore, invention of recombinant strains capable of enhancing the productivity and rate of fermentation will also be highlighted.

Werner (1986) reported *Zymomonas mobilis* as a novel bacterium for ethanol production from sugarcane juice and sugarcane syrup by single-stage fermentation process in microaerophilic conditions. Employing this bacterium yielded up to 94%–98% hydrolyzed monomeric sugar with 95%–98% conversion efficiency, also, the process generated 95.5 g/L ethanol as an end product in 24–30 h. Further, separation and extraction of used microbial inoculum was reported to be carried out from the fermented broth without the need of centrifugation and membrane filtration. The inventors claimed that ethanol produced from this process possessed high commercial value, while the by-product can be used as a carbon source for algal growth. Furthermore, they have also established the dependency of the whole process on raw material used and appealed that accomplishment of high concentration of end product formation is not completely dependent on the substrate quality, as there is not much difference found while comparing sucrose and rotting sugarcane as substrate. They have also added that this process is particularly suited for industrial application with minimum precautions. EP2005/1130085B1 provided information about a new heat-tolerant strain of *Kluyveromyces marxianus* as the fermenting microorganism. The inventors adopted a hydrothermal steam explosion pretreatment strategy aiming at the delignification of lignocellulosic biomass followed by simultaneous saccharification and fermentation as the process of ethanol production. In a similar context, Kumar et al. (2009) also described *Kluyveromyces* sp. as an industrial model for bioethanol production owing to its capability to grow and ferment sugar at a higher temperature (37 °C–55 °C). Further, this strain was also observed to utilize a wide variety of monomeric sugar obtained from pretreatment of lignocellulosic biomass. Furthermore, this strain was also observed to utilize a wide variety of monomeric sugar family obtained from pretreatment of lignocellulosic biomass (e.g., sugarcane bagasse, cassava, potato, and corn). The inventors also interrelated this observation to the process of ethanol production by *Kluyveromyces* sp. IIPE453. The inventors also observed the same during bioethanol production using the recombinant strain *Kluyveromyces* sp. IIPE453. Hence, keeping the robustness, cost-effectiveness ability to ferment a wide variety of sugar, this patent initiated a possibility for the utilization of thermotolerant *Kluyveromyces* sp. in industrial bioethanol production.

Furthermore, various metabolically engineered microorganisms are also employed for the enhanced production of bioethanol owing to their engineered biosynthetic pathways. Recombinant strains provide various advantages over innate strains such as enhanced bioethanol production, less or no accumulation of several inhibitory intermediates. Also the transformants formed using metabolic engineering, provide robust system with credible capability to express a large set of external genes with excellent efficiency. A number of patents have been put forward for the utilization of genetically modified microorganisms, among which a few have been briefly described in this section.

Geng and Zhang (2015) disclosed genome shuffling approach to generate recombinant microorganism for bioethanol production. The inventors produced a hybrid strain ScF2, taking Pichia stipitis and *Saccharomyces cerevisiae* as parent strains and carried out sugar utilization test in all the strains. They observed an improved fructose, xylose, maltose, and cellobiose utilization by ScF2 in comparison to its parent strain; decreased glucose and raffinose utilization capability than *S. cerevisiae*; and less mannose, sucrose, and lactose utilization than *P. stipitis*. Additionally, ScF2 was observed to have a similar sugar utilization pattern as *P. stipitis* for most of the sugars found in lignocellulosic biomass. They have observed an enhanced xylose consumption and fermentation efficiency in case of ScF2 strain in initial elevated concentration of xylose present in the medium with maximum 51 g/L of ethanol with initial xylose concentration of 150 g/L. While ScF2 strain utilized glucose at a very slower rate, the rate of ethanol production was noted to be higher than *P. stipitis*. The inventors also performed simultaneous saccharification and fermentation of oil palm empty fruit bunch hydrolysate and observed a sharp decrease in the hydrolysis rate from 118 h to 46 h for ScF2 as compared to *P. stipitis* (54 h).

Srienc et al. (2014) presented a comparison between the efficiency of bioethanol production by different microorganism as well as their genetically modified version to relate the improvement in downstream processing underlying genetic modification. The inventors took various wild-type yeast such as *Saccharomyces cerevisiae, Kluyveromyces* sp., *Pichia* sp., or *Yarrowia* sp. and bacterial strains such as *Escherichia coli* into consideration and generated several mutants by mutating different genes responsible for C5 and C6 sugar utilization capacity to compare the ethanol production efficiency as well as the inhibitor tolerance capacity. They observed more tolerance against acetic acid and furfural and resistance against other acidic-type inhibitors in case of AFF01 (mutant) than that of its parent *E. coli* strain TCS083. Additionally, mutant strain also overtook the fermentation efficiency, productivity, and yield in the presence of acetic acid or furfural.

Deng et al. (2006) provides information regarding the redirection of carbon flux in a cellulolytic recombinant microorganism to achieve enhanced ethanol yield. The inventors have employed *Clostridium thermocellum, Clostridium cellulolyticum, Clostridium clariflavum, Clostridium phytofermentans, Clostridium acetobutylicum, Caldicellulosiruptor bescii*, and *Caldicellulosiruptor saccharolyticus*. They observed enhanced ethanol production owing to the redirected carbon flux. They postulated that faster ethanol production rate can be achieved by employing continuous or serial dilution cultures for a preferred feedstock.

Selection of appropriate raw material and fermentative microorganism leads to a successful fermentation process. There are three main stages of fermentation process such as pretreatment of lignocellulosic biomass, hydrolysis of polymeric sugar into their monomeric form, and conversion of monomeric sugar units to ethanol by fermentative microorganisms, which fall into the category of mainstream strategies for biofuel production.

13.3.2 MAINSTREAM STRATEGIES

This section involves the patents covering selection of pretreatment strategies. Pretreatment is a key step in fermentation process as it breaks the crystallinity of cellulose and disrupts the rigidity of cellulose, hemicellulose, and lignin network, which in turn helps in enhanced enzymatic hydrolysis. A number of patents have been filed regarding the recent development in pretreatment strategies that have been put forward by various researchers with each passing day around every corner of the world. Hence, a detailed review regarding the patents granted for pretreatment strategies employed might be helpful for researchers not only for designing the experiments but also for discovering new secondary methods for the same. Moreover, patents published in reference to the saccharification of pretreated biomass and the novel fermentation strategies for conversion of monomeric sugar to ethanol have also been included in this section.

Schall and Farahani (2017) describes a pretreatment strategy that includes the application of ionic liquid (IL) at various temperatures ranging from 120°C to 160°C, which leads to rapid and efficient enzyme degradation owing to the generation of more amorphous cellulose, thereby increasing the conversion of cellulose and hemicellulose to glucose and xylose. The authors hypothesized that IL incubation of lignocellulosic biomass required high temperatures in order to unravel the crystalline structure of cellulose. The authors enclosed a modified method that includes an oxidative preprocessing step at temperature nearly equal to the room temperature for lignin oxidation following the IL incubation step. This process leads to a drop in IL incubation temperature range from 120°C to 160°C to 75°C and facilitates partial deconstruction of crystalline cellulose, hemicellulose, as well as lignin and forms easily hydrolysable amorphous cellulose and hemicellulose even in low temperature with higher digestibility yields as compared to that of elevated temperature range. In a similar context, AU2013/204042B2 employed IL pretreatment for lignocellulose conversion to sugar, which substantially improves the sugar yield and efficacy. The authors observed slow reaction rate and low ethanol yield by employing this approach.

Hodge et al. (2015) encloses a type of alkaline pretreatment that not only aids lignin solubilization but also anticipates the resemblance of solubilizing lignin with its native structure such that almost 35% of α-carbon present in solubilized lignin is oxidized from hydroxyl to carbonyl. The inventors described different methods of pretreatment using catalytic approaches for plant cell wall degradation. This process includes the diffusion of a homogeneous catalyst such as copper (II) 2,2-bipyridine ethylene diamine through nanoscale openings within the cell walls to facilitate biomass deconstruction. Homogeneous catalyst comprises of one or more metals; two metals connecting through ligands or a multi-ligand metal complex along with an oxidant generates an oxidation reaction to act as catalyst to change the biomass composition.

Qiulong et al. (2015) disclosed a simple, clean, and pollution-free irradiation pretreatment technique carried out at room temperature with minimal generation of inhibitory product. This pretreatment strategy includes the implication of decay of high-energy radioactive isotope γ-ray Co6t3 to lignocellulose pretreatment to degrade the physiochemical structure of lignocellulosic biomass. The authors claimed that irradiation treatment results in lesser degree of crystallinity in cellulose and degradation of hemicellulose, cellulose, and lignin molecular chains, thereby making the enzymatic degradation significantly effective. However, low reducing sugar concentration and ethanol yield encouraged the authors to employ this pretreatment strategy using batch and fed-batch approaches following the addition of cellulase and hemicellulase for subsequent enzymatic degradation, and observed enhanced hydrolysis rate resulted in high reducing sugar concentrations in fed batch process.

Disappointedly, even with the application of most advanced pretreatment strategy, cost-effective production of fermentable sugar is a myth, which makes the ethanol production less profitable. According to the U.S. Patent No 5,196,069, the degree of transformation of lignocellulosic polysaccharides does not exceed 77%–84% of soluble carbohydrates, while the enzyme concentration required for converting polysaccharides into fermentable monomeric forms is way too large. Along with the pretreatment using different advanced techniques, it is essential to choose proper hydrolysis approaches for optimum conversion of polymeric sugars into monomers. Patents in the context of hydrolysis and degradation of cellulose and hemicellulose into glucose and other monomeric hexose and pentose employing cellulase and hemicellulase from a variety of sources can give the researcher a clear idea regarding the future advancement of the process.

Qiu et al. (2018) disclosed a full bacterium method for saccharification of cellulose and hemicellulose. This method involves a two-step process in which a cellulosome bacterial strain is constructed first under anaerobic condition followed by its addition to pretreated biomass. The inventors claimed that this process is cost-effective and less time-consuming and can be employed for industrialized process in future. In a similar context, Qiu et al. (2018) provides a slight modification in the previous patent to improve full bacterium saccharification efficiency by controlling the pH of the saccharifying medium. This process includes formation of a cellulosome strain in seed culture medium in anaerobic condition until mid-log phase followed by saccharification of seed liquor under specific controlled pH to obtain monomeric sugar. The inventors observed 15–120 g/L liquid glucose and claimed that this method substantially reduced the extent of saccharification. They also claimed this process to be efficient for industrialization purpose with definite high sugar yield. Qiu et al. (2018) also employed a cellulosome bacterium in order to avoid high enzyme cost and low ethanol yield. The authors employed the cellulosome for pretreatment, saccharification, and fermentation process using a wide variety of raw material. The inventors obtained ~90% of saccharification efficiency followed by ethanol production up to 20–98 g/L, ensuring quite high ethanol production. They also claimed this process to be good for industrialization.

Headman et al. (2016) disclosed a process where saccharification and/or fermentation is performed at a temperature below the initial gelatinization temperature in the presence of glucoamylase and alpha-amylase, protease, and/or cellulolytic enzyme. Saccharomyces cerevisiae was employed as the fermenting microorganism. The inventors found that under the above-mentioned condition, S. cerevisiae produced higher ethanol yield and lower glycerol production compared to the conventional fermentation condition.

Lali et al. (2012) disclosed a process of enzymatic hydrolysis using a multi-enzyme, multi-step system. The process involves two steps of saccharification by two different set of enzymes where both the steps involve enzymes from at least one group and perform saccharification. Auxiliary enzymes such as amylases, proteases, lipases, and glucuronidases can be optionally added to both or one of the two steps for increased rate of sugar hydrolysis. The authors claimed to produce fermentable sugar with high efficiency from cellulose, hemicellulose, or the mixture of sugars present in lignocellulosic biomass in very less time period with much higher conversion rate.

Furthermore, along with the appropriate selection of pretreatment and hydrolysis, electing suitable fermentation strategies leads to optimal conversion of lignocellulosic sugars into ethanol by fermentative microorganisms.

Patents covering various fermentation strategies employed worldwide for different lignocellulosic material may lead the researchers to understand the whole bioethanol production process in a better way.

Bradley and Kearns (2009) depicted a process of bioethanol production from corn cob and switch grass, employing solid-state fermentation in rectangular growth chambers made up of mild steel or plastic panels. The authors described a method of cellulase/hemicellulase enzyme production, resulting in substantial cost-reduction in biofuel generation. The whole culture harvested for purifying enzymes is used for the conversion of polymeric lignocellulosic sugar to monomeric ones followed by fermentation combined in a single unit to prevent the inhibition of hydrolytic enzymes by the reaction products. The whole SSF process takes up to 3–6 days, resulting in the production of 2%–4.5% of dilute ethanol stream.

Jung-Gon and Hwan disclosed a fermentation strategy for simultaneous saccharification of polymeric sugar and fermentation of sugar into bioethanol by a process called simultaneous saccharification and fermentation. The authors provided a three-step process comprising of (1) pretreatment of biomass, (b) saccharification at a temperature of 25°C–30°C, and (3) fermentation of the separated supernatant at a temperature of 30°C–70°C resulting in bioethanol production. Microorganisms employed in the whole fermentation process are then separated from the broth followed by distillation of fermentation broth to separate bioethanol.

Belanger et al. (2007) presented an integrated process for the production of ethanol from woody plant biomass. The process initiated with the simultaneous and continuous flow of plant biomass from one direction and aqueous ethanol from the opposite direction. Further, the process continued with the removal of ethanol from biomass mixture followed by subjecting the mixture to continuous flow of water at high temperature and pressure in order to hydrolyze xylose. The last step comprised of the hydrolysis of cellulose present in the biomass.

Brevnova et al. (2011) disclosed an integrated method for degrading various lignocellulosic biomass and fermenting the pentose sugar and converting them into ethanol by employing a full suite of yeast strain or strains that are capable of hydrolyzing cellulose and starch simultaneously, producing less glycerol, utilizing reduced amount of enzyme during fermentation process, and increasing ethanol yield. The inventors also focused on the full suite of proteins expressed in yeast strain or strains that are involved in enzymatic activity of various steps of fermentation process, so that they could be co-expressed in yeast cells to provide synergistic action on lignocellulosic biomass (Table 13.1).

TABLE 13.1 Patents in Biofuel

Patent No.	Patent Title	Year	Description
WO2019/025449A1	Methods of defibrillating cellulosic substrates and producing celluloses using a new family of fungal lytic polysaccharide monooxygenases (LPMO)	2019	Disclosed a physiochemical pretreatment method for defibrillating cellulose containing lignocellulose for the production of nanocelluloses using fungal LPMO
CN2013/104540956A	Methods of processing lignocellulosic biomass using single-stage autohydrolysis and enzymatic hydrolysis with C5 bypass and post-hydrolysis	2012	Disclosed hydrothermal pretreatment as a method to process lignocellulosic biomass. They acquire a two stages process starting with pretreatment of lignocellulosic biomass to low severity in a single-stage pressurized hydrothermal pretreatment followed by enzymatic hydrolysis to release xylose retained in the solid state
CN2013/105518158A	Alkaline treatment of lignocellulosic biomass	2013	Disclosed a method and instrumentation for ionic liquid pretreatment and mild alkaline treatment which can be easily hydrolysed by enzymes
US2013/8551747	Process for producing bioethanol from lignocellulosic plant raw material	2008	Disclosed a pretreatment process especially for cellulose differentiation from the lignocellulosic biomass. They employed pretreatment in the presence of a mixture of formic acid and water, at a temperature range 95–110 °C followed by separating solid phase containing cellulose form liquid phase containing formic acid, lignin, and hemicelluloses at atmospheric pressure
US 2012/022.0005	Method for producing ethanol from lignocellulosic biomass	2009	Focused mainly on lignin removal. Disclosed a three step pretreatment process commencing with steam explosion treatment followed by a flash treatment and lignin removal step by ethanol emersion
US 2012/0329116	Pretreatment of lignocellulosic biomass through removal of inhibitory compounds	2012	Focused on hemicellulose degradation into fibres along with the removal of inhibitory compounds toxic for enzymatic hydrolysis
US2009/OO7504245	Biomass conversion to alcohol using ultrasonic energy	2009	Acquired an ultrasonic pretreatment strategy for delignification of lignocellulosic biomass

TABLE 13.1   (*Continued*)

Patent No.	Patent Title	Year	Description
US2003/6555350B2		2003	Invented a type of wet oxidation or steam explosion pretreatment process in which the whole system is designed in a way that allows the reaction water to be reused in order to reduce the water usage
US2010/7,674,608 B2			Disclosed a type of ionic liquid (IL) pretreatment strategy that would result in partial softening of cellulose followed by regeneration of amorphous cellulose by its subsequent contact with antisolvent. The amorphous cellulose can then be collected from the IL/antisolvent mixture by simple mechanical pretreatments like filtration or centrifugation
US522153. US5536325			Described a two staged process involving the initial step of mild acid pretreatment resulting in hemicellulose depolymerisation into xylose and other monosaccharides followed by cellulose depolymerization into glucose in final one
US5366558			Described a mild hydrolysis process resulting in substantial hemicellulose hydrolysis without any potential damage to cellulose fibers. The solid residue containing cellulose and lignin is subjected to fine grinding which is separated by mild acid pretreatment
US4706903			Pretreatment strategy describing separation of cellulose and lignin by fine grinding followed by acid hydrolysis under tougher conditions until the glucose solution is obtained
US5366558	Method of treating biomass material		Described a method of simultaneous production of glucose by enzymatic hydrolysis by a group of cellulolytic enzymes and fermentation of monomeric glucose into ethanol in presence of fermenting microorganism. This method claimed to improve the ethanol yield

TABLE 13.1 *(Continued)*

Patent No.	Patent Title	Year	Description
Azuma et al. (1984) and US5196069	Apparatus and method for cellulose processing using microwave pretreatment		Proposed a method of microwave pretreatment for enhanced enzymatic hydrolysis of polysaccharides present in lignocellulosic biomass. Such pretreatment claimed to generate about 77%–84% of reducing sugar from total sugar present in lignocellulose stock
US 1979/4237226A			Designed an impact flow-through continuous type reactor for mechanical pretreatment of oak, poplar wood, newspaper and corn straw. The inventors claimed to get fine structured biomass
US2015/0176034 A1	Method for viscosity reduction in co-fermentation ethanol processes		Disclosed a method of pretreatment for reduction of viscosity of processed biomass stream during fermentation
EP2011879A2	Process for producing ethanol		Disclosed a combined biochemical and synthetic conversions process as the pretreatment strategy resulting in high yield ethanol production with concurrent generation of high value co-products
WO2016/110867	Preparation of ethanol by continuous fermentation process		Disclosed a method of bioethanol production by continuous fermentation of starch containing feedstock by arranging the fermenters continuously. Feedstock was provided at the beginning of the process and the product was recovered at the end without any need to discontinue the process for preparation or recovery action
US2013/8551758B2	Substrate-selective co-fermentation process		Disclosed a biological method for independent and co-consumption of sugar or a mixture of sugars present in lignocellulosic biomass along with removing the inhibitory compounds from hydrolysate
JP2012/5935020B2	New production methods of ethanol		Disclosed a method of ethanol production in the presence of a substance having a fermentation inhibiting effects of ethanol

TABLE 13.1 (Continued)

Patent No.	Patent Title	Year	Description
US20120036768A1	High consistency enzymatic hydrolysis for the production of ethanol		Disclosed a method of enzymatic hydrolysis that is rich with fibre consistency of enzymatic hydrolysis mixtures. This consistency can be achieved by exposing the hydrolysis mixture to a group of enzyme followed by thickening the mixture and hydrolyzing it again thereby resulting in highly concentrated fermentable slurry.
WO2009100042A2	Systems and processes for producing biofuels from biomass		Disclosed a process of simultaneous saccharification and fermentation process for converting lignocellulosic biomass keeping the systems and processing cost economic
CA2804738A1	Method and system for processing a biomass for producing biofuels and other products		Disclosed a process of autolysis of lignocellulosic biomass for the production of lipid and nonlipid products in two separate layers which can be further used to produce biofuel. This invention also depicted a processing apparatus useful for bioethanol production
JP2012519500A	Production of fermentation end products from *Clostridium* (*Clostridium*) species	2012	Disclosed a method of enhanced production of bioethanol employing *C. phytofermentans* as fermenting microorganism in fed batch culture in order to improve the fermentation performance. The whole process is being carried out in presence and/or low pH of the fatty acid-containing compounds
CN103328642B	Integrated process for the production of biofuels	2012	Depicts an integrated fermentation process comprising of cellulolytic degradation followed by ethanologenic fermentation for biofuel production.
US8571689B2	Systems and processes for producing biofuels from biomass	2007	Disclosed a process of batch fermentation for biofuel production. The authors determined the optimal mode of fermentation that the whole system takes upon a temporal control horizon specifying biofuel and/or sugar concentration over batch fermentation process
WO2016110867A1	Preparation of ethanol by continuous fermentation process	2015	Disclosed a process for ethanol production by continuous fermentation of biomass by employing a set of fermenters organised to get the continuous end product

Furthermore, a number of patents such as US Patent Publication 2012/0290363, US Patent Publication 2012/0290362, US Patent Publication 2012/0290344, US Patent Publication 2012/0290273, US Patent Publication 2012/0290267, US Patent Publication 2012/0290221, US Patent Publication 2012/0290220, US Patent Publication 2012/0290205, US Patent Publication 2012/0287270, and US Patent Publication 2012/0287270 have been filed concerning monitoring the level of greenhouse gas emission and sustainability for economic production of bioethanol.

13.4   CONCLUSION

In this chapter, various patents published since decades concerning different steps of biofuel production from lignocellulosic biomass as well as other cellulose-containing feedstock have been briefly described in order to enlighten the recent developments in the field of biofuel production. Besides the patents on feedstocks, patents on processes of bioethanol production from lignocellulosic biomass such as the pretreatment and fermentation strategies have also been elaborated. A brief outlook toward the small scale as well as industrial grade developments in the bioenergy field will help us in achieving more economical and sustainable feedstock as well as processes for biorefinery-based lignocellulosic biofuels.

KEYWORDS

- **Patent**
- **biofuel**
- **lignocellulosic biomass**
- **innovation**

REFERENCES

Ahring, B.K.; Thomsen, A.B. Method for processing lignocellulosic material. US2003/ 6,555,350 B2, 2003.

Antolin, R.A.; Valenzuela Romero, M.N.; Alonso Martinez, B.; Molist, R.D.; Garcia Encinas, R.; Gutierrez Montero, M.A.; Vidal, J.Y.; Montes Garcia, L.; Lopez Mac, J.; Marquez

Pinuela, A.; Vazquez Garcia, M. Method of monitoring sustainability of bioproducts. US2012/0290363, 2012.

Azuma, J.I.; Tanaka, F.; Koshijima, T. Enhancement of enzymatic susceptibility of lignocellulosic wastes by microwave irradiation. *Journal of Fermentation Technology* 1984 62(4), 377–384.

Belanger, H.; Watson, J.D.; Yao, J.-L.; Prestidge, R.; Lough, T.J.; Macfarlane, A.; Farid, M.M.; Chen, J.; Elton, C. Process for the production of biofuel from plant materials. CA2007/ 2651628C, 2007.

Bradley, C.; Kearns, R. Process of producing ethanol using cellulose with enzymes generated through solid state culture. US20090117634A1, 2009.

Brevnova, E.; Mcbride, J.E.; Wiswall, E.; Wenger, K.S.; Caiazza, N.; Hau, H.; Argyros, A.; Agbogbo, F.; Rice, C.F.; Barrett, T.; Bardsley, J.S.; Foster, A.; Warner, A.K.; Mellon, M.; Skinner, R.; Shikhare, I.; Haan, R.D.; Gandhi, C.V.; Belcher, A.; Rajgarhia, V.B.; Froehlich, A.C.; Deleault, K.M.; Stonehouse, E.; Tripathi, S.A.; Gosselin, J.; Chiu, Y.-Y.; Haowen, X.U. Yeast expressing saccharolytic enzymes for consolidated bioprocessing using starch and cellulose. US2011/9206444B2, 2011.

Brink, D.L. Method of treating biomass material. US5536325, 1994.

Brink, D.L. Method of treating biomass material. Apparatus for the hydrolysis and disintegration of lignocellulosic. US5366558.

Brink, D.L.; Merriman, M.M.; Mixon, D.A. US4706903, 1987.

Cullingford, H.S.; George, C.E.; Lightsey, G.R. Apparatus and method for cellulose processing using microwave pretreatment. US5196069.

Deng, Y.; Olson, D.G.; van Dijken, J.P.; Shaw, A.J.; Argyros, A.; Barrett, T.; Caiazza, N.; Herring, Lebanon, C.D.; Rogers, S.R.; Agbogbo, F. Engineering microorganisms to increase ethanol production by metabolic redirection. US2006/9803221B2, 2006.

Deshpande, G.B.; Kulkarni, M.A.; Pandurang, R.S.; Kumbhar, P.S.; Bhalewadikar, H.; Patil, D.P.; Budgujar, M.B. Preparation of ethanol by continuous fermentation process. WO2016/ 110867, 2016.

Dottori, F.A. Pretreatment of lignocellulosic biomass through removal of inhibitory compounds. US2012/0329116, 2012.

Eiteman, M.A.; Altman, E. Substrate-selective Co-fermentation Process. US2013/8551758B2, 2013.

Gaddy, J.L.; Arora, D.K.; Ko, C.-W.; Phillips, J.R.; Basu, R.; Wikstrom, C.V.; Clausen, E.C. Methods for increasing the production of ethanol from microbial fermentation. US2007/ 7285402, 2007.

Geng, A.; Zhang, W. Xylose fermenting yeast constructed using a modified genome shuffling method. US2015/0176028A1, 2015.

Ghansham, B.D.; Mahesh, A.K.; Pandurang, R.S.; Pramod, S.K.; Abhishek, B.; Deepak, P.P.; Mahesh, D.B. Preparation of ethanol by continuous fermentation process. WO2016/110867A1, 2016.

Grethlein, H.E. Process for pre-treating cellulosic substrates and for producing sugar therefrom. US4237226, 1979.

Headman, J.; Saunders, J.; Craig, J.; Stevens, M.; SHIHADEH, K.; Victor, P.; John, A.P.; BELL, L. Processes of producing ethanol using a fermenting organism. WO2016/138437, 2016.

Hecht, G.; Stern, M.E.; Brazzell, R.K. Tissue irrigating solutions. US5221537, 1993.

Hodge, D.B.; Hegg, E.L.; Li, Z.; Mathrubootham, V.; Bhala, A.; Bansal, N. Multi-ligand metal complexes and methods of using same to perform oxdative catalytic pretreatment of lignocellulosic biomass. US2015/0352540A1, 2015.

Jung-Gon, P.; Hwan, H.J. Method for producing bio-ethanol using simultaneous saccharification and fermentation with the supernatant of the waste of culture broth. KR101043443B1, 2009.

Kinley, M.T.; Waukee, I.A.; Krohn, B.; Brandon, F.L. Biomass conversion to alcohol using ultrasonic energy. US2009/OO7504245, 2009.

Kirby, K.D.; Plenty, L.; Elsternwick, C.J.M. Production of ethanol by fermentation. US1984/4490469, 1984.

Kondo, A.; Hasunuma, S. New production methods of ethanol. JP2012/5935020B2, 2012.

Kumagai, C.; Taniyama, N.; Nakamura, Y.; Tokushima-shi. Method for producing ethanol from lignocellulosic biomass. US2012/022.0005, 2012.

Kumar, A.D.; Kumar, S.; Dutt, S.C.; Chand, D. Novel strain and a novel process for ethanol production from lignocellulosic biomass at high temperature. US2009/0226993, 2009.

Lali, A.M., et al. Method for production of fermentable sugars from biomass. US2012/0115192, 2012.

Loftus, T.P. Jr. Drive circuit for power switches of a zero-voltage switching power converter. US5268830, 1993.

Lomovsky, O.I.; Korolev, K.G.; Politov, A.A.; Bershak, O.V.; Lomovskaya, T.F.. Method of producing bioethanol from lignocellulose. WO2009/005390, 2009.

Method for determining emissions of greenhouse gases (GHG) in the production of bio-products. US2012/0290344, 2012.

Method of measurement of emissions of greenhouse gas industry related bio-products. US2012/0290362, 2012.

Perdices, B.; Dominguez, O.; Miguel, J.; Garcia, C. Procedure for the production of ethanol from lignocellulosic biomass using a heat-tolerant yeast. EP2005/1130085B1, 2005.

Qiu, C.; Yajun, L.; Bin, L.; Yinang, F. The method that ethyl alcohol is prepared using lignocellulosic. CN2018/109055439A, 2018.

Qiu, C.; Yajun, L.; Shiyue, L.; Yinang, F. Improve the full bacterium method for saccharifying of lignocellulosic saccharification efficiency. CN2018/109097417A, 2018.

Qiu, C.; Yajun, L.; Kuan, Y.; Shiyue, L.; Renmin, L.; Yingang, F. Full bacterium method for saccharifying for lignocellulosic. CN2018/109055459A, 2018.

Qiulong, H.; Xingyao, X.; Xiaojun, S.; Keqin, W.; Liang, C.; Qingming, L.; Kang, W. The method of irradiation pretreatment kind of fed-batch and to achieve a high concentration of the enzymatic hydrolysis of lignocellulosic substrates. CN2015/104651429B, 2015.

Ramos, K.; Santos, D.; Desai, P.; Shrestha, P.; Woods, R.R.; Le, S. Method for viscosity reduction in co-fermentation ethanol processes. US2015/0176034A1, 2015.

Schall, C.A.; Farahani, S.V. Enhancement of lignocellulose saccharification via a low temperature ionic liquid pre-treatment scheme. US2017/0009265A1, 2017.

Srienc, F.; Gilbert, A.; Trinh, C.; Unrean, P. Genetically-engineered ethanol-producing bacteria and methods of using. US2014/8,623,622B2, 2014.

Srienc, F.; Gilbert, A.; Trinh, C.; Unrean, P. Genetically-engineered ethanol-producing bacteria and methods of using. US2014/8623622B2, 2014.

System and method for calculating greenhouse gas emissions in the production of raw material for obtaining bioproducts. US2012/0290273, 2012.

System and method for measuring GHG emissions in bioproduct production processes. US2012/0290267, 2012.

System and method for measuring GHG emissions associated to bioproduct industry. US2012/0290221, 2012.

System and method for calculating greenhouse gas emissions in the production of raw material for obtaining bioproducts. US2012/0290220, 2012.

System and method for measuring GHG emissions in bioproduct production processes. US2012/0290205, 2012.

System and method for measuring GHG emissions associated to by-product industry. US2012/0287270, 2012.

Varanasi, S.; Schall, C.A.; Dadi, A.P. Saccharifying cellulose. US2010/7,674,608B2, 2010.

Verser, D.; Eggeman, T. Process for producing ethanol. US2013/6509180, 2013.

Werner, D.H. Conversion of sucrose to ethanol using the bacterium *Zymomonas mobilis.* WO1986/04357, 1986.

PART III
Utilization of Bioresources in Agriculture

CHAPTER 14

Effective Utilization of Palm Mesocarpic Husk: An Agrowaste in Producing Major Phenolics and Other Chemicals

SWAPAN KUMAR GHOSH[1*], and KINGSUK DAS[2]

[1]Post Graduate Department of Botany, Molecular Mycopathology Lab. Biocontrol Unit, Ramakrishna Mission Vivekananda Centenary College (Autonomous), Rahara, 24 Parganas (North), West Bengal 700118, India

[2]Post Graduate Department of Botany, Serampore College, Serampore, Hooghly, West Bengal 712201, India

*Corresponding author. E-mail: gswapan582@gmail.com

ABSTRACT

All people globally know palm trees and palm fruits. Some popular palms are edible such as date palm (*Phoenix*) and coconut (*Cocos nucifera*), and these are utilized commercially. The dried mesocarpic husks of palms (*Cocos, Borassus, Areca*, etc.) are one of the major agricultural waste materials in developing Third World countries. Each year, it amounts millions to several tons. Apart from jute, coconut mesocarpic husk (coir) is processed and used by the rope-making industry, but majority of it remains unutilized. These unutilized husks and fibrous materials increase unnecessary burdens and promote air pollution. Apart from rope-making industry, chief uses and possible utilities of palm mesocarpic husks are not well investigated. Several scientists reported the compositional and structural characteristics of residual biomass and found that major portion of the waste material is carbohydrate and phenolic in nature. Another research group discovered that several principal phenolic compounds (4-hydroxybenzoic acid [4-HBA], ferulic acid, *p*-coumaric acid, etc.) are accumulated in dried mesocarpic husk. The phenolic, 4-HBA is used as a dietary antioxidant. Other phenolic acids have commercial values. From the studies and researches of these research groups, it is prudent that useful utilization of palm agro-industrial waste is feasible. More investigations are going on for effective application of palm agrowaste.

14.1 INTRODUCTION

The economy of many countries is mainly based on agriculture. As a result, tons and tons agricultural by-products are piling up in the environment as agrowaste. Data from table (Table 14.1) show the total agricultural area and total agrowaste of some countries (Nagendran, 2011). Sadh et al. (2018) classified agro-industrial wastes into two categories—agriculture residues and industrial residues. They further divided agriculture residues into field residues and process residues. Field residues include the matter present in the field after the process of crop harvesting, for example, leaves, stalks, seed pods, and stems left in the fields, while the process residues include the matter existing even after the crop is processed into alternative valuable resources (Sadh et al., 2018). Recently, agricultural waste management (AWM) has become an issue of concern for policymakers (Hai et al., 2010). AWM will be done keeping in mind that wastes may be converted to potential resources rather than undesirable and unwanted ones, without polluting air, water, and land resources. This will require better use of technology and incentives, a change in philosophy and attitudes, and better approaches to AWM. The 3Rs approach, that is, reducing waste and reusing and recycling resources and products (3Rs), leads to efficient minimization of waste generation and management (Obi et al., 2016). Some agrowastes are used in many purposes such as rice husk or bran for rice oil production, rice straw as fodder, thatching, mushroom cultivation, biofertilizers etc.; similarly, wheat straws are used for the same or other purposes. The very innovative uses of vegetable fibers and other agricultural by-products, for example, the ashes of the coconut shell, saw dust, sugarcane bagasse, groundnut shell, oil palm shell, rice husk ash, etc are cement in concrete (Somarriba et al., 2018) as a partial replacement for coarse and fine aggregates (Li et al., 2015), admixtures (Akulova et al., 2016), cement replacement (Ramesh et al., 2017), and even in ultrahigh-strength concrete (Ramos et al., 2014), and lightweight concrete (Anbazhagan and Gopinath, 2017). Other uses include fibers for concrete reinforcement (Sethunarayanan et al., 1989), roof tile production (Jose et al., 2009; Saravanan et al., 2017), and asphalt (Ramadhansyah et al., 2016) in the construction industry for house-building in different countries like India (Akulova et al, 2016), Belarus (Obilade, 2014), Malaysia (Alengaram et al., 2013), Brazil (Jose et al., 2009), and Russia (Asasutjarit et al., 2007). Sukan et al. (2014) produced poly(3-hydroxybutyrate) from orange peel waste. Other agrowastes are also used for many purposes. However, many agrowastes are unutilized and are piled in the environment, and they are rotted like tons of maize stem, cobs, plantain stem, etc. Agrowastes contain many chemical constituents like cellulose, lignin (insoluble), sugar, amino acids, organic acids (soluble), fats, oil waxes, resins, pigments, proteins, and minerals (Caprara et al., 2011; Subba Rao, 1993). They are very good sources of fibers, cellulose, and are utilized in the production of pulp for the paper industry, cellulose derivatives for production of biodegradable plastic, and films for photography. The raw cellulose may be converted to cellulosic derivatives, which include cellulose nitrate, acetate, and xanthenes. Cellulose nitrate has great utilization in lacquer coating for decorative and protective purposes, in photographic industry, etc. The agrowastes from pineapple, citrus orange, banana, watermelon, and papaya were subjected to solid-state fermentation and were used to produce biofertilizer, which was then applied in vegetable plantation (Lim and Matu, 2015). Some agrowastes are burned for waste management but burning releases huge amounts of smoke that causes air pollution. So eco-friendly approaches are needed for waste management as in order to reuse and minimize waste through the recycling process. Recently, research has aimed at converting some agrowastes to commercially useful chemicals such as phenolic compounds, cellulose, and methane (Abdullah et al., 2011; Chakraborty et al., 2006; Suzuki et al., 1998). Palmae family with 210 genera and more than 4000 species (Hutchinson, 1959; Lawrence, 1967) gives us fibrous fruits whose mesocarps are reused as rope or other purpose but some scientists extracted some phenolic chemicals like 4-hydroxybenzoic acid, ferulic acid, *p*-coumaric acid, etc. Some members of this family are presented in Figure 14.1. In this chapter, we will discuss a brief regarding mesocarp agrowaste from Palmae family and their innovative and sustainable management by extracting and purifying phenolic and other chemicals and their utilization.

14.2   ARECACEAE/PALMAE FAMILY

14.2.1   DISTRIBUTION

The members of this family are native to tropical and subtropical climates, as they favor moist and hot climates but can grow in different habitats. The highest number of species is found in Colombia. The main palm growing areas include South America, the Caribbean, and areas of the South Pacific and Southern Asia, but some species are native to desert areas such as the Arabian peninsula and parts of northwestern Mexico (Sethunarayanan et al., 1989). Palms inhabit a variety of ecosystems. More than two-thirds of palm species live in humid moist forests (VPE, 2006).

TABLE 14.1 Total Agricultural Cultivated Area and Total Agrowaste of Some Countries (Nagendran, 2011)

Country	Total Cultivated Area (million hectares)	Total Agrowaste Production (million tonnes/annum)
China	552.83	56.20
Australia	425.44	nr
USA	411.16	nr
India	179.90	nr
Indonesia	48.50	08.65
Nigeria	34.00	nr
Germany	16.90	nr
Egypt	03.53	27.00

NOTE: nr =data not reported.

FIGURE 14.1   Some members of Arecaceae/Palmae (a) *Cocos nucifera* (Coconut palm), (b) *Borassus flabellifer* (Palmyra palm), (c) *Caryota urens* (Fishtail palm), (d) *Areca catechu* (Betel nut palm) (e) *Roystonia regia* (Royal palm) (f) *Phoenix sylvestris* (Wild Date palm) (g) *Thrinax radiata* (Florida Thatch Palm), (h) *Hyphaene thebaica* (Branched palm).

14.2.2 CHARACTERISTIC

Habit—shrubs, trees, or vines, and a solitary shoot ending in a crown of evergreen leaves that are either palmately ("fan-leaved") or pinnately ("feather-leaved") compound and spiral. The inflorescence—spadix or spike surrounded by one or more spathes. The flowers—unisexual or bisexual, white, and radically symmetric. The sepals and petals—three each and may be distinct or joined at the base. The stamens—six in number, with filaments that may be separated attached to each other, or attached to the pistil at the base. The fruit—a single seed drupe or berry-like (Ramos et al., 2014). The fruits of some members are presented in Figure 14.2.

FIGURE 14.2 Fruits of some members of Arecaceae (a) *Cocos nucifera* (b) *Borassus flabellifer* (c) *Areca catachu* (d) *Phoenix paludosa* (Hita) (e) *Areca triandra* (f) *Rhapis excelsa* (g) *Veitchia merrillii* (h) *Kentia belmoreana* (i) *Caryota urens* (j) *Lodoicea maldivica* (double coconut) (k) *Nypa fruticans,* (l) *Arenga* sp. m) *Hyphaene thebaica.*

14.2.3 SUBFAMILY

A revised Moore's classification as proposed by Uhl and Dransfield (1987) showed six subfamilies, namely Coryphoideae, Calamoideae, Nypoideae, Ceroxyloideae, Arecoideae, and Phytelephantoideae.

14.3 MESOCARP AGROWASTE FROM PALMAE

In India and in Southeast Asia, mesocarpic husk material of palm fruits is the major agro-industrial waste, which amounts to millions of tons every year. The availability and huge generation of oil palm agrowastes in Malaysia lead to develop value-added products from them; for example, empty fruit bunches are being treated by physical and chemical processes to extract oil and cellulose (Abdullah et al., 2011), which can be good sources of biopolymers such as polyhydroxyalkanoates (PHAs) and poly-lactate (PLA), as these bioplastics have the same properties as petroleum derivatives used for packaging material (www.biobasics.gc.ca/English/view.asp?x=790#biotech; 2008). Lignocellulose composite material from agrowaste of oil palm, having the strength of 28–40 MPa, can be used as ligament replacement fibers (Abdullah et al., 2011). In our country, the unutilized husks and fibrous materials of mesocarpic portions of palms increase unnecessary burden of free environment and promote air pollution. A portion of waste material of coconut (coir) is processed and used in the rope-making industry. The bulk part of this material is remixed, unused, and their utilities are unexplored.

14.3.1 PHENOLIC CHEMICALS FROM MESOCARP AGROWASTE

Extraction, separation, purification and commercialization of chemical compounds locked in unutilized mesocarps of different members of Areca-ceae family have done some workers, as reported by Dey et al. (2003) that dried mesocarpic husk of Coconut (*Cocos nucifera*) contains good amount of phenolic compounds (13.0 mg/g dry weight). They also detected 4-Hydoxybenzoic acid (4-HBA) and ferulic acids by thin-layer chroma-tography (TLC). Profiling C6–C1 phenolic metabolites in *C. nucifera* has been done (Dey et al., 2005). Following this research, Chakraborty et al. (2006) studied mesocarps of 22 members (18 genera) of Arecaceae for their phenolic contents and they estimated 4-HBA from 1.00–5.6 mg/g dry weight and also ferulic acid ranges from 0.1–2.9 mg/g dry weight. Mesocarp of *Areca catechu* contains maximum amount of 4-HBA (5.0) followed by *Caryota urens* (3.6), *C. nucifera* (3.4), *Borassus flabellifer* (3.1) etc. but the lowest amount was detected in *Roystonea regia* (1.0 mg/gm dry weight). The maximum amount of ferulic acid was detected (2.9 mg/g of dry weight) in mesocarp of *Caryota mitis*.

According to Suzuki et al. (1998), the presence of 4-HBA in coconut husk was indicated. Similar studies carried out on *Phoenix dactylifera* confirmed the presence of ferulic acid, 4-coumaric acid, and 4-HBA bound to the cell walls of date palm leaves and four wall-bound phenolic acids, namely, 4-HBA, 4-coumaric acid, ferulic acid, and sinapic acid, in the root organs of the plant (Modafar et al., 2000). It is to be noted that mesocarps of fruits of all members of Palmae are not waste, some are directly consumed by us as food (e.g., mesocarp of date palm). The chemical structures of 4-HBA, ferulic acid, and *p*-coumaric acid are given in Figure 14.3.

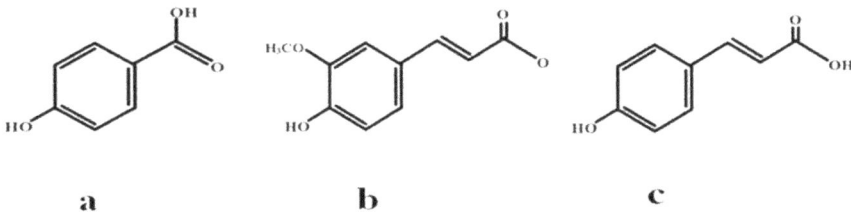

FIGURE 14.3   (a) Chemical structure of 4-hydroxybenzoic acid, (b) ferulic acid, and (c) *p*-coumaric acid.

14.3.2   OTHER CHEMICALS FROM MESOCARPS OF PALMS

Cellulosic polymers such as di and tri acetates have been reported to be produced from oil palm (*Elaeis guineensis*), raffia palm (*Raphia hookeri*), and fruit fibers from coconut fruit (*C. nucifera*). Udonne et al. (2006) showed that the cellulose contents of *Elaeis guineensis* and *Raphia hookeri* are 61.1% and 39%, respectively, and diacetate derivatives of cellulose in branch and fruit fibers of oil palm are 41.2% and 17.85%, respectively, and those for triacetate derivatives are 52% and 51%, respectively. In case of *Raphia*, diacetate and triacetate derivatives produced are 20.8% and 46.6%, respectively (Umoren et al., 2004). *C. nucifera* showed no significant quantity of cellulose, but derived diacetate and triacetate contents are 18.80% and 54%, respectively (Browning et al., 1997; Casey et al., 2000; John et al., 2001).

14.4 CHEMICAL EXTRACTIONS FROM MESOCARP

Extraction procedure of phenolic compounds are discussed below.

14.4.1 *SAPONIFICATION OF CELL WALL MATERIAL FOR RELEASE OF PHENOLIC ACIDS*

The mesocarpic part collected from palm fruits must be first cleaned with detergent such as Surf (1.5% w/v), washed by thorough distilled water, dried, and crashed to powder. Then it is treated with an alkaline solvent from mild to high concentration to release the cell-wall-bound phenolics or for the release of ether and ester-linked phenolics (Parr et al., 1996, 1997). The highly alkaline solution is then acidified well followed by extraction with ethyl acetate for separation of organic phase and aqueous phase. The organic phase-containing phenolic extract is to be collected, evaporated to dryness by rotary evaporator and redissolved again in an organic solvent for further analysis.

14.4.2 *THIN-LAYER CHROMATOGRAPHY AND ULTRAVIOLET ANALYSIS OF PHENOLIC ACIDS*

Thin layer chromatography (preparative) performed using (20 × 20 cm^2) glass plate with 0.3 mm layers of microcrystalline cellulose slurry (0.25 g/mL conc.) and then concentrated phenolic extracts are loaded by capillary in base line drawn previously in few spots onto TLC plates (Chakraborty et al., 2006; Dey et al., 2003). The TLC is run in mobile phase of solvent/solvents within the TLC glass chamber till the mobile phase reaches the upper line drawn on plate. The TLC plates will be developed in 2% formic acid as essentially described by French et al. (1976). After taken out from chambers, the glass plates are dried by an air drier and viewed under a dual wavelength (260 nm–340 nm) *ultraviolet* (UV) lamp. The bands are scrapped, eluted separately in an organic solvent, and the extracts are centrifuged. The purity is checked by an *ultraviolet*–visible (UV–VIS) spectrophotometer. The absorption profiles of the phenolics are determined by differences in absorption maxima (λ_{max}) and minima (λ_{min}) (Waldron et al., 1996). The identification of the sample is done by comparison of the absorption spectra of the sample with those of standard (Parr et al., 1997). The absorption at 225 nm is recorded for quantifying 4-HBA and that for ferulic acid at 310 nm. Further results are confirmed by available methodologies (Sachan et al., 2004).

14.4.3 *HIGH-PERFORMANCE LIQUID CHROMATOGRAPHY ANALYSIS AND MASS SPECTROMETRY ANALYSIS*

The separation of the phenolic acids or other chemicals is generally done by high-performance liquid chromatography (HPLC) using C_{18} column equipped with a dual absorbance detector set at 254 nm and 310 nm. An isocratic linear water system of methanol with a particular flow rate is used to elute the phenolic compounds (Sachan et al., 2004). The data are analyzed on a particular software system. The identity of each phenolic compound is confirmed by comparing retention time and UV spectra with those of the standard compounds. For further analysis, HPLC-purified samples are subjected to gas chromatography and mass spectrometry (GC–MS). Analysis by GC–MS provides the peaks of phenolic compounds, and by library search such as Prediction of Activity Spectra for Substances (PASS)/National Bureau of Standard (NBS)/National Institute for Standard and Technology (NIST) or other, we can identify phenolic or other chemicals and their bioactive functions.

14.5 UTILIZATION OF PHENOLIC COMPOUNDS

Ferulic acid shows a strong antioxidant property by donating hydrogen atom from phenolic hydroxyl group, and it is effective as a free radical scavenger that can cause DNA damage, cancer, and accelerate cell aging (Aragno et al., 2000). It is a very effective UV absorber and it gives a considerable protection to skin against UVB-induced damages (Saija et al., 2000). It exhibits significant anti-inflammatory activity (Kikuzaki et al., 2002). Ferulic acid along with ascorbic acid and vitamin E acts as a stress reliever by preventing thymine dimers in skin (Lin et al., 2005). It is also used in skin whitening process and taken as oral supplements, inhibiting melanin production. Ferulic acid is well-established as an antidiabetic. It is used as an effective agent for retention of food color (e.g., color of green peas), for prevention of discoloration of green tea (Masaji et al., 1999), and as a growth enhancing agent. Ferulic acid can act as a precursor of vanillin as reported by several research groups (Hua et al., 2007; Mathew et al., 2005). Ferulic acid diminishes lipid peroxidation and prevents DNA single strand rupture, and inactivation of certain proteins (Hirose et al., 1999). Reports are available about use of ferulic acid as food preservative. It is used to preserve oranges and inhibits autoxidation of linseed oil (Tsuchiya et al., 1975). Yang et al. (2007) reported that it increases the stability of cytochrome c and it inhibits apoptosis. In matrix-assisted laser desorption/ionization mass spectrometry analysis, ferulic acid is frequently used as a matrix for protein (Beavis et al., 1989).

The use of 4-HBA is multifold. 4-HBA has an important role in estrogenic activity in both in vitro and in vivo conditions (Khetan, 2014). According to Tomas-Barberan and Clifford (2000), 4-HBA is an effective dietary antioxidant. In many plant-based foods, 4-HBA is a natural component. Moreover, 4-HBA is an important preservative for cosmetics, food, and drugs (McQualter et al., 2005). It is general fact that the genome and environmental factors manifest its metabolite expression in plants. According to Sumner et al. (2003), several metabolites may be overexpressed, which are used as phytochemical markers. The chemotaxonomic studies of different taxa are dependent on biomarkers. It is reported that 4-HBA could be considered as a biomarker of Palmae mesocarp wall due to its predominance. Bate-Smith (1962) considered phenolic compounds as good chemotaxonomic markers. Johnson et al. (2002) used flavonoid, which is used as chemotaxonomic marker in case of cultivated Amazonian coca and hydroxycinnamic acids for using as biomarker of Italian blood orange juices (Rapisarda et al., 1999). High accumulation of 4-HBA in Palmae mesocarp wall certainly will be recognized as a phytochemical marker of Palmae.

As 4-HBA along with ferulic acid have notable applications in day-to-day life, further detailed investigations and advanced techniques may be applied for best utilization of these useful phenolic acids. Similarly, cellulose pulp derived from unused agricultural wastes could be used in paper-making industry. Cellulose acetate derivatives (diacetate and triacetates) can be used in plastic production, photographic filmmaking, magnetic tape production, and electrical part production.

14.6 CONCLUSION

After foregoing discussion, agrowastes are now recognized as excellent sources for many practical industrial uses and household making. No doubt the chemical extraction and purification from mesocarpic agrowaste of the Arecaceae family are new lines of waste management; however, literature on this is scant. The phenolic substances or compounds extracted from agrowastes of mesocarp are being utilized in many purposes and also as taxonomic chemomarkers of this family. Although few researchers have done their research work in this field, agrowastes of other family members may be the next line of research by new researchers. This may be of great interest to the existing palm industry as a new economic model. This technological transfer to manage agrowastes of other crops might be encouraged.

KEYWORDS

- **agrowaste**
- **palm**
- **mesocarpic husk**
- **phenolics**

REFERENCES

Abdullah, M.A.; Nazir, M.S.; Wahjoedi, B.A. *Development of value-added biomaterials from oil palm agro-wastes*. In: 2nd International Conference on Biotechnology and Food Science IPCBEE, IACSIT Press, Singapore 2011, vol. 7, pp. 32–35.

Akulova, M.V.; Isakulov, B.R.; Dzhumabaev, M.D.; Imanbekova, A.M. Getting binder of cement, fly ash and sludge, activated by the method of complex electromechanical activation for use in light arbolit concrete. *Naykovedenie* 2016, 3, 1–9.

Alengaram, U.J.; Al Muhit, B.A.; bin Jumaat, M.Z. Utilization of oil palm kernel shell as lightweight aggregate in concrete—A review. *Construction and Building Materials* 2013, 38, 161–172.

Anbazhagan, A.; Gopinath, L. Light weight concrete using coconut shell SSRG. *International Journal of Civil Engineering* 2017, 6, 73–75.

Aragno, M.; Parola, S.; Tamagno, E.; Brignardello, E.; Manti, R.; Danni, O.; Boccuzzi, G. Oxidative derangement in rat synaptosomes induced by hyperglycaemia: Restorative effect of dehydroepiandrosterone treatment. *Biochemical and Pharmacology* 2000, 60, 389–395.

Asasutjarit, C.; Hirunlabh, J.; Khedari, J.; Charoenval, S.; Zeghmati, B. Development of coconut coir–based lightweight cement board. *Construction and Building Materials* 2007, 21, 277–288.

Bate-Smith, E.C. The phenolic constituents in plants and their taxonomic significance. *Botanical Journal of the Linnean Society* 1962, 58, 95–173.

Beavis, R.C.; Chait, B.T.; Fales, H.M. Cinnamic acid derivatives as matrices for ultraviolet laser absorption mass spectrometry of proteins. *Rapid Communications in Mass Spectrometry* 1989, 3(12), 432–435.

Browning, B.L. *The Chemistry of Wood*, 6th Ed., Interscience Publishers Inc., New York, 1997, pp. 56–63.

Caprara, C.; Colla, L.; Lorenzini, G.; Santarelli, C.; Stoppiello, G.; Zanella, D. Development of a model for technical-economical feasibility analysis of biomass and mud gasification plants. *International Journal of Energy Technology* 2011, 3, 1–6.

Casey, J.P. *Pulp and Paper Chemistry and Chemical Technology*, John Wiley and Sons, U.S.A., 2000, vol. 3(1), pp. 150–152.

Chakraborty, M.; Das, K.; Dey, G.; Mitra, A. Unusually high quantity of 4-hydroxybenzoic acid accumulation in cell wall of palm mesocarps. *Biochemical Systematics and Ecology* 2006, 34, 509–513.

Dey, G.; Chakraborty, M.; Mitra, A. Profiling C_6–C_1 phenolic metabolites in *Cocos nucifera. Journal of Plant Physiology* 2005, 162, 375–381.

Dey, G.; Sachan, A.; Ghosh, S.; Mitra, A. Detection of major phenolic acids from dried mesocarpic husk of mature coconut by thin layer chromatography. *Industrial Crops and Products* 2003, 18, 171–176.

French, C.J.; Vance, C.P.; Towers, G.H.N. Conversion of *p*-coumaric acid to *p*-hydroxybenzoic acid by cell free extract of potato tuber and *Polyporus hispidus. Phytochemistry* 1976, 15, 564–566.

Hai, H.T.; Tuyet, N.T.A. *Benefits of the 3R approach for agricultural waste management (AWM) in Vietnam.* Under the Framework of joint Project on Asia Resource Circulation Policy Research Working Paper Series. Institute for Global Environmental Strategies supported by the Ministry of Environment, Japan, 2010.

Hirose, M.; Takahashi, S.; Ogawa, K.; Futakuchi, M.; Shirai, T. Phenolics: blocking agents for heterocyclic amine-induced carcinogenesis. *Food and Chemical Toxicology* 1999, 37, 985–992.

Hua, D.; Ma, C.; Song, L.; Lin, S.; Zhang, Z.; Deng, Z.; Xu, P. Enhanced vanillin production from ferulic acid using adsorbent resin. *Applied Microbiology and Biotechnology* 2007, 74, 783–790.

Hutchinson, J. *The Families of Flowering Plants*. Clarendon Press, UK, 1959.

John, W.S.; Hearle, S.; Calvin, R.; Woodings, C.R. Fibers related to cellulose. In *Regenerated Cellulose Fibers*; Woodings, C., Ed.; Woodhead Publishing Ltd., USA, 2001, 352p.

Johnson, E.L.; Schmidt, W.F.; Cooper, D. Flavonoids as chemotaxonomic markers for cultivated *Amazonian coca. Plant Physiology and Biochemistry* 2002, 40, 89–95.

Jose, A.R.; Sergio, F.S.; Gustavo, H.D.; Holmer, S. Jr. Agricultural wastes as building materials: Properties, performance and applications. In: *Building Materials: Properties, Performance and Applications*; Cornejo, D.N., Haro, J.L., Eds.; Nova Science Publishers, Inc., 2009, vol. 9, pp. 1–44.

Khetan, S.K. *Endocrine Disrupters in the Environment*. Willey, New York, 2014.

Kikuzaki, H.; Hisamoto, M.; Hirose, K.; Akiyama, K.; Taniguchi, H. Antioxidant properties of ferulic acid and its related compounds. *Journal of Agricultural and Food Chemistry* 2002, 50, 2161–2169.

Lawrence, G.H.M. *Taxonomy of Vascular Plants*, Indian Ed., McMillan Company, New York, 1967, p. 823.

Li, G.; Xu, X.; Chen, E.; Fan, J.; Xiong, G. Properties of cement–based bricks with oyster–shells ash. *Journal of Cleaner Production* 2015, 91, 279–287.

Lim, S.F.; Matu, S.U. Utilization of agro-wastes to produce biofertilizer. *International Journal of Energy and Environment Engineering* 2015, 6, 31–35.

Lin, F.H.; Lin, J.Y.; Gupta, R.D.; Tournas, J.A.; Selim, M.A; Monteiro-Riviare, N.A.; Grichnik, J. M.; Zielinski, J.; Pinnel, S.R. Ferulic acid stabilizes a solution of Vitamins C and E and doubles its photoprotection of skin. *Journal of Investigative Dermatology* 2005, 125(4), 826–832.

Mathew, S.; Abraham, T.E. Studies on the production of feruloyl esterase from cereal brans and sugar cane bagasse by microbial fermentation. *Enzyme and Microbial Technology* 2005, 36, 565–570.

McQualter, R.B.; Chong, F.C.; Meyer, E.V.D.; Drew, K.; Michael, G.O.; Walton, N.J.; Viitanen, P. V.; Brumbley, S.M. Initial evaluation of sugarcane as a production platform for *p*-hydroxybenzoic acid. *Plant Biotechnology Journal* 2005, 3, 29–41.

Modafar, C.E.; Tantaoui, A.; Boustani, E.E. Changes in cell wall-bound phenolics compounds and lignin in roots of date palm cultivars differing in susceptibility to *Fusarium oxysporum* f. sp. *albedinis*. *Journal of Phytopathology* 2000, 148, 405–408.

Nagendran, R. *Agricultural waste and pollution*. In *Waste: A Handbook for Management*, Letcher, T.M., Valero, D.A., Eds.; Academic Press, 2011, pp. 341–355.

Obi, F.O.; Ugwuishiwu, B.O.; Nwakaire, J.N. Agricultural waste concept, generation, utilization and management. *Nigerian Journal of Technology* 2016, 35(4), 957–964.

Obilade, I.O. Experimental study on rice husk as fine aggregate in concrete. *International Journal of Engineering and Science* 2014, 3, 9–14.

Parr, A.J.; Ng, A.; Waldron, K.W. Ester-linked phenolic components of carrot cell walls. *Journal of Agricultural and Food Chemistry* 1997, 45(7), 2468–2471.

Parr, A.J.; Waldron, K.W.; Ng, A.; Parker, M.L. The wall-bound phenolics of Chinese water chestnut (Eleocharis dulcis). *Journal of the Science of Food and Agriculture* 1996, 71(4), 501–507.

Ramadhansyah, P.J.; Nurfatin, A.M.; Siti, N.A.J.; Norhafizah, M.; Norhidayah, A.H.; Dewi, S.J. Use of coconut shell from agriculture waste as fine aggregate in asphaltic concrete. *Journal of Engineering and Applied Science* 2016, 11, 7457–7462.

Ramesh, K.V.; Goutham, R.; Siva Kishore, I. An experimental study on partial replacement of bagasse ash in basalt concrete. *International Journal of Civil Engineering and Technology* 2017, 8, 335–341.

Ramos, T.; Matos, A.M.; Sousa–Coutinho, J. Strength and durability of mortar using cork waste ash as cement replacement. *Materials Research* 2014, 17, 893–907.

Rapisarda, P.; Tomaino, A.; Lo Cascio, R.; Bonina, F.; De Pasquale, A.; Saija, A. Antioxidant effectiveness as influenced by phenolic content of fresh orange juices. *Journal of Agricultural and Food Chemistry* 1999, 47, 4718–4723.

Sachan, A.; Ghosh, S.; Mitra, A. An efficient isocratic separation of hydroxycinnamates and their corresponding benzoates from plant and microbial source by HPLC. *Biotechnology and Applied Biochemistry* 2004, 40, 197–200.

Sadh, P.K.; Duhan, S.; Duhan, J.S. Agro-industrial wastes and their utilization using solid state fermentation: A review. *Bioresources and Bioprocessing* 2018, 5, 1.

Saija, A.; Tomaino, A.; Trombetta, D.; De Pasquale, A.; Uccella, N.; Barbuzzi, T.; Paolino, D.; Bonina, F. In vitro and in vivo evaluation of caffeic and ferulic acids as topical photoprotective agents. *International Journal of Pharmaceutics* 2000, 199, 39–47.

Saravanan, J.; Imthiyas Ahamed, S.; Muniyasamy, X.; Muthu Ganesh, P.; Rawther Ibrahim A.S. Low cost roofing tiles using agricultural wastes. *International Journal of Civil Engineering* 2017, 4, 71–75.

Sethunarayanan, R.; Chockalingam, S.; Ramanathan, R. Natural fiber reinforced concrete *Transportation Research Record* 1989, 1226, 57–60.

Somarriba, L.N.; Sokolova, Ermakova, E.V.; Rynkovskaya, M. *A Review of agro-waste materials as partial replacement of fine aggregate in concrete*. IOP Conf. Series: Materials Science and Engineering, 2018; 371, 012012 doi:10.1088/1757–899X/371/1/012012.

Subba Rao, N.S. *Biofertilizer in Agriculture and Forestry*, 3rd Ed., International Science Publisher, New York, 1993.

Sukan, A.; Roy, I.; Keshavarz, T. Agro-industrial waste materials as substrates for the production of poly (3-hydroxybutyric acid). *Journal of Biomaterials and Nanobiotechnology* 2014, 5, 229–240.

Sumner, L.W.; Mendes, P.; Dixon, R.A. Plant metabolomics: Large-scale phytochemistry in the functional genomics era. *Phytochemistry* 2003, 62, **817–836.**

Suzuki, S.; Rodriguez, E.B.; Iiyama, K.; Saito, K.; Shintani, H. Compositional and structural characteristics of residual biomass from tropical plantations. *Journal of Wood Science* 1998, 44(1), 40–46.

Tomas-Barberan, F.A.; Clifford, M.N. Dietary hydroxybenzoic acid derivatives—Nature, occurrence and dietary burden. *Journal of the Science of Food and Agriculture* 2000, 80, 1024–1032.

Tsuchiya, T.; Takasawa, M. Oryzanol, ferulic acid, and their derivatives as preservatives. *Japanese Kokai* 1975, 07, 518–521.

Udonne, A.D. *Preparation of cellulose derivatives from some agricultural wastes.* B.Sc Thesis. University of Uyo, Nigeria, 2006, pp. 41–48.

Uhl, N.W.; Dransfield, J. *Genera Palmarum: A Classification of Palms Based on the Work of Harold E. Moore, Jr.*, Allen Press, Lawrence, Kansas, 1987.

Umoren, S.A.; Umoudoh, A.J.; Akpabio, U.D. Production of cellulosic polymers from agricultural wastes. *Bulletin of Pure and Applied Sciences* 2004, 23, 9–13.

Virtual Palm Encyclopedia (VPE). http://www.plantapalm.com (July 19, 2006).

Waldron, K.; Parr, A.J.; Ng, A.; Ralph, J. Cell wall esterified phenolics dimmers: Identification and quantification by reverse phase high performance liquid chromatography and diode array detection. *Phytochemical Analysis* 1996, 7, 305–312.

Yang, F.; Zhou, B.R.; Zhang, P.; Zhao, Y.F.; Chen, J.; Liang, Y. Binding of ferulic acid to cytochrome c enhances stability of the protein at physiological pH and inhibits cytochrome c-induced apoptosis. *Chemical–Biological Interactions* 2007, 170, 231–243.

CHAPTER 15

CRISPR/Cas9: An Invention for Plant Breeding and Agricultural Sustainability

DEBASHRITA MITTRA and SABUJ SAHOO*

Post Graduate Department of Biotechnology, Utkal University, Vani Vihar, Bhubaneswar, Odisha 751004, India

Corresponding author. E-mail: sabujbiotech@utkaluniversity.ac.in.

ABSTRACT

Viruses are considered to be a common threat to the cellular life as well as to bacteria and archaea. Of the vital mechanisms that have been developed to restrict the virus infection, Clustered Regularly Interspaced Short Palindromic Repeat (CRISPR) is a sequence-specific adaptive immunity that fundamentally enhances our understanding of virus–host interaction. CRISPR represents a family of DNA sequences contained in prokaryotes. The sequences contain snippets of DNA from viruses. The immunity that is based on CRISPR allows the cell to remember, recognize, and clear infections. The plant genome editing which proves to be efficient depends on the induction of double-stranded breaks through site-specified nucleases (NUCs) that initiate the process of DNA repair, which is either based on homologous recombination or nonhomologous end joining. The CRISPR-Cas9 system was recognized as the metamorphose genetic tool because of its simpler framework and due to the broad range of adaptability and applications. Various upgradations in this multipurpose genome-editing technology include options for various genetic manipulations such as generating knockouts, making precise modifications, multiplex genome engineering, and activation and repression of target genes, which have proved to be of great help in plant breeding and agricultural sustainability. The review encompasses an overview of the role of CRISPR in prokaryotic defense mechanism, its prospective role in generating mutagenized plant line, and the applications of this novel interference pathway in crop improvement. The CRISPR/Cas9 system warrants a promise in facilitating both forward and reverse genetics, thereby enhancing research in crops that lack genetic resources and aid in agricultural sustainability.

15.1 INTRODUCTION

With the advent of this new era and the involvement of science in it, people are leading a comfortable and substantial life. They have their access to good plant products, which are essential in daily life including crops, cosmetics, medicines, etc. Despite all these, the plants are facing a lot of threat regarding their existence and productivity. The threats include climatic change, soil erosion, pathogenic attack, and excessive use of chemicals. Plants/crops fall victim to these environmental threats and, as a result, they start deteriorating and their productivity decreases. To meet these challenges, researchers have developed new genome-editing tools, with which they can insert, delete, and replace genes to the plant system in order to increase their efficiency in productivity. So, these threats are taken care of by the advancements in genome-editing tools. The first-generation genome-editing tools include mega-NUCs, zinc-finger nucleases (ZFNs), and transcription activator-like effector nucleases (TALENs), which involve tedious procedures to achieve target specificity and are labor-intensive and time-consuming. ZFNs and TALENs have made it possible for molecular biologists to target any gene of interest. Still these tools are expensive and time-consuming, as they involve complicated steps that require protein engineering. The second-generation genome-editing tool involves CRISPR/Cas9, which includes simple designing and cloning methods, with Cas9 being available to be used with different guide RNAs (gRNAs) targeting multiple sites in the genome. This is a promising tool which can modify genes without rendering corresponding plants as classical genetically modified organism, paving the way to its implementation in agricultural biotechnology.

15.2 CRISPR: THE GENOME-EDITING WONDER

The NUCs that are engineered consist of a nonspecific NUC domain, which is fused with a sequence-specific DNA-binding domain. The NUCs which are fused can accurately cleave the targeted gene, and with the help of nonhomologous end joining (NHEJ) or homology directed repair (HDR), the breaks could be repaired; hence, it is referred to as the "genome editing"

(Gaj et al., 2013). Mega-NUC technologies, like ZFNs and TALENs, were used in the first-generation genome editing, which involves unvaried procedures to achieve target specificity, includes intensive labor, and consumes time, whereas the second-generation genome-editing techniques along with CRISPR/Cas9 involve easier designs, and the execution methodologies are more time-consuming and cost-effective. The research in animal and plant biology has been transformed by the discovery of the CRISPR/Cas9 gene-editing system, and its genome editing utility was first demonstrated in mammalian cells in Jinek et al. (2012).

The CRISPR genome-editing system is more straightforward than ZFNs and TALENs, and a gRNA of about 20 nucleotides complementary to the DNA stretch within the target gene is involved. CRISPR term was first coined in 2002 (Jansen et al., 2002). The steps involved for the cleavage of CRISPR require (1) a synthetic (gRNA) sequence that is short and made up of 20 nucleotides, which binds to the target DNA and (2) 3–4 bases that are cleaved by Cas9 NUC enzyme after the protospacer adjacent motif (PAM), generally 50 NGG (Jinek et al., 2012). The canonical PAM is the sequence 5'-NGG-3' where "N" can be any nucleobase followed by "G" guanine. gRNAs can transport Cas9 for gene editing throughout the genome. However, gene editing is done at the site only at which the Cas9 NUC comprises two domains: (a) RuvC-like domains and (b) an HNH domain, where each domain cuts one strand of DNA. Due to the development of this methodology, CRISPR technology has been widely implemented in plants and animals. Since the last half decade, the plant genomes have been edited by CRISPR/Cas9 technology, and as compared to ZFNs and TALENs, these are easy to use.

CRISPR/Cas9 system is a simple inexpensive and multifaceted tool for genome editing, resulting in a groundswell of research based on the technique, which has been known as "THE CRISPR CRAZE" (Pennisi, 2013). It is the most widely used system, type II CRISPR/Cas9system from *Streptococcus pyogenes* (Jinek et al., 2012). These systems are a part of the adaptive immune system of bacteria and archae, protecting them against the invading nucleic acids such as viruses by cleaving the foreign DNA in a sequence-dependent manner.

Site-specific modification is achieved by a single-guide RNA (sgRNA) (usually about 20 nucleotides), which is complementary to a target gene or locus and is anchored by a PAM. Cas9 NUC cleaves the targeted DNA to generate double-strand breaks (DSBs), which are subsequently repaired by NHEJ or HDR mechanisms. NHEJ may also introduce indels that cause frame shift mutations by disrupting gene functions; when combined with double or multiplex gRNA design, NHEJ involves targeted chromosome deletions, whereas HDR can be engineered for target gene correction, gene replacement, and gene knock-in (Figure 15.1).

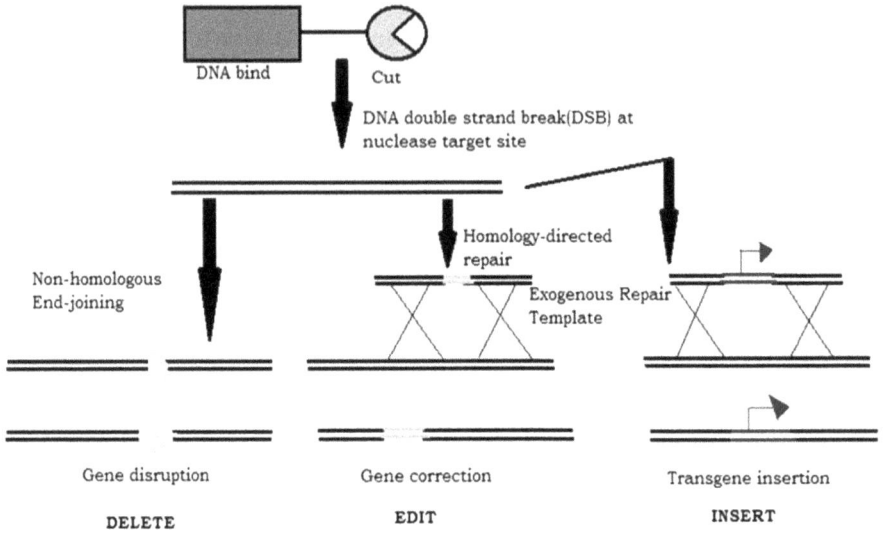

FIGURE 15.1 DNA repair facilitated by CRISPR-Cas. The repair without using a template occurs through the nonhomologous end joining (NHEJ) pathway and can be used to disrupt the function of a gene, effectively deleting it. Repair by using a template through the homology-directed repair (HDR) pathway which can enable precise alterations and insertions from the template DNA sequence into the genome.

15.2.1 COMPONENTS OF CRISPR-CAS9 BACTERIAL SYSTEM

The three components of a bacterial CRISPR-Cas9 system are a Cas9 NUC, CRISPR RNA (crRNA), and trans-activating crRNA. The two components, crRNA and trRNA, form a complex by base pairing, which ultimately guides Cas9 to the target sites complementary to 20-nucleotide sequence of crRNA (Jinek et al., 2012). CRISPR functions along with the Cas genes that flank CRISPR loci in the genome. Cas9 is a DNA endonuclease that has two NUC domains, HNH and RuvC, for introducing DSB in the target DNA. The complementary strand is cleaved by the HNH domain of Cas9, while the identical strand is cleaved by the RuvC domain. A short conserved sequence, known as PAM, is required by Cas9, which follows 3' end of the crRNA complementary sequence immediately (Mojica et al., 2009). For target specificity and molecular capabilities of Cas9, the CRISPR-Cas9 system relies on RNA sequences, which makes it successful.

15.2.1.1 CAS9 NUCLEASE

A conserved core and a bi-lobed architecture with adjacent active sites and two nucleic acid-binding grooves were revealed with the help of Cas9 endonuclease of different subtypes. The lobes comprise of a large globular recognition (REC) lobe, which is connected to a small NUC lobe. The REC lobe is known to be a Cas9-specific functional domain which comprises of two domains, REC1 and REC2, and also consists of a long α-helical arginine-rich domain, referred to as the "bridge helix." The NUC lobe includes two NUC domains, RuvC and HNH, and a PAM-interacting domain (PI domain) (Nishimasu et al., 2014).

15.2.1.2 CRISPR RNA

The transcription of the CRISPR-Cas loci to generate an RNA protein guide complex follows a simple process in most organisms, which also displays several type-specific differences. All systems transcribe the CRISPR locus and then process the RNA with Cas ribo-NUC and form a CRISPR ribonucleoprotein (crRNP) complex (Pul et al., 2010).

15.2.1.3 TRANS-ACTIVATING CRRNA

It is a small trans-encoded RNA, which was first discovered in human pathogen, *S. pyogenes*. The maturation of crRNA and the processing pathway in type II systems is clearly illustrated by *S. pyogenes*, which significantly differs. It has been stated that the process requires an extra tracrRNA molecule, which depends on the housekeeping RNase III. Then there is a complementary pairing of the repeat sequences within the pre-crRNA primary transcript, which forms an RNA duplex, pre-crRNA-tracrRNA, that is recognized and cleaved by RNase III in the presence of Cas9 protein. Cas9 is bound to the crRNA and tracrRNA into a ternary complex which helps in executing silencing in the effector complex of *S. pyogens* (Jinek et al., 2012).

15.2.2 TYPES/SUBTYPES AND THE STRUCTURES OF DIFFERENT CAS PROTEINS

Variations in CRISPR-Cas systems are attributed to prokaryote-infecting viruses' interaction. The Cas system is primarily classified into two classes based on the composition of effector modules (protein complexes), namely Class I (most abundant 90%) and II CRISPR-Cas, and distributed among archaeal (87%) and bacterial (50%) genomes; Class I CRISPR-Cas is based on the composition of effector modules (protein complexes) with designated functions, whereas in Class II CRISPR-Cas, the effector is associated with a single multidomain protein that helps in the integration of a viral protospacer into the bacterial genome. The effector proteins involved in the target REC and cleavage steps are generally differ right through types.

The CRISPR-Cas systems of bacterial and archaeal adaptive immunity are due to direct repeat arrays separated by unique spacers and multiple CRISPR-associated (Cas) small sets of core genes that encode proteins, which act as a go-between stage of the CRISPR response. Different Cas gene types/subtypes present in CRISPR-Cas loci (Table 15.1) are essential and responsible for the defense and other ancillary functions. The functional role of most genes is still unknown. The CRISPR-Cas systems mediate immunity to assaulting genetic elements via a three-stage process, namely adaptation, expression, and interference.

15.2.3   MECHANISM OF CRISPR/CAS SYSTEM

Three genes are involved in the mechanism of CRISPR/Cas system, one for encoding Cas9 NUC and two noncoding RNA genes for tracrRNA and pre-crRNA (Deltcheva et al., 2011; Jansen et al., 2002; Jia et al., 2013). The pre-crRNA contains a 20-nucleotide sequence corresponding to a viral DNA signature, which is called the proto-spacer sequence and is a part of the CRISPR repeat sequence. The second RNA molecule is called tracrRNA, and it is also encoded at the CRISPR locus. The pre-crRNAs are transcribed from the CRISPR region and are further processed into mature forms by tracrRNA (Anders et al., 2014; Chylinski et al., 2013). Multiple crRNAs and tracrRNAs can be packaged together to form sgRNA. The sgRNA, joined together with Cas9, is inserted into a plasmid to transfect the target cells.

15.2.4   MECHANISM OF CRISPR-CAS SYSTEM IS SUBDIVIDED INTO THREE FUNCTIONAL PROCESSES

1. Spacer acquisition
2. CRISPR RNA (pre-crRNA and tracrRNA) maturation
3. Target interference

TABLE 15.1 Classification of Types/Subtypes and the Structures of Different Cas Proteins

Gene Name	Type/Subtype	Examples of Structures and Structural Features
Cas1	Type I, Type II, Type III	PDB: 3GOD, 3LFX, 2YZS. Unique fold with two domains
Cas2	Type I, Type II, Type III	PDB: 2IVY, 2I8E and 3EXC RRM (ferredoxin) fold
Cas3	Type I	PDB: 3S4L and 3SKD
Cas4	Type (I-A, I-B, I-C, I-D, II-B)	PDB: 4IC1
Cas5	Type I	PDB: 3KG4; 3VZI; 3VZH, Two domains of RRM (ferredoxin) fold, the C-terminal domain is deteriorated in many cas5 protein of type I, RAMP super family
Cas6	Type I, Type III	PDB: 2XLJ, 1WJ9, 3I4H, 4C8Z, 4D2D Two domains of RRM (feeredoxin) fold. RAMP super family
Cas7	Type I	PDB: 3PS0, 4N0L RRM (ferredoxin) fold with subdomain, RAMP super family
Cas8	I-A	PDB: 4AN8
Cas9	Type II	PDB: 4OGC, 4OO8, 4CMP Cas9 has several subdomains and adapts a two lobed general structure. Beyond two catalytic nuclease domains its subdomain does not appear to be similar to other known protein structure
Cas10	I-D	PDB: 3UNG, 4DO2 Two domains homologous to palm domain polymerases and cyclases, both belonging to RRM (ferredoxin) fold

### 15.2.4.1	SPACER ACQUISITION

In response to the virus or plasmid assault, bacteria confront the invader by copying and incorporating new spacers from the invading genetic elements into the CRISPR locus. CRISPR loci in bacterium consist of spacers (viral or plasmid DNA inserted into CRISPR locus protospacers) leading to type II adaptive immune systems. On subsequent assault, Cas9 NUC attaches to tracerRNA-crRNA and guides Cas9 NUC to the invading protospacer sequence.

### 15.2.4.2	CRRNA MATURATION

The functional crRNAs are prepared from the transcript of the repeat-spacer array where the tracrRNA hybridizes at the repeat region of the pre-crRNA with the aid of Cas9.

### 15.2.4.3	TARGET INTERFERENCE

Here, the crRNA-tracrRNA duplex directed the cleavage of invading foreign genetic elements by Cas9 NUCs, and these NUCs induce a DSB to destroy the invading viral genome (Tsai et al., 2014). Only those viral sequences are recognized by Cas9 which are complementary to the guide crRNA, where target sequence is directly 5′ to a PAM with the sequence NGG where "N" can be any nucleotide. However, absence of PAM NGG sequences at the CRISPR locus prevents the RNA-guided nuclease from cleaving the CRISPR locus itself (Miao et al., 2013).

## 15.3	CRISPR-CAS9: A BOON FOR PLANT GENETIC MATERIAL

### 15.3.1	ROLE OF CRISPR IN PLANT IMPROVEMENT

The CRISPR/Cas system can be utilized for introducing desirable changes like targeted single and multiple gene knockouts of determined genes in plants (Brouns et al., 2008) and introducing single-nucleotide polymorphisms (SNPs) into a gene of interest (Voytas, 2013) for improving economic traits.

Gene regulatory elements are also studied by expressing affinity or fluorescent tagged proteins at their native loci in the genome (Belhaj et al., 2013; Kim et al., 2014) for molecular stacking of multiple pest resistance genes into plants (D'Halluin and Ruiter, 2013). The CRISPR/Cas system has been incorporated into many plant species using multiple sgRNAs for various functional studies including *Arabidopsis thaliana*, tobacco (*Nicotiana benthamin*), sweet orange (*Citrus sinesis*), rice (*Oryza sativa*), wheat (*Triticum aestivum*), sorghum (*Sorghum bicolor*), maize(*Zea mays*), and liverwort (*Machantia polymorpha*) (Jinek et al., 2012).

A CRISPR/Cas9 toolkit has been developed, which contains various sgRNAs, and its potential is tested for multiplex genome editing by validating the biallelic mutations for TRY and CPC genes in Arabidopsis and zmHKT1 genes in transgenic maize (Cong et al., 2013). Plant architecture, grain dormancy, and seed dispersed are the important targets for the improvement of crops in *Brassica oleracea* and *Hordeum vulgare*. Targeted mutagenesis has also been successfully applied to edit soybean AlS1 gene to obtain chlorsulfuron herbicide-resistant plants (Li et al., 2015).

The precise editing of genes is required for crop improvement for their functional characterization. The CRISPR-Cas9 system has the appropriate ability to meet this demand. Transcriptional regulation change, epigenetic modification (Hu et al., 2018), and microscopic visualization of specific genome loci (Chen et al., 2013) are well understood by using the CRISPR-Cas9 system. Gene replacement, knockout, chromosomal deletions, and raising marker-free GM plants are also achieved using CRISPR-Cas9 technology.

15.3.2 ROLE OF CRISPR IN PLANT BREEDING AND GENOME EDITING

There are many bacteria which protect themselves from the invasion of viruses using adaptive immune systems called CRISPR-Cas. The CRISPR-Cas system recognizes genetic sequences specific to that of invaders and cuts them. CRISPR-Cas can be repurposed as a molecular scissor, which can be used to make cuts at specific locations in a plant genome. Subsequent repair by the cell's repair mechanism introduces precise change at the specific cut site. This advanced plant breeding tool represents tremendous promise and potential. It also facilitates precision crop breeding by working with the native characteristics available within the crop, a process called genome editing. This system can also be applied to produce transgenic crops. DuPont Pioneer being a world leader in CRISPR-Cas advanced breeding applications in agriculture, is developing its first CRISPR-Cas-enabled commercial product, a next generation of waxy corn, to be the pioneer agricultural product in the market developed with CRISPR-Cas. CRISPR-Cas has numerous potential agricultural applications, which include improvement to yield, disease resistance, drought tolerance, and output characteristics. In CRISPR-Cas, the Cas9 protein facilitates genome editing by functioning as a precise and programmable molecular scissor which cuts the DNA at specific locations. Following the cut of the target DNA sequence, the CRISPR-Cas system takes advantages of the naturally occurring cellular DNA repair mechanism to delete genes, edit genes, or insert genes.

CRISPR-Cas-facilitated HDR can also be used for introducing characteristics from nonnative source (i.e., transgenes). CRISPR-Cas has an ability to do this in a targeted manner, which provides a significant advantage over transgenic products that are currently on market. Pioneer has established a CRISPR-Cas advanced breeding platform to develop seed products for greater environmental resiliency, productivity, and sustainability. CRISPR-Cas has numerous potential agricultural applications including improvements beneficial to the end user such as output characteristics and nutritional content. In a recently published research paper, Pioneer scientists have described the first application of CRISPR-Cas to improve a corn plant's own ability to withstand drought stress (Shi et al., 2016).

CRISPR-Cas has been used to target a gene identified for its innate ability to promote drought tolerance. Other Pioneer publications related to CRISPR-Cas-advanced plant breeding include reports demonstrating the efficiency and flexibility of the CRISPR-Cas system in both corn (Svitashev et al., 2015; Svitashev et al., 2016) and soybean (Li et al., 2015). Conventional plant breeding is not sufficient to meet increasing food demands and other environmental challenges. On the contrary, CRISPR technology is erasing barriers to genome editing and could revolutionize plant breeding. CRISPR/Cas9 is the most recent, characterized, and rapidly developing genome-editing technology that has been successfully utilized in a wide variety of organisms.

Gene drive technology is being researched and developed for the purpose of reducing or eliminating human, ecological, or agricultural pest populations, or immunizing other desirable or endangered species against pest and disease. The ability of human beings to alter populations within ecosystems through genetic engineering raises issues associated with biodiversity and conservation, which in turn may affect the abilities of current and future generations to use and enjoy the benefits of the natural world. Applications in genome editing in plants are as follows:

- Targeted mutagenesis or gene knockout
- Multiplexing: editing of multiple loci simultaneously
- Gene replacement and trait stacking

- Gene regulation: RNA interference and activation
- Visualization of loci in large genomes
- Generation of marker-free transgenics.

Targeted modifications of genomes are attributed to genome editing by sequence-specific NUCs. CRISPR-Cas9 has transfigured the recent developments in plant genome editing and its application against plant pathogens to combat disease resistance in plants. Applications of the CRISPR-based genome-editing approach in plants are given in Table 15.2.

TABLE 15.2 Application of CRISPR-based Genome Editing Approach in Plants

Plant Species	Target Gene/Type of Modification	References
Oryza sativa (Rice)	OsERF922 (ethylene responsive factor)/Indels in ORF	Zhou et al., 2015
Nicotiana benthamiana (Tobacco)	BeYDV	Baltes et al., 2015
Zea mays (Maize)	ZmIPK gene in maize protoplast	Liang Z et al., 2014
Triticum aestivum (Bread Wheat)	TaMLO-AL, TaMLO-B1 and TaMLOD1	Wang et al., 2014
Arabidopsis thaliana (Mustard)	ds DNA of virus (A7, B7, and C3 regions)	Ji et al., 2015
Nicotiana tabacum (Tobacco)	NtPDS and NtPDR6 used for targeted mutagenesis	Gao J et al., 2015
Solanum lycopersicum (Tomato)	SLAGO7 coding sequence was targeted	Brooks C et al., 2014
Solanum tuberosum (Potato)	StIAA2 coding sequence	Wang S et al., 2015
Glycine max (Soybean)	Knocking out GFP transgene and modifying nine endogenous loci	Jacob et al., 2015
Cucumis sativus (Cucumber)	elF4E (eukaryotic translation initiation)	Chandrasekharan et al., 2016

The simpler structure and suitable construction procedure of CRISPR/Cas system make it more economical and effective than other site-specific NUCs. It has large application in developing better quality and quantity of crops to provide food security. CRISPR/Cas9-mediated editing of *SBEIIb* creates high amylose rice (Sun et al., 2017), and higher content of amylose increases resistant starch in rice. High resistant starch in cereals is good for health, and it reduces the chances of noninfectious diseases such as heart diseases and diabetes (Regina et al., 2006). Editing of genes *GW2*, *GW5*, and *TGW6* in rice improved the grain weight of rice (Xu et al., 2016). Generation of parthenocarpic tomato was achieved by editing the *SlIAA9* gene in *Solanum lycopersicum* (Ueta et al., 2017). *RIN* gene, the prime regulator of tomato-ripening encodes the MADS-box transcription factor, which regulates fruit ripening that is edited by this technique to inhibit the tomato ripening (Ito et al., 2015). This method was employed to evaluate starch quality in potato by means of editing gene encoding *granule-bound starch synthase* (*GBSS*), enzyme responsible for amylose synthesis (Andersson et al., 2017). *Camelina sativa* having increased oil content, drought- and salt-tolerant soybean, white button mushroom (*Agaricus bisporus*) with antibrowning properties, waxy corn with starch composed exclusively of amylopectinare are other few examples of CRISPR/Cas-mediated genome editing in food crops. The high efficiency of gene editing and convenient handling make this system widely applicable in crop improvement, eradicating diseases, and therapeutics. This can be used on mammalian system only after increasing the specificity and reducing the off-target mutations.

15.3.3   CRISPR IN IDENTIFICATION AND CONFIRMATION OF MUTATION

CRISPR is the most sophisticated, cost-efficient, less time-consuming, and easily engineered technique that orchestrates its practice and implications in clinical approaches for gene editing in plants, animals, microbes, etc. for cleaving of the site-specific DNA. CRISPR works by inducing DSBs in the DNA by an NUC called Cas at target genomic loci through gRNA, which are joined by NHEJ, an error-prone mechanism, and can result in frameshift insertion or deletion (Indel). An important challenge is the confirmation of mutation after editing through ZFN, TALEN, CRISPR/Cas techniques.

There can be four types of mutations, namely no mutation, heterozygous mutation (one allele is mutated), biallelic mutation (both alleles are mutated but the sequence of each allele is distinct), or homozygous mutation (the same mutation in both the alleles). The gene-editing technique can cause on-target or off-target mutations.

The detection and confirmation of on-target and off-target mutations are a serious concern. The on-target mutation can be detected by mismatch detection assays, mismatch cleavage assays, high-resolution melting analysis, and heteroduplex mobility assays. Most of the techniques to detect the on-target mutations are based on PCR and the ice-cold PCR (Milbury et al., 2011), which does not favor the amplification of the dominant wild-type sequence but improves the detection of rare mutant sequences in chimeric clones.

Mutations related to the loss of primer binding can be detected by qPCR by determining and characterizing the mutations; large deletions are detected by amplified fragment length polymorphism. In fluorescent PCR capillary gel electrophoresis, the indel sites present in genome can be amplified by PCR using fluorophore-labeled primers, and the resulting labeled amplicons are resolved by capillary gel electrophoresis; mutations are detected by comparing the differences of mobility with the wild-type amplicon.

The off-target mutation can be detected by amplification and sequencing of preselected-off-target sites, whole exome sequencing, genome sequencing, digenome sequencing, direct in situ breaks labeling, enrichment on streptavidin, and next-generation sequencing (BLESS). Linear amplification-mediated high-throughput genome-wide translocation sequencing (LAM-HTGTS), GUIDE-seq, and Digenome-seq are other methods for off-target confirmation (Zischewski et al., 2017). BLESS is the technique used to confirm the off-target mutations caused by sequence-specific NUCs; its detection is unbiased, and it makes genome-wide REC of DSBs possible. LAM-HTGTS is employed to know genomic translocations caused by end-joining between genomic DSBs.

CRISPR/Cas cleavage in vitro can be used to detect low-frequency DNA mutation of long fragment deletion, short fragment deletions, and SNP. Thus, the CRISPR/Cas cleavage in vitro can be used as a convenient mutation detection technique due to its high sensitivity (Jia et al., 2018).

15.4 PERSPECTIVES OF CRISPR

CRISPR, the gene-editing technique derived from the bacterial adaptive immune system, already been applied in both plants and animals, is having a high success rate. Being the most advanced technique, this needs some further modifications for improved results. Further studies on the exact mechanism of dissociation of Cas protein from the gRNA and its recycling, the structure of Cas with reference to the PAM sequence, delivery of multiple gRNA at once, effectiveness of mutations in future generations, and rate of occurrence of off-target mutation and its elimination are needed for improvement in agricultural traits and food crops (Ismail et al., 2019).

For the production of transgenic plants in which the Agrobacterium method is not applicable, CRISPR can be used with optimized transformation process (Pandey et al., 2019). Advanced CRISPR technique can provide protection to the crop plants from viral, bacterial, and fungal attacks by early detection (Pandey et al., 2019). CRISPR can satisfy the food requirement of growing population using artificial/synthetic genes to remove the incompatibility in traditional plant breeding. Recently developed CRISPR/ Cpf1-causing phenotype-linked mutations in the AT-rich region[1] can be used to develop biotic/abiotic stress-tolerant crop plant species, producing staple food for human beings. A big challenge in the way of success of CRISPR in plant gene editing is deficient knowledge about the complete genome sequence of the plant, which can be modified further for societal benefit. The creation of a worldwide data base having all the information about CRISPR will lay concrete on the way for more discoveries in the agricultural field and food production.

15.5 CONCLUSION

Since half a decade, this technology has been used in many plant systems for functional studies and combating biotic and abiotic stresses as well as to improve other agronomic traits. Several modifications have been made to this technology, which have eventually led to an increase in the on-target efficiency. CRISPR/Cas9-based genome editing will gain popularity. This tool will be an important one to obtain suitably edited plants, which will help achieve the zero-hunger rate as well as will maintain food for the growing human population. Compared to other tools used for genetic engineering, CRISPR-Cas9 is precise, cheap, remarkably powerful, and easy to use. It has rapidly become the most powerful gene-editing tool among researchers in fields such as human biology, agriculture, and microbiology.

CRISPR will transform our world in various major ways, namely through correction of the genetic errors that cause disease, by elimination of microbes that cause disease, by resurrecting species, by creating new, healthier foods, and by eradication of the planet's most dangerous pest. The CRISPR-Cas gene editing has been proven to be more promising in the field of agricultural research. Scientists from Cold Spring Harbor Laboratory in New York have used this tool to increase the yield of tomato plants. The lab has developed a method to edit the genes that determine tomato size, branching architecture, and ultimately, shape of the plant for a greater harvest. Regardless of advancements in CRISPR-Cas systems, the specificity of Cas9/gRNA targeting is still challenging, and it is hard to pin down the prediction of Cas9 off-targets, resulting in limited functional impact. Designing of a novel Cas9 protein with enhanced specificity by genome-wide binding and cutting can be achieved by employing genetic engineering tools.

---

[1] Regions with high content of adenine and thymine residues.

ACKNOWLEDGMENTS

We acknowledge the resources provided by the Post Graduate Department of Biotechnology, Utkal University and Center for Environment, Climate Change & Public Health, Utkal University under RUSA 2.0 grants of University Grants Commission, for the support.

KEYWORDS

- CRISPR-Cas system
- viral DNA snippets
- adaptive immunity
- plant genome editing
- agricultural sustainability

REFERENCES

Anders, C.; Niewoehner, O.; Duerst, A.; Jinek, M. Structural basis of PAM-dependent target DNA recognition by the Cas9 endonuclease. *Nature* 2014, 513(7519), 569–573.

Andersson, M.; Turesson, H.; Nicolia, A.; Fält, A.S.; Samuelsson, M.; Hofvander, P. Efficient targeted multiallelic mutagenesis in tetraploid potato (*Solanum tuberosum*) by transient CRISPR-Cas9 expression in protoplasts. *Plant Cell Reports* 2017, 36(1), 117–128.

Barrangou, R. CRISPR-Cas systems and RNA-guided interference. *Wiley Interdisciplinary Reviews: RNA* 2013, 4(3), 267–278.

Belhaj, K.; Chaparro-Garcia, A.; Kamoun, S.; Nekrasov, V. Plant genome editing made easy: targeted mutagenesis in model and crop plants using the CRISPR/Cas system. *Plant Methods* 2013, 9(1), 39.

Brouns, S.J.; Jore, M.M.; Lundgren, M.; Westra, E.R.; Slijkhuis, R.H.; Snijders, B.; Dickman, M.J.; Makarova, K.S.; Koonin, E.V.; Oost, J.v.d. Small CRISPR RNAs guide antiviral defense in prokaryotes. *Science* 2008, 321(5891), 960–964.

Chen, M.; Licon, K.; Otsuka, R.; Pillus, L.; Ideker, T. Decoupling epigenetic and genetic effects through systematic analysis of gene position. *Cell Reports* 2013, 3(1), 128–137.

Chylinski, K.; Le Rhun, A.; Charpentier, E. The tracrRNA and Cas9 families of type II CRISPR-Cas immunity systems. *RNA Biology* 2013, 10(5), 726–737.

Cong, L.; Ran, F.A.; Cox, D.; Lin, S.; Barretto, R.; Habib, N.; Hsu, P.D.; Wu, X.; Jiang, W.; Marraffini, L.A.; Zhang, F. Multiplex genome engineering using CRISPR/Cas systems. *Science* 2013, 339(6121), 819–823.

Deltcheva, E.; Chylinski, K.; Sharma, C.M.; Gonzales, K.; Chao, Y.; Pirzada, Z.A.; Eckert, M.R.; Vogel, J.; Charpentier, E. CRISPR RNA maturation by trans-encoded small RNA and host factor RNase III. *Nature* 2011, 471, 602–607.

D'Halluin, K.; Ruiter, R. Directed genome engineering for genome optimization. *International Journal of Developmental Biology* 2013, 57(6–8), 621–627.

Gaj, T.; Gersbach, C.A.; Barbas, C.F. ZFN, TALEN, and CRISPR/Cas-based methods for genome engineering. *Trends in Biotechnology* 2013, 31(7), 397–405.

Hu, L.; Xiao, P.; Jiang, Y.; Dong, M.; Chen, Z.; Li, H.; Hu, Z.; Lei, A.; Wang, J. Transgenerational epigenetic inheritance under environmental stress by genome-wide DNA methylation profiling in cyanobacterium. *Frontiers in Microbiology* 2018, 9, 1479.

Ismail, Eş.; Gavahian, M.; Marti-Quijal, F.J.; Lorenzo, J.M.; Khaneghah, A.M.; Tsatsanis, C.; Kampranis, S.C.; Barba, F.J. The application of the CRISPR-Cas9 genome editing machinery in food and agricultural science: Current status, future perspectives, and associated challenges. *Biotechnology Advances* 2019, 37(3), 410–421.

Ito, Y.; Nishizawa-Yokoi, A.; Endo, M.; Mikami, M.; Toki, S. CRISPR/Cas9-mediated mutagenesis of the RIN locus that regulates tomato fruit ripening. *Biochemical and Biophysical Research Communications* 2015, 467(1), 76–82.

Jansen, R.; Embden, J.D.; Gaastra, W.; Schouls, L.M. Identification of genes that are associated with DNA repeats in prokaryotes. *Molecular Microbiology* 2002, 43, 1565–1575.

Jia, C.; Huai, C.; Ding, J.; Hu, L.; Su, B.; Chen, H.; Lu, D. New applications of CRISPR/Cas9 system on mutant DNA detection. *Gene* 2018, 641, 55–62.

Jiang, W.; Maniv, I.; Arain, F.; Wang, Y.; Levin, B.R.; Marraffini, L.A. Dealing with the evolutionary downside of CRISPR immunity, bacteria and beneficial plasmids. *PLoS Genetics* 2013, 9(9), e1003844.

Jinek, M.; Chylinski, K.; Fonfara, I.; Hauer, M.; Doudna, J.A.; Charpentier, E. A programmable dual-RNA-guided DNA endonuclease in adaptive bacterial immunity. *Science* 2012, 337(6096), 816–21.

Jinek, M.; East, A.; Cheng, A.; Lin, S.; Ma, E.; Doudna, J. RNA-programmed genome editing in human cells. *elife* 2013, 2, e0047.

Kim, K.; Vinayagam, A.; Perrimon, N. A rapid genome-wide microRNA screen identifies miR-14 as a modulator of Hedgehog signaling. *Cell Reports* 2014, 7(6), 2066–2077.

Li, D.; Song, J.Z.; Li, H.; Shan, M.-H.; Liang, Y.; Zhu, J.; Xie, Z. Storage lipid synthesis is necessary for autophagy induced by nitrogen starvation. *FEBS Letters* 2015, 589(2), 269–276.

Miao, J.; Guo, D.; Zhang, J.; Huang, Q.; Qin, G.; Zhang, X.; Wan, J.; Gu, H.; Qu, L.-J. Targeted mutagenesis in rice using CRISPR-Cas system. *Cell Research* 2013, 23(10), 1233–1236.

Milbury, C.A.; Li, J.; Liu, P.; Makrigiorgos, G.M. COLD-PCR, improving the sensitivity of molecular diagnostics assays. *Expert Review of Molecular Diagnostics* 2011, 11(2),159–169.

Mojica, F.J.; Díez-Villaseñor, C.; García-Martínez, J. et al., Short motif sequences determine the targets of the prokaryotic CRISPR defense system. *Microbiology* 2009, 155(Pt 3), 733–740.

Nishimasu, H.; Ran, F.A.; Hsu, P.D.; Konermann, S.; Shehata, S.I.; Dohmae, N.; Ishitani, R.; Zhang, F.; Nureki, O. Crystal structure of Cas9 in complex with guide RNA and target DNA. *Cell* 2014, 156, 935–949.

Pandey, P.K.; Quilichini, T.D.; Vaid, N.; Gao, P.; Xiang, D.; Datla, R. Versatile and multifaceted CRISPR/Cas gene editing tool for plant research. *Seminars in Cell and Developmental Biology* 2019, S1084–9521(18)30115–0.

Pennisi, E. The CRISPR craze. *Science* 2013, 341(6148), 833–836.

Pul, U.; Wurm, R.; Arslan, Z.; Gieben, R.; Hofmann, N.; Wagner, R. Identification and characterization of *E. coli* CRISPR-Cas promoters and their silencing by H-NS. *Molecular Microbiology* 2010, 75(6), 1495–1512.

Regina, A.; Bird, A.; Topping, D.; Bowden, S.; Freeman, J.; Barsby, T.; Kosar-Hashemi, B.; Li, Z.; Rahman, S.; Morell, M. High-amylose wheat generated by RNA interference improves indices of large-bowel health in rats. *Proceedings of the National Academy of Sciences* 2006, 103(10), 3546–3551.

Shi, Y.; Chen, X.; Elsasser, S. et al. Rpn1 provides adjacent receptor sites for substrate binding and deubiquitination by the proteasome. *Science* 2016, 351(6275), doi:10.1126/science. aad9421.

Sun, Y.; Jiao, G.; Liu, Z.; Zhang, X.; Li, J.; Guo, X.; Du, W.; Du, J.; Francis, F.; Zhao, Y.; Xia, L. Generation of high-amylose rice through CRISPR/Cas9-mediated targeted mutagenesis of starch branching enzymes. *Frontiers in Plant Science* 2017, 8, 298.

Svitashev, S.; Schwartz, C.; Lenderts, B.; Young, J.K.; Mark, C.A. Genome editing in maize directed by CRISPR-Cas9 ribonucleoprotein complexes. *Nature Communications* 2016, 7, 13274.

Svitashev, S.; Young, J.K.; Schwartz, C.; Gao, H.; Falco, S.C.; Cigan, A.M. Targeted mutagenesis, precise gene editing, and site-specific gene insertion in maize using Cas9 and Guide RNA. *Plant Physiology* 2015, 169(2), 931–945.

Tsai, S.Q.; Wyvekens, N.; Khayter, C. et al. Dimeric CRISPR RNA-guided FokI nucleases for highly specific genome editing. *Nature Biotechnology* 2014, 32(6), 569–576.

Ueta, R.; Abe, C.; Watanabe, T.; Sugano, S.S.; Ishihara, R.; Ezura, H.; Osakabe, Y.; Osakabe, K. Rapid breeding of parthenocarpic tomato plants using CRISPR/Cas9. *Scientific Reports* 2017, 7(1), 507.

Voytas, D.F. Plant genome engineering with sequence- specific nucleases. *Annual Review of Plant Biololgy* 2013, 64, 327–350.

Xu, R.; Yang, Y.; Qin, R.; Li, H.; Qiu, C.; Li, L.; Wei, P.; Yang, J. Rapid improvement of grain weight via highly efficient CRISPR/Cas9-mediated multiplex genome editing in rice. *Journal of Genetics and Genomics* 2016, 43(8), 529–532.

Zischewski, J.; Fischer, R.; Bortesi, L. Detection of on-target and off-target mutations generated by CRISPR/Cas9 and other sequence-specific nucleases. *Biotechnology Advances* 2017, 35, 95–104.

CHAPTER 16

Self-Sustained Ramie Cultivation in Tripura: A Source of Multiple Commercial Uses for Adoption as an Alternative Livelihood

SHAON RAY CHAUDHURI[1*], BASANT KUMAR AGARWALA[2,3],
SUNIL K. SETT[4], PRIYASANKAR CHAUDHURI[2], PIYALI PAUL[1],
GOURAV BHATTACHARJEE[1], SUMONA DEB[1], SUKANYA CHOWDHURY[1],
PURNASREE DEVI[1], SINCHINI BARMAN[1], MANDAKINI GOGOI[1],
TETHI BISWAS[1,5], PURABI BAIDYA[1], ABHISPA BORA[1],
AMRITA CHAKRABORTY[1], CHAITALI CHANDA[5], SAURAV SAHA[1],
AJOY MODAK[1], GAUTAM DAS[1], PRIYA SARKAR[1], RONALD JAMATIA[1],
AMITAVA MUKHERJEE[6], ASHUTOSH KUMAR[1],
ASHOKE RANJAN THAKUR[7], MATHUMAL SUDARSHAN[8],
RAJIB NATH[9], LEENA MISHRA[10], INDRANIL MUKHERJEE[5],
GAUTAM BOSE[10], AMARPREET SINGH[11], and RANJAN KUMAR NAIK[12]

[1]Department of Microbiology, Tripura University, Suryamaninagar, West Tripura, Tripura 799022, India

[2]Department of Zoology, Tripura University, Suryamani Nagar, West Tripura, Tripura 799022, India

[3]Tripura State Pollution Control Board, Parivesh Bhawan, Pandit Nehru Complex, Agartala, West Tripura, Tripura 799006, India

[4]Department of Jute and Fibre Technology, Calcutta University, Kolkata, West Bengal 700019, India

[5]Centre of Excellence in Environmental Technology and Management, Maulana Abul Kalam Azad University of Technology, Haringhata, Nadia, West Bengal 741249, India

[6]Centre for Nanobiotechnology, Vellore Institute of Technology, Vellore, Tamil Nadu 632914, India

*[7] School of Science, Sister Nivedita University, New Town,
West Bengal 700156, India*

*[8] Trace Element Laboratory, Inter University Consortium, Kolkata,
West Bengal 700098, India*

*[9] Department of Agronomy, Bidhan Chandra Krishi Viswavidyalaya,
Mohanpur, West Bengal 741252, India.*

*[10] Division of Mechanical Processing, ICAR-National Institute of Natural
Fibre Engineering and Technology, Kolkata, West Bengal 700040, India*

[11] Ramie Research Station, Sorbhog, Barpeta, Assam 781317, India

*[12] ICAR-Central Research Institute for Jute & Allied Fibres, Barrackpore,
Kolkata, West Bengal 700120, India*

[*] Corresponding author. E-mail: shaonraychaudhuri@tripurauniv.in

ABSTRACT

Ramie, the king of natural fibers, has an important place in the global textile market. Many countries have climate suitable for growing Ramie including the north eastern region of India, but the raw material used globally comes from China, Indonesia, Philippines, Korea, Vietnam, and Japan. This study reports the implementation of organic Ramie cultivation in Tripura (23.9408N and 91.9882°E) using indigenously developed microbial biofertilizer. The decorticated waste was converted into vermicompost within 77 days of incubation with healthy growth of *Eisenia fetida*. The vermicompost positively impacted the height (p-value 0.01), basal diameter (p-value 7.72E−05), fiber yield (p-value 1.56E−07), and moisture content (1.38E−06) of Ramie fiber when compared to control (unfertilized condition) as well as chemical fertilizer treatment. The gum from the decorticated fiber was an excellent feed for bacterial growth and hence is of importance for biotechnology companies. The leaf extract with antibacterial property could be effectively used for green synthesis of the more potent bactericidal silver nanoparticles. The above findings indicated the prospect of putting the nonagricultural lands into eco-friendly Ramie cultivation through organic farming, thereby making it a zero pollution practice with generation of raw material for textile, biotechnology, and pharmaceutical industries.

16.1 INTRODUCTION

Ramie (*Boehmeria nivea* [*B. nivea*]) cultivation and its applications have made Hansan in Korea a World Heritage site. Literature suggests that Ramie, commonly known as China Grass, is one of the most preferred sources of textile fibers used for royal families since ages. The earliest evidence of its use is as mummy cloths in Egypt (5000–3000 BC). Its antimicrobial property has been explored in Egypt (Mummy Cloths) as well as in China (burial shrouds) since ancient times (Jose et al., 2016; Pandey, 2007a). It was much later that Ramie leaf extract was reported to have antioxidant, antidiabetic, antihypertensive, and anticancer properties (Mitra et al., 2013; Nho et al., 2010), albeit with limited scientific documentation. Ramie has immense applications in terms of fiber-based products, medicinal properties of its leaf and rhizome, edible nature of its leaves for human as well as animal consumption, to name a few. Microbial degumming of Ramie fiber has been the topic of investigation for many groups (Bruhlmann et al., 2000; Jose et al., 2016; Liang et al., 2015; Ray et al., 2014; Zhang et al., 2013). Upon blending with cotton and silk, the fiber is used as apparels, furnishing, towels, canvass, filter cloths, and so on (Gupta et al., 2015). The fiber waste is suitable for manufacturing currency notes and cigarette paper, while the woody waste has an option of being used for making plywood and fiber board (Mitra et al., 2013).

Ramie cultivation requires humid climate without water logging with an annual rainfall of about 1500–2500 mm. Despite a substantial portion of the globe having the requisite growth conditions for Ramie, its cultivation is confined only to certain areas of Asia like China, Indonesia, Philippines, Korea, Vietnam, Japan, and it is imported by other countries including India which harbors more than 19 germplasms of Ramie with 5 high-yielding varieties, of which only one (R1411) is commercially cultivated. Among the producers, China with 72,934 hectare (ha) area under Ramie cultivation (as in 2011) was the major producer (124,000 tons) of the fiber with an efficiency of 1700 kg/ha (Kim et al., 2014) production. According to the optimum climatic condition, Ramie can grow well in some parts of West Bengal and in northeastern states including Assam, Tripura, Arunachal Pradesh, Nagaland, and Meghalaya, to name a few. However, only some places of Assam and North Bengal grow Ramie, that too covering only 100 ha area (Mitra et al., 2013). Meghalaya has been actively involved in Ramie cultivation for the past few years. Ramie fiber is of high economic value, with the raw fiber costing about INR 100–115/kg, while the chemically degummed fiber costs around INR 230/kg. Cultivation of Ramie does not need very fertile land to start with but requires addition of fertilizer and micronutrients after the onset of harvesting for maintaining high production of the crop. Once planted, the rhizome goes on producing fiber for 16 years with harvesting done up to six times a year from the 2nd year of plantation. In order to maintain this high production rate, routine fertilization along with addition of the decorticated waste as mulch is practiced. The first bottleneck for large-scale cultivation of Ramie, in spite of its high demand and economic value, is the fertilizer requirement. Besides the cost factor, the problem with chemical fertilizer application is the utilization of only 12%–30% of the applied fertilizer while the rest is washed out with agricultural runoff. It makes chemical fertilizer-based agriculture economically less effective while polluting the environment (Ray Chaudhuri et al., 2017). The second bottleneck in large-scale Ramie cultivation is the management of the large amount of waste generated as a result of the decortication of the fiber. Every kilogram of fiber generation produces about 500 g of decorticated waste (dust of leaves and stem). The amount of decorticated Ramie waste produced annually per hectare of land is about 1874–2481.5 kg. The fate of this waste is either application as mulch on the Ramie field or dumping into landfill, which is expensive. Application of biocompost alone cannot maintain the high level of growth of Ramie; hence, additional fertilization is unavoidable. Application of Ramie decorticated waste for other value-added product generation would make the cultivation of Ramie profitable and pollution free.

The decorticated fiber retains about 25%–28% gum (Pandey, 2007a, 2007b), which needs to be further removed in an optimized manner so as to ensure luster with increased strength of the final finished fiber. The gum in turn acts as a microbial feed (Banerjee et al., 2018). Effective removal of gum from Ramie determines its appropriate end-uses (Jiang et al., 2017). Research is being carried out to improve the texture and quality of the finished fiber for enhancing the commercialization potential of the finished product. Microbial enzymes have always found their ways in industrial applications (Beg et al., 2001; De Souza and Magalhaes, 2010; Hoondal et al., 2002; Kuhad et al., 2011). In recent times, microbial enzymes have become the workhorse of textile industries for bio-polishing of raw materials or products for increasing the luster and strength of the final product. The use of microbial enzymes ensures that on one hand there is little chemical residue which would pollute the environment, while on the other hand, the fabric surface becomes smoother and more lustrous (Pandey, 2007a) due to use of enzymes for degumming. It also significantly reduces pilling, softens the fabric, and provides a smooth appearance to the fabric. Upon blending with synthetic fiber, the biochemically treated fibers lead to the formation of strong threads that can be weaved in fabric with enhanced luster and strength (Figure 16.1a and b). This report is a compilation of the work done by the Microbial Technology Group at Tripura University (23.9408°N and 91.9882°E) (India) on different aspects of Ramie cultivation and processing to propose it as a self-sustainable, profit earning solution for the entrepreneurs adopting Ramie cultivation.

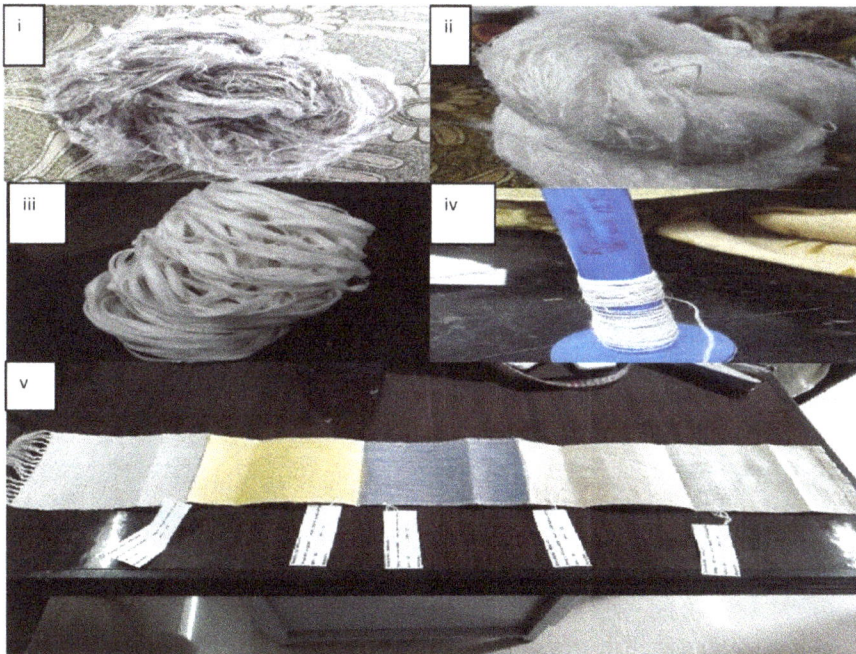

(a)

(b)

FIGURE 16.1   (a) Processing of Ramie fiber: (i) unopened degummed fiber; (ii) opened degummed fiber; (iii) sliver; (iv) thread of Ramie with viscose; (v) blended fabric using linen/cotton and above mentioned thread, (b) Images of fabric produced from Ramie yarn observed under stereo dissecting zoom microscope (Leica) at 12× magnification.

16.2   RAMIE DECORTICATED WASTE AS SUBSTRATE FOR VERMICULTURE

One of the major problems associated with Ramie cultivation is maintaining the production of Ramie fiber while managing the waste. Vermicomposting, the process of organic waste degradation with the help of the microbial community residing in the earthworm intestine as well as added from outside (as cow dung), is an effective way of solid waste management in which the final product matures faster than that of traditional composting. However, the composition of the decorticated waste has an important impact on it being suitable as a substrate for vermicomposting. In order to test the potential of Ramie decorticated waste for vermicomposting and vermiculture, the elemental composition of the leaves was analyzed through energy dispersive X-ray fluorescence (EDXRF) to reveal the presence of the following elements: phosphorus (558 mg/kg), sulfur (4.4 g/kg), chlorine (2.55 g/kg), potassium (12.02 g/kg), calcium (23.56 g/kg), manganese (15.86 mg/kg), iron (176.61 mg/kg), copper (5.68 mg/kg), and zinc (15.92 mg/kg). High calcium content makes it suitable for earthworm feeding, as earthworms need continuous supply of calcium for formation of cast through which they increase the porosity and improve structure, aeration, drainage, and moisture-holding capacity of soil (https, //www.dpi.nsw.gov.au/agriculture/soils/biology/earthworms). Humus-rich casting with 80% sulfur makes the soil more fertile (Aladesida et al., 2014). Casting is known to contain potassium, iron, zinc, phosphorous, calcium, manganese, and magnesium (Ayoola and Olayiwola, 2014). An essential component of casting is humic acid, mainly from plant origin, which preferentially binds elements like calcium, iron, potassium, sulfur, as well as phosphorous that in turn are released as per requirement for plant growth in soil. The presence of these elements indicates Ramie leaf to be a preferred substrate for growth of earthworm. There are other components in the leaf which, beyond a concentration, might have an inhibitory effect on growth of earthworm. Qualitative analysis of the leaf extract revealed the presence of alkaloids, flavonoids, phenols, glycosides, tannins, carbohydrates, steroids, and proteins, while absence of terpenoids, saponins, and anthocyanins. The presence of alkaloids and glycosides makes them unsuitable for growth of earthworm unless degraded during pre-composting. *Pseudomonas putida*, Seriatia sp., *Pseudomonas aeruginosa*, *Rhodococcus* sp., and *Staphylococcus aureus* present in cow dung (10^{11} cells/g of cow dung) are known to degrade alkaloids and glycosides (https, //microbewiki.kenyon.edu/index.php/Purine alkaloid degradation; Girija et al., 2013). On the other hand, presence of steroid helps in the development of earthworm (Romano et al., 2015). On performing the quantitative analysis, the leaf extract revealed the presence of 1.89 mg/g of protein and 1.99 mg/g of phenol. Both protein and phenols are essential components of an earthworm (Aldarraji et al., 2013). All these indicate Ramie leaf to be a suitable substrate for vermiculture (*Eisenia fetida* [*E. fetida*]), provided the alkaloids and glycosides are reduced through pre-composting before using it as a substrate for vermiculture. As a result of pre-composting, the substrate changes and becomes palatable for the earthworm, leading to a significant increase in earthworm biomass (Kostecka and Kaniuczak, 2008; Mistry et al., 2015).

The validation of the prediction of Ramie decorticated waste as a suitable substrate for vermiculture was carried out experimentally in two different scales (200 g, 3.25 kg). For the small-scale trial, 7 days of precomposting was allowed, while for the large-scale experiment, 24 days were required. In both the experiments, 25 and 200 adult earthworms (*E. fetida*) were added, respectively, after pre-composting. The increase in biomass over a period of 30 days was highly significant (p-value 2.04×10^{-5}). The cocoon number (p-value 0.25) as well as earthworm density (p-value 0.25) also increased with time but not to a significant extent. The data indicate a 6.5-fold increase in earthworm density within a span of 30 days. It was observed that there was no significant change in the content of available organic carbon (p-value 0.27), phosphorous (p-value 0.183), pH (p-value 0.38), available nitrogen (p-value 0.346), and total nitrogen (p-value 0.063) in the final product as compared to decorticated waste, while a significant change in available potassium (p-value 0.0053) was observed in the vermicompost after 30 days of incubation. The validation of data was done using paired two-sample t-test. The statistics was expected to improve on further optimization of the process during scale-up trials. The large-scale trial carried out using 66% and 100% cow dung for pre-composting was maintained for 57 days after releasing the earthworm. The substrate still had unutilized nutrients on the day of biomass measurement. The biomass which was the total weight of all cocoons, juveniles, and adult earthworms per 3.35 kg of vermicompost indicated 1.76-fold increase in production with 100% of cow dung when compared to 66%, but 0.7-fold less weight of the adult earthworms. However, the significant part was the cocoon production in Ramie decorticated waste. From the current study, it could be extrapolated that *E. fetida* produced approximately 18×10^4 to 2.36×10^5 cocoon/m^3, while *Perionyx excavatus* produced 3,600 cocoon/m^3 of tank. The reason for mentioning the approximation is that there might be spatial variation in cocoon count within the tank, but random sampling in large number ensures minimal error. As per literature, *E. fetida* reproduces at a slow rate (0.35 to 0.5 cocoons/day) when compared to

Perionyx excavatus (1.1 to 1.4 cocoons/day) (Bhattacharjee and Chaudhuri, 2002; Hallatt et al., 1990). In spite of the fact that *E. fetida* produces less number of cocoons than *Perionyx excavates*, Ramie decorticated waste showed much higher cocoon production in case of the former proving this waste to be a much better substrate for earthworm culture (vermiculture). A comparative study of vermicomposting of leaves of rubber, cashew nut, pineapple, and decorticated waste of Ramie at small scale (100 g substrate each) reconfirmed Ramie decorticated waste to be a suitable substrate for vermiculture. Further, EDXRF analysis of the pre- and post-vermicompost samples revealed that most of the elements present were in less amount in the vermicompost, indicating that the earthworm had utilized the element during vermicomposting and had survived in a healthy manner in the feed provided. This proved Ramie waste to be a suitable substrate for vermiculture. Hence, producing a byproduct (substrate for vermiculture) of economic value besides vermicompost for organic farming.

In order to create a zero-pollution environment during Ramie cultivation, it is essential to convert the waste into a useful product that can be utilized completely. Ramie decorticated waste could be used efficiently for mushroom cultivation after adequate precomposting and sterilization (for 15 min at a pressure of 15 pounds/inch^2). During cultivation, precautions need to be taken to prevent growth of unwanted fungus and insects as it is more enriched than rice straw, which is conventionally used as substrate for growing mushroom.

Vermicompost and substrate for mushroom as well as vermiculture could be the three useful byproducts of Ramie decorticated waste.

16.3   ORGANIC FARMING OF RAMIE

Another bottleneck in Ramie cultivation is the maintenance of fertility of the soil from the second year of plantation while yielding five to six harvests a year. The soil has to be repleted with manure to ensure retention of fertility of the soil. Usually, the mulch (leaves) is added to the soil as amendment. But it takes time for them to get decomposed to a form that could be picked up by the plants. Tripura has substantial unutilized land and is geographically cut off from the rest of the mainland. It receives the requisite rainfall to sustain Ramie cultivation. The living expenses are extremely high because of the elevated transportation cost. The principal occupation of the people is cultivation and weaving. The possibility of Ramie cultivation being an alternative livelihood for the local residents with access to land is currently being explored to boost the textile industry of the state in particular and the country in general.

In a country like India, the need of the hour is to work on real-life problems, bearing in mind one's expertise and the available resources. A major drawback of using chemical fertilizers for sustaining cultivation is that only 12%–30% of the applied fertilizer is utilized by the plants while the rest leaches out of the soil into surrounding water bodies, hence polluting the environment and making the process economically nonfavorable due to loss of essential plant growth nutrients. The other option is organic farming with sustained production of fiber. The vermicompost developed using Ramie decorticated waste was used for cultivation of Ramie and mung bean (*Vigna radiata* var. MEHA) during field and pot trials, respectively. It demonstrated positive influence on growth in terms of decreased plant height (*p*-value 0.0035), root length (*p*-value 2.23E−08), and nodulation (*p*-value 0.0021) and increased yield (*p*-value 0.0389) and carbohydrate content (*p*-value 0.0402) of mung bean seeds. Though the nutritional quality (in terms of plant growth nutrients) of vermicompost was similar to conventional compost, it could sustain productivity of healthy mung bean cultivation. The microbial population (4.7×10^6 cells/g of vermicompost) could also help in retaining soil quality during cultivation of mung bean. These factors could be responsible for the improved performance of the vermicompost during mung bean cultivation.

This vermicompost significantly increased the number of stems per rhizome (*p*-value 1.4×10^{-2}), average height of stem (*p*-value 1.2×10^{-4}), basal diameter (*p*-value 3.89×10^{-3}), and fiber yield (*p*-value 2.31×10^{-3}) of Ramie (*B. nivea*) when compared to control (without fertilizer) and chemical (fertilized with NPK) treatment within 6 months of application, with no significant change in soil N-P-K ratio, pH, and carbon content. When compared to chemical fertilizer alone, the average height of stem (*p*-value 0.03), basal diameter (*p*-value 0.01), and fiber yield (*p*-value 0.001) were significantly higher in vermicompost treatment in spite of no significant increase in the number of stems per rhizome (*p*-value 0.5). Improvement in the agronomic parameters is an essential feature of any plant growth trial. Here the trial was carried out using one rhizome at a time in replicates of 10. The tensile strength, tenacity, fineness, brightness, and yellowness of the fibers obtained from these three treatments did not show any significant change ensuring higher yield with maintained quality with vermicompost application in case of Ramie cultivation. The Ramie fiber gum that is removed during fiber processing works as a microbial growth supplement for strains of *Pseudomonas* sp. and *Bacillus* sp. (Banerjee et al., 2018).

Other biofertilizers developed by the group were tested for sustaining Ramie cultivation at Suryamaninagar, West Tripura in a 196.42 m² land. Standard 4 m² plots were taken for testing the effect of different fertilizers on yield of Ramie fiber. The tested fertilizers were chemical (NPK in the ratio of 20:15:15 kg/ha), microbial biofertilizer (1.47×10^{10} cells) (Ray Chaudhuri et al., 2017), and vermicompost (from Ramie decorticated waste) (2.8 kg in 4 m²) applied only one time during zero day of experiment. After 40 days of growth, the fiber yield per stem for control (tap water alone), chemical fertilizer with tap water, biofertilizer with tap water, and vermicompost with tap water grown set were 0.73 g, 1.59 g, 1.50 g, and 1.39 g, respectively. The same plots were used for a second round of cultivation, with no further addition of fertilizers to check the soil fertility condition after one round of cultivation. After 54 days of growth, the fiber yield per stem for control (tap water alone), chemical fertilizer with tap water, biofertilizer with tap water, vermicompost with tap water grown set were 0.58 g, 0.91 g, 1.05 g, and 1.01 g, respectively. It shows that the fertility of the soil is maintained even during the second harvest. The data generated by the group shows successful conversion of decorticated waste into vermicompost, and the work done shows a positive impact of its application on agronomic parameters, fiber yield, and gum yield of Ramie. In order to make Ramie cultivation a self-sustained, zero pollution process, we attempted to find other uses of the Ramie waste. The gum from Ramie fiber was found to work as a bacterial feed (Banerjee et al., 2018), while the fiber processing itself requires a lot of chemicals, leading to generation of wastewater that requires proper treatment (Chakraborty et al., 2018).

These observations enhance the possibility of increasing self-sustained Ramie cultivation in barren land providing alternative livelihood to the land owners, whereby they could use the land to generate raw material for the textile industry along with vermiculture, vermicompost, mushroom, as well as microbial feed. Through this approach, there is complete conversion of waste into value-added products, making Ramie cultivation a zero pollution agricultural practice with generation of byproducts.

16.4   RAMIE LEAF EXTRACT AS ANTIMICROBIAL AGENT

Ramie has multiple applications of its different parts. The current section explores the silver nanoparticle (Ag-NP)-forming ability of the Ramie leaf extract leading to its antimicrobial activity, which is more potent when compared to few other locally available medicinal plants (mentioned below).

In order to understand the reason behind the enhanced antimicrobial potential of Ramie phytochemicals, the qualitative and quantitative analysis were done for leaf extracts of each variety. Qualitative analysis confirmed the presence of proteins, phenols, alkaloids, flavonoids, and carbohydrates in the water extracts of all of the plant leaves except thankuni (*Centella asiatica*) which lacked phenol and Shiuli (*Nyctanthes abor-tristis*) and Sajne (*Moringa oleifera*) which lacked flavonoid. Anthocyanin and terpenoids were found to be absent in all the leaf extracts. Saponin was absent only in Ramie; glycoside was absent in lemon (*Citrus limon*), aparajita (*Clitoria ternatea*) and thankuni; tannin was absent in thankuni while steroid was present in Ramie and Shiuli. Among the components present in most of the extracts, alkaloids, flavonoids, phenols, and tannins are antimicrobial in nature. Terpenoids, glycosides, saponins, tannins, and flavonoids are anti-diarrheal in nature while phenols, tannins, and alkaloids are known to be antihelminthic in nature (Tiwari et al., 2011). These components are responsible for the medicinal property of the leaves of the above-mentioned plants. Protein and phenolic contents of the phytochemicals from the above-mentioned plants showed protein content to be highest in Ramie (1.51 mg/g of leaves) as compared to the other five plants, that is, Sajne (0.73 mg/g of leaves), Shiuli (0.8 mg/g of leaves), thankuni (0.7 mg/g of leaves), lemon (1.11 mg/g of leaves), and aparajita (0.77 mg/g of leaves). In case of phenolic content, it was again found to be highest in Ramie (1.6 mg/g of leaves) as compared to the other four plants, that is, Sajne (0.85 mg/g of leaves), Shiuli (0.76 mg/g of leaves), lemon (1.29 mg/g of leaves), and aparajita (1 mg/g of leaves), while absent in thankuni.

Researchers reported protein to act as reducing agent and phenol as a stabilizing agent in Ag-NP generation (Mittal et al., 2014; Salleh et al., 2003). From the data, it is clear that Ramie contains the highest amount of protein and phenolics, which played a key role as reducing and stabilizing agents, respectively, for synthesis of Ag-NP. The statistical analysis was done using ANOVA where p-values were found to be 1.21×10^{-4} and 2.6×10^{-7}, respectively, for protein and phenolics, which indicates a statistically significant difference in the amount of protein and phenolics among the plants. The Ramie leaf extract was further found to be a potent source for green synthesis of Ag-NPs from silver nitrate solution and was more efficient (4.55 mg/g of wet leaf) in terms of nanoparticle generation as compared to Shiuli (0.67 mg/g wet leaf), aparajita (0.92 mg/g wet leaf), thankuni (0.91 mg/g wet leaf), lemon (0.48 mg/g wet leaf), and Sajne (0.38 mg/g wet leaf). Characterization of the nanoparticles was done by ultraviolet–visible spectroscopy (UV–VIS), scanning electron microscopy (SEM) with energy dispersive X-ray analysis (Figure 16.2a) and Fourier transform infrared spectroscopy (FTIR)

(Figure 16.2b). The nanoparticles were in the range of 15–125 nm, with majority having sizes between 35 nm and 40 nm (Figure 16.2c). Production of nanoparticles was optimized at 50 °C and pH 6.8. The components in the leaf extract which might be responsible for nanoparticle generation were proteins, flavonoids, and terpenoids (as reducing agents) as well as phenolics and alkaloids (as capping and stabilizing agents). These nanoparticles were tested for their antimicrobial activity against well-characterized strains of *Bacillus* sp. under planktonic and biofilm states, revealing little effect of the nanoparticles on preformed biofilms (Figure 16.3a and b). Minimum inhibitory concentrations (MICs) of the nanoparticles against *Bacillus* sp. (MCC0008) and *Bacillus* sp. (MTCC8995) under planktonic condition were 100 µg/mL and 250 µg/mL, respectively. This reveals the potential of the Ramie leaf extract for Ag-NP synthesis and is the first scientific documentation in case of Ramie leaves.

16.5 FUTURE SCOPE TO ADOPT RAMIE CULTIVATION AS AN ALTERNATIVE LIVELIHOOD OPTION

Available cited scientific literature states that 100 ha of land in India is under Ramie cultivation, of which 90 ha is in Northeast India, producing 1600 to 2400 kg/ha/year of fresh fiber, which amounts for 150–200 kg/day (dry weight) and hence 73,000 kg/year fiber production. In Philippines, Ramie is grown as a mixed crop in coconut plantations. In India, it is grown with pineapple, rubber, highland paddy, and areca nut. The cost of cultivation of Ramie/ha in the first year as per literature is about INR 37,259 while the return is about INR 80,500 (cost of Ramie fiber INR 115/kg). The investment in the second year is around INR 50,400 while the return is about INR 207,000. The expenditure for preparation of land, planting material, and planting cost is only during the first year. Weed management, staging/ harvesting, and irrigation double during the second year commensurate with the production. The labor charge and the decortication charges also double during the second year, but when Ramie cultivation becomes the alternative livelihood for the land owners who themselves with their family get involved in the *cultivation*, savings also occur in the labor cost. Ramie grows round the year in irrigated land, while from April to September, it grows in areas with an annual rainfall of 1500–3000 mm. The crop needs to be harvested every 45 days. The decortication is carried out at a rate of 400–450 kg/h processing 2400 kg/day. Hence, crop from 1 ha land would require about 12 days for decortication. The vermicomposting of the decorticated waste

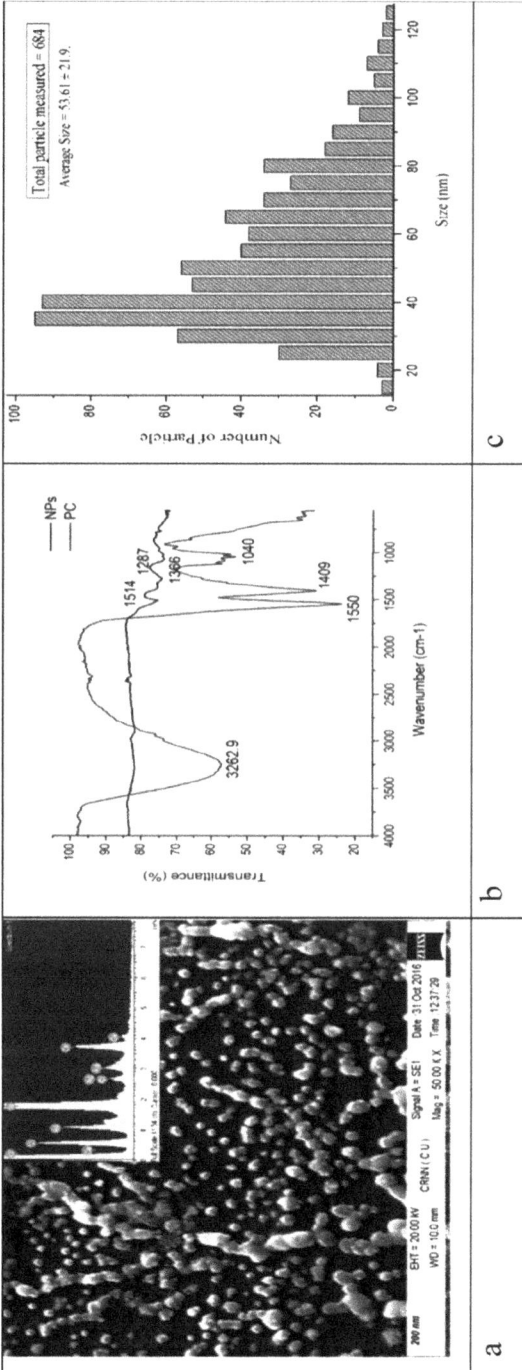

FIGURE 16.2 Silver nanoparticle (Ag-NPs) synthesis using Ramie leaf phytochemical. (a) SEM image of Ramie phytochemical synthesized Ag-NPs. The inset EDX image shows silver to be present in these particles, (b) FTIR spectra of Ramie Ag-NP and phytochemical. NPs means Ramie Ag-NP while phytochemical is represented as PC. FTIR analysis for both the samples was done using standard protocol. In the case of synthesized NPs the peak intensity at 3263, 1550, and 1408 cm^{-1} was observed to be significantly decreased, with slight red shift, suggesting that hydroxyl and COO- functional groups could be largely involved in the synthesis of NPs. From the data it can be concluded that reductive synthesis of the NPs was accomplished due to flavonoids and polyphenol compounds present in the phytochemicals. (c) Bar diagram showing the size distribution of nanoparticle. Size distribution curve was plotted using the data from 684 particles.

will follow harvesting. Simultaneously, fertilization would be done after harvesting using the vermicompost prepared from the decorticated wastes of previous harvests. The process continues round the year producing raw material that could be sold at the cost of INR 100–115/kg. Upon being planted, the rhizome (Rs 25/kg) goes on producing the crop for 16 years. The chemically processed fiber decreases in weight by one-third and is sold at a cost of INR 230/kg. This results in round-the-year generation of revenue for the Ramie farmers.

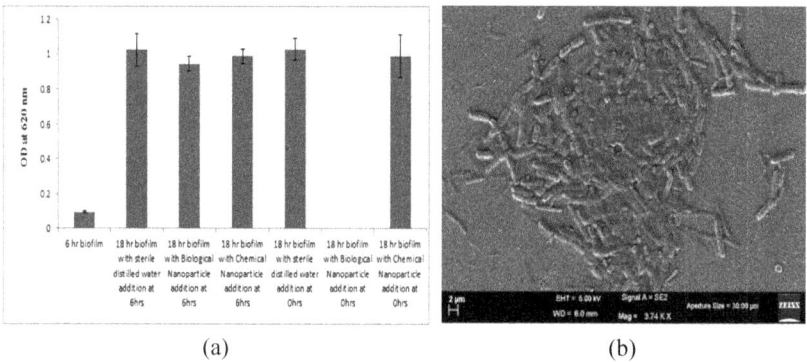

(a)   (b)

FIGURE 16.3 (a) Bar diagram showing the effect of Ag-NPs on bacterial biofilm of MCC0008. Biofilm of MCC0008 was formed by growing the culture in LB at staionary condition and at 37 °C for 6 h. At 6 h, one set was used for crystal violet staining while others were inoculated with nanoparticles (from biological as well as chemical origin). However, the control was inoculated with same amount of sterile distilled water and again incubated for 12 h under similar condition. After incubation biofilm was stained with crystal violet, excess stain was washed with distilled water and the bound stain was eluted with ethanol and the absorbance was measured as per standard procedure. The result was documented. (b) Field emission scanning electron microscopic (FESEM) image showing the effect of Ag-NPs on preformed bacterial biofilm. Biofilm was formed by growing the culture in LB broth at staionary condition and at 37 °C for 6 h. At 6 h Ramie phytochemical synthesized Ag-NPs was added to the culture at 500 μg/mL concentration. It was again incubated for 12 h under similar condition. After incubation biofilm was coated with platinum and observed under FESEM. The 6 h old preformed biofilm on the glass surface did not get destroyed.

Ramie mulch, consisting of short fibers along with other natural fibers, is used as a means of crop improvement as green manure, which is not detrimental to the environment. However, it takes time for the mulch to get degraded in soil before it contributes to the soil fertility. It also has a potential of being used for bioethanol production due to its high biomass yield. After extraction, Ramie fiber can also be used for producing paper of superior quality. It can be used as feedstock for cattle and goats. Vermicomposting can be done in a portion of the farm, requiring manpower for the formation of vermi bed, watering for moisture maintenance, stirring, harvesting, sieving, and packaging/application. This would provide work round the year for the families getting involved with Ramie cultivation as an alternative livelihood along with generation of a byproduct, (vermicompost) which is sold at approximately INR 9/kg. In addition, the decorticated waste being a suitable substrate for vermiculture, the overproduced earthworms could also be sold (INR 3.45/individual) as it has application in multiple fields. The constant monetary growth of vermicomposting in recent years has made it viable for adoption in the cooperative sector.

16.6 CONCLUSION

The state of Tripura in northeastern India has a climate suitable for Ramie cultivation. The economy of the state is strongly influenced by its textile industry. Tripura has substantial unutilized land and is geographically cut off from the rest of the mainland. The living expenses are extremely high because of the elevated transportation cost. The principal occupation of the people is cultivation and weaving. The possibility of Ramie cultivation as an alternative livelihood for the local residents with access to land has been explored to boost the textile industry of the state in particular and the country in general on a small scale, with discrete unit operation at a laboratory scale. Each component reported in this study reveals the potential of sustained Ramie cultivation as a source of raw material for various commercial ventures, making its adoption a viable alternative livelihood option for places like Tripura. Hence, the next step should be a scaled-up trial in an integrated approach at the state level, which would lay the foundation for future roadmap of organic farming of such economically beneficial crops.

ACKNOWLEDGMENTS

The authors acknowledge Tripura University for providing the laboratory infrastructure and field trial facility for conducting the experiments; Department of Biotechnology, Government of India (GOI) for funding the project under the Twinning scheme; Ministry of Human Resource and Development, GOI for funding under the FAST scheme (Centre of Excellence in Environmental Technology and Management, Maulana Abul Kalam Azad University of Technology, West Bengal); Late Mr. Sourav

Chakraborty, Ms Manjila Gupta and Mr. Arindam Roy for their assistance; Prof Lalit Mohan Gantayet for reviewing the manuscript and providing his valuable input. The corresponding author would like to thank Carl Zeiss for the Microscopic Facility at Calcutta University. The facility at ICAR-National Institute of Natural Fibre Engineering and Technology was used for fiber quality analysis; EDXRF analysis was carried out at UGC DAE Consortium at Kolkata, India.

AUTHORS' CONTRIBUTIONS

Piyali Paul, Gourav Bhattacharjee, and Sumona Deb contributed equally to this project.

KEYWORDS

- Ramie
- silver nanoparticles (Ag-NPs)
- vermiculture
- organic farming
- zero pollution cultivation

REFERENCES

Aladesida, A.A.; Owa, S.O.; Dedeke, G.A.; Adewoyin, O.A. Prospects and challenges of vermiculture practices in southwest Nigeria. *African Journal* of *Environmental Science and Technology* 2014, 8, 185–189.

Aldarraji, Q.M.; Halimoon, N.; Majid, N.M. Antioxidant activity and total phenolic content of earthworm paste of *Lumbricus rubellus* (red worm) and *Eudrilus eugenia* (African night crawler). *Journal of Entomology and Nematology* 2013, 5, 33–37.

Ayoola, P.B.; Olayiwola, A.O. Trace elements and major minerals evaluation of earthworm casts from a selected site in Southwestern Nigeria. *ARPN Journal of Agricultural and Biological Science* 2014, 9, 216–218.

Banerjee, S.; Gupta, M.; Roy, A.; Chakraborty, A.; Ray Chaudhuri, S. Ramie (*Boehmeria nivea*) Gum: a natural feed to sustain and stimulate the growth of bacteria. *Journal of Bacteriology and Mycology* 2018, 5, 1–3.

Beg, Q.K.; Kapoor, M.; Mahajan, L.; Hoondal, G.S. Microbial xylanases and their industrial applications: a review. *Applied Microbiology and Biotechnology* 2001, 56, 326–338.

Bhattacharjee, G.; Chaudhuri, P.S. Cocoon production, morphology, hatching pattern and fecundity in seven tropical earthworm species—a laboratory-based investigation. *Journal of Biosciences* 2002, 27, 283–294.

Bruhlmann, F.; Leupin, M.; Erismann, K.H.; Fiechter, A. Enzymatic degumming of ramie bast fibers. *Journal of Biotechnology* 2000, 76, 43–50.

Chakraborty, A.; Bhowmik, A.; Jana, S.; Bharadwaj, P.; Das, D.; Das, B.; Debnath, P.; Agarwala B.K.; Ray Chaudhuri, S. Evolution of waste water treatment technology and impact of microbial technology in pollution minimization during natural fiber processing. *Current Trends in Fashion Technology & Textile Engineering* 2018, 3, 1–4.

de Souza, P.M.; Magalhaes, P.O.E. Application of microbial α-amylase in industry—a review. *Brazilian Journal of Microbiology* 2010, 41, 850–861.

Girija, D.; Deepa, K.; Xavier, F.; Antony, I.; Shidhi, P.R. Analysis of cowdung microbiota—a metagenomic approach. *Indian Journal of Biotechnology* 2013, 12, 372–378.

Gupta, M.; Roy, A.; Banerjee, S.; Kapoor, R.; Adhikari, B.; Thakur, A.R.; RayChaudhuri, S. *Bacillus* sp MCC2138: a potential candidate for microbial degumming of Ramie. *International Journal of Fiber and Textile Research* 2015, 5, 39–43.

Hallatt, L.; Reinecke, A.J.; Viljoen, S.A. Life cycle of the oriental compost worm Perionyx excavatus (Oligochaeta). *South African Journal of Zoology* 1990, 25, 41–45.

Hoondal, G.S.; Tiwari, R.P.; Tewari, R.; Dahiya, N.; Beg, Q.K. Microbial alkaline pectinases and their industrial applications: a review. *Applied Microbiology and Biotechnology* 2002, 59, 409–418.

Jiang, W.; Han, G.; Zhang, Y.; Liu, S.; Zhou, C.; Song, Y.; Zhang, X.; Xia, Y. Monitoring chemical changes on the surface of kenaf fiber during degumming process using infrared microspectroscopy. *Scientific Reports* 2017, 7, 1–8.

Jose, S.; Rajna, S.; Ghosh, P. Ramie fibre processing and value addition. *Asian Journal of Textile* 2016, 7, 1–9.

Kim, A.R.; Kang, S.T.; Jeong, E; Lee, J.J. Effects of Ramie leaf according to drying methods on antioxidant activity and growth inhibitory effects of cancer cells. *Journal of the Korean Society of Food Science and Nutrition* 2014, 43, 682–689.

Kostecka, J.; Kaniuczak, J. Vermicomposting of duckweed (*Lemna Minor* L.) biomass by *EiseniaFetida*(Sav.) earthworm. *Journal of Elementology* 2008, 13, 571–579.

Kuhad, R.C.; Gupta, R.; Singh, A. Microbial cellulases and their industrial applications. *Enzyme Research* 2011, 2011, 1–10.

Liang, C.; Gui, X.; Zhou, C.; Xue, Y.; Ma, Y.; Tang, S.Y. Improving the thermoactivity and thermostability of pectate lyase from *Bacillus pumilus* for ramie degumming. *Applied Microbiology and Biotechnology* 2015, 99, 2673–2682.

Mistry, J.; Mukhopadhyay, A.M.; Baur, G.N. Status of N P K in vermicompost prepared by two common weed and two medicinal plants. *International Journal of Applied Sciences and Biotechnology* 2015, 3, 193–196.

Mitra, S.; Saha, S.; Guha, B.; Chakrabarti, K.; Satya, P.; Sharma, A.K.; Gawande, S.P.; Kumar, M.; Saha, M. Ramie: the strongest bast fibre of nature. Technical Bulletin No. 8/2013, Central Research Institute for Jute and Allied Fibres, ICAR. 2013. doi:10.13140/2.1.3519.5842.

Mittal, J.; Batra, A.; Singh, A.; Sharma, M.M. Phytofabrication of nanoparticles through plant as nanofactories. *Advances in Natural Sciences: Nanoscience and Nanotechnology* 2014, 5, 1–10.

Nho, J.W.; Hwang, I.G.; Kim, H.Y.; Lee, Y.R.; Woo, K.S.; Hwang, B.Y.; Chang, S.J.; Lee, J.; Jeong, H.S. Free radical scavenging, angiotensin I-converting enzyme (ACE) inhibitory, and in vitro anticancer activities of ramie (*Boehmeria nivea*) leaves extracts. *Food Science and Biotechnology* 2010, 19, 383–390.

Pandey, S.N. Ramie fibre: Part I. Chemical composition and chemical properties. A critical review of recent developments. *Textile Progress* 2007a, 39, 1–66.

Pandey, S.N. Ramie fibre: Part II. Physical fibre properties. A critical appreciation of recent developments. *Textile Progress* 2007b, 39, 189–268.

Ray Chaudhuri, S.; Mishra, M.; De, S.; Samal, B.; Saha, A.; Banerjee, S.; Chakraborty, A.; Chakraborty, A.; Pardhiya, S.; Gola, D.; Chakraborty, J.; Ghosh, S.; Jangid, K.; Mukherjee, I.; Sudarshan, M.; Nath, R.; Thakur, A.R. Microbe-based strategy for plant nutrient management. In *Biological Wastewater Treatment and Resource Recovery*; Farooq, R., Ahmed, Z., Eds.; Intech, Croatia, 2017; pp. 38–55.

Ray, D.P.; Satya, P.; Mitra, S.; Banerjee, P.; Ghosh, R.K. Degumming of ramie: challenge to the queen of fibres. *International Journal of Bioresource Science* 2014, 1, 37–41.

Romano, M.C.; Jimenez, P.; Brito, C.M.; Valdez, R.A. Parasites and steroid hormones synthesis, their role in the parasite physiology and development. *Frontiers in Neuroscience* 2015, 9, 1–5.

Salleh, A.B.; Ghazali, F.M.; Rahman, R,N,Z,A.; Basri, M. Bioremediation of petroleum hydrocarbon pollution. *Indian Journal of Biotechnology* 2003, 2, 411–425.

Tiwari, P.; Kumar, B.; Kaur, M.; Kaur, G.; Kaur, H. Phytochemical screening and extraction—a review. *Internationale Pharmaceutica Sciencia* 2011, 1, 98–106.

Zhang, C.; Yao, J.; Zhou, C.; Mao, L.; Zhang G; Ma, Y. The alkaline pectate lyase PEL168 of *Bacillus subtilis* heterologously expressed in *Pichia pastoris* is more stable and efficient for degumming ramie fiber. *BMC Biotechnology* 2013, 13, 1–9.

CHAPTER 17

Azolla: A Viable Resource of Biofertilizer and Livestock Feed for Sustainable Development of Farming Community

UPENDRA KUMAR*, SNEHASINI ROUT, MEGHA KAVIRAJ,
SWASTIKA KUNDU, HARI NARAYAN, HIMANI PRIYA, and A. K. NAYAK

*Microbiology Laboratory, Crop Production Division,
ICAR-National Rice Research Institute, Cuttack, Odisha 753006, India*

Corresponding author. E-mail: ukumarmb@gmail.com.

ABSTRACT

Azolla is a genus of small aquatic floating fern that generally grows and multiply on the surface of the lakes, fresh water ponds and various kinds of streams. It has seven distinct species, that is, *A. caroliniana, A. filiculoides, A. mexicana, A. microphylla, A. rubra, A. pinnata,* and *A. nilotica,* which belongs to two subsections, namely, Euazolla and Rhizosperma. The usage and application of this fern is very versatile in agricultural and industrial fields mainly as biofertilizer, compost, livestock feed, bioaccumulator of heavy metals, waste water treatments, biofuel and so on, and therefore, it is designated as "green gold mine." It is also accepted widely in farming community because of its fast-growing nature, high nutritional value, and nitrogen-fixing ability. The presence of symbiotic cyanobionts in dorsal leaf pockets of *Azolla* make it enables to fix atmospheric nitrogen and thus serves as potential biofertilizer in agriculture. Besides rapid growth rate of *Azolla*, its high nutritional value also enables it a suitable and sustainable livestock feed. Owing to the various importance of *Azolla*, the present chapter describes its prospect as biofertilizer and livestock feed for sustainable agriculture.

17.1   INTRODUCTION

Azolla is a genus of aquatic small floating fern and grouped into two sub-sections, that is, Euazolla and Rhizosperma. Section Euazolla has five species, namely *Azolla caroliniana*, *A. filiculoides*, *A. mexicana*, *A. microphylla*, and *A. rubra*, and section Rhizosperma has two species—*A. pinnata* and *A. nilotica*. (Kumar and Nayak, 2019; Kumar et al., 2019). Wagner (1997) tagged *Azolla* as a "green gold mine" and eventually in 2006, Carrapiço called it a "superorganism" (Carrapiço, 2010). *Azolla* is acknowledged by the farming community due to its fast-growing nature, high nutritional value, and nitrogen (N)-fixing ability. There is a well-known fact that long-term use of inorganic N fertilizers makes the soil acidic (Bouman et al., 1995; Kumar et al., 2017a, 2018a). Therefore, farmers are trying to adopt organic fertilizers that provide natural nutrients to their crop, thus improving the soil quality (Bhardwaj et al., 2014; Sharma et al., 2019). *Azolla* biofertilizer is one of the best organic fertilizers, which help to increase plant productivity by fixing the atmospheric N into ammonia (Kumar et al., 2017b). Moreover, *Azolla*-cyanobionts have co-evolved together and are associated symbiotically and able to fix around 1100 kg of N ha^{-1} year^{-1} (Yadav et al., 2014; Zahran, 1999). *Azolla* biofertilizer is also termed as "Azobiofer" (Mian, 2002).

Livestock production has been another vital field of agriculture since long and considered as backbone of Indian economy, particularly a source of income in rural areas. Due to lack of land availability, there is an intense shortage of feed and fodder; hence, to fulfill the demand of livestock products and their increasing population, a search for a suitable substitute is need of the hour (Table 17.1). Moreover, maintaining off-season milk production is also a serious concern; hence, many researchers advocated various unconventional feed and fodder (Gouri et al., 2012; Khatun et al., 1999; Tamang and Samanta, 1993). Cottonseed cake is an expensive traditional supplement, which could not simply afforded by the farmers as a regular protein source and it is also not sufficient enough to fulfill the required need of nutrients of animals. Therefore, many workers reported that *Azolla* might be serve as one of the acceptable, as well as nutritional, substitutes for sustainable livestock (Chatterjee et al., 2013; Khare et al., 2016) and poultry feed (Alalade and Iyayi, 2006; Chichilichi et al., 2015). Keeping the above viewpoints, in the present chapter, we describe the beneficial role of *Azolla* as a biofertilizer and as livestock feed for sustainable agriculture.

TABLE 17.1 Comparison of Relative Biomass Production of *Azolla* and Different Fodders

Name of fodder	Annual Production Biomass (MT/ha)	Dry Matter Content (MT/ha)	Protein Content (%)
Azolla	1000	80	24
Hybrid Napier (CO$_3$)	250	50	4
Kolakattao grass	40	8	0.8
Lucerne (*Medicago sativa*)	80	16	3.2
Cowpea (*Vigna sinensis*)	35	7	1.4
Subabul (*Leucaena leucocephala*)	80	16	3.2
Sorghum (*Sorghum vulgare*)	40	3.2	0.6

SOURCE: Adapted from Katole et al. (2013).

17.2 CULTIVATION OF *AZOLLA* AND ITS APPLICATION

Azolla cultivation has been generally initiated in a small nursery with an area of 20 m^{-2} maintaining water level about 5–10 cm. The area must be under shade where no direct sunlight comes in contact with the chamber. Around 0.4–0.5 kg m^{-2} fresh *Azolla* can be inoculated. Phosphate fertilizer (4–20 kg P$_2$O$_5$ ha^{-1}) is applied in the nursery for better growth and development of *Azolla*. To avoid pests, an insecticide such as furadon (3% a.i. as a carbofuran) can be sprayed. *Azolla* multiplies fast and becomes double within 3 days of inoculums by vegetative propagation. *Azolla* mat is formed above the water surface within 2–3 weeks. Overgrown *Azolla* is harvested at regular intervals to avoid over population and inoculated to other fields or used as livestock or poultry feed.

For Azobiofer, *Azolla* can be applied as (a) incorporation of dried *Azolla* in soil, (b) incorporation of fresh *Azolla* after transplantation of crop with 5–10 cm water level in the field (Mian, 2002; Singh and Singh, 1986), or (c) making a compost pit with fresh *Azolla*, then applied to the roots of crops like manure (Anand and Pereira, 2006), whereas for cattle livestock feed, two common ways are (a) by feeding fresh *Azolla* mixed with other commercial feeds or (b) by feeding dry *Azolla* mixed with other dried straw fodders. For aquaculture, *Azolla* can be given as supplementary feeding material and given directly to the fish in the pond.

17.3   PERPETUAL RELATIONSHIP BETWEEN *AZOLLA* AND CYANOBIONTS

Endophytic cyanobacteria are present inside the air cavity of the dorsal leaf lobe of *Azolla* and both partners are involved in carrying out photosynthesis as *Azolla* contains chlorophylls a and b and carotenoids, whereas its cyanobionts contain chlorophyll a, phycobiliproteins, and carotenoids (Figure 17.1). Thus, they successfully exhibit carbon dioxide fixation and produce intermediates and are finally able to convert them into sucrose as a major product. However, N-fixation is performed only by cyanobacteria, as nitrogenase enzyme is present only in its heterocysts which catalyze di-nitrogen into ammonia. Ammonium produced through N-fixation is assimilated by the enzymatic process involving glutamine synthetase, glutamate synthase, and glutamate dehydrogenase, and this process is performed by both the partners (Silver and Schröder, 1982).

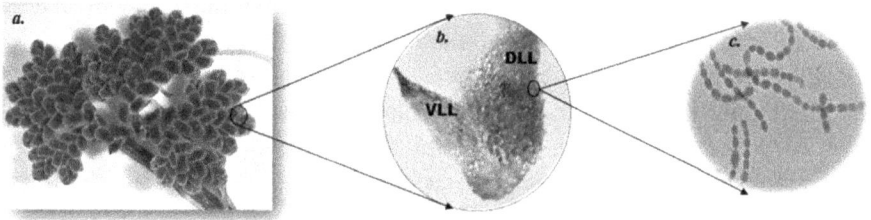

FIGURE 17.1   Overall view of *Azolla* and its cyanobionts: (a) *Azolla* frond; (b) Dorsal (DLL) and ventral (VLL) leaf lobe of *Azolla*; (c) Cyanobacteria present inside the leaf cavity of *Azolla* (adapted from Kumar and Nayak, 2019).

17.4   PROSPECTS OF *AZOLLA* AS A BIOFERTILIZER

17.4.1   *AZOBIOFER IN RICE CULTIVATION*

Azobiofer is used in the paddy field either as a monocrop or as an intercrop (Table 17.2). The use of azobiofer is efficiently used in the countries of Asia (Yadav et al., 2014). The utilization of azobiofer as a monocrop applied in the rice field expanded the yield by 112% compared to the control method during the decrepit season. Whereas its application as intercrop increases the rice yield by 23%, and if applied as a both monocrop and intercrop in wheat it increases the yield by 216% (Ripley et al., 2003; Roy, 2016a). *Azolla* has quick decomposition rate in soil and thus, it speeds up the N uptake

TABLE 17.2 Role of *Azolla* as Biofertilizer in Agricultural Sector

Azolla species	Crop	Used as	Total Nitrogen Fixed	Experimental Details — Role	References
A. pinnata	Rice	Dual crop	75.5%–98.2%	*Azolla* combined with reduced dose of urea suggested that it can substitute up to 25% of nitrogen	Yao et al., 2018
A. pinnata	Rice	Dual crop	71%	Deep placement of urea combined with *Azolla* reduces N loss and improves fertilizer N recovery	Yao et al., 2018
A. pinnata	Rice	Manure	Not specified	Incorporation of fresh *Azolla* at 20 t ha^{-1} and 5 t ha^{-1} for compost, increased the soil available P, plant P content and tiller number	Setiawati et al., 2018
A. filiculoides	Rice	Compost	Not specified	Under water deficit conditions, application of *Azolla* compost provided grain yield of an average of 13.8% higher than that of the nonamended control	Razavipour et al., 2018
All species	Coffee	Compost	Not specified	*Azolla* acts as a cheap source of nitrogen in coffee	Titus and Pereira, 2006
A. filiculoides	Wheat	Manure	Not specified	*Azolla* increases the nutrient availability in soil by decomposition, resulting into increase of plant growth and grain yield	Ripley et al., 2003
A. microphylla	Rice	Dual crop	Not specified	*Azolla* acts as biofertilizer	Rudsamee, 2008
Azolla (not specified)	Rice	Manure	82.4%	Grain yield results indicated that growing and incorporating one layer of *Azolla* may save 40% urea	Akhtar et al., 2002
A. pinnata	Mushroom	Manure	Not specified	Provides more lignin to mushroom growth	Jain et al., 1988
A. pinnata	Rice	Dual crop	59.5% in India 65.1% in Thailand 65.9% in Vietnam 61.2% in Bangladesh	*Azolla* showed better result as compared to that of BGA. *Azolla* increased the grain yield by 24.8 to 32.3%	Singh and Singh, 1987

feasibility of the rice plant (Akhter et al., 2002; Rudsamee, 2008). The rapid multiplicative nature and fast decomposing rate of *Azolla* made it an essential observable factor to utilize as a green-manure-cum-biofertilizer in the rice field (Setiawati et al., 2018; Singh and Singh, 1987; Yao et al., 2018a, 2018b). The azobiofer basal application (10–12 t ha^{-1}) enriched soil nitrogen up to 60 kg ha^{-1} and cut down about 30–35 kg of N fertilizer requirement of the rice crop. Weed suppression in the rice field by forming a thick mat of *Azolla* is another advantage of it (Cheng et al., 2010; Janiya and Moody, 1984). The use of azobiofer increases the rice yield by 20%–30% (Raja et al., 2012). Besides, the successful use of azobiofer under paddy cultivation in Asia is now also spreading in the rest of the world like Italy (Bocchi and Malgioglio, 2010) and Africa (Carrapiço et al., 2002).

17.4.2 AZOBIOFER IN OTHER CROP CULTIVATIONS

Azobiofer has been incorporated successfully in other crop cultivations, mainly spinach and coffee. Organic compost of *Azolla* as N fertilizer was used to observe the growth, production, and quality (nitrate content) of spinach (Suratno and Asyim, 2018). In coffee cultivation, *Azolla* was used as a biofertilizer in the form of dried and fresh green compost materials (Anand and Pereira, 2006; Titus and Pereira, 2006).

17.5 NUTRITIONAL VALUE OF AZOLLA

Azolla is a high-yielding aquatic plant, containing 20%–37% protein (Kathirvelan et al., 2015; Kumar et al., 2019; Singh, 1980; Sreemannaryana et al., 1993; Subudhi and Singh, 1978; Van Hove, 1989). However, the protein content that can be digested of *Azolla* is 56.6% (Tamany et al., 1992), thereby limiting the higher incorporation of *Azolla* in poultry diets. The following amino acids, such as 0.47%–0.53% leucine and lysine, 0.11%–0.17% methionine, 0.53%–0.55% threonine, and 0.14%–0.15% tryptophan, are present in *Azolla*. Arginine and valine, the predominant essential amino acids, are also found in *Azolla*, while sulfur-containing amino acids are deficient (Alalade and Iyayi, 2006; Leterme et al., 2009). *Azolla* also contains 0.8%–6.7% crude fat, namely, polyunsaturated fatty acids (PUFAs) and omega-3 and omega-6 fatty acids (Ali and Leeson, 1995; Buckingham et al., 1978; Fujiwara et al., 1947; Querubin et al., 1986, 1987; Sreemannaryana et al., 1993; Subudhi and Singh, 1978;). Neutral detergent fiber (NDF) and crude fiber contents in *Azolla* are in the range of 42%–62% and 14%–16%, respectively (Basak et al., 2002; Cherryl et al., 2014; Querubin et al., 1987). N-free extract in *Azolla* is 31% (Ali and Leeson, 1995; Basak et al., 2002; Querubin et al., 1986, 1987) and ash content is 15%–16% (Basak et al., 2002; Buckingham et al., 1978). Lower acid detergent fiber (ADF) and NDF in *Azolla* demonstrate a superior proficiency of utilization of nonruminant animals. *Azolla* contains several mineral properties, namely calcium (1.16%), phosphorus (1.29%), potassium (1.25%), and magnesium (0.25%), while sodium, manganese, iron, copper, and zinc were 23.79 ppm, 174.42 ppm, 755.73 ppm, 16.74 ppm, and 87.59 ppm, respectively. Additionally, it also contains 3.4%–0.55% chlorophyll and 0.26% iron substance on dry matter basis. Apart from *Azolla* being a vital source of plant proteins, fresh *Azolla* is also an important source of provitamin A and β-carotenes (Kathirvelan et al., 2015; Lejeune et al., 2000). *Azolla*, being carotene-rich ranging from 300 to 600 ppm, can be positioned among carotene-rich plants, though nutrient composition varies among different species of *Azolla* (Table 17.3).

17.5.1 AZOLLA AS A LIVESTOCK FEED FOR RUMINANTS

In India, due to lacking inadequate land for fodder production, farmers have started to use *Azolla* as a better supplement of livestock feed due to its low cost production and maintenance technology (Akhud et al., 2017; Chandewar et al., 2018; Kumar et al. 2019; Mathur et al., 2013; Srinivas Kumar et al., 2012; Tawasoli et al., 2018). In South India, low-cost milk production technology is supposed to be taken up widely by dairy farmers particularly those areas where insufficient land are available for fodder production (Pillai et al., 2005). *Azolla* is a valuable green feed supplement having high content of protein and mineral contents which are ideally suited for livestock feed at a very low-input cost dairy production (Gowda et al., 2015; Indira et al., 2009; Katole et al., 2013; Parashuramulu et al., 2013). One report depicted that supplementation of fresh *A. microphylla* at the rate of 1.5 kg^{-1} day^{-1} animal^{-1} in cross breed growing heifers made them healthy for longer period (Khare et al., 2016). *A. microphylla* at the rate of 25% with regular meal also acts as a useful alternate of mustard cake protein and has shown beneficial impact on Jalauni lambs (Das et al., 2017). Utilization of nutritional aspects of *Azolla* as feed supplementation for other large ruminantsis also revealed by other workers (Chatterjee etal.,2013; Trivedi et al., 2005). The nutrient composition of *Azolla* increases the milk yield in cattle and feed conversion ratio. Singh (1980) reported that in crossbred heifer's digestibility and body weight increased with intake of *Azolla* as a feed substitute and saved 20%–25% of costly commercial feed.

TABLE 17.3 Nutrients Composition of Different Species of *Azolla*

Nutrient Parameters	Unit	*Azolla* spp.					
		AP	AF	AR	AMI	AME	AC
Dry matter (DM)	% as fed	5.44	6.84	6.5	6.94	4.1	4.48
Moisture (%)		94.56	93.15	93.5	93.05	95.9	95.5
Crude protein	% DM	21.77	22.65	23.13	25.86	22.65	18.56
Ether extract	% DM	3.43	3.82	2.9	3.90	4.65	2.98
Total sugar	% DM	7.66	7.53	7.62	7.16	7.39	6.27
Total antioxidants (ABTS)	% DM	53.2	55.25	56.14	76.87	68.93	63.72
Neutral detergent fibre (NDF)	% DM	55.12	42.04	40.34	29.23	42.31	51.12
Total ash	% DM	25.2	30.8	33.6	40.4	24.4	40.2
Organic matter	% DM	74.8	69.2	66.4	59.6	75.6	59.8
Minerals							
Ca	g/kg DM	6.21	10.41	14.25	16.73	18.02	22.10
P (g/kg)	g/kg DM	9.28	7.26	7.17	6.46	10.25	7.88
Fe	mg/kg DM	2680.7	8907.0	8916.6	13276.8	5623.3	14,104.2
Cu	mg/kg DM	18.8	34.4	29.8	47.8	30.9	27.1
Zn	mg/kg DM	33.1	29.2	33.9	54.7	51.7	67.6
Mn	mg/kg DM	205.3	288.1	281.8	313.2	350.7	378.3
B	mg/kg DM	18.92	38.48	37.48	42.42	25.32	46.11
Ni	mg/kg DM	6.51	15.26	12.79	23.57	13.62	23.32

AP: A. pinnata; AF: *A.filiculoides;* AR: *A. rubra;* AMI: *A. microphylla*; AME: *A. mexicana* AC: *A. caroliniana.*

SOURCE: Adapted from Kumar et al. (2015, 2019).

17.5.2 AZOLLA AS A FEED FOR NON-RUMINANTS OR OTHER LIVESTOCK

Azolla is also feed for nonruminants such as fish, swine, and poultry (Alcantara and Querubin, 1985; Ara et al., 2015; Bhattacharyya et al., 2016; Castillo et al., 1981; Cherryl et al., 2013) (Table 17.4). Utilization of green *Azolla* as a fodder at low quantity in the poultry diets showed better yield performance in terms of total protein contents and pigmentation of egg yolk

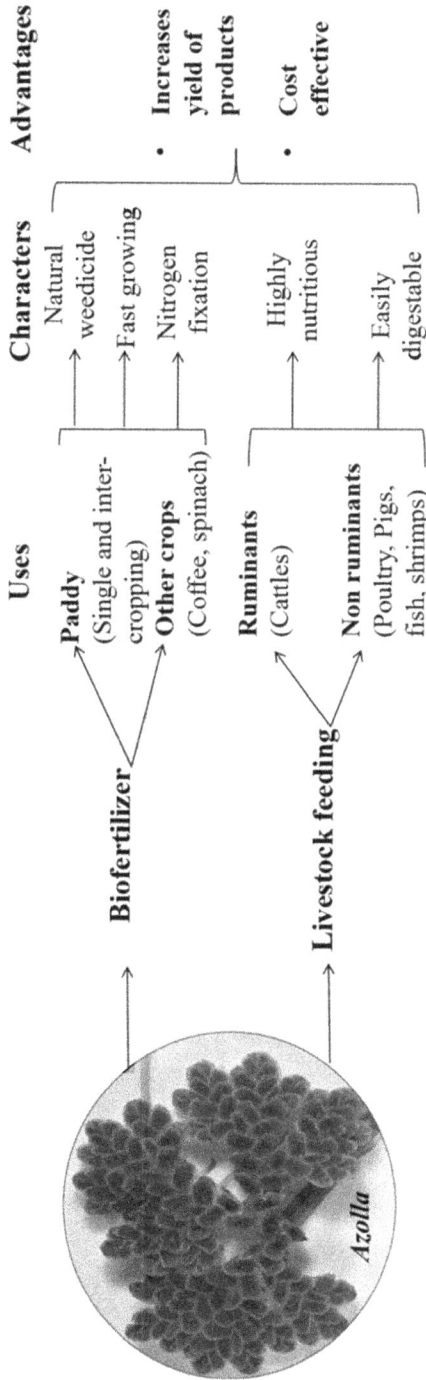

FIGURE 17.2 Beneficial role of *Azolla* in context with biofertilizer and livestock feed

and skin (Maurice et al., 1984). It is also a potential feed supplement for chickens and could replace 20% of commercial feed as diets of growing chickens (Subudhi and Singh, 1978). *Azolla* at the 5% inclusion level is highly economical in broiler chickens as per recent reports by Ara et al. (2015). Use of *Azolla* as a sole source of poultry feed may be unsuitable because of its low digestibility (Buckingham et al., 1978; Khatun, 1996), but it can be used in the combination of other feed ingredients such as rice bran, wheat bran, and maize. In Vietnam, *Azolla* is reported to be used as feed for domestic fowl (Minh, 2005). Moreover, no mortality in birds was reported during experimentation on broilers fed with *Azolla* meal (Basak et al., 2002). It is accounted to have potential as a feedstuff for egg-type chicks, and no mortality is observed in layers fed with *A. pinnata* (Alalade and Iyayi, 2006). Contrasting reports also indicating in broiler chickens which have been fed with *Azolla*-based diets regularly, but did not show any kind of significant difference with nonfeeders (Ara et al., 2015; Dhumal et al., 2009). Without any adverse effect on growth and health of ducks, *Azolla* could replace 20% of crude protein of soybeans diets and feed conversion ratio and could fetch higher net profit compared to conventional poultry diets (Acharya et al., 2015; Becerra et al., 1995; Sujatha et al., 2013). Escobin (1987) also reported that fresh *Azolla* does not compromise on the production efficiency in egg and meat-producing ducks. Researchers also found that birds gained their body weight after intake of approximately 5%–15% of *Azolla* in their diet (Basak et al.2002; Cambel, 1984; Querubin et al., 1986). The weight gain variations could be due to strain variations of *Azolla* meal at different levels as well as the nutrient components used along with the metabolic activity of the used experimental animal. *Azolla* is rich in cysteine while deficient in other sulfur-containing amino acids and lysine like other aquatic plants, this opens the ground for *Azolla* being a better food supplement for fishes (Cruz-Velásquez et al., 2014; Das et al., 2018; Fiogbe et al., 2004; Shiomi and Kitoh, 2001). It is also used as a substitute for defatted soybean meal diet of juvenile black tiger shrimp (*Penaeus monodon*) (Sudaryono, 2006). The considerable increasing in absorption rate on the experimental fish may be due to the ample of essential amino acids found in *Azolla* spp. (Sithara and Kamalaveni, 2008). Report indicated that the body weight of grass and common carps are increased by 174 and 35.8 g fish^{-1}, respectively, after feeding 30% *Azolla* as diet supplement (Ninawe, 2003; Spotte, 1979). Hence, to attend the optimum level of growth and survival of fish, the fish diets should have comparatively high amount of protein than other commercial feeds. Due to the presence of high concentration of protein, *Azolla* is also supposed to be a good substitute of feed material for rabbit (Anitha, 2015;

TABLE 17.4 Role of *Azolla* as a Livestock Feed

Azolla Species	Fed by	As Nutrient Source	Role	Reference
pinnata	Crossbred cows	—	Increased body weight and feed intake	Chandewar et al., 2018
Azolla (not specified)	Buffalo calves	—	Increased dry matter intake and growth rate	Akhud et al., 2018
A. pinnata	Thai Silver Brab (*Barbonymus gonionotus*)	Protein and carbohydrate	Increased quality production	Das et al., 2018
A. pinnata	Broiler chicks	—	Increase body weight, FCR (feed consumption ratio) and performance index	Kumar et al., 2018b
A. pinnata	Vanraja breed of old chicks	Protein	Improved growth, feed conversion efficiency and immune system	Tawasoli et al., 2018
A. pinnata	Commercial broilers	—	Improved weight gain	Bhattacharya et al., 2016
A. pinnata	Hariana heifers	—	Increased body weight and nutrient digestibility	Roy et al., 2016b
A. microphylla	Rabbits	Protein	Increased weight gain and FCR	Anitha, 2015
A. pinnata	White pekin broilers	—	Higher efficiency, FCR and gain in body weight.	Acharya et al., 2015
Azolla (not specified)	Commercial broilers	Protein	Higher erythrocyte catalase with better immune response	Chichilichi et al., 2015
A. filiculiodes	Amazonian fish (*Cachama blanca* and *Piaractus branchypomus*)	Carbohydrate	Improved growth performance	Cruz-Velasquez et al., 2014
A. microphylla	Crossbred cattle	—	Increased productivity in case of growth, milk and meat	Chaterjee et al., 2013
A. pinnata	Cow and buffalo	—	Increase in buffalo milk yield	Mathur et al., 2013
A. pinnata	Ruminants	Protein	High digestibility of dry matter	Parashuramulu et al., 2013

TABLE 17.4 *(Continued)*

Azolla Species	Fed by	As Nutrient Source	Role	Reference
A. pinnata	Murrah buffalo bulls	Protein	Increased digestibility	Srinivas et al., 2013
A. pinnata	Ducks	—	Increase in egg weight	Sujatha et al., 2013
A. pinnata	Crossbred pigs	—	Protein source, higher cost per kg gain	Cherryl et al., 2013
A. filiculoides	Pigs	Minerals and essential amino acids	Increase in weight of colons and enhanced digestive secretions	Leterme et al., 2009
Azolla (not specified)	Buffalo calves	Protein	Increased body weight	Indira et al., 2009
A. pinnata	Lambs	—	Increased body weight	Trivedi et al., 2005
A. pinnata	Egg-type chicks	Protein	Improved growth performance	Alalade and Iyayi, 2005
A. microphylla	Rabbit	Fiber	Increased body weight	Lebas et al., 2004
A. microphylla	Fish (*Oreochromis niloticus*)	Protein	Exhibited weight gain	Fiogbe et al., 2004
A. pinnata	Broilers	Protein	Increased body weight and higher energy efficiency	Basak et al., 2002
A. filiculiodes	Fish (*Tilapia nilotica*)	—	Increased weight gain and Tilapia feed replacement	Shiomi and Kitoh, 2001
A. microphylla	Poultry	—	Significant increase in body weight gain	Brue and Latshaw, 1985
A. pinnata	Commercial broiler chickens	—	Increase in weight gain and feed consumption	Ali and Leeson, 1995

Lebas, 2004). *Azolla* is reported to be low in digestible energy for growing piglets but it is a good source of minerals and essential amino acids for them (Leterme et al., 2009). However, it could only replacement partially the protein source for growing-fattening pigs (Becerra et al., 1995).The major limitation of *Azolla* as a supplementary food diet is having relatively high quantity of acid detergent fiber and lignin contents, that is why, preferably its fewer intakes is observed by nonruminant animals (Tamany et al., 1992 and Basak et al., 2002).

17.6 CONCLUSION AND FUTURE PROSPECTS

Azolla serves as a potential azobiofer in rice, as it is symbiotically associated with N-fixing cyanobacteria. Combination of azobiofer with minimal recommended chemical fertilizer gave higher yields, better plant growth, and other parameters, minimizing the hazardous effect of chemical fertilizer, mainly urea. *Azolla* is considered as an economic and efficient feed supplement for different groups of animals including both ruminant and nonruminant. *Azolla* is an excellent source of proteins, carbohydrate, minerals, and essential amino acids, and a well-accepted feed for cattle, goats, sheep, and chicken.

Despite of several advantages, azobiofer and azofeed are still underutilized. Therefore, we must take initiative to make its popularization among the farmers and making them aware of its advantages as biofertilizer and livestock feed. Though *Azolla* is a popular biofertilizer for rice growers, its use in other crops is very limited. The status of soil health in field after incorporation of *Azolla* in combination with other traditional fertilizers is also limited. As *Azolla* has invasive nature, it moves into canals and other water bodies where eutrophication typically occurs, which causes algal blooms and harms the aquatic ecosystem as well as human life. Hence, a robust management practice should be developed to minimize overgrowth of *Azolla* in an aquatic environment. Sporocarp germination of *Azolla* under unfavorable conditions may serve as one of the options to mitigate this challenge. Molecular and metagenomic approaches may also be explored to enhance the understanding of symbiosis of *Azolla*-cyanobionts, which may help to boost diazotrophic potential, so that it can supplement chemical nitrogenous fertilizer for agricultural crop.

Azolla has potential to serve as an azofeed; hence, an attempt could be made for making *Azolla* in a pellet form with known composition of nutrients, which will serve as a probiotic medicine for the animal husbandry and helps to boost its immunity and productivity. Research could also be performed in fish, rabbits, and pigs to obtain more detailed information about their growth, feed consumption ratio, body weight, and other profitability parameters. An enhanced emphasis is requisite to increase nutrient use efficiency in ruminant yield production systems and having a special role in the animal diet value since they can produce human edible protein from human inedible protein such as grass, crop residues, and agro-industrial by-products. Overall, *Azolla* may play a significant role in doubling the farm income of India without affecting the soil health and environment.

KEYWORDS

- *Azolla*
- biofertilizer
- livestock feed
- sustainable agriculture

REFERENCES

Acharya, P.; Mohanty, G.P.; Pradhan, C.R.; Mishra, S.K.; Beura, N.C.; Moharana, B. Exploring the effects of inclusion of dietary fresh *Azolla* on the performance of White Pekin broiler ducks. *Veterinary World* 2015, 8(11), 1293.

Akhter, S.; Mian, M.H.; Kader, M. A.; Begum, S.A. Combination of *Azolla* and urea nitrogen for satisfactory production of irrigated Boro rice (BRRI Dhan 29). *Journal of Agronomy* 2002, 1(3), 127–130.

Akhud, M.W.; Ingole, A.S.; Atkare, V.G.; Khupse, S.M.; Deshmukh, S.V. Effect of feeding different levels of concentrate replace with *Azolla* on Nagpuri buffalo calves. *Journal of Soils and Crops* 2017, 27(2), 105–108.

Alalade, O.A.; Iyayi, E.A. Chemical composition and the feeding value of *Azolla* (*Azolla pinnata*) meal for egg-type chicks. *International Journal* of *Poultry Science* 2006, 5(2), 137–141.

Alcantara, P.F.; Querubin, L.J. Feeding value of *Azolla* meal for broilers [Philippines]. Philippine *Journal of Veterinary and Animal Sciences* 1985, 11(3–4), 1–8.

Ali, M.A.; Leeson, S. The nutritive value of some indigenous Asian poultry feed ingredients. *Animal Feed Science and Technology* 1995, 55(3–4), 227–237.

Anand, T.; Pereira, G.N. *Azolla* as a biofertilizer in coffee plantations. *Communications in Soil Science and Plant Analysis* 2006, 36(13–14), 1737–1746.

Anitha, K.C. Effect of supplementary feeding of *Azolla* on performance of broiler rabbits (PhD dissertation). Karnataka Veterinary, Animal and Fisheries Sciences University, Bidar, 2015.

Ara, S.; Adil, S.; Banday, M.T.; Khan, M.A. Feeding potential of aquatic fern *Azolla* in broiler chicken ration. *Journal of Poultry Science and Technology* 2015, 3, 15–19.

Basak, B.; Pramanik, M. A. H.; Rahman, M. S.; Tarafdar, S. U.; Roy, B. C. *Azolla* (*Azolla pinnata*) as a feed ingredient in broiler ration. *International Journal of Poultry Science* 2002, 1(1), 29–34.

Becerra, M.; Preston, T.R.; Ogle, B. Effect of replacing whole boiled soya beans with *Azolla* in the diets of growing ducks. *Livestock Research for Rural Development* 1995, 7(3), 32–38.

Bhardwaj, D.; Ansari, M.W.; Sahoo, R.K.; Tuteja, N. Biofertilizers function as key player in sustainable agriculture by improving soil fertility, plant tolerance and crop productivity. *Microbial Cell Factories* 2014, 13(1), 66.

Bhattacharyya, A.; Shukla, P. K.; Roy, D.; Shukla, M. Effect of *Azolla* supplementation on growth, immune competence and carcass characteristics of commercial broilers. *Journal of Animal Research* 2016, 6(5), 941.

Bocchi, S; Malgioglio, A. *Azolla-Anabaena* as a biofertilizer for rice paddy fields in the Po Valley, a temperate rice area in Northern Italy. *International Journal of Agronomy* 2010, 2010, 1–5.

Bouman, O.T.; Curtin, D.; Campbell, C.A.; Biederbeck, V.O.; Ukrainetz, H. Soil acidification from long-term use of anhydrous ammonia and urea. *Soil Science Society of America Journal* 1995, 59(5), 1488–1494.

Buckingham, K.W.; Ela, S.W.; Morris, J.G.; Goldman, C.R. Nutritive value of the nitrogen-fixing aquatic fern *Azolla filiculoides*. *Journal of Agricultural and Food Chemistry* 1978, 26(5), 1230–1234.

Cambel, I.M. *Growth Performance of Broilers Fed with Varying Levels of Azolla Meal. University of Southern Mindanao,* Kabacan, North Cotabato, *Philippines, 1984,* p. 66.

Carrapiço, F. *Azolla* as a superorganism: its implication in symbiotic studies. In *Symbioses and Stress*, Seckbach J.; Grube M. Eds.; Springer, Dordrecht, The Netherlands, 2010, pp. 225–241.

Carrapiço, F.; Teixeira, G.; Diniz, M.A. *Azolla* as a biofertilizer in Africa: a challenge for the future. *Revista de Ciências Agrárias* 2002, 23, 120–138.

Castillo, L.S., Gerpacio, A.L.; Sd, P.F. Exploratory studies on *Azolla* and fermented rice hulls in broiler diets [study conducted in the Philippines]. In Annual Convention of Philippine Society of Animal Science (PSAS). University of the *Philippines* Los Baños, College, Laguna, Metro Manila, Philippines, 1981, pp. 13–14.

Chandewar, D.; Rathod, K.; Mohale, D. Effect of green *Azolla* (*Azolla pinnata*) feeding on productive performance of crossbred cows. *Trends in Biosciences* 2018, 11(6), 745–747.

Chatterjee, A; Sharma P; Ghosh, MK; Mandal, M; Roy, K. Utilization of *Azolla microphylla* as feed supplement for crossbred cattle. *International Journal of Agriculture Food Science & Technology* 2013, 4, 207–214.

Cheng, W.; Sakai, H.; Matsushima, M.; Yagi, K.; Hasegawa, T. Response of the floating aquatic fern *Azolla* filiculoides to elevated CO_2, temperature, and phosphorus levels. *Hydrobiologia* 2010, 656(1), 5–14.

Cherryl, D. M.; Prasad, R. M.V.; Jayalaxmi, P. A. Study on economics of inclusion of *Azolla pinnata* in swine rations. *International Journal of Veterinary Science and Medicine* 2013, 1, 50–56.

Cherryl, D.M.; Prasad, R.M.V.; Rao, S.J.; Jayalaxmi, P.; Rao, B.E. Effect of inclusion of *Azolla pinnata* on the haematological and carcass characteristics of crossbred large white Yorkshire pigs. *Veterinary World*, 2014, 7(2), 78.

Chichilichi, B.; Mohanty, G. P.; Mishra, S. K.; Pradhan, C. R.; Behura, N. C.; Das, A.; Behera, K. Effect of partial supplementation of sun-dried *Azolla* as a protein source on the immunity and antioxidant status of commercial broilers. *Veterinary World*. 2015, 8(9), 1126.

Cruz-Velásquez, Y.; Kijora, C.; Agudelo-Martínez, V.; Schulz, C. Inclusion of fermented aquatic plants as feed resource for cachama blanca, *Piaractus brachypomus*, fed low-fish meal diets. *Orinoquia*. 2014, 18, 229–236.

Das, M.M.; Agarwal, R.K.; Singh, J.B.; Kumar, S.; Singh, R.P.; Kumar, S. Nutrient intake and utilization in lambs fed azolla microphylla meal as a partial replacement for mustard cake in concentrate mixture. *Indian Journal of Animal Nutrition* 2017, 34, 45–49.

Das, M.; Rahim, F.; Hossain, M. Evaluation of fresh *Azolla pinnata* as a low-cost supplemental feed for Thai Silver Barb *Barbonymus gonionotus*. *Fishes* 2018, 3(1), 15.

Dhumal, M.V., Siddiqui, M.F., Siddiqui, M.B.A. and Avari, P.E. Performance of broilers fed on different levels of *Azolla* meal. *Indian Journal of Poultry Science* 2009, 44(1), 65–68.

Escobin, Jr, R.P. Fresh *Azolla* (*Azolla microphylla* kaulfuss) as partial replacement to palay-snail-shrimp based ration for laying mallard and growth-fattening muscovy ducks. *University* of the *Philippines Los Baños*, College, Laguna, Philippines, 1987.

Fiogbé, E.D.; Micha, J.C.; Van Hove, C. Use of a natural aquatic fern, *Azolla microphylla*, as a main component in food for the omnivorous–phytoplanktonophagous tilapia, *Oreochromis niloticus* L. *Journal of Applied Ichthyology* 2004, 20(6), 517–520.

Gouri, M.D.; Sanganal, J.S.; Gopinath, C.R.; Kalibavi, C. Importance of *Azolla* as a sustainable feed for livestock and poultry—a review. *Agricultural Reviews* 2012, 33(2), 93–103.

Gowda, N.K.S.; Manegar, A; Verma, S.; Valleesha, N.C.; Maya, G.; Pal, D.T.; Suresh, K.P. *Azolla* (*Azolla pinnata*) as a green feed supplement for dairy cattle—an on-farm study. *Animal Nutrition and Feed Technolog* 2015, 15, 283–287.

Indira, D.; Rao, K. S.; Suresh, J.; Naidu, K. V.; Ravi, A. *Azolla* (*Azolla pinnata*) as feed supplement in buffalo calves on growth performance. *Indian Journal of Animal Nutrition* 2009, 26(4), 345–348.

Janiya, J.D.; Moody, K. Use of *Azolla* to suppress weeds in transplanted rice. *Inter. J. Pest Manag.* 1984, 30(1), 1–6.

Kathirvelan, C.; Banupriya, S.; Purushothaman, M.R. *Azolla*—an alternate and sustainable feed for livestock. *International Journal of Environmental Science and Technology* 2015, 4(4), 1153–1157.

Katole, S.B.; Kumar, P.; Patil, R.D. Environmental pollutants and livestock health a review. *Veterinary Research International* 2013, 1(1), 1–3.

Khare, A.; Chatterjee, A.; Mondal, M.; Karunakaran, M.; Ghosh, M.K.; Dutta.; T.K. Effect of supplementing *Azolla microphylla* on feed intake and blood parameters in growing crossbred female calves. *Indian Journal of Animal Nutrition* 2016, 33, 224–227.

Khatun, M.A. Utilization of *Azolla* (*Azolla pinnata*) in the diet of laying hen (M.Sc. thesis). Department of Poultry Science, *Bangladesh Agricultural* University, Mymensingh, *Bangladesh,* 1996.

Khatun, A.; Ali, M.A.; Dingle, J.G. Comparison of the nutritive value for laying hens of diets containing *Azolla* (*Azolla pinnata*) based on formulation using digestible protein and digestible amino acid versus total protein and total amino acid. *Animal Feed Science and Technology* 1999, 81(1–2), 43–56.

Kumar, U.; Nayak, A.K.; Panneerselvam, P.; Kumar, A.; Mohanty, S.; Shahid, M.; Sahoo, A.; Kaviraj, M.; Priya, H.; Jambhulkar, N.N., Dash, P.K. Cyanobiont diversity in six *Azolla* spp. and relation to *Azolla*-nutrient profiling. *Planta*. 2019, 1–13.

Kumar, U.; Nayak, A.K. Azolla germplasms at NRRI, conservation, characterization and utilization. NRRI Research Bulletin No. 19., 2019, p. 68.

Kumar, U.; Nayak, A.K.; Shahid, M.; Gupta, V.V.S.R.; Panneerselvam, P.; Mohanty, S.; Kaviraj, M.; Kumar, A.; Chatterjee, D.; Lal, B.; Gautam, P.; Tripathi, R.; Panda, B.B. Continuous application of inorganic and organic fertilizers over 47 years in paddy soil alters the bacterial community structure and its influence on rice production. *Agriculture, Ecosystems and Environment* 2018a, 262, 65–75.

Kumar, U.; Panneerselvam, P.; Govindasamy, V.; Vithalkumar, L.; Senthilkumar, M.; Banik, A.; Annapurna, K. Long-term aromatic rice cultivation effect on frequency and diversity of diazotrophs in its rhizosphere. *Ecological Engineering* 2017b, 10, 227–236.

Kumar, U.; Shahid, M.; Tripathi, R.; Mohanty, S.; Kumar, A.; Bhattacharya, P.; Lal, B.; Gautam, P.; Raja, R.; Panda, B.B.; Jambhulakar, N.N.; Shukla, A.K.; Nayak, A.K. Variation of functional diversity of soil microbial community in sub-humid tropical rice-rice cropping system under long-term organic and inorganic fertilization. *Ecological Indicators* 2017a, 73, 536–543.

Kumar, M.; Dhuria, R. K.; Jain, D.; Nehra, R.; Sharma, T.; Prajapat, U. K.; Kumar, S.; Siyag, S. S. Effect of inclusion of sun dried *Azolla* (*Azolla pinnata*) at different levels on the growth and performance of broiler chicks. *Journal of Animal Science* 2018b, 8(4), 629–632.

Lebas, F. Reflections on rabbit nutrition with a special emphasis on feed ingredients utilization. In Proceedings of the 8th World Rabbit Congress, Puebla, Mexico, September 7–10, 2004, pp. 686–736.

Lejeune, A.; Peng, J.; Le Boulengé, E.; Larondelle, Y.; Van Hove, C. Carotene content of *Azolla* and its variations during drying and storage treatments. *Animal Feed Science and Technology* 2000, 84(3–4), 295–301.

Leterme, P.; Londono, A.M.; Munoz, J.E.; Súarez, J.; Bedoya, C.A.; Souffrant, W.B.; Buldgen, A. Nutritional value of aquatic ferns (*Azolla filiculoides* and *Salvinia molesta*) in pigs. *Animal Feed Science and Technology* 2009, 149(1–2), 135–148.

Liu, C.W.; Sung, Y.; Chen, B.C.; Lai, H.Y. Effects of nitrogen fertilizers on the growth and nitrate content of lettuce (*Lactuca sativa* L.). *International Journal of Environmental Research and Public Health* 2014, 11(4), 4427–4440.

Mathur, G.N.; Sharma, R.; Choudhary, P.C. Use of *Azolla* (*Azolla pinnata*) as cattle feed supplement. *Journal of Krishi Vigyan* 2013, 2(1), 73–75.

Maurice, D.V.; Jones, J.E.; Dillon, C.R; Weber, J.M. Chemical composition and nutritional value of Brazilian elodea (*Egeria densa*) for the chick. *Poultry Science* 1984, 63(2), 317–323.

Mian, M.H. Azobiofer: a technology of production and use of *Azolla* as biofertiliser for irrigated rice and fish cultivation. Biofertilisers in action. Rural Industries Research and Development Corporation, Canberra, 2002, pp. 45–54.

Minh, D.V. Effect of supplementation, breed, season and location on feed intake and performance of scavenging chickens in Vietnam. *Acta Universitatis Agriculturae Sueciae* 2005, 101.

Ninawe, A.S. Need to promote sewage fed fish culture as ecofriendly production technology. *Dimensions of Environmental Threats* 2003, 166.

Parashuramulu, S.; Swain, P.S.; Nagalakshmi, D. Protein fractionation and in vitro digestibility of *Azolla* in ruminants. *Online Journal of Animal and Feed Research* 2013, 3(3), 129–132.

Pillai, P.K.; Premalatha, S.; Rajamony, S. *Azolla*: a sustainable feed substitute for livestock. *Leisa India*, 2002, 4(1), 15–17.

Querubin, L.J.; Alcantara, P.F.; Luis, E.S.; Princesa, A.O. Chemical composition and feeding value of *Azolla* in broiler ration. *Animal Products Technology* 1987, 13, 46.

Querubin, L.J.; Alcantara, P.F.; Luis, E.S.; Princesa, A.O. Chemical composition and feeding value of *Azolla* in broiler ration. *Philippines Journal of Veterinary and Animal Sciences* 1986, 12, 65.

Raja, W.; Rathaur, P.; John, S.A.; Ramteke, P.W. Azolla-Anabaena association and its significance in supportable agriculture. *Hacettepe Journal of Biology and Chemistry* 2012, 40(1), 1–6.

Ripley, B. S.; Kiguli, L. N.; Barker, N. P.; Grobbelaar, J. U. *Azolla filiculoides* as a biofertiliser of wheat under dry-land soil conditions. *South African Journal of Botany* 2003, 69(3), 295–300.

Roy, D.; Pakhira, M.; Bera, S. A review on biology, cultivation and utilization of *Azolla*. *Advances in Life Sciences* 2016a, 5, 11–15.

Roy, D.; Kumar, V.; Kumar, M.; Sirohi, R.; Singh, Y.; Singh, J. K. Effect of feeding *Azolla pinnata* on growth performance, feed intake, nutrient digestibility and blood biochemicals of *Hariana heifers* fed on roughage based diet. *Indian Journal of Dairy Science* 2016b, 69(2), 190–196.

Rudsamee, W. *Classification and nitrogen fixation efficiency analysis of Azolla species in rice fields* (MSc. Biotech. Dissertation). Suranaree University of Technology, Nakhon Ratchasima Province, Thailand, 2008.

Setiawati, M.R.; Damayani, M.; Herdiyantoro, D.; Suryatmana, P.; Anggraini, D.; Khumairah, F.H. The application dosage of *Azolla pinnata* in fresh and powder form as organic fertilizer on soil chemical properties, growth and yield of rice plant. *AIP Conference Proceedings*, 2018, 1927(1), 030017.

Sharma, S.; Padbhushan, R.; Kumar, U. Integrated nutrient management in rice–wheat cropping system: an evidence on sustainability in Indian subcontinent through meta-analysis. *Agronomy* 2019, 9(2), 71.

Shiomi, N.; Kitoh, S. Culture of *Azolla* in a pond, nutrient composition, and use as fish feed. *Soil Science and Plant Nutrition* 2001, 47(1), 27–34.

Silver, W.S.; Schröder, E.C. Eds. Practical application of *Azolla* for rice production. Proceedings of an International Workshop, Mayaguez, Puerto Rico, Springer Science & Business Media, November 17–19, 1982.

Singh, A. L.; Singh, P. K. Comparative study on *Azolla* and blue-green algae dual culture with rice. *Israel Journal of Botany* 1987, 36(2), 53–61.

Singh, A.L.; Singh, P.K. Comparative studies on different methods of *Azolla* utilization in rice culture. *Journal of Agricultural Science* 1986, 107(2), 273–278.

Singh, Y.P. *Feasibility, nutritive value and economics of Azolla anabaena as an animal feed* (M.Sc. thesis). G.B. Pant University, Pantnagar, Uttar Pradesh, India, 1980.

Sithara, K.; Kamalaveni, K. Formulation of low-cost feed using *Azolla* as a protein supplement and its influence on feed utilization in fishes. *Current Biotica*, 2008, 2(2), 212–219.

Spotte, S. *Marine aquarium keeping*. John Wiley & Sons, New York, 1993.

Sreemannaryana, D.; Ramachandraiah, K.; Sudarshan, K.M.; Romanaiah, N.V.; Ramaprasad, J. Utilization of *Azolla* as a rabbit feed. *Indian Veterinary Journal* 1993, 70, 285–286.

Srinivas Kumar, D.; Prasad, R.M.V.; Raja Kishore, K.; Raghava Rao, E. Effect of Azolla (*Azolla pinnata*) based concentrate mixture on nutrient utilization in buffalo bulls. *Indian Journal of Animal Research* 2012, 46(3), 268–271.

Subudhi, B.P.R.; Singh, P.K. Nutritive value of the water fern *Azolla pinnata* for chicks. *Poultry Science*, 1978, 57(2), 378–380.

Sudaryono, A. Use of *Azolla* (*Azolla pinnata*) meal as a substitute for defatted soybean meal in diets of juvenile black tiger shrimp (*Penaeus monodon*). *Journal of Coastal Development* 2011, 9(3), 145–154.

Sujatha, T.; Kundu, A.; Jeyakumar, S.; Kundu, M. S. *Azolla* supplementation: feed cost benefit in duck ration in Andaman Islands. *Tamil Nadu Journal of Veterinary and Animal Sciences* 2013, 9(2), 130–136.

Suratno, S.; Asyim, M. The effect of organic composes of *Azolla* and nitrogen fertilizer towards growth, production and quality spinach (*Amaranthus* sp.). In Proceedings of the 1st International Conference on Food and Agriculture, Bali, Indonesia, October 20–21, 2018, Vol. 2, pp. 440–444.

Tamang, Y.; Samanta, G. Feeding value of *Azolla* (*Azolla pinnata*) an aquatic fern in Black Bengal goats. *Indian Journal of Animal Sciences* 1993, 63 (2), 188–191.

Tamany, Y.; Samanta, G.; Chakraborty, N.; Mondal, L. Nutritive value of *Azolla* (*Azolla pinnata*) and its potentiality of feeding in goats. *Environment and Ecology* 1992, 10, 755–756.

Tawasoli, M.J.; Kahate, P.A.; Shelke, R.R.; Chavan, S.D.; Shegokar S.R. Effect of feeding *Azolla* (*Azolla pinnata*) meal on feed intake and feed conversion efficiency of Vanraja poultry birds. Int *Journal of Agricultural Science* 2018, 10(14), 6733–6736.

Titus, A.; Pereira, G.N. *Azolla* as a biofertilizer in coffee plantations. 2006. https://ecofriend-lycoffee.org/azolla-as-a-biofertilizer-in-coffee-plantations/ (accessed February 1, 2006).

Trivedi, M.M.; Parnerkar, S.; Patel, A.M. Effect of feeding non-conventional creep mixtures on growth performance of pre-weaned lambs. *International Journal of Agriculture and Biology* 2005, 7(2), 175–179.

Wagner, G.M. *Azolla*: a review of its biology and utilization. *Botanical Review* 1997, 63(1), 1–26.

Yadav, R.K.; Abraham, G.; Singh, Y.V.; Singh, P.K. Advancements in the utilization of *Azolla-Anabaena* system in relation to sustainable agricultural practices. Proceedings of the Indian National Science Academy 2014, 80(2), 301–316.

Yao, Y.; Zhang, M.; Tian Y.; Zhao, M.; Zeng, K.; Zhang, B.; Zhao, M.; Yin, B. *Azolla* biofertilizer for improving low nitrogen use efficiency in an intensive rice cropping system. *Field Crops Research* 2018a, 216, 158–164.

Yao, Y.; Zhang, M.; Tian, Y.; Zhao, M.; Zhang, B.; Zeng, K.; Zhao, M.; Yin, B. Urea deep placement in combination with *Azolla* for reducing nitrogen loss and improving fertilizer nitrogen recovery in rice field. *Field Crops Research* 2018b, 218, 141–149.

Zahran, H. H. Rhizobium-legume symbiosis and nitrogen fixation under severe conditions and in an arid climate. *Microbiology and Molecular Biology Reviews* 1999, 63(4), 968–989.

PART IV

Utilization of Bioresources in the Environment

CHAPTER 18

Bioremediation of Arsenic-Contaminated Environment Using Bacteria: An Update

RANJAN KUMAR MOHAPATRA[1*], RITESH PATTNAIK[1],
BIRENDRA KUMAR BINDHANI[1], HRUDAYANATH THATOI[3], and
PANKAJ KUMAR PARHI[2*]

[1]*School of Biotechnology, KIIT Deemed to be University, Bhubaneswar, Odisha 751024, India*

[2]*Department of Chemistry, Fakir Mohan University, Balasore, Odisha 756089, India*

[3]*Department of Biotechnology, North Orissa University, Baripada, Odisha 757003, India*

Corresponding author. E-mail: parhipankaj@gmail.com; mranjankumar.biotech@gmail.com.

ABSTRACT

Recently, environmental pollution due to arsenic contamination has become a global concern. Rapid industrialization and urbanization activities along with technological advancement are continuously increasing the heavy metal (including arsenic) load in the environment. Arsenic contamination in soil, water, air, and sediment environment from various anthropogenic and geogenic sources is a major environmental issue in many regions of the world including India owing to its hazardous nature in the biological ecosystem. The content of arsenic(III/V) causes severe human health and mortality problems to millions of people worldwide annually when present at an eminent concentration (>0.05 mg/L) level. Arsenic is generally nonbiodegradable, but it can be transformed to its less toxic form through oxidation–reduction or methylation reaction and can be immobilized through sorption, complexation, and precipitation processes. Bioremediation of arsenic using bacteria is now gaining increasing global attention due to its greater advantages over conventional physicochemical methods. During arsenic bioremediation, bacterial biomass can interact with arsenic metal ions by passive and/or active bioremediation mechanisms including extracellular adsorption/entrapment/precipitation, intracellular absorption, methylation, and oxidation–reduction process. Indigenous arsenic-resistant bacteria (heterotrophs and chemo-litho-autotrophs) present in the arsenic-contaminated area developed resistance against arsenic toxicity, which were substantially used for arsenic detoxification processes. This chapter provides insight into recent information about the global arsenic pollution, arsenic toxicity, and successful implementation of arsenic-resistant bacteria isolated from both saline and nonsaline sources in an environmentally safe and cost-effective manner. It also gives a clear understanding about the possible mechanisms involved in the bacterial arsenic bioremediation process.

18.1 INTRODUCTION

In the present century, generation of huge amounts of inorganic (e.g., toxic heavy metals) and organic (e.g., petroleum hydrocarbons, phenolic compounds, fertilizers, and pesticides) contaminants due to intense industrialization, mining, and urbanization activities has shot up the environmental pollution (Akhtar et al., 2013). Among the toxic heavy metals, environmental pollution caused by arsenic contamination is now considered to be vital due to its severe toxicity and potential health risk. In Southeast Asia region, basically in India (West Bengal, Chhattisgarh, Bihar, Telangana, and Uttar Pradesh) and Bangladesh, humans and animals are regularly exposed to elevated levels of arsenic pollution through contaminated drinking water and food crops (Dey et al. 2016; Satyapal et al. 2018). arsenic (As) is a metalloid that comes under group "V" element of the periodic table and thus considered as a heavy metal (Satyapal et al., 2016). In nature, arsenic is found in four different oxidation states, that is, +5, +3, 0, and -3. Among these, pentavalent (As^{+5}; arsenate) and trivalent (As^{+3}) are the most commonly existing inorganic forms of arsenic (Satyapal et al., 2018). Alternatively, some organic arsenicals, which are derived from pesticides, herbicides, and preservatives, are also present in the environment (Satyapal et al., 2018). Both pentavalent (+5) and trivalent (+3) oxidation forms of arsenic are poisonous; however, As(III) is found to be 1000-fold more toxic than As(V) due to its high mobility (Dey et al., 2016; Satyapal et al., 2018). Among both inorganic and organic forms of arsenic, some commonly available inorganic trivalent arsenic compounds such as sodium arsenite, arsenic trioxide, and arsenic trichloride are more poisonous causing multiple hazards to the living organisms by producing the reactive oxygen species (ROS) and damaging the DNA (Adeniji, 2004). Excessive arsenic consumption beyond the permissible limit (0.01 mg/L) can create multiple acute and chronic health disorders such as skin cancer, gastrointestinal problems, cardiovascular failure, anemia, weakness, and lethargy (Dey et al., 2016; Mohapatra et al., 2017a). Heavy metals including arsenic cannot be easily degraded by the natural phenomenon and therefore can persist in the environment for a long time, resulting in bioaccumulations and biomagnifications (Chaudhary et al., 2014; Voica et al., 2016). Arsenic remediation can be achieved by either transforming the more toxic forms to relatively less toxic form or immobilizing it through applying various physicochemical and biological methods (Aryal and Liakopoulou-Kyriakides, 2015). Physicochemical remediation processes are not expedient due to high operational cost, high energy consumption, low efficiency, and generation of secondary pollutants (Voica et al., 2016), whereas bioremediation of arsenic using arsenic-tolerant bacteria has been widely accepted as the most potential alternative for cleanup of arsenic pollution in an efficient, economic, and environment-friendly way (Mohapatra et al., 2017a; Voica et al., 2016). Bacteria residing in an arsenic-contaminated area can develop resistance toward arsenic toxicity by applying a variety of inherent mechanisms such as precipitation, volatilization, cell surface sorption, intracellular accumulation, and ATP-mediated efflux system (Ahalya et al., 2003; Das et al., 2008; Naik et al., 2012). Hence, these native bacteria have the capacity to remove arsenic ions, and their removal efficiency can be enhanced by providing suitable abiotic factors such as, pH, temperature, salinity, and nutrients (Mohapatra et al., 2019; 2017a).

This chapter deliberates the recent updates of arsenic bioremediation using arsenic-resistant bacteria, factors affecting the bioremediation process, and the possible mechanisms associated with the arsenic decontamination process.

18.2 SOURCES OF ARSENIC CONTAMINATION

The main sources of arsenic contamination in the environment are from both natural and anthropogenic process that resulted wide distribution of arsenic compound in soil, water, air and crops (Satyapal et al., 2016) (Figure 18.1). Origination of arsenic in the environment (soil, water, sediment, and air) takes place from both natural (geogenic) and anthropogenic sources (Satyapal et al., 2018). Natural geogenic sources such as volcanic eruption, weathering, fossil fuels, minerals, arsenic-bearing parent rock/sedimentary rock, etc. are the key origin points of arsenic flow to the environment (Mohapatra et al., 2017a), whereas various anthropogenic activities like agriculture, mining, smelting, refining, electroplating, coal combustion, painting, and chemical manufacturing have released huge amount of arsenic to the environment (Akhtar et al., 2013; Dey et al., 2016; Mohapatra et al., 2017a). Manufacturing practices including agricultural chemicals (e.g., pesticides, herbicides, fertilizers, wood preservatives, etc.), dying materials, and medical products are the major sources of arsenic contamination (Mohapatra et al., 2017a; Vishnoi and Singh, 2014).

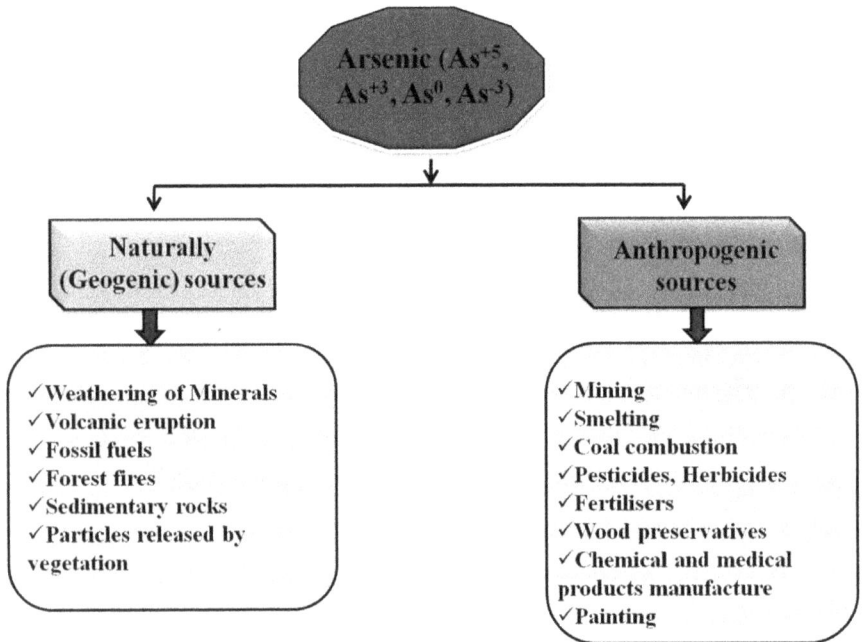

FIGURE 18.1   Sources of arsenic contamination in the environment.

18.3   ARSENIC TOXICITY AND POTENTIAL HEALTH HAZARDS

Humans along with animals are exposed to the arsenic when consumed arsenic contaminated water for drinking and taking food cropping in an arsenic contaminated field or irrigated with arsenic contaminated water (Dey et al., 2016; Satyapal et al., 2018). In 1997, the maximum permissible limit for arsenic contamination in drinking water was 0.01 mg/L as per World

Health Organization (WHO) (Aksornchu et al., 2008; WHO, 2011). But some developing countries like Indian and Bangladesh are still trying to stay with the earlier permitted limit, that is, 0.05 mg/L (Dey et al., 2017; Khan et al., 2000). The acute symptoms of arsenic poisoning, as a consequence of consuming arsenic above the permissible limit (i.e., 0.05 mg/L), are skin discoloration, skin thickening, and ultimately, skin cancer (Banerjee et al. 2011; Dey et al., 2017; Khan et al. 2000). Arsenic accumulation in human body, causes several detrimental health effects by affecting digestive system, respiratory system, nervous system, and circulatory system, including skin and muscle (Dey et al., 2016; Satyapal et al., 2016, 2018). Arsenic consumption beyond the permissible limit (0.01 mg/L) can generate multiple acute and chronic health disorders (Dey et al., 2016). Trivalent inorganic form of arsenic compounds such as arsenic trioxide, sodium arsenite and arsenic trichloride can cause neurotoxicity of both peripheral and central nervous systems, including sensory changes, muscle tenderness, followed by progressive weakness from the proximal to distal muscle groups (Adenjii, 2004; Klaassen, 2019). Similarly, pentavalent inorganic forms of arsenic such as arsenic pentoxide, arsenic acid, and arsenates can also affects enzyme activity of humans (Klaassen, 2019). Trivalent or pentavalent forms of organoarsenic compound, especially methylated form, cause biomethylation and potential health hazards to humans, animals, and other living organisms by entering the food chain (through soil, food, water, air) (Mateos et al., 2006; Adenjii, 2004).

Excessive exposure of arsenic even in low concentration creates various acute health problems such as skin itching, hyperkeratosis, fever, melanosis, gastrointestinal disorder (e.g., nausea, anorexia, stomach irritation, abdominal pain, enlarged liver and spleen), weight loss, loss of appetite, weakness, anemia, lethargy, granulocytopenia, cardiac arrhythmia, and cardiovascular failure (Ahsan et al., 2012; Cavalca et al., 2013; Dey et al., 2016; Mohapatra et al., 2017a; Taran et al., 2013). Long-term exposure of arsenic can cause chronic respiratory disorders, lung irritation, immune-suppression, arsenicosis, sensory loss, changes in skin epithelium, and cancer due to DNA damage (Adenjii, 2004; Mohapatra et al., 2017a). Intake of high arsenic dose can cause infertility, fatal health problems, miscarriages in women, hypertension, type-II diabetes, brain damage, and cardiovascular problems including coronary artery disease, peripheral vascular disease, and atherosclerosis (Mohapatra et al., 2017a). Arsenic can be entered into the cells of the living organisms by following the similar mechanism as phosphate transported into the cell due to their structural analogy resulting in disruption of phosphorylation reaction (Shrestha et al., 2008). As(III) can also cease the protein function by binding to sulfhydryl groups of cysteine residues (Cavalca et al., 2013).

The cellular accumulation of arsenic can hinder succinic dehydrogenase activity and deactivate oxidative phosphorylation, which can result in the inhibition of mitochondrial ATPase activity and NAD reduction (Klaassen, 2019). Hence, arsenic exposure can inhibit mitochondrial respiration, leading to a decrease in cellular ATP synthesis (Klaassen, 2019).

18.4 RECENT SCENARIO OF ARSENIC CONTAMINATION

Among all toxic metalloids, arsenic is the most prevalent poisonous and toxic metalloid that contaminates the environment, especially in groundwater system. Arsenic poisoning due to consumption of pumped-out groundwater through a number of tube wells by millions of people for drinking purpose was reported as the greatest mass poisoning in human history by WHO (Dey et al., 2017; Vaughan, 2006). Recently, various human anthropogenic activities have upshot arsenic contamination in drinking water all over the world, especially in Southeast Asia region. According to a report, it was estimated that more than 6 million people in West Bengal state (India) (Anyanwu and Ugwu, 2010; Dey et al., 2016) and 46 million people in Bangladesh are at risk from drinking water due to arsenic poisoning (Bachate et al., 2009; Dey et al., 2016). In some regions in India (e.g., West Bengal, Chhattisgarh, Bihar, Telangana [Hyderabad], and Uttar Pradesh), the arsenic contamination in groundwater is now in alarming condition. In West Bengal, 12 districts are found to be having arsenic contamination above the WHO permissible limit and among them, 6 districts such as Murshidabad, Burdwan, Nadia, Hoogly, 24 parganas (North) and 24 parganas (South) are severely affected (Chakraborty et al., 2009). About 44.4% of the total population and 38.4% of the total area of West Bengal are affected by the arsenic poisoning. The average arsenic concentration (i.e., 0.02 mg/L) in groundwater is found to be higher than the prescribed standard limit, with a wide range of variation, that is, 0.05 to 3.7 mg/L. The presence of elevated levels of arsenic in groundwater in West Bengal is mainly due to the geochemical reactions comprising oxidation of arsenic-rich pyrites or reductive dissolution of arsenic-rich iron hydroxides, which takes place during the ground water abstraction.

In Chhattisgarh, some parts between Dongargaon and Mohla (~500 sq. km.) in Rajnandgaon district are affected by the arsenic contamination with varying concentration range. In Hyderabad, the industrial area of Patancheru and around Khazipally is mostly affected by the arsenic contamination due to the effluent disposal by more than 40 agrochemicals and pesticide-manufacturing industries present in that vicinity (Prasad, 2011). The arsenic concentration in groundwater in Gangetic plain of Bihar is found to be higher than the permissible limit, which may be due to the natural release of arsenic from the Holocene sediments containing clay and slit (Satyapal et al., 2018).

18.5 ARSENIC REMEDIATION APPROACHES

Due to high degree of toxicity and solubility, it is very much essential to eradicate arsenic contamination for maintaining social wellness. There are varieties of conventional methods that have been used for the removal of arsenic from the contaminated aqueous system, such as membrane filtration, precipitation, reverse osmosis, coagulation, oxidation–reduction, and adsorption (Bahar et al., 2012; Dey et al., 2016, 2017). These physicochemical methods have several disadvantages like high operational cost, generation of harmful secondary waste, working limitations, and low efficiency (Das and Dash, 2014; Mohapatra et al., 2017a, 2017b). In contrast to the physicochemical remediation process, microbial bioremediation is proven as a potential alternative for eco-friendly approach for cleanup of the arsenic contaminated area because of its low operation cost, less energy requirements, nonproduction of toxic by-products, and high removal efficiency of arsenic concentration (Clausen, 2000; Das and Dash, 2014; Dey et al., 2017; Voica et al., 2016). Microorganisms having unique metal-resistant genotypic characteristic features toward arsenic tolerance can potentially be used for arsenic bioremediation exercise (Das and Dash, 2014). Complex structure of the cell membrane and distinctive metabolic activity of the microorganisms can made them a potential agent for diminishing the arsenic toxicity through transforming it to less toxic form during the cellular metabolic process and/or remove them by cellular biosorption/accumulation phenomenon (Das and Dash, 2014; Dey et al., 2016; Mohapatra et al., 2017a).

18.6 ARSENIC BIOREMEDIATION USING BACTERIA

Bioremediation process is a natural incident in which biological organisms (microorganisms or plants) play an important role in either removal of toxic metal ions through biosorption or transformation of toxic metals to their less toxic form. Among all forms of bioremediation, microbial bioremediation using metabolically diverse bacteria is now emerging as an efficient metal removal technique for detoxification of the arsenic contaminated environment. Bacteria possess multiple bioremediation potentials and hence are confirmed to be beneficial agents for cleanup of the toxic pollutants from both environmental and economic points of view (Das and Dash. 2014; Gupta et al., 2016). Nowadays, the goal of potential bioremediation of toxic pollutants (including metals/metalloids) is achieved using native bacteria isolated from contaminated sites and stimulating their detoxification ability by providing them optimum levels of proper nutrients and other essential chemicals needed for their metabolism (Das and Dash, 2014). Nevertheless, studies regarding the use of suitable non-native microbes as well as genetically engineered microbes suited for bioremediation of specific contaminated sites are still being conducted (Das and Das, 2014).

18.6.1 ARSENIC BIOREMEDIATION BY SALINE BACTERIA

Most of the polluted environments are characterized by elevated or low temperature, alkaline or acidic pH, high pressure, and high salt concentration. Salt-tolerant bacteria, mainly native marine bacteria that are suitably adapted to most adverse environment conditions (i.e., varying temperature, pH, salinity, conductance, water currents), naturally possess complex characteristic features of adaptation. The bacterial population isolated from the marine sources are believed to be better utilized in bioremediation of toxic metals/metalloids under saline condition. Application of indigenous marine bacteria for in situ bioremediation of arsenic without any genetic manipulation in any adverse conditions is the key advantage. Salt-tolerant arsenic resistant bacteria can be isolated from water, soil, sediments of the marine ecosystem, mangroves associated with the marine habitats, normal flora of marine organisms, deep-sea hydrothermal vents, and also saline industrial effluents (Dash et al., 2013; Mohapatra et al., 2017a; Naik et al., 2012). Marine sediments are more often contaminated with various toxic heavy metals (Mohapatra and Panda, 2017) and, hence, benthic bacteria attached to sediment particles show more resistance toward toxic metals (Mohapatra et al., 2016; Naik et al. 2012). Deep-sea hydrothermal vent liquids also contain various toxic metals and are likely to be a storehouse of native bacteria adapted to toxic metals/metalloids (Devika et al., 2013; Naik et al., 2012). Coral reef and marine fauna are also frequently exposed to the toxic compounds from industrial and mine effluents discharged into the marine ecosystem through river and, hence, become a source of finding arsenic-resistant bacteria (Naik et al., 2012). Various potential salt-tolerant bacteria such as *Bacillus* sp. and *Aneurinibacillus* sp. (Dey et al., 2016), *Halobacterium* sp. and *Natronobacterium* sp. (Williams et al., 2013), *Halorcula* sp. (Taran et al., 2013), *Halomonas marina, Alteromonas malleoli, Marinomonas communis*, and *Vibrio alginolyticus* (Takeuchi et al., 2007) were isolated from different arsenic contaminated saline environmental sources, as reported by various researchers, for bioremediation/detoxification of the toxic As(III/V), which are presented in Table 18.1.

18.6.2   ARSENIC BIOREMEDIATION BY NONSALINE BACTERIA

A variety of arsenic resistant bacteria isolated from nonsaline environment such as *Pseudomonas* sp. (Satyapal et al., 2018), *Bacillus* sp. (Biswas and Sarkar, 2019; Dey et al., 2016; Roychowdhury et al., 2018), *Paenibacillus* sp. and *Escherichia coli* (Vishnoi and Singh, 2014), *Brevibacillus* sp. (Banerjee et al., 2013), *Micrococcus* sp. (Paul et al., 2018), *Actinobacteria* sp., *Microbacterium* sp., *Rhizobium* sp. (Paul et al., 2014), *Haemophilus* sp. (Ike et al., 2008), *Kytococcus* sp., *Staphylococcus* sp. (Roychowdhury et al., 2018), *Aneurinibacillus* sp. (Dey et al., 2016), and *Corynebacterium* sp. (Ghodsi et al., 2011) have been successfully applied for arsenic bioremediation, as reported by various researchers, which are given in Table 18.1.

18.7   CLASSIFICATION OF THE BACTERIAL ARSENIC BIOREMEDIATION PROCESS

Arsenic bioremediation using bacteria can be broadly classified into two main types, one is on the basis of bacterial metabolism and the other is cellular location involved in arsenic removal.

18.7.1   ON THE BASIS OF BACTERIAL CELLULAR METABOLISM

On the basis of cellular metabolism, bacterial bioremediation can be further classified as metabolism-dependent (active method)- and metabolism-independent (passive method) process.

Metabolism-dependent bioremediation is carried out only by viable (live) bacterial cells in which the arsenic ions are transported across the cell membrane during cellular metabolism, resulting in intracellular accumulation (Ahalya et al., 2003). During metabolism, independent bioremediation, arsenic removal can be achieved through the physicochemical interactions between the arsenic ions and the bacterial cell wall ligands. Several cell wall components like lipids, proteins, and polysaccharides of the bacterial biomass containing many functional groups like amino, carboxyl, sulfate, and phosphate help in arsenic

TABLE 18.1 Arsenic(III/V) Bioremediation by Various Native Arsenic Resistant Bacteria Isolated from Saline and Nonsaline Sources

Name of As-resistant Bacteria	Isolation Source	Tolerance	pH/NaCl	References
Arsenic Bioremediation by Saline Bacteria				
Bacillus sp. KM02	Ground water alkaline nature, Burdwan, WB, India	As(III): 500 mg/L, As(V): 4500 mg/L	NaCl: 8%–10% pH:7.23	Dey et al., 2016
Aneurinibacillus aneurinilyticus				
Halobacterium saccharovorum, Halobacterium salinarium, Natronobacterium gregoryi	Solar salt pan, Rajakkamangalam coast, India	As: 0.001 mM	–	Williams et al., 2013
Halorcula sp. IRU1	Hypersaline Urmia lake, Iran	As(III): 90 mg/L	NaCl: 25% pH: 8.0	Taran et al. 2013
Halomonas marina IAM 14107	NCIMB, Aberdeen, Scotland	As: 730 mg/L	NaCl: 25% pH: –	Takeuchi et al., 2007
Alteromonas macleodii IAM 12914	IAM, Tokyo, Japan	As: 310 mg/L		
Marinomonas communis IAM12914		As: 210 mg/L		
Vibrio alginolyticus NCIMB		As: 510 mg/L		
Arsenic Bioremediation by Nonsaline Bacteria				
Pseudomonas extremorientalis	Gangtic plane, Bihar	As(III): 13–15 mM As(V): 200 mM	pH: 7.0	Satyapal et al., 2018
Micrococcus sp. KUMAs15	As-contaminated fields, Nadia, West Bengal	–	–	Paul et al., 2018
Bacillus sp. BAS-1	Shallow aquafier, Bhojpur, Bihar	As(III): 70 mM As(V): 1000 mM	–	Biswas and Sarkar, 2019
Haemophillus, Micrococcus, Bacillus sp.	Soil sample	As(III): 1500 mg/L	pH 7.0–10.0	Ike et al., 2008
Actonobacteria BAS123i, Microbacterium CAS905i, Pseudomonas CAS912i, Rhizobium CAS934i	As-contaminated ground water, West Bengal	As(III): 10 mM As(V): 450 mM	pH: 5.0–10.0	Paul et al., 2014

TABLE 18.1 *(Continued)*

	Arsenic Bioremediation by Saline Bacteria			
Name of As-resistant Bacteria	Isolation Source	Tolerance	pH/NaCl	References
Bacillus sp., *Micrococcus* sp., *Kytococcus* sp., *Staphylococcus* sp	Fly ash pond	As(III): 30 mM As(V): 72 mM	pH: 7.0–8.0 NaCl: 3%–5%	Roychowdhury et al., 2018
Bacillus selenatarsenatis SF-1	Effluent drain sediment	—	pH: 8.0	Yamamura et al., 2008
Bacillus sp. KM02, *Aneurinibacillus aneurinilyticus* BS-1	Arsenic affected ground water, Burdwan, West Bengal	As(III): 550 mg/L As(V): 4500 mg/L	pH: 7.0 NaCl: 4–6%	Dey et al., 2016
Bacillus macerans, Bacillus megaterimand, Corynebacterium vitarumen	Arsenic contaminated soil, Isfahan, Iran	As(III): 128mM	pH: 7.0	Ghodsi et al., 2011
Bacillus cereus W2	Contaminated soil, Miyazaki prefecture, Japan	As(III): 50 mg/L	pH: 7.0	Miyatake and Hayashi, 2011
Bacillus subtilis, Paenibacillus macerans, Bacillus megaterium, Bacillus pumilius, Escherichia coli	Arsenic contaminated soil, Lakhmipurkheri and Unnao, Uttarpradesh	As(III): 40 m M	—	Vishnoi and Sing, 2014
Brevibacillus brevis	Contaminated soil, Nadia, West Bengal	As(III): 500 mg/L As(V): 1000 mg/L	—	Banerjee et al., 2013

components. The binding process involves many rapid reversible mechanisms like physical adsorption, ion exchange, complexation, and precipitation (Mohapatra et al., 2017a). Physical adsorption takes place by the effect of van der Waals' forces of attraction and electrostatic forces of interaction between the bacterial cell wall components and the arsenic ions in aqueous medium (Ahalya et al., 2003). In case of ion-exchange mechanism, the exchange of metal ions including arsenic (As^{5+}, As^{3+}) takes place with the naturally occurring cellular ions like K^+, Na^+, Ca^{2+}, and Mg^{2+} present in the cell wall matrix (Ahalya et al., 2003; Mohapatra et al., 2017a). Bioremediation of arsenic can also be achieved through formation of complexes during the interaction between arsenic and the active functional sites (Mohapatra et al., 2017a). Bacteria can also produce complexing/chelating agents like oxalic acid, citric acid, fumaric acid, gluonic acid, lactic acid, and malic acid,. which may help in chelation and formation of metal/metalloid complexes. Precipitation may also occur either dependent or independent of cellular metabolism. As a defense mechanism, the bacterial cell reacts with the toxic components in the vicinity and produces compounds that help in precipitation.

In cellular metabolism-independent mechanism, precipitation occurs as a consequence of the chemical interaction between the toxic arsenic components and the bacteria (Ahalya et al., 2003). Bacteria also possess a variety of adaption, survival, and bioremediation mechanisms like precipitation (as phosphates, sulfides, and carbonates), volatilization (via methylation/ethylation/reduction), ATP-mediated effluxing, intracellular bioaccumulation (mediated by metallothionein proteins), cell surface adsorption, and surface sequestration in extracellular polymeric substances (EPSs) (Mohapatra et al., 2017a) for highly contaminated environments.

18.7.2   ON THE BASIS OF BACTERIAL CELLULAR LOCATION

On the basis of the cellular location of the bacteria which actively take part in the bioremediation process, the bioremediation can be categorized as extracellular accumulation/precipitation, cell surface sorption/precipitation, and intracellular accumulation. In active bioremediation (with live bacteria), bio-sorption/-accumulation of toxic compounds like in various cellular regions is a multi-step process. The pollutant first adsorbs to the cell surface by interacting with the cell surface ligands, resulting in extracellular accumulation/precipitation. Then the adsorbed pollutant (metal/metalloid ions) transports across the cell membrane and enters into the cytoplasm, resulting in intracellular accumulation (Das et al., 2008).

18.8 MECHANISM OF ARSENIC BIOREMEDIATION USING BACTERIA

Bioremediation of toxic metalloids like arsenic using bacteria mainly comprises of two major techniques such as biotransformation and biosorption (Karigar and Rao, 2011). The bacteria used in arsenic bioremediation process having various arsenic-processing features such as uptake, reflux, surface adsorption, intracellular assimilation, immobilization, complexion, precipitation, volatilization, ion-exchange, and oxidation–reduction, which are described in Figure 18.2 (Mohapatra et al., 2017a; Stelting et al., 2012).

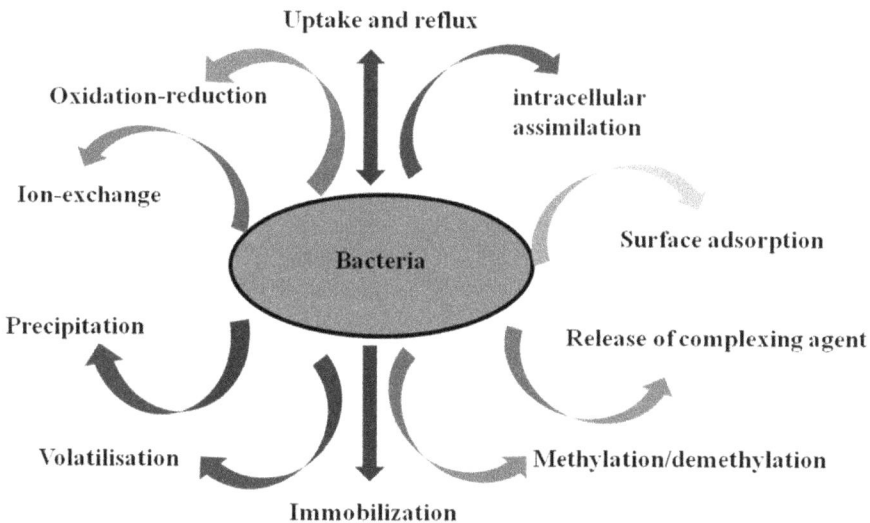

FIGURE 18.2 Different mechanisms associated with bacterial arsenic bioremediation process.

18.8.1 MECHANISM OF ARSENIC BIOTRANSFORMATION USING BACTERIA

In the bacterial biotransformation process, bacteria can decrease the toxicity of the arsenic compounds using them in their metabolic processes as energy sources, and transform them through energy-yielding oxidation–reduction reactions by utilizing oxygen, carbon dioxide (CO_2), nitrates, sulfate acetate, lactate, and glucose as electron acceptors/donors (Akhtar et al., 2013; Dey et al., 2017). It was reported that bacteria having heavy metal-resistant capacity can chemically transform toxic metals/metalloids through their common cellular metabolism comprising a variety of pathways like oxidation–reduction, methylation–demethylation, and precipitation (Dey et al., 2017; Silver and Phung, 2005). Bacteria can exploit arsenic in their metabolic process either as an electron acceptor for anaerobic respiration or as an electron donor to support chemoautotrophic fixation of CO_2 into cell carbon (Akhtar et al., 2013; Stolz and Oremland, 1999). Dissimilatory arsenate-reducing bacteria use arsenate as an electron acceptor and reduce it to arsenite. Chemoautotrophic arsenite oxidizer bacteria use CO_2 as a carbon source and arsenite as an electron source, oxidizing it to arsenate for energy, whereas heterotrophic arsenite oxidizers use oxygen as an electron acceptor for oxidize arsenite to arsenate (Akhtar et al., 2013). For detoxification of arsenic, transformation of As(III) to its less toxic form As(V) through bacterial oxidation process is gaining interest as a cost-effective and an environmentally safe method than the conventional oxidation process using oxygen, hydrogen peroxide, chlorine, and ozone (Dey et al., 2016). Bacteria have developed different arsenic detoxification mechanisms like arsenite methylation arsenite oxidation to transform more toxic arsenite to less toxic arsenate (Dey et al., 2016; Qin et al., 2006). A special type of enzyme is present in the protoplasm of arsenic oxidizing bacteria called "arsenic oxidase," which helps in oxidation of arsenite to form (Andreoni et al., 2012; Dey et al., 2016).

Arsenic tolerance in bacteria is mediated by the gene products of the *ars* operon containing *arsA*, *arsB*, *arsC*, *arsD*, and *arsR*. *arsA* and *arsB* encode for an arsenic efflux pump (i.e., *ArsAB* ATPase), *arsC* encodes for an arsenate reductase. *arsD* acts as arsenic chaperone that transfers arsenite from the glutathione-bound complexes to *ArsA* subunit of the *ArsAB* complex and activates it, and *arsR* encodes for a transcriptional regulator (Arsène-Ploetze et al., 2010; Satyapal et al., 2016). Several genes like *aoxR*, *aoxB*, and *aoxC* are also involved arsenite oxidation (Satyapal et al., 2018).

18.8.2 *MECHANISM OF ARSENIC BIOSORPTION USING BACTERIA*

Bacterial biosorption process involves sorption of the metal/metalloid ions (sorbate) present in an aqueous solution on to the surface of a solid bacterial biomass (biosorbent). The biosorbent biomass attracts the contaminant ions for binding due to their higher affinity toward charged ions which mainly depends upon the chemical constituents of the bacterial cell wall (Mohapatra et al., 2017a). The degree of biosorption affinity of the bacterial biomass is evaluated according to the distribution of ions between the solid biosorbent and liquid (aqueous) phases, and the process continues till equilibrium is established between the amount of contaminant-bound biosorbent and free ions in the solution (Das et al., 2008). Transition metal ions having free d-orbital can coordinate three to eight ligands and often exhibit an octahedral coordination (Aryal and Liakopoulou-Kyriakides, 2013). Thus, the three-dimensional network structure of surface peptidoglycan layer present in bacterial cell wall can be suitable for the binding of these transitional metals like arsenic (Fomina and Gadd, 2014). EPSs of the bacterial cell play a key role in the binding and adsorption of the toxic arsenic ions. EPSs comprising of various organic substances like peptidoglycan, phospholipids, lipopoly-saccharides, proteins, and teichoic and teichuronic acids are responsible for the binding of arsenic ions (Din et al. 2014; Vijayaraghavan and Yun 2008). Several functional groups such as carboxylic, amino, thiol, hydroxyl, and hydrocarboxylic are present in the bacterial cell surface and actively take part in the binding process (Mohapatra et al., 2017a, 2017b, 2017c). The cell wall of Gram-positive bacteria contains a thick layer of peptidoglycan, which exhibits lower levels of surface complexation due to the heavily cross-linked peptidoglycan layer, whereas Gram-negative bacteria contain a highly permeable porous outer membrane and expose most of their cellular components (lipopolysaccharides, phospholipids, and proteins) due to the lack of thick peptidoglycan layer (Joo et al. 2010; Mohapatra et al., 2017a).

18.9 APPROACHES FOR ENHANCED ARSENIC BIOREMEDIATION

Efficiency of arsenic bioremediation can be enhanced by applying various physical–chemical–biological techniques. Modification of bacterial cell wall, use of bacterial consortia, and genetic modification of the bacterial strain are the three major approaches described below.

18.9.1 MODIFICATIONS OF BACTERIAL CELL WALL FOR IMPROVED BIOSORPTION

Enhanced biosorption of toxic metals/metalloids can be achieved by modi-fying the cell surface–binding sites by applying chemical and biological (molecular) techniques, which results in high affinity toward contaminants (Dash et al., 2013; Mohapatra et al., 2017a). Chemical modification of bacterial cell wall by treating with various inorganic and organic substances, such as NaOH, KOH, $(Na)_2CO_3$, NH_4OH, TritonX-100, C_2H_5OH, CH_3OH,

H_2SO_4,Na_2CO_3, $(NH_4)_2SO_4$, HCl, polyacrylic acid, chloroform-methanol, acetone, and toluene, results in the rupture of bacterial cell wall and/or formation of additional binding sites, which ultimately enhances the biosorption capacity (Mohapatra et al., 2017a; Mao et al. 2013). Biological modification of bacterial strains can be done by applying genetic engineering and protein engineering techniques.

18.9.2   *ARSENIC BIOREMEDIATION USING BACTERIAL CONSORTIA*

The degree of arsenic bioremediation can be enhanced by using consortia of a group of arsenic- resistant bacteria and also using genetically modified bacteria (Das and Dash, 2014; Mohapatra et al., 2017a). The use of bacterial consortia instead of single bacterial cultures has increased the bioremediation efficiency dramatically due to their synergistic metabolism. The arsenic bioremediation process becomes much more faster and efficient with the use of bacterial consortia due to the fact that (1) the metabolic intermediate of one bacterium can be utilized by another for efficient degradation and (2) the development of suitable entrap methods for efficient biosorption, etc. (Das and Dash et al., 2014).

18.9.3   *ARSENIC BIOREMEDIATION USING GENETICALLY MODIFIED BACTERIA*

Bacterial arsenic bioremediation can also be regulated by genes or plasmids present inside the bacterial cell (Guo et al., 2001). Therefore, in order to increase the bioremediation potential and metabolic activities of any bacteria, insertion of certain functional genes into their genome is essential. The above phenomenon can be achieved through the insertion of new genes and/or insertion of a new plasmid into the bacterial genomic materials, resulting in the alteration of metabolic pathways such as ion transport, chemotaxis, and overexpression of metal-binding proteins (e.g., mercury binding protein). (Das and Dash, 2014; Pieper and Reineke, 2000). Due to considerable developments in the field of genetic engineering, genetically engineered bacteria for bioremediation of toxic substances like arsenic are available. However, till date, limited reports are available regarding the genetic manipulation of arsenic resistant bacteria to achieve the goal of enhanced arsenic bioremediation (Mohapatra et al., 2017a).

18.10 FACTORS AFFECTING ARSENIC BIOREMEDIATION USING BACTERIA

Several physicochemical and biological factors like pH, temperature, biomass concentration, contact time, initial concentration of pollutants, and the interfering co-contaminants could affect microbial bioremediation of toxic elements like arsenic (Mohapatra et al., 2017a, 2017b, 2017c).

18.10.1 EFFECT OF pH

pH is an important factor, which significantly affects the chemistry of the cell surface functional groups of bacterial biomass. Biosorption capacity for some cationic heavy metals increases with increase in the pH owing to the more negative binding sites exposed on the biomass surface. On the other hand, the binding sites of the microbial cell wall are blocked by hydrogen ions at lower pH, which would hinder binding of cations (Mohapatra et al., 2017a). At lower pH, deprotonation of the cell surface functional groups occurs, resulting in the formation of positively charged active sites favoring the binding of anionic metal ions (Mohapatra et al., 2017a). Arsenic is an anion having -3 charges, but it can also form cations with $+3$ or $+5$ charge and hence, binding takes place in both acidic and alkaline pH according to the charge content of the arsenic molecule.

18.10.2 EFFECT OF TEMPERATURE

Temperature is also one among the important factors affecting bioremediation. When temperature increases, the bioremediation is usually accelerated due to the increase of the surface molecule movement and kinetic energy of the solute (Mohapatra et al., 2017a). However, some binding sites available for metal ion binding may also be destroyed at extreme temperatures. Bacterial sorption of metals/metalloids is normally dependent on the exothermic or endothermic nature of the interactions between the metal/metalloid ions and the bacterial cells.

18.10.3 EFFECT OF BACTERIAL BIOMASS

Bioremediation of arsenic ions also depends on the concentration of the bacterial biomass applied during bioremediation process. Application of more bacterial biomass leads to increased biosorption efficiency owing to the availability of surplus binding sites. Up to a certain limit of increase in bacterial biomass dosage, the rate of bioremediation remains steady, due to the saturation of free cellular binding sites.

18.10.4   EFFECT OF CONTACT TIME

Contact time duration is the resident time existing between the bacterial cell and the toxic arsenic ions. The contact time is one of the important factors that influences the biosorption efficiency. The contact time at which the biosorption reached its maximum and the entire sorption–desorption processes occurred between the cell surface and ions remain saturated, which is considered as an equilibrium time (Mohapatra et al., 2107a, 2019).

18.10.5   EFFECT OF INITIAL ARSENIC CONCENTRATION

The initial concentration of the arsenic ions being used for bioremediation could also affect the bioremediation rate. During biosorption process, the initial concentration of arsenic ions has a major role as it provides the necessary driving force to overcome the mass transfer resistance between the aqueous and solid phases (Mohapatra et al., 2017a). Higher biosorption rate can be obtained in case of lower initial arsenic concentration as compared to the high initial concentration, which may be due to the availability of more free bacterial binding sites for enhanced interaction with arsenic molecules (Mohapatra et al., 2017a). The very high or low concentration level of the arsenic ions is influencing the bioremediation process in both the ways, one is at higher concentration level it can create toxicity to the bacterial cell whereas, in case of lower concentration level, efficiency of the cellular enzymes reduced to recognize the arsenic ions (as substrate) (Pearce et al., 2003).

18.10.6   EFFECT OF INTERFERING IONS (CO-CONTAMINANTS)

Bioremediation of arsenic containing waste may be hampered due to the influence co-contaminants (metal ions including both cations and anions) present in the waste due to competition phenomena for biosorption sites. The decrease of biosorption efficiency in the presence of multiple charged metallic ions may be due to the competition of both metal ions (cations and anions) for the same binding sites on the cell surface (Aryal and Liakopoulou Kyriakides, 2015). In some cases, metal cations as competent ions may increase biosorption of anionic species by enhancing the additional binding sites (Aryal and Liakopoulou Kyriakides, 2011).

18.11 *IN SITU* AND *EX SITU* APPROACHES FOR BACTERIAL ARSENIC BIOREMEDIATION

In situ bacterial bioremediation involves application of bacterial treatment to clean up the arsenic contaminant at the site of contamination (Vanloocke et al., 1975). In situ bioremediation technique involves transformation, immobilization, and/or separation of toxic arsenic compound from the bulk waste (Vanloocke et al., 1975). The optimization of in situ bioremediation process requires the amalgamation of knowledge from both scientific and engineering disciplines. Many *in situ* bioremediation practices generally employed involve biosparging, bioventing, bioaugmentation, biopiling, etc., for soil decontamination (Das and Dash, 2014). *Ex situ* process involves the removal of arsenic contaminated waste from the site of contamination for bacterial bioremediation involving transformation/biosorption/detoxification and then restoring the site with detoxified soil (Das and Dash, 2014). A common example of *ex situ* bioremediation is composting, in which the organic wastes are collectively dumped and subjected to degradation by soil microbes, usually at elevated temperatures (Das and Dash, 2014). The *in situ* technique is cost-effective and has less impact on the ecosystem as compared to the *ex situ* technique (Vanloocke et al., 1975).

18.12 CONCLUSION AND FUTURE PROSPECTIVE

In the 21st century, the environmental pollution due to arsenic contamination is emerging as a major threat to biological ecosystem. arsenic bioremediation using arsenic-resistant bacteria has now became a one-stop solution, which is required globally for decontamination of arsenic-contaminated environment due to its several advantages over other remediation processes. Among various microorganisms employed in arsenic bioremediation, native bacteria residing in arsenic-contaminated sites can adapt to arsenic toxicity and can potentially be used in bioremediation process. In addition to nonsaline bacteria, marine bacteria can also have greater advantages over nonmarine bacteria, as they are more adaptable to quickly changing noxious environment. Though numerous studies have been carried out for arsenic bioremediation using a variety of bacteria isolated from several contaminated areas, still there is a need for exploring highly potential bacterial diversity through combined molecular and metabolic approaches for potential arsenic bioremediation. The arsenic bioremediation using variety of bacteria is doubtlessly take as greener approach but there is also some limitations such as complete detoxification of the highly arsenic contaminated environment under adverse physicochemical conditions not possible effectively by the isolated bacteria till date. Hence, in future, there is need for employing promising genetic engineering technologies to produce more potential genetically engineered microorganisms and optimizing their enzyme production, metabolic pathways, and growth conditions, which will be highly useful for in situ bioremediation of arsenic-contaminated environment.

KEYWORDS

- **arsenic contamination**
- **arsenic-resistant bacteria**
- **bioremediation**
- **biosorption**
- **bio-oxidation**

REFERENCES

Adeniji, A. *Bioremediation of arsenic, chromium, lead, and mercury*. National network of environmental management studies fellow. U.S. Environmental Protection Agency Office of Solid Waste and Emergency Response Technology Innovation Office, Washington, DC, 2004, pp. 14–19.

Ahalya, N.; Ramachandra, T.V.; Kanamadi, R.D. Biosorption of heavy metals. *Research Journal of Chemistry and Environment* 2003, 7, 71–79.

Ahsan, N.; Faruque, K.; Shamma, F.; Islam, N.; Akhand, A.A. Arsenic adsorption by bacterial extracellular polymeric substances. *Bangladesh Journal of Microbiology* 2012, 28, 80–83.

Akhtar, M.S.; Chali, B.; Azam, T. Bioremediation of arsenic and lead by plants and microbes from contaminated soil. *Research in Plant Sciences* 2013, 1, 68–73.

Aksornchu, P.; Prasertsan, P.; Sobhon, V. Isolation of arsenic-tolerant bacteria from arsenic-contaminated soil. *Songklanakarin Journal of Science and Technology* 2008, 30, 95–102.

Andreoni, V.; Zanchi, R.; Cavalca, L.; Corsini, A.; Romagnoli, C.; Canzi, E. Arsenite oxidation in Ancylobacter dichloromethanicus As3–1b strain, detection of genes involved in arsenite oxidation and CO_2 fixation. *Current Microbiology* 2012, 65, 212–218.

Anyanwu, C.U.; Ugwu, C.E. Incidence of arsenic resistant bacteria isolated from a sewage treatment plant. *International Journal of Basic and Applied Science* 2010, 10, 64–78.

Arsène-Ploetze, F.; Koechler, S.; Marchal, M.; Coppée, J.Y.; Chandler, M.; Bonnefoy, V.; Brochier-Armanet, C.; Barakat, M.; Barbe, V.; Battaglia-Brunet, F.; Bruneel, O. Structure, function and evolution of the *Thiomonas* spp. genome. *PLoS Genetics* 2010, 6, e1000859.

Aryal, M.; Liakopoulou-Kyriakides, M. Binding mechanism and biosorption characteristics of Fe(III) by *Pseudomonas* sp. cells. *Journal of Water Sustainability* 2013, 3, 117–131.

Aryal, M.; Liakopoulou-Kyriakides, M. Bioremoval of heavy metals by bacterial biomass. *Environmental Monitoring and Assessment* 2015, 187, 4173.

Aryal, M.; Liakopoulou-Kyriakides, M. Equilibrium, kinetics and thermodynamic studies on phosphate biosorption from aqueous solutions by Fe(III)-treated *Staphylococcus xylosus* biomass: common ion effect. *Colloids and Surfaces A: Physicochemical and Engineering Aspects* 2011, 387, 43–49.

Bachate, S.P.; Cavalca, L.; Andreoni, V. Arsenic-resistant bacteria isolated from agricultural soils of Bangladesh and characterization of arsenate-reducing strains. *J. Appl. Microbial.* 2009, 107, 145–156.

Bahar, M.M.; Megharaj, M.; Naidu, R. Arsenic bioremediation potential of a new arsenite-oxidizing bacterium *Stenotrophomonas* sp. MM-7 isolated from soil. *Biodegradation* 2012, 23, 803–812.

Banerjee, S.; Datta, S.; Chattyopadhyay, D.; Sarkar, P. Arsenic accumulating and transforming bacteria isolated from contaminated soil for potential use in bioremediation. *Journal of Environmental Science and Health Part A* 2011, 4, 1736–1747.

Banerjee, S.; Majumdar, J.; Samal, A.C.; Bhattachariya, P.; Santra, S.C. Biotransformation and bioaccumulation of arsenic by *Brevibacillus brevis* isolated from arsenic contaminated region of West Bengal. *IOSR Journal of Environmental Science, Toxicology and Food Technology* 2013, 3, 1–10.

Biswas, R.; Sarkar, A. Characterization of arsenite-oxidizing bacteria to decipher their role in arsenic bioremediation. *Preparative Biochemistry and Biotechnology* 2019, 49, 30–37.

Cavalca, L.; Corsini, A.; Zaccheo, P.; Andreoni, V.; Muyzer, G. Microbial transformations of arsenic: perspectives for biological removal of arsenic from water. *Future Microbiology* 2013, 8, 753–768.

Chakraborty, D.; Das, B.; Rahman, M.M.; Chowdhury, U.K.; Biswas, B.; Goswami, A.B.; Nayek, B.; Pal, A.; Sengupta, M.K.; Ahmed, S.; Hossain, A.; Basu, G.; Roychowdhury, T.; Das, D. Status of groundwater arsenic contamination in the state of West Bengal, India: a 20-year study report. *Molecular Nutrition & Food Research* 2009, 53, 542–551.

Chaudhary, A.; Salgaonkar, B.B.; Braganca, J.M. Cadmium tolerance by haloarchaeal strains isolated from solar salterns of Goa, India. *International Journal of Bioscience, Biochemistry and Bioinformatics* 2014, 4, 1–6.

Clausen, A. Isolating metal-tolerant bacteria capable of removing copper, chromium, and arsenic from treated wood. *Waste Management and Research* 2000, 18, 264–268.

Das, N.; Vimala, R.; Karthika, P. Biosorption of heavy metals—an overview. *Indian Journal of Biotechnology* 2008, 7, 159–169.

Das, S.; Dash, H.R. Microbial bioremediation: a potential tool for restoration of contaminated areas. In *Microbial Biodegradation and Bioremediation*; Das, S., Ed.; Elsevier, Waltham, MA, 2014, Vol. 1; pp. 1–21.

Dash, H.R.; Mangwani, N.; Chakraborty, J.; Kumari, S.; Das, S. Marine bacteria: potential candidates for enhanced bioremediation. *Applied Microbiology and Biotechnology* 2013, 97, 561–571.

Devika, L.; Rajaram, R.; Mathivanan, K. Multiple heavy metal and antibiotic tolerance bacteria isolated from equatorial Indian Ocean. *International Journal of Microbiology Research* 2013, 4, 212–218.

Dey, U.; Chatterjee, S.; Mondal, N.K. Investigation of bioremediation of arsenic by bacteria isolated from an arsenic contaminated area. *Environmental Processes* 2017, 4, 183–199.

Dey, U.; Chatterjee, S.; Mondal, N.K. Isolation and characterization of arsenic-resistant bacteria and possible application in bioremediation. *Biotechnology Reports* 2016, 10, 1–7.

Din, M.I.; Hussain, Z.; Mirza, M.L.; Shah, A.T.; Athar, M.M. Adsorption optimization of lead(II) using *Saccharum bengalense* as a non-conventional low cost biosorbent; isotherm and thermodynamics modeling. *International Journal of Phytoremediation* 2014, 16, 889–908.

Fomina, M.; Gadd, G.M. Biosorption: current perspectives on concept, definition and application. *Bioresource Technolog* 2014, 160, 3–14.

Ghodsi, H.; Hoodaji, M.; Tahmourespour, A.; Gheisari, M.M. Investigation of bioremediation of arsenic by bacteria isolated from contaminated soil. *African Journal of Microbiology Research* 2011, 5, 5889–5895.

Guo, X.J.; Fujino, Y.; Kaneko, S.; Wu, K.; Xia, Y.; Yoshimura, T. Arsenic contamination of groundwater and prevalence of arsenical dermatosis in the Hetao plain area, Inner Mongolia, China. *Molecular and Cellular Biochemistry* 2001, 222, 137–140.

Gupta, A.; Joia, J.; Sood, A.; Sood, R.; Sidhu, C.; Kaur, G. Microbes as potential tool for remediation of heavy metals: a review. *Journal of Microbial and Biochemical Technology* 2016, 8, 364–72.

Ike, M.; Miyazaki, T.; Yamamoto, N.; Sei, K.; Soda, S. Removal of arsenic from groundwater by arsenite-oxidizing bacteria. *Water Science and Technology* 2008, 58, 1095–1100.

Joo, J.H.; Hassan, S.H.A.; Oh, S.E. Comparative study of biosorption of Zn^{+2} by *Pseudomonas aeruginosa* and *Bacillus cereus*. *International Biodeterioration & Biodegradation* 2010, 64, 734–741.

Karigar, C.S.; Rao, S.S. Role of microbial enzymes in the bioremediation of pollutants: a review. *Enzyme Research* 2011, 2011, 1–11. doi:10.4061/2011/805187.

Khan, A.H.; Rasul, S.B.; Munir, A.; Habibuddowla, M.; Alauddin, M.; Newaz, S.S.; Hussan, A. Appraisal of a simple arsenic removal method for groundwater of Bangladesh. *Journal of Environmental Science and Health* 2000, 35, 1021–1041.

Klaassen, C.D. *Casarett & Doull's Toxicology: The Basic Science of Poisons*, 9th ed.; McGraw-Hill, New York, 2019.

Mao, J.; Won, S.W.; Yun, Y.S. Development of poly (acrylic acid)-modified bacterial biomass as a high-performance biosorbent for removal of Cd(II) from aqueous solution. *Industrial & Engineering Chemistry Research* 2013, 52, 6446–6452.

Mateos, L.M.; Ordóñez, E.; Letek, M.; Gil, J.A. *Corynebacterium glutamicum* as a model bacterium for the bioremediation of arsenic. *International Microbiology* 2006, 9, 207–215.

Miyatake, M.; Hayashi, S. Characteristics of arsenic removal by *Bacillus cereus* strain W2. *Resources Processing* 2011, 58, 101–107.

Mohapatra, R. K.; Panda, C. R. Spatiotemporal variation of water quality and assessment of pollution potential in Paradip port due to port activities. *Indian Journal of Geo Marine Sciences*. 2017, 46(07), 1274–1286.

Mohapatra, R.K.; Parhi, P.K.; Patra, J.K.; Panda, C.R.; Thatoi, H.N. Biodetoxification of toxic heavy metals by marine metal resistant bacteria: a novel approach for bioremediation of the polluted saline environment. In *Microbial Biotechnology*; Patra, J.K.; Vishnuprasad, C.N.; Das, G., Eds.; Springer, Singapore, 2017a, Vol. 1, pp. 343–376.

Mohapatra, R.K.; Parhi, P.K.; Thatoi, H.; Panda, C.R. Bioreduction of hexavalent chromium by *Exiguobacterium indicum* strain MW1 isolated from marine water of Paradip Port, Odisha, India. *Chemistry and Ecology* 2017b, 33, 114–130.

Mohapatra, R.K.; Pandey, S.; Thatoi, H.; Panda, C.R. Reduction of chromium (VI) by marine bacterium *Brevibacillus laterosporus* under varying saline and pH conditions. *Environmental Engineering Science* 2017c, 34, 617–626.

Mohapatra, R.K.; Pandey, S.; Thatoi, H.N.; Panda, C.R. Screening and evaluation of multi-metal tolerance of chromate resistant marine bacteria isolated from water and sediment samples of Paradip port, Odisha Coast. *Journal of Advances in Microbiology* 2016, 2, 135–147.

Mohapatra, R.K.; Parhi, P.K.; Pandey, S.; Bindhani, B.K.; Thatoi, H.; Panda, C.R. Active and passive biosorption of Pb (II) using live and dead biomass of marine bacterium *Bacillus xiamenensis* PbRPSD202: kinetics and isotherm studies. *Journal of Environmental Management* 2019, 247, 121–134.

Naik, M.M.; Pandey, A.; Dubey, S.K. Bioremediation of metals mediated by marine bacteria. In *Microorganisms in Environmental Management*; Satyanarayana, T.; Johri, B.N.; Prakash, A. Eds.; Springer, Dordrecht, 2012, Vol. 1, pp. 665–682.

Paul, D.; Poddar, S.; Sar, P. Characterization of arsenite-oxidizing bacteria isolated from arsenic-contaminated groundwater of West Bengal. *Journal of Environmental Science and Health Part A* 2014, 49, 1481–1492.

Paul, T.; Chakraborty, A.; Islam, E.; Mukherjee, S.K. Arsenic bioremediation potential of arsenite-oxidizing *Micrococcus* sp. KUMAs15 isolated from contaminated soil. *Pedosphere* 2018, 28, 299–310.

Pearce, C.I.; Lloyd, J.R.; Guthrie, J.T. The removal of colour from textile wastewater using whole bacterial cells: a review. *Dyes and Pigments* 2003, 58, 179–196.

Pieper, D.H.; Reineke, W. Engineering bacteria for bioremediation. *Current Opinion in Biotechnology* 2000, 11, 262–270.

Prasad, M.N.V. *A State-of-the-Art Report on Bioremediation, Its Applications to Contaminated Sites in India*. Ministry of Environment, Forests and Climate Change, Government of India, 2011.

Qin, J.; Rosen, B.P.; Zhang, Y.; Wang, G.; Franke, S.; Rensing, C. Arsenic detoxification and evolution of trimethylarsine gas by a microbial arsenite S-adenosylmethionine methyltransferase. *Proceedings of the National Academy of Sciences* 2006, 103, 2075–2080.

Roychowdhury, R.; Roy, M.; Rakshit, A.; Sarkar, S.; Mukherjee, P. Arsenic bioremediation by indigenous heavy metal resistant bacteria of fly ash pond. *Bulletin of Environmental Contamination and Toxicology* 2018, 101, 527–535.

Satyapal, G.K.; Mishra, S.K.; Srivastava, A.; Ranjan, R.K.; Prakash, K.; Haque, R.; Kumar, N. Possible bioremediation of arsenic toxicity by isolating indigenous bacteria from the middle Gangetic plain of Bihar, India. *Biotechnology Reports* 2018, 17, 117–125.

Satyapal, G.K.; Rani, S.; Kumar, M.; Kumar, N. Potential role of arsenic resistant bacteria in bioremediation: current status and future prospects. *Journal of Microbial and Biochemical Technology* 2016, 8, 256–258.

Shrestha, R.A.; Lama, B.; Joshi, J.; Sillanpää, M. Effects of Mn(II) and Fe(II) on microbial removal of arsenic(III). *Environmental Science and Pollution Research* 2008, 15, 303–307.

Silver, S.; Phung, L.T. Genes and enzymes involved in bacterial oxidation and reduction of inorganic arsenic. *Applied and Environmental Microbiology* 2005, 71, 599–608.

Stelting, S.; Burns, R.G.; Sunna, A.; Visnovsky, G.; Bunt, C.R. Immobilization of *Pseudomonas* sp. strain ADP: a stable inoculant for the bioremediation of atrazine. *Applied Clay Science* 2012, 64, 90–93.

Stolz, J.F.; Oremland, R.S.; Bacterial respiration of arsenic and selenium. *FEMS Microbiology Reviews* 1999, 23, 615–627.

Takeuchi, M.; Kawahata, H.; Gupta, L.P.; Kita, N.; Morishita, Y.; Ono, Y.; Komai, T. Arsenic resistance and removal by marine and non-marine bacteria. *Journal of Biotechnology* 2007, 127, 434–442.

Taran, M.; Safari, M.; Monaza, A.; Reza, J.Z.; Bakhtiyari, S. Optimal conditions for the biological removal of arsenic by a novel halophilic archaea in different conditions and its process optimization. *Polish Journal of Chemical Technology* 2013, 15, 7–9.

Vanloocke, R.; De Borger, R.; Voets, J.P.; Verstraete, W. Soil and groundwater contamination by oil spills; problems and remedies. *International Journal of Environmental Studies* 1975, 8, 99–111.

Vaughan, D.J. Arsenic. *Elements* 2006, *2*, 71–75.

Vijayaraghavan, K.; Yun, Y.S. Bacterial biosorbents and biosorption. *Biotechnology Advances* 2008, 26, 266–291.

Vishnoi, N.; Singh, D.P. Biotransformation of arsenic by bacterial strains mediated by oxido-reductase enzyme system. *Molecular and Cellular Biology* 2014, 60, 7–14.

Voica, D.M.; Bartha, L.; Banciu, H.L.; Oren, A. Heavy metal resistance in halophilic bacteria and archaea. *FEMS Microbiology Letters* 2016, 363, 146.

WHO, E.F. Guidelines for drinking-water quality. *WHO Chronicle* 2011, 38, 104–108.

Williams, G.P.; Gnanadesigan, M.; Ravikumar, S. Isolation, identification and metal tolerance of halobacteial strains. *Indian Journal of Geo Marine Sciences*. 2013, 42(03), 402–408.

Yamamura, S.; Watanabe, M.; Kanzaki, M.; Soda, S.; Ike, M. Removal of arsenic from contaminated soils by microbial reduction of arsenate and quinone. *Environmental Science & Technology* 2008, 42, 6154–6159.

CHAPTER 19

Potential Applications of Coal Fly Ash: A Focus on Its Fungistatic Properties and Zeolite Synthesis

SANGEETA RAUT

Centre for Biotechnology, School of Pharmaceutical Sciences, Siksha 'O' Anusandhan (Deemed to be University), Bhubaneswar, Odisha 751003, India, E-mail: rautsan.bio@gmail.com.

ABSTRACT

Coal fly ash (CFA), an industrial by-product derived from coal combustion in thermal power plants, has increased, generating environmental problems. It is one of the most complex anthropogenic materials and results in a waste of recoverable resources. Its improper disposal has become an environmental concern. The need of the hour is to develop methods for the utilization of fly ash (FA). This chapter summarizes studies concerning the characterization and potential applications of FA on the basis of the numerous studies published on the subject worldwide over the past few decades. The chapter first describes the generation, physicochemical properties, and hazards of CFA at the global level, and also focuses on its current and potential applications, including its use in agricultural applications, soil stabilization, the cement and concrete industries, zeolite synthesis, etc. It also deals with an evaluation of the antifungal effect of CFA. Among the various methods proposed for the reuse of FA, conversion to zeolite offers the greatest benefits; the process diverts ash waste materials from disposal sites and transforms them into useful secondary products for applications ranging from environmental mitigation to catalysis. Finally, the advantages and disadvantages of these applications, the mode of FA utilization worldwide, and directions for future research are discussed.

19.1   INTRODUCTION

Coal is the second major fossil fuel source for energy production. It is largely distributed worldwide, with approved reserves of approximately 1000 billion tones in total. Despite the Paris climate agreement, coal will still play a key role in power generation in the foreseeable future, considering the growing demands for energy, particularly in the developing countries. Globally, with around 28% of the total generated energy reported in 2016, coal has the second rank among other energy sources (BP Statistical review, 2016). The latest report published in 2017 by the US Energy Information Administration (EIA) evaluates the worldwide coal consumption from 1990 to 2040 (International Energy Outlook, 2017). The EIA outlook reported that a coal consumption of around 160 quadrillion Btu (QBTU) (a quadrillion Btu roughly equals the amount of energy in 45 million tons of coal) would be required from 2015 to 2040. The consumption rate in China and the United States will decrease, but it will be compensated by India. China will still remain the largest consumer of coal, with a partial reduction in 2040 (about 73 QBTU), while India's coal consumption will grow at an average rate of 2.6% per year from 2015 to 2040. It is estimated that the overall coal demand for the power sectors in India will increase from 672 million tonnes (Mt) in 2017 to 827–1277 million tonnes (Mt) by 2030. The consumption of coal in Organisation for Economic Co-operation and Development (OECD) countries will decrease by an average of 0.6% per year during 2015–2040, due to the growing competitions for natural gas and renewable sources. In other places, such as Africa, the Middle East, and non-OECD Asia, the rate of consumption will gradually enhance until 2040 (International Energy Outlook, 2017). In addition, the EIA predicts that the amount of coal utilization for the generation of power will remain stable, and then drop slightly, due to the replacement of coal by natural gas and renewable and nuclear power in the OECD countries. EIA's projections show that the amount of electricity generation will grow till 2040 (with a growth rate of 2% annually), implying that coal will constitute a significant portion in this growth (International Energy Outlook, 2017).

The continuous demands for coal lead to more coal extraction and, subsequently, more coal ash (CA) production (Figure 19.1a and b). Coal-fed power plants generate tons of fly ash (FA), bottom ash, and polycyclic aromatic hydrocarbons, whose accumulation in landfills leads to destructive effects on living organisms and ecosystems (Figure 19.1c and d). Coal fly ash (CFA) represents 65%–95% of the total ash generated. It is composed of mineral mixture consisting of various oxides, such as SiO_2, Al_2O_3, and metal oxides, which can be alkaline, alkaline earth, and transition metal oxides, that make it potentially toxic. The major parameters affecting CFA quality are the chemical composition of the coal burned and the combustion conditions, that is, the rate of oxidation and pulverization.

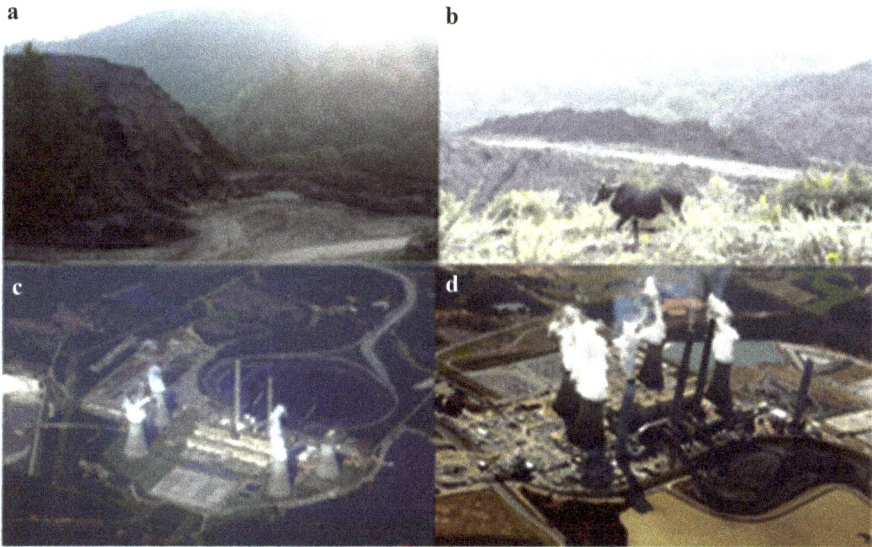

FIGURE 19.1 (a) Production of coal ash in a coal mine site (The mountain contains coal ash more than two million tons) and (b) The negative effects of CFA accumulation on the environment (destruction of natural ecosystem), Savadkouh, Mazandaran, Iran, (c) CFA emission into the air and environment by coal-fired power plant in Georgia, and (d) Accumulation of coal fly ash in the landfills around Georgia coal-fired power plant (https://en.wikipedia.org/wiki/Plant_Scherer).

The world annual rate of CFA generation in 2012 was reported to be around 800 million tons, from which China, India, US, and EU generated 500, 140 and 115 million tons, respectively (Belviso, 2018a).

Therefore, the direct use of CFA leads to hazardous effects on the ecosystem (Huggins et al., 2016). The chemical mineralogical composition of CFA allows its usage in many fields. The utilization rates of CFA in the United States, EU, India, and China were estimated to be around 50%, 90%, 60%, and 67%, respectively (Yao et al., 2015). Notably, a considerable amount of CFA is used for various applications, especially in construction (Xu and Shi, 2018) and civil engineering (Aljerf, 2015; Hoy et al., 2016), where it stands as a substitute to cement in concrete (Ebrahimi et al., 2017).

Due to the existence of several components in it and larger surface area, its application in amendments of soil-improving texture, bulk density, water-holding capacity, and nutrients has been appreciated (Ram and Masto, 2014; Shaheen et al., 2014). Also, it is used as a precursor to prepare zeolites (Fukasawa et al., 2017) and mesoporous silica (Li and Qiao, 2016). It is also used as an adsorbent to capture CO_2, SO_2, NO_2, and mercury (Ji et al., 2017; Wang et al., 2016; Yang et al., 2017) from the air. Moreover, its applications include the generation of FA-based geopolymers (Boke et al., 2015; Zhuang et al., 2016), recovery of valuable metals from CFA (Ding et al., 2017; Sahoo et al., 2016), and synthesis of FA polymer composites for various purposes like preparation of electromagnetic interference-shielding materials (Lu et al., 2015; Revanasiddappa et al., 2018). One of the potential applications of FA is its use as fungicides.

Until now, the rate of FA utilization compared to total ash production is very low. Conversion of this waste (FA ~ threat to the environment) to a valuable product in the form of zeolite can be an attractive option for efficient CO_2 capturing. Furthermore, it is a more affordable and economically viable solution than land disposal. Some other emerging applications of FA zeolites, such as molecular sieves, ion exchangers, gas adsorbents, and catalysts, have also opened up new approaches to protect the environment.

19.2   OPPORTUNITIES AND CHALLENGES OF COAL FLY ASH

It is predicted that with the development of the world economy and increase in demand for energy, large amounts of CFA are produced by power generation or industrial coal combustion. Moreover, there is no comprehensive plan to utilize this generated CFA for the fabrication of value-added products. Due to the lack of adequate use, researchers have been working on the possibility of CFA conversion into a series of effective products, such as zeolites, catalysts, photocatalysts, and geopolymers. It has been found that CFA contains a large amount of aluminosilicate, and also the high price of commercial types of mentioned products, CFA can be used for the production of competitive products with lower cost. This is a golden opportunity for both coal factories and coal-fired power plants to sustain and use lucrative plans by establishing an autonomous or lateral unit for the valorization of CFA. The implementation of beneficial plans in both coal factories and coal-fired power plants not only helps them to get rid of upcoming challenges of CFA (CFA environmental problems), but also increases their incomes by their selling CFA-based products to increase their market through competing with similar commercial products. In fact, the most suitable way for the generation of clean energy from fossil fuels like coal is the renovation of available technologies in the related industries for the reduction of air and water pollutants. Controlling the dose of gaseous and aqueous pollutants is crucial to prevent their negative effects on the aquatic organism, ecosystem, and particularly the human health that can be carried out by cost-effective and efficient products like CFA-based adsorbents with acceptable reusability properties in comparison with the similar commercial products. Therefore, investing on research and development projects can provide remarkable income for coal industries, besides solving the global environmental crises of CFA.

19.3 APPLICATIONS OF COAL FLY ASH

19.3.1 ENVIRONMENTAL USES OF FLY ASH

Although FA has many applications, the most prominent ones are in environmental remediation as adsorbent for gas and wastewater treatment (Apiratikul and Pavasant, 2008; Hui et al., 2005; Querol et al., 2002a), consumer goods as detergent builders (Hui and Chao, 2006a), and industrial processes as catalysts (Wang et al., 2008). Carbon capture, utilization, and storage is one of the proposed technologies for reducing global CO_2 emissions, and it presents many opportunities for FA utilization, since it is cheaply available, and could help in reducing the environmental risks associated with FA disposal. In some cases, it might be possible to use carbonated FA as a construction material or additive. The most prevalent applications and properties of CFA are reported in Figure 19.2.

19.3.2 BIO-CORROSION AND BIODETERIORATION IN CONSTRUCTION MATERIALS

The bioactivity of living microorganisms on building materials is a severe problem, because of the likely effect on the utility of properties and reduction in the service life of structures due to biocorrosion and subsequent biodeterioration (Cwalina, 2008; George et al., 2012). The biocorrosion of concrete, plasters, and mortars is caused by the bioactivity of different microorganisms such as bacteria, yeast (one-cell fungi), molds (filamentous fungi), microscopic fungi, lichens, mosses on the exteriors and interiors of buildings and structures, or nostocs, algae (primitive plants) seen at sewage water treatment plants and in water and sewage pipelines, cooling towers, dams, maritime structures, bridges, and tanks. The microbes colonize the surfaces of concrete and also the pores, capillaries, and microcracks, causing damage through biodeterioration (George et al., 2012). Biocorrosion is normally connected with a formation of biofilms. The biofilms affect the overall biodeterioration of building materials (Morton, 2003). Microbial metabolites like biogenic carbon dioxide (CO_2), hydrogen sulfide (H_2S), ammonia (NH_3), and mineral and organic acids are the main drivers of biodeterioration, reacting with the components of concrete and mortar (George et al., 2012). The acidic metabolites react mainly with calcareous components such as portlandite $Ca(OH)_2$ crystals and C-S-H gels available in the form of tobermorite $C_5S_6H_5$ and xonotlite C_5S_5H, as well as with the carbonate aggregates consisting of limestone and dolomite.

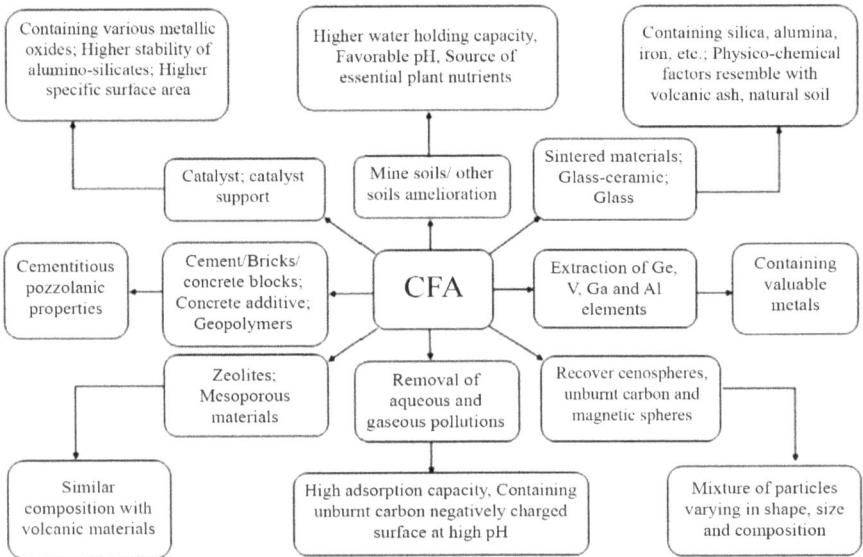

FIGURE 19.2   Prevalent applications and properties of CFA.

The increase in corrosion products causes a tremendous expansion of concrete, which is ultimately manifested by total destruction (George et al., 2012; House and Weiss, 2014; Morton, 2003). The antifungal effect of CFA products means that it has an inhibiting effect on the growth and reproduction of fungi by 100%, however, without lethal action, so the material's surface is not invaded by fungi.

The use of materials and by-products of CFA for enhancing the anti-microbial properties of construction materials is very useful to prevent the presence, growth, and reproduction of microorganisms, and so to decrease the biocorrosion and biodeterioration commonly caused by the bioactivity of microorganisms. In spite of these potential pathways for CFA utilization, the use of this material in zeolite synthesis is minimal, and large volumes are currently disposed in landfills. It can be advantageous to extend the application of CFA.

19.4 OVERVIEW OF ZEOLITE SYNTHESIS FROM ASH

Zeolites are microporous, crystalline, hydrated aluminosilicates characterized by a three-dimensional network of tetrahedral $(SiAl)O_4$ units that form a system of interconnected pores. In the event where the FA contains low amounts of unburned carbon, materials such as zeolites can be synthesized. Due to its high content of silicon and aluminum compounds, FA provides an ideal base material for the synthesis of zeolites. The aluminum ion produces a negative net charge, which is balanced by the presence of an extra cation in the framework (Figure 19.3). These zeolites are synthesized via different techniques. The purity of the synthesized zeolites from the above techniques often varies between 40% and 99%, depending on the conversion technique used and on the composition of FA. When FA is used as a precursor for zeolite, the synthesis follows the pathway shown in Figure 19.4. The type and amount of zeolite formed results from the combination of different conditions and parameters such as temperature, pressure, solution alkalinity, activation solution-to-FA ratio, and the formation process. The second part of this chapter investigates the major methods that have been used to synthesize zeolites from combustion residue materials and explores the effects of each of the aforementioned parameters and the advantages and limitations of each synthesis process. The main applications of these synthetic products are also briefly discussed.

19.4.1 SYNTHESIS PROCESSES

Dissolution begins the synthesis process but does not complete before the subsequent stages begin. When a sufficient Si and Al concentration is reached in the solution, the condensation and nucleation processes begin on the surface of the FA aluminosilicate particles, effectively stopping the dissolution step. This explains why in most FA zeolites crystals can be found

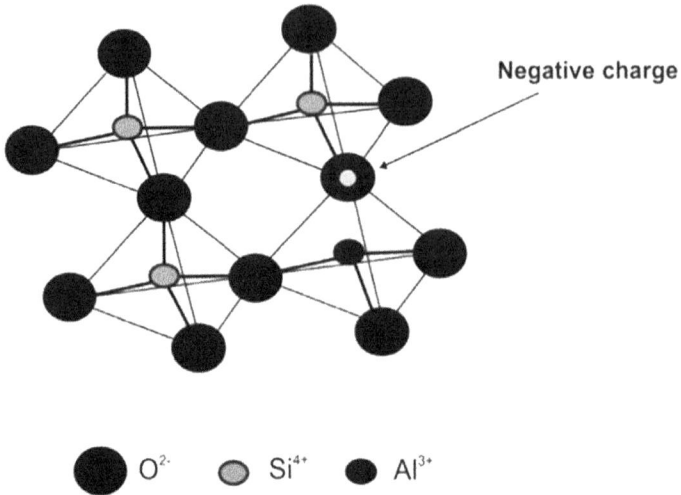

FIGURE 19.3   Zeolite framework of tetrahedral $[SiO_4]_4$ with the Si–Al substitution yielding a negative charge (adapted from Querol et al., 2002).

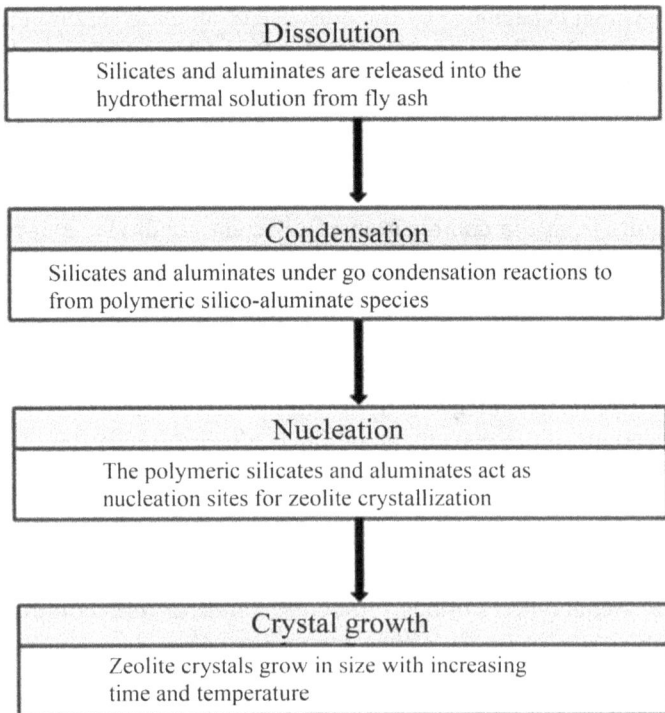

FIGURE 19.4   Zeolite formation pathway from fly ash.

aggregated on top of the larger FA particles. The low purity of the zeolite product can be attributed to the high content of inert components in the FA, such as CaO and Fe_2O_3, which do not participate in zeolitization reaction or in undissolved the silico-aluminate phase. In order to resolve these issues and produce high-quality zeolites from FA, Jha and Narain have proposed a three-step activation (TSA) process. In this process, there is three times repetition of the hydrothermal step of recycling the produced zeolite with the alkaline supernatant solution. The TSA process produced enhanced cation-exchange capacity (CEC) of the FA zeolites (843 mEq/100 g), in comparison to the conventional methods (388 mEq/100 g for step-1 activation and 250 mEq/100 g for step-2 activation) (Hollman et al., 1999). The supernatant solution also has considerably reduced the concentrations of Si, Al, Na, and heavy-metal species, hence making their disposal comparatively safer. The cost analysis of the proposed method revealed that TSA was cheaper in comparison to both one-step and two-step activation processes, since the open reflux system employed for TSA was cheaper than the closed reflux system used for the one- and two-step activations (Jha and Singh, 2016). A similar fusion technique was proposed where the crystalline phases in the FA were further broken down by re-fusing the fusion product with NaOH. Although the TSA-fusion process enhanced zeolite yield, its performance was lower than the TSA-hydrothermal process.

The overall purity of the synthesized zeolites will also be lowered in the presence of nonreacting phases or components like CaO and Fe_2O_3 in the FA. Therefore, to obtain high-purity zeolites, the nonreacting FA phases need to be removed by a pretreatment process. FA pretreatment is usually done by either of the following ways:

1. *Acid digestion*: FA is agitated in a concentrated acid solution (3M HNO_3, H_2SO_4, or HCl) at a temperature between 60 °C and 120 °C for an extensive period. This treatment leaches out CaO, Fe_2O_3, and undesired phases that are soluble in the acid, which results in FA with much less impurities because of the dissolution of alumina during the digestion. An increase in the Si-to-Al ratio of the FA is also observed after pretreatment.

2. *Magnetic separation*: This involves the implementation of strong magnets for the removal of unwanted Fe_2O_3 particles from FA, resulting in higher purity zeolites obtained from the FA. After the pretreatment, an improvement in the crystallinity of zeolites synthesized from the FA has also been observed (Hollman et al., 1999; Rayalu et al., 2000).

Figure 19.5 shows a simplified scheme of the CFA zeolitization mechanism based on Al and Si dissolution, geopolymer formation, crystalline structure nucleation, and, finally, zeolite crystal growth.

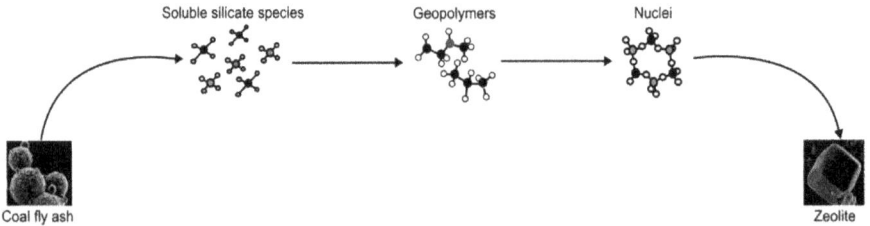

FIGURE 19.5   Schematic of the coal fly ash zeolitization mechanism (adapted from Bukhari et al., 2015).

The main zeolite synthesis processes include (1) the conventional hydrothermal process; (2) the alkaline fusion-assisted hydrothermal process; and (3) multistep treatment. In this section, the literature on each approach is analyzed, focusing on the processes and parameters that influence the type of zeolite formed from ash and the main drawbacks of each method.

19.4.1.1   THE HYDROTHERMAL PROCESS

The first comparative result of alkaline-treated coal combustion ash was published by Mondra Gon. The experiments were carried out with different concentrations of sodium hydroxide (from 2.0 to 13.0 M) at various temperatures (90–100 °C) and for different periods of time (8–48 h) using distilled water. The outcome of the experiments showed that the amorphous material in the ash is altered into crystalline phase, which takes the form of zeolite X, even though zeolite P and hydroxyl sodalite are also produced. Also, in order to lessen the concentrations of iron and alkaline oxides present on the outer surfaces of the ash particles, the FA was treated with acid (HCl or H_2SO_4). The acid treatment was shown to improve zeolite synthesis, maybe due to huge concentration of silica and aluminum in the acid-treated samples, while adversely affecting the absorptive properties of the final products. Later, an analysis of hydrothermal treatment behavior of two different CFA samples was carried out (Singer and Berkgaut, 1995). This study concluded that with a high CEC, the solid residues can be converted into a new material characterized by ~50% zeolite material. These authors validated the results formerly obtained by Catalfamo et al. (1993). They showed that the yield of zeolite is affected by the presence of mullite and, subordinately, quartz. An analysis of temperature effect on the formation of zeolite in a Na_2O-K_2O-Al_2O_3-SiO_2-H_2O system was performed by Lin and His (1995), which revealed that under changing incubation temperatures, different phases of zeolite are synthesized; zeolite P was synthesized at a lower temperature (~70 °C), analcime at 150 °C, hydroxysodalite was formed when the temperature was as high as 170 °C, and crinite was the only zeolite forming at the highest temperature of 200°C.

The conventional hydrothermal process has been depicted with the help of a flowchart in Figure 19.6. Thus, it can be concluded that (1) reactive Al_2O_3 contributes to zeolite formation only in small amounts, and (2) during a typical hydrothermal process, the amounts of reactive elements are limited. This leads to lower zeolite synthesis, and the main disadvantages of the hydrothermal process are due to the presence of unreacted FA in the final product.

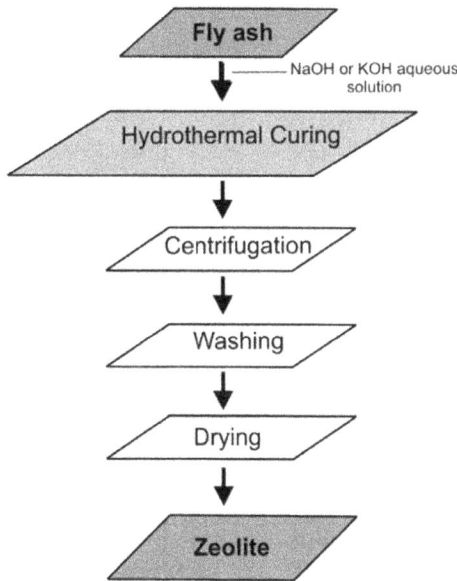

FIGURE 19.6 Flowchart of the conventional hydrothermal process (adapted from Jha et al., 2011).

19.4.1.2 *THE ALKALINE FUSION-ASSISTED HYDROTHERMAL PROCESS*

Querol and his co-researchers published their first paper on zeolite synthesis from CFA by alkaline hydrothermal methods in 1995, and related papers later on (Querol et al., 1997, 1999, 2001). They made an analysis on the synthesis of zeolite as a function of FA-to-solution ratio, solution chemistry, and reaction time. They explained that upon the reaction time, the FA-to-solution ratio and the chemical and mineralogical composition of the raw materials are the factors upon which the volume, type, and crystalline size of the newly formed minerals rely (Querol et al., 1995). They found that the main reactive phase of FA in zeolite synthesis is aluminosilicate, and also confirmed the low reactivity of mullite. In following studies, Querol et al. (1997) drew a distinction between two different zeolite crystal growth processes, which included:

1. Those involving pure alkaline activation of the aluminosilicate glass and occurring under short periods of activation at low temperatures; and

2. Those involving the dissolution of SiO_2 and Al_2O_3 from quartz, mullite, and aluminosilicate glass, occurring under long periods of activation at high temperatures.

The authors showed that the reason for the differences in types of synthesis of zeolites using fly ash with similar SiO_2/Al_2O_3 ratios as well as the same activation conditions was the differences in the mineralogical composition (in the samples with same chemical composition). Querol et al. (1999) also studied the physical, chemical, and mineralogical properties of 14 Spanish FA samples and carried out pilot plant-scale experiments (Querol et al., 2001) in order to elucidate the prime attributes influencing the employment of FA for synthesis of zeolite. An analysis on the effect of aging and seeding on zeolite synthesis from CFA was performed by Zhao et al. (1997). Their data revealed that aging is crucial for the dissolution of Si and Al into basic solutions, forming ring-like structures, and, subsequently, for zeolite formation. Selective induction occurs due to seeding during the synthesis of a specific zeolite in the absence of other impurities, and this skips the processes of induction and nucleation. Many other studies have been documented on the optimum condition for zeolite formation using CFA; a majority of these show that the newly formed zeolite types are dominated by faujasite, sodalite and hydroxysodalite, P zeolite, and A-type zeolite (Franus et al., 2014; Gross-Lorgouilloux et al., 2010; Ma et al., 1998; Moutsatsou et al., 2006; Steenbruggen and Hollman, 1998; Tanaka et al., 2003, 2004a, 2004b; Vucinic et al., 2003; Waek et al., 2008). The typical scanning electron microscopy images of zeolites have been shown in Figure 19.7. Tanaka et al. (2002) study showed Na–A zeolite formation as a single phase when the

SiO_2/Al_2O_3 molar ratio of the CFA solutions was modified in the range of 1.0 $\leq SiO_2/Al_2O_3 \leq 2.0$, although Na–X zeolite got crystallized at $SiO_2/Al_2O_3 \geq$ 2.5. A study by Murayama et al. (2002) on the mechanism of synthesis of zeolite showed that the changes in the concentrations of Al^{3+} and Si^{4+} in the liquid phase during hydrothermal reaction is determined by the dissolution of amorphous aluminosilicates (characteristic of FA composition). With the increase in concentration of these elements, aluminosilicate gel is formed by the condensation of ions, a type of preformation of zeolite crystal, which is more reactive with rise in temperature. The aluminosilicate gel begins to quickly deposit on the particle surface like a large flake and gets transformed into zeolite crystals with the onset of the condensation reaction.

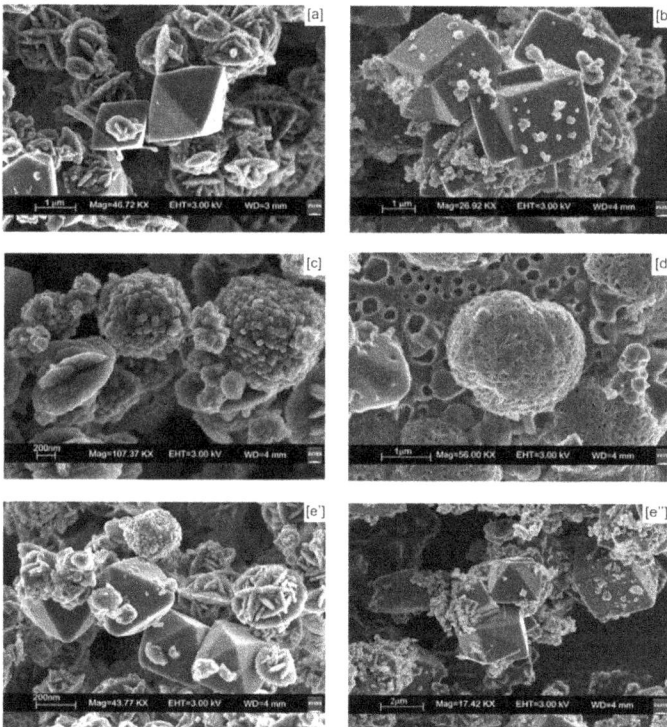

FIGURE 19.7 SEM images showing the typical morphology of (a) faujasite, (b) A-type zeolite, (c) Na-P zeolite, (d) sodalite, and (e′) and (e″) zeolite intergrowth.

The dissolution–precipitation process was proposed by studying the mechanism of controlling CFA conversion into zeolites by means of hydrothermal alkaline treatment. The authors demonstrated that the formation of hydroxyl sodalite takes place using silica-lean fly ash at high concentrations of NaOH, whereas crystallization of Na-P zeolites require moderate concentrations of NaOH with high and low concentrations of Si and Al, respectively.

The following paragraph illustrates some of the improved conventional hydrothermal methods. The recent publications that used biomass ash for zeolite synthesis, documented the synthesis of analcime and other phases via the hydrothermal treatment of biomass ash from a fluidized-bed forest combustor (Jimenez et al., 2017), documented synthesis of ZSM-5 zeolite by the implementation of FA from the pulp and paper industries and using bagasse ash as a silica source for the synthesis of Na-Y zeolite (Worathanakul and Tobarameeku, 2016).

19.4.1.3   *THE MULTISTEP TREATMENT METHOD*

Several variations have been documented on conventional and prefusion hydrothermal methods focusing on controlling single-zeolite synthesis or increasing the amount of mineral formation using CFA as the raw material. There have been description of processes which favor the synthesis of a more stable zeolite via nucleation, basing on the supplementation of external silica, observing higher reactivity in the silica-enriched mixture (Gross-Lorgouilloux et al., 2010).

Other researchers have studied a two-step method for zeolite synthesis, which involves an initial silicon and aluminum-extraction stage and a second zeolite hydrothermal synthesis step (El-Naggar et al., 2008; Hollman et al., 1999; Hui and Chao, 2006b; Matlob et al., 2012; Moreno et al., 2002, 2004; Tanaka et al., 2006). The methods including step-wise changes in temperature during hydrothermal treatment have also been carried out (Hui and Chao, 2006).

The molten-salt method proposed by Park et al. (2000) is one of the most recognized among the nonconventional methods of zeolite synthesis. The researchers used molten–salt mixtures (which included NaOH–NaNO$_3$, NaOH–KNO$_3$, and NH$_4$F–NH$_4$NO$_3$) to create sodalite and cancrinite from CFA, without adding water. This method employed NaOH, KNO$_3$, or NH$_4$NO$_3$ as bases or mineralizers, whereas NaNO$_3$, KNO$_3$, or NH$_4$NO$_3$ acted as stabilizers.

There are also studies describing the formation of nanocrystalline silicalite-1 from a biomass silica source as hand analyses and discuss the zeolite synthesis through a multistep method (Pengthamkeerati et al., 2015).

19.4.1.4   KEY FACTORS AND ISSUES: A BRIEF SUMMARY

The technical and general issues regarding synthesis of zeolite via conventional and prefusion hydrothermal processes have been long debated in the community. They have been finding solutions in both multistep and nonconventional methods. Various cases show that these treatments are well-known hydrothermal methods that have been modified. A general characteristic feature of these processes is an increase in the concentration of Si and Al in the solution that engenders a lower amount of residual, unreacted CFA material, as well as selected "pure zeolite" synthesis. The major issues related to these treatments are huge costs, complex techniques, and prolonged preparatory periods. Figures 19.8a and b illustrate the "two-step process" and "molten salt method," respectively.

19.4.2   END USES OF SYNTHETIC ZEOLITES

The zeolites synthesized from CFA have diverse and well-recognized applications. A study by Belviso et al. (2015a) reveals that there is a decrease in the potential toxicity of the raw material after the conversion of CFA to zeolite; also, there is less mobility of elements such as Cd, Pb, Ni, and Cr in the synthetic products because of the entrapment of these elements during the formation of new mineral structures. The above-mentioned fact might be the reason for the wide applications of FA. The successful deployment of synthetic CFA products in environmental pollution, catalytic processes, and selecting promising innovative applications has been outlined hereafter. This section focuses on the documentation that transformation of waste material to useful products is a feasible process.

19.4.2.1   ENVIRONMENTAL POLLUTION

Remediation of mine water and treatment of acid mine drainage have been employing zeolites produced from CFA (Rayalu et al., 2000; Rios et al., 2008; Vadapalli et al., 2010). Moreover, zeolites have also been used in the elimination of ammonium from contaminated solutions (Zhang et al., 2007, 2011), as well as in making polluted water free from heavy metals (Belviso et al., 2010a; Medina et al., 2010; Miyaji et al., 2010; Moreno et al., 2001a; Querol et al., 2001; Scott et al., 2001; Visa et al., 2012). A study shows the efficacy in the removal of all metals from mine water by zeolite formed from FA, and,

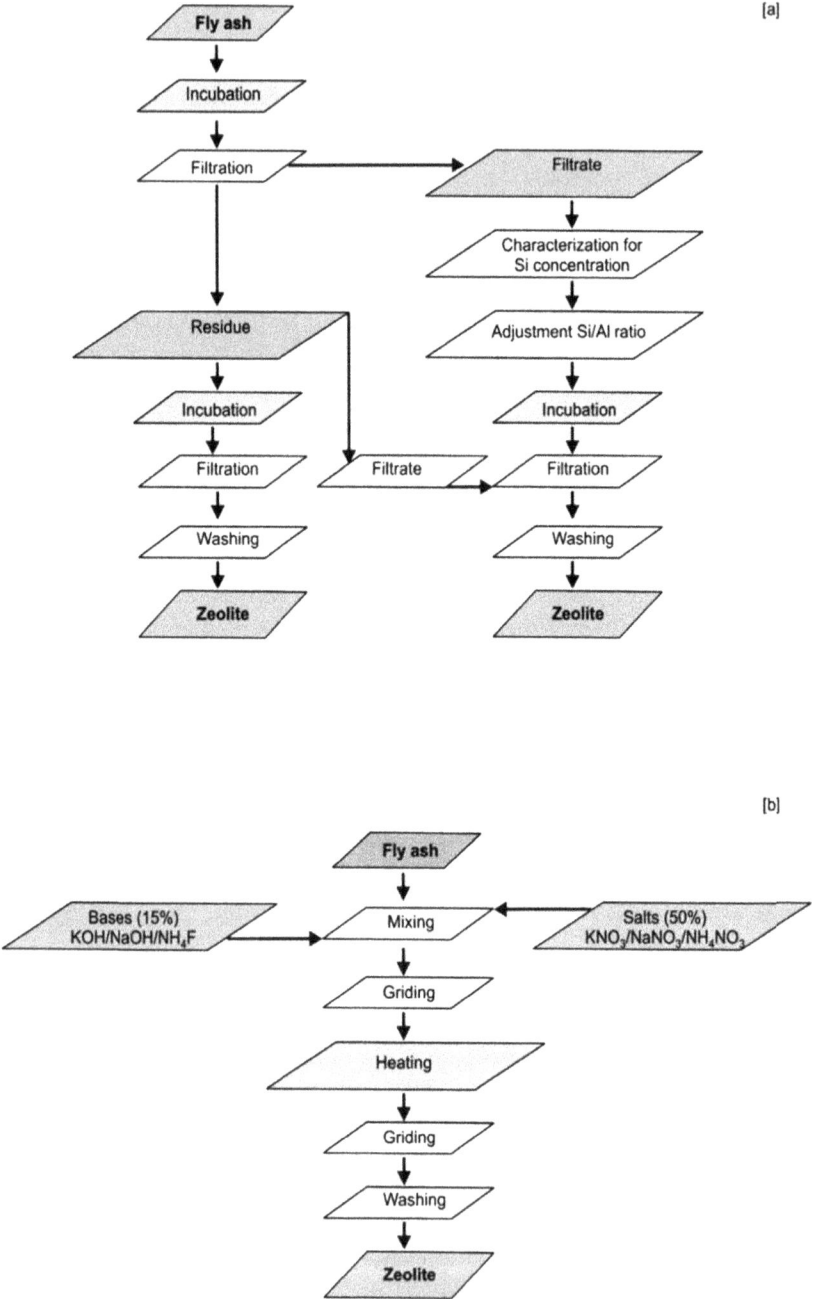

FIGURE 19.8   Flowchart of (a) the two-step process and (b) the molten salt method process (adapted from Jha et al., 2011)

subsequently, increased doses leading to improved removal. Chalupnik et al. (2013) demonstrated their results on the application of synthetic zeolite in the purification of wastewater from mines, which included a cost-effective passive treatment to eliminate radium isotopes and other constant pollutants like barium, iron, and manganese. Somerset et al. (2008) illustrated the reduction in lead and mercury concentrations in mine wastewater by 95% and 30%, respectively, on application of synthetic zeolites. Experiments on the capability of CFA-derived zeolite in removing ammonium from aqueous solutions concluded that synthetic faujasite is extremely proficient as an ion exchanger, as the kinetic Ho's pseudo-second-order model indicated (Zhang et al., 2011). Furthermore, thermodynamic studies show that an exothermic reaction takes place when ammonium is taken up during the synthesis of zeolite. Wu et al. (2006) explored the simultaneous elimination of ammonium and phosphate by FA synthetic products, and also revealed that efficiency in the remediation of polluted water improved through the presaturation of synthetic zeolite with a suitable cation such as Al^{3+}, Ca^{2+}, or Fe^{3+}. Another study showed that the use of H_2SO_4 treated synthetic zeolite greatly improve the removal efficacy of both ammonium and phosphate (Zhang et al., 2007). Metal absorption by zeolite synthesized from CFA was first scrutinized by Lin and His (1995). Some authors have reported the use of altered FA as an adsorbent for lead in contaminated aqueous solutions (Woolard et al., 2000). They investigated the efficacy of zeolite synthesized from FA as absorbents for lead in contaminated waters, and demonstrated the higher efficiency of Na-P1 zeolite than hydroxyl sodalite. The authors who proposed that the rate of lead-ion removal was influenced by zeolite type and morphology established these results (Scott et al., 2001). An investigation was made by Lee et al. (2003) on the absorption efficiency and selectivity of lead metals with the use of zeolite synthesized from FA as a function of initial metal concentration, pH of the solution, adsorption isotherm, and rate. After the onset of the experiments, nearly 90% of lead ions were found to be absorbed within 30 min and reached equilibrium within 2 h. To evaluate the capability of hydroxyl sodalite formed from FA for absorption of lead ions in aqueous media, Golbad et al. (2017) performed adsorption isotherm studies and found out that with increase in contact time, the capacity of absorption also increased. They observed a removal efficiency of 98.1% for a solid-to-liquid ratio of 3.0 g/dm^3, with an initial lead concentration of 100 mg/dm^3. Analyses on adsorption capacity of synthetic zeolite for trivalent chromium have also been performed (Hui et al., 2005; Penilla et al., 2006). A previous study demonstrated the removal mechanism of chromium by zeolitic FA involving the ion exchange and precipitation as chromium hydroxide (Wu et al., 2008).

Furthermore, the crucial role of pH has been observed, because the efficiency of toxic element removal reached 100% when pH values were above the range of the solution. This occurred due to the $Cr(OH)_3$ precipitation on the absorbent surfaces. Finally, a recent investigation has also been carried out on the kinetic behavior of indigo carmine dye adsorption by zeolite formed from CFA (deCarvalho et al., 2011).

In order to mitigate polluted soils, various remediation technologies have been developed. Among these technologies, many studies have demonstrated positive results when zeolites produced from FA were added to polluted soil (Moreno et al., 2001b; Querol et al., 2006). Column experiments have been performed by Lin et al. (1998), wherein they added 0.5–16 g of zeolite synthesized from CFA to 100 g Cd-contaminated soil samples, and the results showed that leaching of toxic elements was strongly inhibited by the addition of the synthetic product, indicating the capability of zeolite in stabilizing the Cd contents in polluted soil. Moreno et al. (2001a) have shown the immobilization of heavy metals in contaminated soil when zeolite synthesized from FA was added. Some authors, in order to remediate polluted soil, used Na-P1 zeolite synthesized from Spanish CFA. (Querol et al., 2006). Their results indicate that on addition of synthetic zeolite, pH of the soil instigated the precipitation of metal hydroxides which eventually lead to the immobilization of metals. Some papers have also documented the efficacy of zeolite formed from FA in the remediation of contaminated soils characterized by elevated concentrations of toxic elements (Belviso et al., 2010b, 2012; Terzano et al., 2005, 2007).

Synthetic zeolites can also be employed to adsorb gases like CO_2, SO_2, and mercury from exhaust gases. An analysis was carried out on the adsorption capacity of zeolite synthesized from silica extracts from FA to adsorb CO_2 and SO_2, and the sorption values were estimated up to 74 mg/g and 297 mg/g, respectively (Querol et al., 2002b). The authors also expressed that the real potential utilizations of this synthetic product for gas treatment lie in uptake of water vapor and/or SO_2 sorption from restricted water gaseous effluents. The uptake of mercury from a gaseous stream was tested with Na-X and Na-P zeolite formed from CFA by Wdowin et al. (2014). The authors thermally activated the synthetic products and treated them with silver in order to enhance the Hg adsorption efficiency. The data revealed that the performance of the zeolite was enhanced on treatment with silver. This was due to the exceptional Brunauer–Emmett–Teller surface area and combined micropore and mesopore volumes. The authors, later on, demonstrated various crucial factors affecting the mercury removal from the exhaust gases, which included sorbent texture (powder or granulated), exhaust gas flow rate, and contact time and temperature of the experiment (Wdowin et al., 2015). This study confirmed the reduction in the level of mercury in flue gas by Ag-impregnated synthetic Na-X zeolite. It showed long-term mercury breakthrough, ranging from 15% to 40% (these results of capture of mercury were obtained for granulated material).

The utilization of synthetic zeolite has been likewise tried for the adsorption of benzene, toluene, and isomers of xylene (BTX) from gas streams (Bandura et al., 2016), aqueous solutions (Bandura et al., 2017a), and land-based petroleum spills (Bandura et al., 2015, 2017b). A high BTX-sorption capacity in gas streams for Na-X zeolite formed from CFA was shown by Bandura et al. (2016), who demonstrated that this behavior is brought about by the pore structure of the faujasite zeolite group, which is characterized by channels with huge dimensions. Bandura et al. (2017a) studied rather the BTX-removal efficacy of zeolite synthesized from CFA, from aqueous solutions. It was also observed that adsorption by synthetic Na-P1 lead to the elimination of xylenes, toluene, and benzene by 99%, 77%, and 35%, respectively, at equilibrium. The order of removal (xylenes > toluene > benzene) corresponded with declining hydrophobicity and elevated water solubility and molecular weight. Additionally, the data specified that an increase in initial BTX concentration due to the saturation of adsorption sites at higher concentrations of elements leads to a decrease in zeolite-removal efficacy. Comparable experiments have been carried out to utilize synthetic zeolite altered with surfactants. Finally, Bandura et al. (2015) researched on the high sorption limit of Na-P1 and Na-X zeolites for two diesel fuels and one motor oil. The outcomes demonstrate that the zeolite's mesoporous filling plays a predominant role in the sorption process, and that the dispersion of particle size and sorbent surface are conclusive factors in the immobilization of petroleum products.

19.4.2.2   CATALYSIS

Few reports focus on FA-derived zeolite in catalyst application. It was investigated that the synthesis of Na-Y zeolite from Colombian CFA was modified by synthesizing zeolite through Na ion exchange with ammonium and hydrogen ions. By this process, catalytic efficiency of synthesized zeolite is improved in petroleum refinery catalytic-cracking units, which convert high-boiling fractions of petroleum crude to gasoline, diesel, and other products, as suggested by Htay and Oo (2008). The findings and hypothesis of Rios et al. (2012) are in agreement with those of Yamamoto et al. (2007), who explain zeolite ion exchange as being one of the most valuable methods for the preparation of heterogeneous catalysts, since structural cations can be easily exchanged with dissolved cations by zeolites. The authors emphasize that transition metals (and their ionic forms) stabilized in zeolite crystalline lattices have been used for catalyzing a wide range of organic reactions. Investigation of the utilization of FA-derived zeolite as a catalyst for polypropylene pyrolysis was performed by Nam et al. (2004) and Kim et al. (2003). FA collected from pulp and paper industries in Thailand was used as raw material for synthesizing ZSM-5 and also investigated derived zeolites for catalytic fast pyrolysis applications (Vichaphund et al., 2014).

19.4.2.3 *EMERGING APPLICATIONS AND NANOZEOLITE*

In the past few years, lot of interest has been given to the use of FA to form zeolite with particular properties (such as magnetic properties) and/or nanoparticle sizes (Belviso et al., 2015a, 2015b; Chen et al., 2012; Fungaro et al., 2012) for use in innovative applications (Saranya and Selvaraj, 2017). Zeolite with singular magnetic properties was formed by combining magnetite with zeolite synthesized from CA (Yamaura and Fungaro, 2013).

19.5 CONCLUSION

This chapter aims to give an overview of the various peer-review studies on CFA that have been published since 1990. CA is the major solid residue produced by coal combustion in thermoelectric power plants. The rapidly increasing production of combustion wastes over the past few years has provoked researchers to find latest waste applications and innovative methods for the conversion of waste into useful products. CFA utilization has been investigated in many contexts since the first publication in 1985, and the profuse research on FA conversion into zeolitic products has featured prominently. The synthesis of zeolite from CFA is generally carried out in a closed alkaline reaction system by one of three main processes—the conventional hydrothermal process, the alkaline fusion-assisted hydrothermal process, or multistep treatment. Each process can be characterized by its major advantages and disadvantages. However, although the conversion of FA to zeolite constitutes a quite well-documented approach to wisely reducing the amount of combustion residues deposited in landfills, more research is required, primarily regarding the synthesis of nanoparticle zeolites and their potential utilization in medical applications.

Fungistatic ash products can be utilized in a wide range of applications aimed at eliminating fungal growth, from preventive use to repair and reconstruction. Future research must focus on quantifying additional, potential fungistatic effects from minor elements and also on the evaluation of Portland cement composites with fungistatic ash.

KEYWORDS

- **coal fly ash**
- **environmental pollution**
- **fungistatic**
- **zeolite synthesis**

REFERENCES

Aljerf, L. Effect of thermal-cured hydraulic cement admixtures on the mechanical properties of concrete. *Interceram* 2015, 64, 346–356.

Apiratikul, R.; Pavasant, P. Sorption of Cu^{2+}, Cd^{2+}, and Pb^{2+} using modified zeolite from coal fly ash. *Chemical Engineering Journal* 2008, 144, 245–258.

Bandura, L.; Franus M.; Jozefaciuk G.; Franus W. Synthetic zeolites from fly ash as effective mineral sorbents for land-based petroleum spills cleanup. *Fuel* 2015, 147, 100–107.

Bandura, L.; Panek R.; Rotko M.; Franus W. Synthetic zeolites from fly ash for an effective trapping of BTX in gas stream. *Microporous and Mesoporous Materials* 2016, 223, 1–9.

Bandura, L.; Kolodynska D.; Franus W. Adsorption of BTX from aqueous solutions by Na-P1 zeolite obtained from flyash. *Process Safety and Environmental Protection* 2017a, 109, 214–23.

Bandura, L.; Woszuk, A.; Kolodynska, D.; Franus, W. Application of mineral sorbents for removal of petroleum substances: a review. *Minerals* 2017b, 37, 1–25.

Belviso, C. Ultrasonic vs hydrothermal method: different approaches to convert fly ash into zeolite. How they affect the stability of synthetic products over time? *Ultrasonics Sonochemistry* 2018, 43, 9–14.

Belviso, C.; Cavalcante, F.; Fiore, S. Synthesis of zeolite from Italian coal fly ash. Differences in crystallization temperature using seawater instead of distilled water. *Waste Manag.* 2010a, 30, 839–847.

Belviso, C.; Cavalcante, F.; Ragone, P.; Fiore, S. Immobilization of Ni by synthesizing zeolite at low temperatures in a polluted soil. *Chemosphere* 2010b, 78, 1172 –6.

Belviso, C.; Cavalcante, F.; Ragone, P.; Fiore, S. Immobilization of Zn and Pb in polluted soil by in-situ crystallization zeolites from fly ash. *Water, Air, & Soil Pollution* 2012, 223, 5357–5364.

Belviso, C.; Cavalcante, F.; Di Gennaro, S.; Palma, A.; Ragone, P.; Fiore, S. Mobility of trace elements in fly ash and in zeolitised coal fly ash. *Fuel* 2015a, 144, 369–79.

Belviso, C.; Giannossa, L.C.; Huertas, F.J.; Lettino A, Mangone A, Fiore S. Synthesis of zeolites at low temperatures in fly ash-kaolinite mixtures. *Microporous and Mesoporous Materials* 2015b, 212, 35–47.

Boke, N., Birch, G.D., Nyale, S.M., Petrik, L.F. New synthesis method for the production of coal fly ash-based foamed geopolymers. *Construction and Building Materials* 2015, 75, 189–199.

Catalfamo, P.; Corigliano, F.; Primerano, P.; DiPasquale, S. Study of the pre-crystallization stage of hydrothermally treated amorphous aluminosilicates through the composition of the aqueous phase. *Journal of the Chemical Society, Faraday Transactions*1993, 89, 171–175.

Chalupnik, S.; Franus, W.; Wysocka, M.; Gzyl, G. Application of zeolites for radium removal from mine water. *Environmental Science and Pollution Research.* 2013, 20, 7900–7906.

Chen, X.Y.; Khunjar, W.; Jun, Z.; Li, J.L.; Yu, X.; Zhang, Z. Synthesis of nano-zeolite from coal flyash and its potential for nutrient sequestration from an aerobically digested swine wastewater. *Bioresource Technology* 2012, 110, 79–85.

Cwalina, B. Biodeterioration of concrete. *Architecture Civil Engineering Environment* 2008, 1, 133–140.

deCarvalho, D.A.; Fungaro Magdalena, C.P.; Cunico, P. Adsorption of indigo carmine from aqueous solution using coal flyash and zeolite from fly ash. *Journal of Radioanalytical and Nuclear Chemistry* 2011, 289, 617–626.

Ding, J., Ma, S., Shen, S., Xie, Z., Zheng, S., Zhang, Y. Research and industrialization progress of recovering alumina from fly ash: a concise review. *Waste Management* 2017, 60, 375–387.

Ebrahimi, A., Saffari, M., Milani, D., Montoya, A., Valix, M., Abbas, A. Sustainable transformation of fly ash industrial waste into a construction cement blend via CO_2 carbonation. *Journal of Cleaner Production* 2017, 156, 660–669.

El-Naggar, M.R.; El-Kamash, A.M.; El-Dessouky, M.I.; Ghonaim, A. K. Two-step method for preparation of NaA-X zeolite blends from flyash for removal of cesiumions. *Journal of Hazardous Materials* 2008, 154, 963–972.

Franus, W.; Wdowin, M.; Franus, M. Synthesis and characterization of zeolites prepared from industrial fly ash. *Environmental Monitoring and Assessment* 2014, 186, 5721–5729.

Fukasawa, T., Karisma, A.D., Shibata, D., Huang, A.-N., Fukui, K. Synthesis of zeolite from coal fly ash by microwave hydrothermal treatment with pulverization process. *Advanced Powder Technology* 2017, 28, 798–804.

Fungaro, D. A.; Yamaura, M.; Craesmeyer, G. R. Uranium removal from aqueous solution by zeolite from flyash iron oxide magnetic nano composite. *International Review of Chemical Engineering* 2012, 4, 353–358.

George, R.P.; Vishwakarma, V.; Samal, S.S.; Mudali, U.K. Current understanding and future approaches for controlling microbially influenced concrete corrosion: a review. *Concrete Research Letters* 2012, 3, 491–506.

Golbad, S.; Khoshnoud, P.; Abu-Zahra, N. Hydrothermal synthesis of hydroxyl soda-lite from flyash for the removal of lead ions from water. *International Journal of Environmental Science and Technology* 2017, 14, 135–142.

Gross-Lorgouilloux, M.; Soulard, M.; Caullet, P.; Patarin, J.; Moleiro, E.; Saude, I. Conversion of coal flyashes into fauja site under soft temperature and pressure conditions: influence of additional silica. *Microporous and Mesoporous Materials* 2010, 127, 41–49.

Hollman, G.G.; Steenbruggen, G.; Janssen-Jurkovicova, M. A two-step for the synthesis of zeolites from coal fly ash. *Fuel* 1999, 78, 1225–1230.

House, M.W.; Weiss, W.J. Review of microbially induced corrosion and comments on needs related to testing procedures. In *Proceedings of the 4th International Conference on the*

Durability of Concrete Structures. Purdue University, West Lafayette, IN, USA, 2014, pp. 94–103.

Hoy, M., Horpibulsuk, S., Rachan, R., Chinkulkijniwat, A., Arulrajah, A. Recycled asphalt pavement–fly ash geopolymers as a sustainable pavement base material: strength and toxic leaching investigations. *Science of the Total Environment* 2016, 573, 19–26.

Htay, M.M.; Oo, M.M. Preparation of zeoliteY catalyst for petroleum cracking. *World Academy of Science, Engineering and Technology* 2008, 24, 114–120.

Huggins, F.E.; Rezaee, M.; Honaker, R.Q.; Hower, J.C. On the removal of hexavalent chromium from a Class F fly ash. *Waste Management* 2016, 51, 105–110.

Hui, K.S.; Chao, C.Y.H.; Kot, S.C. Removal of mixed heavy metal ions in wastewater by zeolite 4A and residual products from recycled coal fly ash. *Journal of Hazardous Materials* 2005, 127, 89–101.

Hui, K.S.; Chao, C.Y.H. Pure, single phase, high crystalline, chamfered-edge zeolite 4A synthesized from coal fly ash for use as a builder in detergents. *Journal of Hazardous Materials* 2006a, 137, 401–409.

Hui, K.S.; Chao, C.Y.H. Effects of step-change of synthesis tem-perature on synthesis of zeolite 4A from coal flyash. *Microporous and Mesoporous Materials* 2006b, 88, 145–151.

International Energy Outlook. U.S. Energy Information Administration. EIA released in September 14, 2017. www.eia.gov/ieo

Jha, B.; Singh, D.N. Fly ash zeolites: innovations, applications, and directions. In: Ochsner, A.; Da Silva, L.F.M.; Altenbach, H. (Eds.); *Advanced Structured Materials*. Springer, Singapore, 2016.

Ji, L., Yu, H., Wang, X., Grigore, M., French, D., Gozukara, Y.M., Yu, J., Zeng, M. CO_2 sequestration by direct mineralisation using fly ash from Chinese Shenfu coal. *Fuel Processing Technology* 2017, 156, 429–437.

Jimenez, I.; Perez, G.; Guerrero, A.; Ruiz, B. Mineral phases synthesized by hydro-thermal treatment from biomass ashes. *International Journal of Mineral Processing* 2017, 158, 8–12.

Kim, S.S.; Kim, J.H.; Chung, S.H. A study on the application of flyash-derived zeolite materials for pyrolysis of polypropylene. *Journal of Industrial and Engineering Chemistry* 2003, 9, 287–293.

Lee, M.G.; Cheon, J.K.; Kam, S.K. Heavy metal adsorption characteristics of zeolite synthesized from fly ash. *Journal of Industrial and Engineering Chemistry* 2003, 9, 174–180.

Li, C.-C.; Qiao, X.-C. A new approach to prepare mesoporous silica using coal fly ash. *Chemical Engineering Journal* 2016, 302, 388–394.

Lin, C.F.; His, H.C. Resource recovery of waste flyash, synthesis of zeolite-like materials. *Environmental Science & Technology* 1995, 29, 1109–1117.

Lin, C.F.; Lo, S.S.; Lin, H.Y.; Lee, Y. Stabilization of cadmium contaminated soils using synthesized zeolite. *Journal of Hazardous Materials* 1998, 60, 217–226.

Lu, N.; Wang, X.; Meng, L.; Ding, C.; Liu, W.; Shi, H.; Hu, X.; Wu, K. Electromagnetic interference shielding effectiveness of magnesium alloy-fly ash composites. *Journal of Alloys and Compound* 2015, 650, 871–877.

Ma, W.; Brown, P.W.; Komarneni, S. Characterization and cation exchange properties of zeolite synthesized from fly ashes. *Journal of Materials Research and Technology* 1998, 13, 3–7.

Matlob, A.S.; Kamarudin, R.A.; Jubri, Z.; Ramli, Z. Using the response surface methodology to optimize the extraction of silica and alumina from coal fly ash for the synthesis of zeolite Na-A. *Arabian Journal for Science and Engineering* 2012, 37, 27–40.

Medina, A.; Gamero, P.; Almanza, J.M.; Vargas, A.; Montoya, A.; Vargas, G.; Izquierdo, M. Fly ash from a Mexican mineral coal. II. Source of W zeolite and its effectiveness in arsenic(V) adsorption. *Journal of Hazardous Materials* 2010, 181, 91–104.

Miyaji, F.; Masuda, S.; Suyama, Y. Adsorption removal of lead and cadmium ions from aqueous solution with coal flyash derived zeolite/sepiolite composit. *Journal of the Ceramic Society of Japan* 2010, 118, 1062–1066.

Moreno, N.; Querol, X.; Lopez-Soler, A.; Andres, J.M.; Janssen, M.; Nugteren, H.; Towler, M.; Stanton, K. Determining suitability of a fly ash for silica extraction and zeolite synthesis. *Journal of Chemical Technology & Biotechnology* 2004, 79, 1009–1018.

Moreno, N.; Querol, X.; Plana, F.; Andres, J.M.; Janssen, M.; Nugteren, H. Pure zeolite synthesis from silica extracted from coal flyashes. *Journal of Chemical Technology & Biotechnology* 2002, 77, 274–279.

Moreno, N.; Querol, X.; Alastuey, A.; Garcia-Sanchez, A.; Soler, L.A.; Ayora, C. Immobilization of heavy metals in polluted soils by the addition of zeolitic materials synthesized from coal flyash. International Ash Utilization Symposium, Center for Applied Energy Research, University of Kentucky. Lexington, KY, 2001a, pp. 1–8.

Moreno, N.; Querol, X.; Ayora, C.; Alastuey, A.; Fernandez-Pereira, C.; Janssen-Jurkovi-cova, M. Potential environmental applications of pure zeolitic material synthesized from fly ash. *Journal of Environmental Engineering* 2001b, 994–1002.

Morton, G. Things that go rot in the night—a review of biodeterioration. *Microbiology Today* 2003, 30, 103–106.

Moutsatsou, A.; Stamatakis, E.; Hatzizotzia, K.; Protonotarios, V. The utilization of Ca-rich and Ca-Si-rich fly ashes in zeolites production. *Fuel.* 2006, 85, 657–663.

Murayama, N.; Yamamoto, H.; Shibata, J. Mechanism of zeolite synthesis from coal fly ash by alkali hydrothermal reaction. *International Journal of Mineral Processing* 2002, 64, 1–17.

Nam, Y.M.; Kim, S.M.; Lee, J.H.; Kim, S.J.; Chung, S.H. A study on the application of fly ash-derived zeolite materials for pyrolysis of polypropylene(II). *Journal* of *Industrial* and *Engineering Chemistry* 2004; 10, 788–793.

Park, M.; Choi, C.L.; Lim, W.T.; Kim, M.C.; Choi, J.; Heo, N.H. Molten-salt method for the synthesis of zeolitic materials. I. Zeolite formation in alkaline molten-salt system. *Microporous and Mesoporous Materials* 2000, 37, 81–89.

Pengthamkeerati, P.; Kraewong, W.; Meesuk, L. Green synthesis of nano-silicalite-1: biomass flyash as asilica source and mother liquid recycling. *Environmental Progress & Sustainable Energy* 2015, 34, 1–6.

Penilla, R.P.; Bustos, G.; Elizalde, S.G. Immobilization of Cs, Cd, Pb, and Cr by synthetic zeolites from Spanish low-calcium coal fly ash. *Fuel* 2006, 85, 823–832.

Querol, X.; Alastuey, A.; Turiel, F.; Lopez-Soler, A. Synthesis of zeolites by alkaline activation of ferro-aluminous flyash. *Fuel* 1995, 74, 1226–1231.

Querol, X.; Plana, F.; Alastuey, A.; Lopez-Soler, A. Synthesis of Na-zeolites from fly ash. *Fuel* 1997, 76, 793–799.

Querol, X.; Umana, J.C.; Alastuey, A.; Bertrana, C.; Lopez-Soler, A. Plana, F. Physico-chemical characterization of Spanish Fly Ashes. *Energy Sources* 1999, 21, 883–898.

Querol, X.; Umaena, J.C.; Plana, F.; Alastuey, A.; Lopez-Soler, A.; Medinacelli, A.; Domingo, M.J.; Garcia-Rojo, E. Synthesis of zeolites from flyash at pilot plant scale. Examples of potential applications. *Fuel* 2001, 80, 857–865.

Querol, X.; Moreno, N.; Umana, J.C.; Alastuey, A.; Hernandez, E.; Lopez-Soler, A.; Plana, F. Synthesis of zeolites from coal flyash: an overview. *International Journal* of *Coal Geology* 2002a, 50, 413–423.

Querol, X.; Moreno, N.; Uman, J.C.; Juan, R.; Hernandez, S.; Fernandez-Pereira, C.; Ayora, C.; Janssen, M.; Garcia-Martinez, J.; Linares-Solano, A.; Cazorla-Amoros, D. Application of zeolitic material synthesized from fly ash to the decontamination of waste water and flue gas. *Journal of Chemical Technology & Biotechnology* 2002b, 77, 292–298.

Querol, X.; Alastuey, A.; Moreno, N.; Alvarez-Ayuo, E.; Garcia-Sanchez, A.; Cama, J.; Ayora, C.; Simon, M. Immobilization of heavy metals in polluted soils by the the addition of zeolitic materials synthesized from coal flyash. *Chemosphere* 2006, 62, 171–80.

Ram, L.; Masto, R. Fly ash for soil amelioration: a review on the influence of ash blending with inorganic and organic amendments. *Earth-Science Reviews* 2014, 128, 52–74.

Rayalu, S.; Meshram, S.U.; Hasan, M.Z. Highly crystalline faujasitic zeolites from fly ash. *Journal of Hazardous Materials* 2000, 77, 123–131.

Revanasiddappa, M.; Swamy, D.S.; Vinay, K.; Ravikiran, Y.; Raghavendra, S. Synthesis, characterization and DC conductivity studies of conducting polyaniline/PVA/Fly ash polymer composites. *AIP Conference Proceedings* 2018, pp. 090070. https://doi.org/10.1063/1.5032917

Rios, C.A.; Williams, C.D.; Roberts, C.L. Removal of heavy metals from acid mined rainage (AMD) using coal fly ash, natural clinker and synthetic zeolites. *Journal of Hazardous Materials* 2008, 156, 23–35.

Sahoo, P.K.; Kim, K.; Powell, M.; Equeenuddin, S.M. Recovery of metals and other beneficial products from coal fly ash: a sustainable approach for fly ash management. *International Journal of Coal Science and Technology* 2016, 3, 267–283.

Saranya, D.; Selvaraj, V. Hydrothermal assisted synthesis of zeolite based nickel deposited poly (pyrrole-co-fluoroaniline)/CuS catalyst for methanol and sulphur fuel cell applications. *Journal of Electroanalytical Chemistry* 2017, 787, 55–65.

Scott, J.; Guang, D.; Naeramitmarnsuk, K.; Thabuot, M.; Amal, R. Zeolite synthesis from coal flyash for the removal of lead ions from aqueous solution. *Journal of Chemical Technology and Biotechnology* 2001, 77, 63–69.

Shaheen, S.M.; Hooda, P.S.; Tsadilas, C.D. Opportunities and challenges in the use of coal fly ash for soil improvements—a review. *Journal of Environmental Management* 2014, 145, 249–267.

Singer, A.; Berkgaut, V. Cation exchange properties of hydrothermally treated coal flyash. *Environmental Science and Technology* 1995, 29, 1748–1753.

Somerset, V.; Petrik, L.; Iwuoha, E. Alkaline hydrothermal conversion of flyash precipitates into zeolites. The removal of mercury and lead ions from wastewater. *Journal of Environmental Management* 2008, 87, 125–131.

Steenbruggen, G.; Hollman, G.G. The synthesis of zeolites from fly ash and the properties of the zeolite products. *Journal of Geochemical Exploration* 1998, 62, 305–309.

Tanaka, H.; Sakai, Y.; Hino, R. Formation of Na-A and X zeolites from waste solutions in conversion of coal fly ash zeolites. *Materials Research Bulletin* 2002, 37, 1873–1884.

Tanaka, H.; Matsumura, S.; Furusawa, S.; Hino, R. Conversation of fly ash into zeolites for ion-exchange applications. *Journal of Materials Science* 2003, 22, 32–35.

Tanaka, H.; Matsumura, S.; Hino, R. Formation process of Na-X zeolites from coal fly ash. *Journal of Materials Science* 2004a, 39, 1677–1682.

Tanaka, H.; Miyagawa, A.; Eguchi, H.; Hino, R. Synthesis of a single-phase Na-A zeolite from coal fly ash by dialysis. *Industrial and Engineering Chemistry Research* 2004b, 43, 6090–6094.

Tanaka, H.; Eguchi, H.; Fujimoto, S.; Hino, R. Two-step process for synthesis of as in-glephase Na-A zeolite from coal flyash by dialysis. *Fuel* 2006, 85, 1329–1334.

Terzano, R.; Spagnuolo, M.; Medici, L.; Vekemans, B.; Vincze, L.; Janssens, K.; Ruggiero, P. Copper stabilization by zeolite synthesis in polluted soils treated with coal flyash. *Environmental Science and Technology* 2005, 20, 6280–6287.

Terzano, R.; Spagnuolo, M.; Medici, L.; Dorrine, W.; Janssens, K.; Ruggiero, P. Microscopic single particle characterization of zeolites synthesized in a soil polluted by copper or cadmium and treated with coal fly ash. *Applied Clay Science* 2007, 35, 128–138.

Vadapalli, V.R.K.; Gitari, W.M.; Ellendt, A.; Petrik, L.F.; Balfour, G. Synthesis of zeolite-P from coal fly ash derivative and its utilization in mine-water remediation. *South African Journal of Science* 2010, 106, 1–7.

Vichaphund, S.; Aht-Ong, D.; Sricharoenchaikul, V.; Atong, D. Characteristic of fly ash derived-zeolite and its catalytic performance for fast pyrolysis of Jatropha waste. *Environmental Technology* 2014, 35, 2254–2261.

Visa, M.; Isac, L.; Duta, A. Flyash adsorbents for multi-cation wastewater treatment. *Applied Surface Science* 2012, 258, 6345–6352.

Vucinic, D.; Miljanovic, I.; Rosic, A.; Lazic, P. Effect of Na_2O/SiO_2 mole ratio on the crystal type of zeolite synthesized from coal fly ash. *Journal of the Serbian Chemical Society* 2003, 68, 471–478.

Waek, T.T.; Saito, F.; Zhang, Q. The effect of low solid/liquid ratio on hydrothermal synthesis of zeolites from fly ash. *Fuel* 2008, 87, 3194–3199.

Wang, F., Wang, S., Meng, Y., Zhang, L., Wu, Q., Hao, J. Mechanisms and roles of fly ash compositions on the adsorption and oxidation of mercury in flue gas from coal combustion. *Fuel* 2016, 163, 232–239.

Wang, S. Application of solid ash-based catalysts in heterogeneous catalysis. *Environmental Science and Technology* 2008, 42, 7055–7063.

Wdowin, M.; Wiatros-Motyka, M.M.; Panek, R.; Stevens, L.A.; Franus, W.; Snape, C. E. Experimental study of mercury removal from exhaust gases. *Fuel* 2014, 128, 415–457.

Wdowin, M.; Macherzynski, M.; Panek, R.; Gorecki, J.; Franus, W. Investigation of the sorption of mercury vapour from exhaust gas by an Ag-Xzeolite. *Clays and Clay Minerals* 2015, 50, 31–40.

Woolard, C.; Petrus, K.; Vander Horst, M. The use of a modified flyash as an adsorbent for lead. *Water SA* 2000, 26, 531–536.

Worathanakul, P.; Tobarameeku, P. Development of Na Y zeolite derived from biomass and environmental assessment of carbon dioxide reduction. *MATEC Web of Conferences* 2016, 62, 1–5.

Wu, D.; Zhang, B.; Li, C.; Zhang, Z.; Kong, H. Simultaneous removal of ammonium and phosphate by zeolite synthesized from fly ash as influenced by salt treatment. *Journal of Colloid and Interface Science* 2006, 304, 300–306.

Wu, D.; Sui, Y.; He, S.; Wang, X.; Li, C.; Kong, H. Removal of trivalent chromium from aqueous solution by zeolite synthesized from coal fly ash. *Journal of Hazardous Materials* 2008, 155, 415–423.

Xu, G., Shi, X. Characteristics and applications of fly ash as a sustainable construction material: a state-of-the-art review. *Resources, Conservation and Recycling* 2018, 136, 95–109.

Yamamoto, T.; Eiadua, A.; Kim, S.I.; Ohmori, T. Preparation and characterization of cobalt cation-exchanged NaX zeolite as catalyst for wastewater treatment. *Journal of Industrial and Engineering Chemistry* 2007, 13, 1142–1148.

Yamaura, M.; Fungaro, D.M. Synthesis and characterization of magnetic adsorbent prepared by magnetite nanoparticles and zeolite from coal fly ash. *Journal of Materials Science* 2013, 48, 93–101.

Yang, J.; Zhao, Y.; Zhang, S.; Liu, H.; Chang, L.; Ma, S.; Zhang, J.; Zheng, C. Mercury removal from flue gas by magnetospheres present in fly ash: role of iron species and modification by HF. *Fuel Processing Technology* 2017 167, 263–270.

Yao, Z.T.; Ji, X.S.; Sarker, P.K.; Tang, J.H.; Ge, L.Q.; Xia, M.S.; Xi, Y.Q. A comprehensive review on the applications of coal fly ash. *Earth-Science Reviews* 2015, 141, 105–121.

Zhang, B.; Wu, D.; Wang, C.; He, S.; Zhang, Z.; Kong, H. Simultaneous removal of ammonium and phosphate by zeolite synthesized from coal fly ash as influenced by treatment. *Journal of Environmental Sciences* 2007, 19, 540–545.

Zhang, M.; Zhang, H.; Xu, D.; Han, L.; Niu, D.; Tian, B.; Zhang, J.; Zhang, L.; Wu, W. Removal of ammonium from aqueous solutions using zeolite synthesized from fly ash by a fusion method. *Desalination* 2011, 271, 111–121.

Zhao, X.S.; Lu, G.Q.; Zhu, H.Y. Effects of ageing and seeding on the formation of zeolite Y from coal fly ash. *Journal of Porous Material* 1997, 4, 245–251.

Zhuang, X.Y.; Chen, L.; Komarneni, S.; Zhou, C.H.; Tong, D.S.; Yang, H.M.; Yu, W.H.; Wang, H. Fly ash-based geopolymer: clean production, properties and applications. *Journal of Cleaner Production* 2016, 125, 253–267.

## CHAPTER 20

Heavy Metal Toxicity and Their Bioremediation Using Microbes, Plants, and Nanobiomaterials

S. K. SAHU* and H. G. BEHURIA

Department of Biotechnology, North Orissa University, Baripada 757003, Odisha, India

Corresponding author. E-mail: rsantoshnou@gmail.com

ABSTRACT

Heavy metals and their derivatives constitute one of the most toxic environmental pollutants that exhibit multiple deleterious effects on biosphere. As physical or chemical procedure for removal of heavy metal is expensive, need special instrumentation and not eco-friendly, bioremediation involving microbes and plants has received considerable interest over the years. Use of nanomaterials synthesized by microbes and plant products in bioremediation of heavy metals is an eco-friendly and cost-effective procedure for detection and their removal from contaminated air, water, and soil. Heavy metal tolerant bacteria like *Bacillus coagulans, Streptococcus mitis,* and *Pseudomonas fluorescens* are capable of tolerating high concentration of toxic heavy metals such as Hg, Cu, Cr, Cd, Zn, Co due to their altered cellular mechanism. Similarly, plants such as *Gundelia tournefortii, Reseda lutea, Scariola orientalis, Eleagnum angustifolia, Noaea Mucronata* are known for their high heavy metal accumulating abilities from soil and water bodies and converting them into nontoxic components. Therefore, nanomaterials synthesized using microbial and plant components may exhibit enhanced ability to detect and detoxify heavy metals from the environments. Biogenic nanoparticle synthesis involves plants, fungi, yeast, actinomycetes, bacteria, etc., or their products to synthesize or fabricate nanoparticles. Biomaterials such as polysaccharides, metallo-thionins, chitin, and polyphenols have shown promising results for preparation of nanomaterials and their use for the effective bioremediation of heavy metals. This chapter describes heavy metal toxicity in the environments, their toxic effects and their bioremediation using plants, bacteria and bionanomaterials.

20.1   INTRODUCTION

Out of 118 known elements, 80 are metals. Living organisms are made from 34 elements. Carbon (C), hydrogen (H), nitrogen (N), oxygen (O), phosphorous (P), sulfur (S), and chlorine (Cl) constitute 99% of a cell, whereas the other 1% is made from trace elements such as cobalt (Co), copper (Cu), iron (Fe), manganese (Mn), molybdenum (Mo), and vanadium (V) (Underwood et al., 1997). Metals such as potassium (K^+), sodium (Na^+), and calcium (Ca^{2+}) that are essential constituents of the cell do not fall into the category of heavy metals. However, other metals such as Cu, Zn, Mn, Fe, etc., constitute heavy metals. These heavy metals either act as cofactors in enzymes or as prosthetic groups of proteins (e.g., hemoglobins, hemocyanins, etc.) that is essentially required for their structure and function. These metals are not toxic to cells at their physiological concentration (Table 20.1). However, other metals those are not a component of living system includes cadmium (Cd), chromium (Cr), Mercury (Hg), lead (Pb), arsenic (As), and antimony (Sb). These metals are not required by cellular enzymes or proteins and are toxic to all known life forms. Heavy metals and their derivatives are among the most hazardous environmental pollutants because of their toxicity, nonbiodegradability, and ability of bioaccumulation. In human, heavy metal toxicity has profound pathophysiological consequences (Gautam et al., 2014; American Physical Society, 1990; Ratnaike, 2003; Satarug and Moore, 2004). Anthropogenic activities such as metallurgical operations, petrochemical refining, and mining activities have led to substantial increase in toxic metals into the environment.

These heavy metals are mainly transported by runoff water and contaminated soil and water bodies downstream to their site of origin. Heavy metals enter into the body of organisms inhabiting the contaminated sites. For example, bacteria and other microbes and plants inhabiting heavy metal contaminated soil exhibit high concentration of these metals. Fishes grown in heavy metal contaminated water exhibit higher concentration of heavy metal in their tissues. Similarly, vegetables grown in fields enriched in heavy metals contain elevated amount of different heavy metals. Consumption of heavy metal contaminated water, aquatic animals, and vegetables constitutes the most important route of heavy metal contamination in human (Medici and Huster, 2017). Once, inside the body, these heavy metals make their route to the organs such as liver, kidney, and brain those act as metabolic hubs and accumulate there to a level that interferes with normal functioning of these organs (Arnich et al., 2012). Ironically, these heavy metals do not produce any obvious symptoms before they affect the function of these organs. In this chapter, we discuss the recent advances to remove heavy metal contamination using nanobiotechnological processes.

TABLE 20.1 Physiological Limit of Heavy Metals

Metal	Physiological Tolerance Limit	Function	References
Co	0.13–0.48 µg/kg bw/day	Important component of vitamin B12 (cobalamin). Promotes RBC formation. Activates enzymes, replaces zinc in some enzymes. It also helps in repair of myelin, which surrounds and protects nerve cells	Arnich et al., 2012
Cu	900 µg/day	Maintains collagen and elastin. It helps to maintain healthy bones, blood vessels, nerves, and immune function. Contributes to iron absorption	Medici, 2017
Zn	40 mg/day	Involved in many reactions of the cellular metabolism, such as antioxidative defense, protein synthesis, carbohydrate metabolism and stability of genetic materials	Fosmire, 1990
Pb	10 µg/dL	No known function	Wani et al., 2005
As	10–50 µg/L	No known function	Ratnaike, 2003
Cd	30 µg/day	No known function	Satarug, 2000
Hg	0.1–0.47 µg/kg/day	No known function	Jones, 1999; Montague, 1998

20.2 MECHANISM OF HEAVY METAL TOXICITY ON LIVING ORGANISM

Heavy metals affect the growth, development, and survival of organisms. Heavy metal contaminated food can seriously deplete some essential nutrients such as Fe, Ca, and Mg in the body. These metals are displaced by heavy metals forming their bound proteins. Lack of these essential nutrients lead to depleted immunological developments, mental retardation, and psychological disabilities. Heavy metals also enhance the cytosolic reactive oxygen species (ROS) in cytosol (Figure 20.1). Heavy metals can be divided into two groups according to accumulation in the cell, redox active (Fe, Cu, Cr, and Co), and redox inactive (Cd, Zn, Ni, Al, etc.). The redox active heavy metals are directly involved in the redox reaction in cells and result in the formation of $O_2^{\bullet-}$ and ${}^{\bullet}OH$ production by Haber–Weiss and Fenton reactions (Sharma and Dietz, 2009; Fosmire, 1990). Exposure to redox inactive heavy metals also results in oxidative stress through indirect mechanisms such as interaction with the antioxidant defense system, disruption of the electron transport chain, or induction of lipid peroxidation. Subsequently, it induces the increase in lipoxygenase activity. Another important mechanism of heavy metal toxicity is the ability of it to bind strongly to O, N, and S atoms (Nieboer and Richardson, 1980). HMs can inactivate enzymes by binding to cysteine residues. For example, Cd binding to sulfhydryl groups of structural proteins and enzymes leads to misfolding and inhibition of activity and interference with enzymatic regulation. Heavy metals such as mercury attached to the selenohydryl and sulfhydryl groups which undergo reaction with methyl mercury results in damaging the tertiary and quaternary protein structure and alter the cellular function (Patrick, 2002). It is also involved with the process of transcription and translation resulting in the disappearance of ribosomes and eradication of endoplasmic reticulum and the activity of natural killer cells. High accumulation of Cd in the body is capable of binding with cystein, glutamate, histidine, and aspartate ligands and can lead to the deficiency of Fe (Boonyapookana et al., 2002). Cd and zinc have the same oxidation states and hence Cd can replace zinc present in metallothionein, thereby inhibiting it from acting as a free-radical scavenger within the cell. Heavy metal such as Pb increases the level of ROS and decreases antioxidants level. Since glutathione exists both in reduced (GSH) and oxidized (GSSG) state, the reduced form of glutathione gives its reducing equivalents ($H^+ + e^-$) from its thiol groups of cystein to ROS in order to make them stable. In the presence of the enzyme glutathione peroxidase, reduced glutathione readily binds with another molecule of glutathione after donating the electron and forms glutathione disulfide (GSSG). The reduced form (GSH) of glutathione accounts for 90% of the total glutathione content and the oxidized form (GSSG) accounts for 10% under normal conditions. Under the condition of oxidative stress, the concentration of GSSG exceeds the concentration of GSH and induces lipid peroxidation (Noctor et al., 2016).

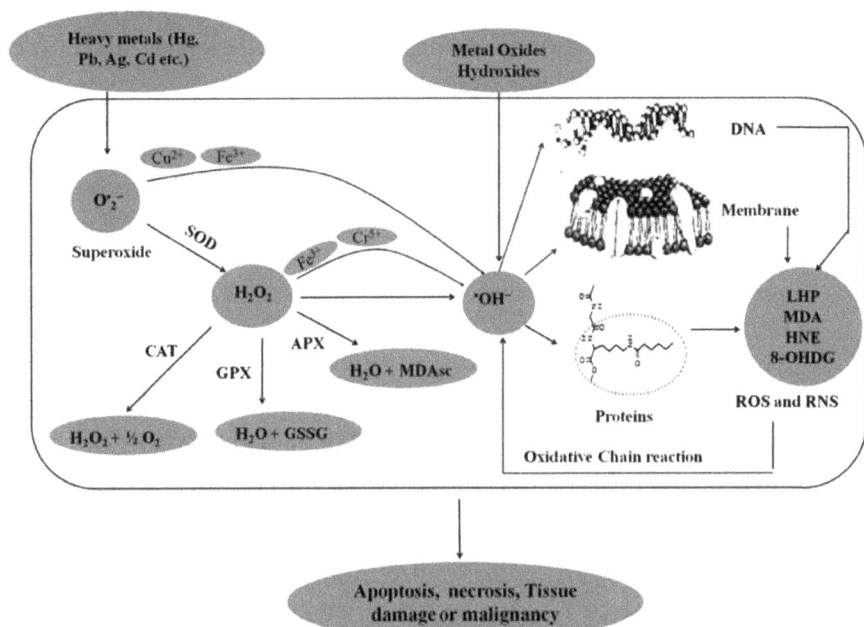

FIGURE 20.1 Mechanism of heavy metal toxicity on bacteria. Heavy metals such as lead (Pb), silver (Ag), and cadmium (Cd) can directly penetrate through the plasma membrane. After entering the cytosol, these heavy metals lead to generation of superoxide ion ($O^{•2-}$). This superoxide ion is acted upon by superoxide dismutase (SOD) to form H_2O_2 or interacts with transition metal ions such as Cu^{2+} and Fe^{3+} to form hydroxyl radical ($^•OH^-$). $^•OH^-$ is highly reactive and leads to oxidation of lipids, proteins, and DNA. The products of this oxidation are reactive oxygen species (ROS) or reactive nitrogen species (RNS), lipid hydroperoxides (LHP), malonal dehyde (MDA), hydroxyl noneals (HNE), and 8-hydroxydeoxyguanosine (8-OHDG). These oxidative intermediates are reactive and lead to the formation of $^•OH^-$ that further damages the cellular components. Cells express antioxidant enzymes such as catalase (CAT), superoxide dismutase (SOD), and glutathione peroxidase (GPX) and ascorbate peroxidase (APX) to reduce the concentration of ROS generated in cytosol. The consequences of heavy-metal-induced cellular oxidative stress are apoptosis, necrosis, tissue damage, and malignancy.

20.3 METHODS OF DETOXIFICATION OF HEAVY METALS

Commonly used procedures for removing heavy metal ions from water and industrial effluents include chemical precipitation, lime coagulation, ion exchange, reverse osmosis, and solvent extraction (Sekhar et al., 1998).

1. *Reverse osmosis.* It is a process in which heavy metal contaminated water is treated through a semipermeable membrane at a pressure

greater than osmotic pressure exerted by the dissolved metal ions in wastewater. The major disadvantage of this method is its high cost to be implemented globally (Ahlya et al., 2003).

2. *Electrodialysis*. In this process, the dissolved heavy metals are separated through the use of semipermeable ion-selective membranes. During this process, application of an electric potential between the two electrodes causes shifting of cations and anions toward cathode and anode, respectively. Because of alternate spacing of cation and anion-selective membranes, cells of concentrated and dilute salts are formed. The disadvantage is the formation of metal hydroxides that tends to precipitate and close the membranes (Ahlya et al., 2003).

3. *Ultrafiltration*. This method is a pressure driven membrane operation that uses porous membranes for removal of heavy metals from contaminated water. The major disadvantage of this process is the generation of sludge.

4. *Ion exchange*. In this process, metal ions from dilute solutions are exchanged with ions held by electrostatic forces on the exchange resin. It is one of the most common methods used for removal of heavy metal ions from contaminating water. The major drawbacks of this method include high cost and partial removal of certain ions (Ahlya et al., 2003). Further, this method uses resins that are toxic to the environment.

5. *Chemical precipitation*. Precipitation of heavy metals is achieved by the addition of coagulants such as alum, lime, Fe salts, and organic polymers. Major disadvantages of this method include use of nonecofriendly chemicals to precipitate the contaminated heavy metals and production of large amount of sludge containing toxic compounds.

Recent advancement in the field of nanobiotechnology has led to the development of highly efficient methods of bioremediation. These bacteria could be directly used for bioremediation methods or could be immobilized on solid substrates depending upon the bioremediation requirement. However, different cell free components of bacteria such as extracellular polymeric substances (EPS), enzymes, and other proteins, polysaccharides and secretory products have given promising results. Further, these bacterial components having high ability of heavy metal detoxifying ability that have been used to manufacture NPs with enhanced ability of heavy metal bioremediation. Smaller particle size enables the development of smaller sensors, which can be deployed more easily into remote locations. Recently,

NMs have been suggested as efficient, cost-effective, and environmentally friendly alternatives to existing treatment materials, in both resource conservation and environmental remediation (Tripathi et al., 2018).

20.4 NMS FOR BIOREMEDIATION OF HEAVY METALS

The term nanotechnology was first coined by Richard Feyman in 1959 and is otherwise termed as the next generation industrial revolution (Feynman, 1959). Recent advances in nanotechnology have led to the development of several NMs for enhanced remediation of heavy metals from contaminated soil and water bodies. In environmental science, NMs are used as bioflocculant, microbial monitoring and detection, and chemical degradation (Liu, 2006).

Using NMs for bioremediation of environmental contaminants is proved to be effective for the following reasons:

1. NPs have high surface to volume ratio per unit mass of the material, hence greatly enhance the bioremediation processes that is a surface-dependent phenomenon.
2. At nanoscale dimension, the material needs very low activation energy because of quantum effect, leading to very low activation energy required for a physical or chemical process associated with NP.
3. Surface plasmon resonance is another phenomenon exhibited by NPs which can be used for the detection of toxic material.
4. NPs can access toxic substances hidden to microbes because of their submicroscopic sizes.

However, the biggest drawback of these NPs is that most are toxic at long-term use. They are mostly synthesized partially or fully from metals or their derivatives. However, recent developments in the field of nanosynthesis have led to the development of bionanomaterials that overcome these toxic effects of metal-based NPs. Hence, many researches are being focused on bionanomaterials that are either completely or partially biological derivatives. NPs have been used for (1) Sensing and detection of hazardous chemical in the environment, (2) Its removal from the environment, (3) Detoxification of the compound, and (4) Prevention of pollution. Nanotechnologies are pervasive solution vectors in our economic environment. It is necessary to develop new methods to assess development for better understanding of nanotechnology-based innovation (Sekhara et al., 1998).

20.5 BIONANOTECHNOLOGY PROCEDURES

NPs (nanoscale particles = NSPs) are atomic or molecular aggregates with dimension between 1 and 100 nm that can drastically modify their physico-chemical properties compared with the bulk material. NPs can be made from a variety of bulk materials and they can act depending on chemical composition, size, or shape of the particles. NPs are broadly divided in two groups of organic and inorganic NPs. Organic NPs include CNPs (fullerenes), while some of the inorganic NPs include magnetic NPs, noble metal NPs (e.g., gold and silver) and semiconductor NPs (e.g., titanium dioxide and zinc oxide).

Ruffini-Castiglione and Cremonini, identified three types of NSPs, natural (e.g., volcanic or lunar dust and mineral composites), incidental (resulting from anthropogenic activity, e.g., diesel exhaust, coal combustion, and welding fumes), and engineered. The latter includes metal-based materials quantum dots, nanogold, nanozinc, nanoaluminium, TiO_2, ZnO, and Al_2O_3. Biological synthesis of NPs has grown markedly to create novel materials that are ecofriendly, cost-effective, and stable with great importance in wider application in the areas of electronics, medicine, and agriculture. Physical and chemical methods are most widely available method for synthesis of NPs. Chemical method of synthesis is advantageous as it takes short period of time for synthesis of large quantity of NPs. However, in this method capping agents are required for size stabilization of the NPs (Kavitha et al., 2013). Moreover, chemicals reagents used generally for NPs synthesis and stabilization are toxic and lead to byproducts that are not environment benign. The nontoxic and ecofriendly procedure for NP synthesis has led to biological approaches of NPs which otherwise termed as green synthesis. Biological approaches of NP synthesis involve different microorganisms such as plants, fungi, yeast, actinomycetes, bacteria, etc., or their products to synthesize or fabricate NPs. These organisms may lead to synthesis of NPs that are either intracellular or extracellular depending upon the site of reduction of the metal ions.

20.6 NMS AS ADSORBENTS OF HEAVY METALS FROM WATER

Industrial wastewater containing toxic heavy metal ions such as Hg^{2+}, Cd^{2+}, Pb^{2+}, Ag^+, Cu^{2+}, etc., could be removed by several mechanisms such as (1) chemical precipitation, (2) reverse osmosis, (3) electrochemical treatment, and (4) ion exchange. Although traditional sorbents could remove heavy metal ions from wastewater, the low sorption capacities and efficiencies limit their application deeply. To solve these defects of traditional sorbents, NMs are used as the novel ones to remove heavy metal ions in wastewater. The purified polysaccharides from plants, animals, and microflora sources were used as reducing and stabilizing agents for the synthesis of NPs (Delbarre-Ladrat et al., 2014). Polysaccharides have hydroxyl and hemiacetal groups, which play a vital role in reduction and stabilization that generate vast chances for their application and probable mass production.

20.7 BIOLOGICALLY DERIVED NMS

20.7.1 *PHYTOCHEMICAL-BASED NPS*

Phytochemical-based NPs are synthesized using phytochemicals alone or chemical materials fabricated with phytochemicals. Advantages of phytochemical-mediated NPs include their ecofriendly method of synthesis, the absence of toxicity, and the presence of natural capping agents. Further, they are easily available, safe to handle, and possess a broad variability of metabolites that may aid in reduction. Taking use of plant tissue culture techniques and downstream processing procedures, it is possible to synthesize metallic as well as oxide NPs on an industrial scale once issues such as the metabolic status of the plant are properly addressed. It is evident from compiled information that effect of NPs varies from plant to plant and depends on their mode of application, size, and concentrations (Manzer et al., 2015). Plants such as *Gundelia tournefortii*, *Centaurea virgata, Reseda lutea, Scariola orientalis, Eleagnum angustifolia*, and *Noaea mucronata* are known for high their heavy metal accumulating abilities. *N. mucronata* belonging to family *Chenopodiaceae* is the best Pb accumulator and also a good accumulator for Zn, Cu, and Ni. *Reseda lutea* is found to be the best Fe accumulator and *Marrubium vulgare* is the best accumulator of Cd.

20.7.2 *BACTERIA-MEDIATED SYNTHESIS OF NPS*

Bacteria are microscale efficient living machines that can synthesize a variety of novel biological products. Further, they could be genetically manipulated for production of specific metabolite in large quantity. Bacterial products such as enzymes, vitamins, polysaccharides, biodegradable polymers, lipopolycaccharides, etc., have been used to synthesize NPs (Figure 20.2). Extracellular secretion of enzymes offers the advantage of producing large quantities of NPs of size 100–200 nm in a relatively pure state, free from other cellular proteins. The special metal binding abilities of the bacterial cells and S-layers make them useful for technical applications in bioremediation and nanotechnology. Many microflorae can produce NPs through both intra and extracellular levels. Synthesis of AgNPs can be performed in various parts of the microbial cells (Narayanan and Sakthivel, 2010). Several lactic acid bacteria such as *Lactobacillus* spp., *Pediococcus pentosaceus,* and *Enterococcus faecium* are able to reduce Ag^+ to AgNPs. Recently, it was found that the AgNPs with a high surface area have more reactivity toward chemicals compounds and are effective tools in treatment of wastewater within a short time period (Kang et al., 2000). Chitosan-stabilized AgNPs combined with advanced oxidation process showed good results in the degradation of various dyes (Santhanalakshmi and Dhanalakshmi, 2012).

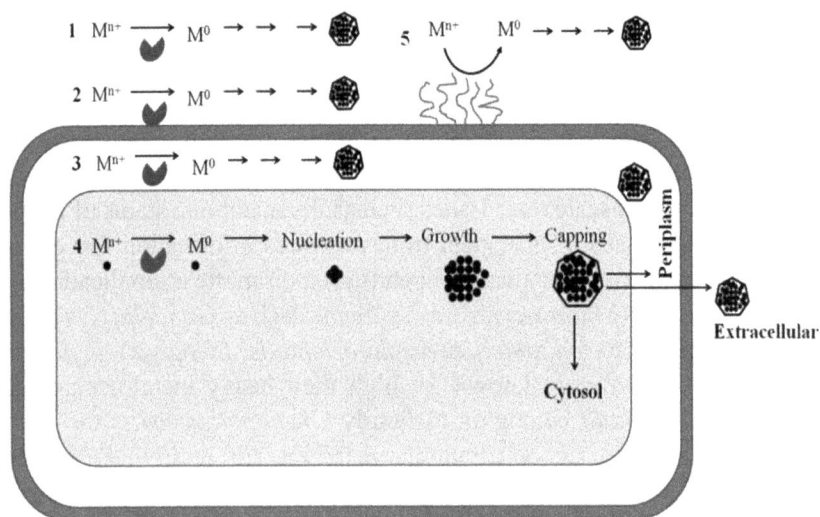

FIGURE 20.2 Mechanism of nanoparticle synthesis by bacteria. Site of nanoparticle synthesis in bacteria by reduction of metal ion (M^+) can be reduced by (1) extracellular protein (e.g., metal reductase) to metallic form (M^0). This metallic form accumulates by a process termed as nucleation in a multistep process that is stabilized by the biological agents such as lipopolysaccharides synthesized by the cell. The site of nucleation for nanoparticles may be on the (2) cell surface, (3) periplasmic space, (4) cytosol, or (5) polymers attached to the cell surface. After synthesis in cytosol, the nanoparticles are either retained in the cytosol or secreted into the periplasm or extracellular medium.

Bacteria could be modified with metal NPs that enhance their bioremediation abilities. For example, palladium, Pd(0) NPs deposited on the cell wall and loaded inside the cytoplasm of *Shewanella oneidensis* can be charged with H$^+$ radicals by adding different substrates such as hydrogen, acetate, and formate as electron donors in a bioreductive assay containing Pd(II). When these charged Pd(0)-deposited *S. oneidensis* cells are brought in contact with chlorinated compounds, the H$^+$ radical on the Pd(0) catalytically react with PCP, resulting in the removal of the Cl molecule from the chlorinated compounds. The green synthesis of AgNPs was performed using EPS from *Leuconostoc lactis* as both, a reducing and stabilizing agent. EPS-AgNPs were spherical in shape, with an average size of 35 nm, exhibited stabilization up to 437.1 °C and the required energy is 808.2 J/g. EPS stabilized AgNPs were found to be efficient in facilitating the degradation process of two industrial textile dyes, methyl orange and congo red. This makes EPS-AgNPs a suitable, cheap, and environment friendly candidate for the biodegradation of harmful textile dyes (Saravanana et al., 2017).

20.7.3   NPS FROM YEAST AND FUNGI

Fungi are an excellent source of extracellular enzymes that greatly influence synthesis of NPs. Compared with bacteria, fungi could be used as a source for the production of larger amounts of monodispersed NPs. This higher yield of NPs using fungi could be attributed to larger quantity of extracellular enzymes and proteins secreted by fungal cells compared to bacteria. For industrial applications, fungi possess certain properties which include high production of specific enzymes or metabolite, high growth rate, easy handling in large-scale production, and low-cost requirement for production procedures which provides advantages over other fungus methods (Vahabi et al., 2011). In synthesis of numerous enzymes and rapid growth with the use of simple nutrients, yeast strains possess certain benefits over bacteria and the synthesis of metallic NPs employing the yeast is being considered (Kumar et al., 2011).

20.7.4   ENZYME AND PROTEIN-BASED NPS

Enzymes are specialized class of protein with high catalytic efficiency. However, these incredible biomolecules exhibit a great degree of variability in their catalytic efficiency, specificities, reaction conditions, and stability. Many enzymes are short lived, being prone to oxidative damage, temperature, heavy metals, and pH of the reaction medium. However, attachment to magnetic FeNPs is an effective way to increase their stability, longevity, and reusability (Steiert et al., 2018). It also enables their easy separation from rest of the components in the reaction medium. Enzymes such as trypsin and peroxidases have been used to uniform core shell magnetic nanoparticles (MNPs). The reported study indicates that the lifetime and activity of enzymes increase dramatically from a few hours to weeks and that MNP-enzyme conjugates are more stable, efficient, and economical for commercial uses (Cordova et al., 2014; Varca et al., 2014).

20.8 CONCLUSION

Most of these heavy metal contaminants accumulate in surface layer of soils and water bodies. So, these heavy metals get into the organisms and biomagnified, resulting in detrimental consequences such as heavy metal toxicity. Bioremediation of these heavy metals by using microbes such as bacteria is an ecofriendly step toward detection and removal of heavy metal contaminants. However, the intervention of nanotechnology into bioremediation procedure has increased the fidelity and effectiveness of the remediation process. NMs synthesize from biological agents are nontoxic, ecofriendly, and much more sensitive that detects and removes the heavy metal contaminants at nanomolar level. Hence, exploration of these technologies for environmental bioremediation processes must be encouraged in both laboratory and large scale sustainable green technology.

KEYWORDS

- heavy metals
- biosorption
- biomaterial
- exopolysaccharide
- reactive oxygen species (ROS)
- bioremediation

REFERENCES

Ahalya, N.; Ramachandra, T.V.; Kanamadi, R.D. Biosorption of heavy metal. *Res. J. Chem. Environ.* **2003**, *7*, 1–23.

American Physical Society, California Institute of Technology, 1992, pp. 1–56.

Arnich, N.; Sirot, V.; Rivière, G.; Jean, J.; Noël, L.; Guérin, T.; Leblanc, J.C. Dietary exposure to trace elements and health risk assessment in the 2nd French total diet study. *Food Chem. Toxicol.* **2012**, *50*, 2432–2449.

Boonyapookana, B.; Upatham, E.S.; Kruatrachue, M.; Pokethitiyook, P.; Singhakaew, S. Phytoaccumulation and phytotoxicity of cadmium and chromium in duckweed Wolffia globosa. *Int. J. Phytoremediation.* **2002**, *4*, 87–100.

Cordova, G.; Attwood, S.; Gaikwad, R.; Gu, F.; Leonenko, Z. Magnetic force microscopy characterization of super paramagnetic iron oxide nanoparticles (SPIONs). *Nano Biomed. Eng.* **2014**, *6(1)*, 31–39.

Gautam, R.K., Sharma, S.K., Mahiya, S., Chattopadhyaya, M.C. Contamination of heavy metals in aquatic media, transport, toxicity and technologies for remediation. *Royal Soc. Chem.* **2014**, *1*, 1–23.

Delbarre-Ladrat, C.;Sinquin, C.; Lebellenger, L.; Zykwinska, A.; Colliec-Jouault, S. Exopoly-saccharides produced by marine bacteria and their applications as glycosaminoglycan-like molecules. *Front. Chem.* **2014**, *2*, 85.

Feynman, R.P. There's Plenty of Room at the Bottom: An Invitation to Enter a New Field of Physics. In *Handbook of Nanoscience, Engineering, and Technology*; CRC Press: Boca Raton, FL, USA, 1959; pp. 3–12.

Fosmire, G.J. Zinc toxicity. *Am J. Clin. Nutr.* **1990**, *51*, 225–227.

Kang, S.F.; Liao, C.H.; Po, S.T. Decolorization of textile wastewater by photo-fenton oxidation technology. *Chemosphere.* **2000**, *41*, 1287–1294.

Kavitha, K.S.; Baker, S.; Rakshith, D.; Kavitha, H.U.; Rao, Y.; Harini, H.C.; B.P.; Satish, S. Plants as green source towards synthesis of nanoparticles. *Int. Res. J. Bio. Sci.* **2013**, *2*, 66–76.

Kumar, D.; Karthik. L.; Kumar, G.; Roa, K.B. Biosynthesis of silver anoparticles from marine yeast and their antimicrobial activity against multidrug resistant pathogens. *Pharmacology.* **2011**, *3*, 1100–1111.

Liu, R.; Munro, S.; Nguyen, T.; Siuda, T.; Suciu, D.; Integrated microfluidic custom array device for bacterial genotyping and identification. *J. Assoc. Lab. Autom.* **2006**, *11*, 360–367.

Manzer, H.; Mohamed, H.S.; Whaibi, A.; Firoz, M.; Mutahhar, Y.; Khaishany, A. Nanotechnology and plant science. In *Nanoparticles and Their Impact on Plant Siddiqui*; M.H., Al-Whaibi, M.H., Mohammad, F., Eds.; Springer, **2015**.

Medici, V.; Huster, D. Animal models of Wilson disease. *Handb. Clin. Neurol.* **2017**, *142*, 57–70.

Narayanan, K.B.; Sakthivel, N. Biological synthesis of metal nanoparticles by microbes. *Adv. Colloid Interface Sci.* **2010**, *156*, 1–13.

Nieboer, I.; Richardson, E. Replacement of the nondescript term 'heavy metals' by a biologically and chemically significant classification of metal ions. *Environ. Poll. Series B, Chem. Phy.* **1980**, *1*, 3–26.

Noctor, G.; Mhamdi, A.; Foyer, C.H. Oxidative stress and antioxidative systems, recipes for successful data collection and interpretation. *Plant Cell Environ.* **2016**, *39*, 1140–1160.

Patrick, L. Mercury toxicity and antioxidants: Part 1: Role of glutathione and alpha-lipoic acid in the treatment of mercury toxicity. *Altern. Med. Rev.* **2002**, *7*, 456–471.

Ratnaike, R.N. Acute and chronic arsenic toxicity. *Postgrad. Med. J.* **2003**, *79*, 391–396.

Santhanalakshmi, J.; Dhanalakshmi, V. Chitosan silver nanoparticles assisted oxidation of textile dyes with H_2O_2 aqueous solution, Kinetic studies with pH and mass effect. *Ind. J. Sci. Technol.* **2012**, *5*, 3834–3838.

Saravanana, C.; Rajesh, R.; Kaviarasanc, T.; Muthukumarc, K.; Kavitakea, D.; Kumar, P.; Shetty, H. Synthesis of silver nanoparticles using bacterial exopolysaccharide and its application for degradation of azo-dyes. *Biotechnol. Rep.* **2017**, *15*, 33–40.

Satarug, S.; Moore, M.R. Adverse health effects of chronic exposure to low-level cadmium in foodstuffs and cigarette smoke. *Environ. Health Perspect.* **2004**, *112*, 1099–1103.

Sekhara, S.K.; Subramanian, S.; Modakc, M.; Natarajan, K.A. Removal of metal ions using an industrial biomass with reference to environmental control. *Int. J. Min. Process.* **1998**, *53*, 107–120.

Sharma, S.S.; Dietz, K.J. The relationship between metal toxicity and cellular redox imbalance. *Trends Plant Sci.* **2009**, *14*, 43–50.

Steiert, E.; Radi, L.; Fach, M.; Wich, P.R. Protein-based nanoparticles for the delivery of enzymes with antibacterial activity. *Macromol. Rapid. Commun.* **2018**, 39, e1800186. doi: 10.1002/marc.201800186.

Tripathi, S.C.; R.; Anuradha, J.; Chauhan, D.C. Nano-bioremediation: Nanotechnology and bioremediation. In *Biostimulation Remediation Technologies for Groundwater Contaminants*; Rathoure, A.K., Ed.; IGI Global, **2018**, pp. 202–219. http://doi:10.4018/978-1-5225-4162-2. ch012 **2018**, 202–219.

Underwood, E.J.; George, L.F.; Garton, A. The incidence of trace element deficiency diseases. *Phil. Trans. R. Soc. Lond. B*, **1997**, *274*.

Vahabi, K.; Mansoori, G.L.; Karimi, S. Biosynthesis of silver nanoparticles by fungus *Trichoderma reesei* (A route for large-scale production of AgNPs). *Insci. J.* **2011**, *1*, 65–79.

Varca, G.H.C.; Caroline, C.F.; Lopes, P.S.; Mathora, M.B.; Grassellic, M.; Lugão, A.B. Radio-synthesized protein-based nanoparticles for biomedical purposes. *Rad. Phy. Chem.* **2014**, *94*, 181–185.

## CHAPTER 21

Spent Coffee Waste Conversion to Value-Added Products for the Pharmaceutical and Horticulture Industry

PRIYA SARKAR[1], TETHI BISWAS[1,2], CHAITALI CHANDA[2], AMRITA SAHA[3], MATHUMAL SUDARSHAN[4], CHANCHAL MAJUMDER[5], and SHAON RAY CHAUDHURI[1*]

[1]Department of Microbiology, Tripura University, Suryamani Nagar, Tripura West 799022, India

[2]Centre of Excellence in Environmental Technology and Management, Maulana Abul Kalam Azad University of Technology, Haringhata, Nadia, West Bengal 741249, India

[3]Department of Environmental Science, Amity Institute of Environmental Sciences, Amity University, Kolkata, West Bengal 700135, India

[4]Trace Element Laboratory, Inter University Consortium, Kolkata, Kolkata Centre, Sector-III/Plot LB-8, Salt Lake, Kolkata, West Bengal 700098, India

[5]Department of Civil Engineering, Indian Institute of Engineering Science and Technology, Shibpur, Botanic Garden, Howrah, West Bengal 711103, India

[*]Corresponding author. E-mail: shaon.raychaudhuri@gmail.com

ABSTRACT

Coffee, a valuable product of world trade, is a popular beverage globally. After petroleum, it is the highest traded commodity. The International Coffee Organization reports that approximately 6 million tons of the solid waste is generated annually, of which spent coffee ground (SCG) forms a major portion. Currently, SCG has a little commercial value being mostly disposed into landfills with a minor fraction being used as compost. Their utility as pollutant absorbent has been investigated by research groups. The SGC instead of being a burden for environment could be used as a valuable raw material for generation of value-added products for pharmaceutical industry and for horticulture. Our research is focused on making coffee processing a zero-pollution industry. The SCG has been effectively used for extraction of caffeine (37.3 mg per gram of SCG) and synthesis of silver nanoparticle (0.112 mg per gram of SCG) exhibiting antimicrobial activity against bacteria in planktonic as well as biofilms state. Both these components are of use to pharmaceutical industry. The extracted SCG functioned as efficient soil conditioner enhancing seed germination significantly (p value = 0.05) owing to its porosity enhancing ability and rich elemental content, revealing its importance in horticulture industry. Hence, this opens up new avenue for applications of SCG, making the coffee industry further eco-friendly.

21.1   INTRODUCTION

Coffee, a pivotal component of world trade, is also an important part of Indian culture. These plants are cultivated mostly in the equatorial regions of Southeast Asia, India, Africa, and America covering over 70 countries. Most common cultivars of coffee are Arabica and Robusta. The ripe berries of Coffea plant (native of tropical Africa) after processing, drying, roasting to varying degrees (based on the desired end product quality), and grinding are finally brewed to produce coffee as a beverage. The remaining solid wastes from cultivation and preparation of coffee are coffee cherry, coffee husks, peel, pulp, and the SCG (Cruz et al., 2012). Large volume of wastewater with high carbon load is generated due to the washing steps, which demands treatment before being discharged. The other wastes produced in relatively low quantities are the defective green beans and the leaves (Blinova et al., 2017). The ever increasing pollution from the industrial activities demands for proper utilization or abatement of wastes, and coffee industry is no exception.

India produced 318,200 metric tons of coffee in 2012–2013, while it generates 7.5 metric tons of dry coffee waste per year (Chinmai et al., 2014; Coffee Market Report, 2018). The spent coffee ground (SCG) is mostly dumped into the landfills. Some scientific reports have assessed their potential for production of biodiesel (Caetano et al., 2014; Blinova et al., 2017), sugars (Mussatto et al., 2011), activated carbon (Kemp et al., 2015), compost (Santos et al., 2017), adsorbent for metal ions (Davila-Guzman et al., 2016), bioethanol (Kwon et al., 2013), substrate for mushroom cultivation

(Kimura, 2013), additive in ceremics (Fonseca et al., 2014), adsorbent for dye (Baek et al., 2010; Roh et al., 2012), source of natural phenolic antioxidants (Mussatto et al., 2011; Zuorro et al., 2012), and biomaterial in the pharmaceutical, food, and polymer industry. The SCG contains large amounts of organic compounds such as cellulose (12.4%), hemicelluloses (39.1%), lignin (23.9%), fat (2.29%), protein (17.44%), and total dietary fibers (60.46%), fatty acids, phenolics, alkaloids, and other polysaccharides that justify its valorization (Ballesteros et al., 2014). Bioactive compounds such as chlorogenic acids, caffeine, caffeoylquinic acids, feruloylquinic acids, and dicaffeoylquinic acids are abundant in coffee leading to the bitter taste and are known to be active antioxidants that may cause health benefits to coffee drinkers (Farah and Donangelo, 2006; Farah et al., 2006; Moeenfard et al., 2014).

The presence of phenolic compounds suggests their potential application for green synthesis of nanoparticles (NPs). From the green chemistry perspective, the preparation of NPs includes the choice of the solvent for the synthesis, reducing agent, and a nontoxic substance for the stabilization of the NPs. NPs are broadly grouped under two categories, namely, organic NPs (carbon NPs) and inorganic NPs, which includes metal NPs {silver (Ag), gold (Au), and platinum (Pt)}. Ag-NPs are mostly smaller than 100 nm and comprise of around 20–15,000 silver molecules known for their antibacterial activity due to their large surface-area-to-volume ratio (Morones-Ramirez et al., 2013; Rai et al., 2009). There are reports of successful production of Ag-NPs from different plant parts such as tea and coffee extract, blueberry, pomegranate, and turmeric extract, which are known for the presence of phenolic compounds, terpenoids, alkaloid, sterols, anthocyanins, flavonoids, tannins, and proteins, which act as capping and stabilizing agents. The factors that affect the biosynthesis of NPs are pH, temperature, concentration of plant extract, and the concentration of metal solution. Factors influencing the NP morphology include concentration and pH of plant extract, as well as temperature during particle synthesis. Ag-NPs synthesized at alkaline pH are smaller and well dispersed than those at acidic pH (Sanghi and Verma, 2009; Iravani and Zolfaghari, 2013).

SCG is a rich source of caffeine, which can be extracted to be used by pharmaceutical industries (Baucells et al., 1993; Cruz et al., 2012). It is the most studied content because of its well-established psychoactive effects and promotion of energy metabolism. Although the caffeine content in coffee waste is lower than that in coffee beans, it is in substantial amount. The extraction depends on the solubility, grind level, roast level, dwell time, brew time, water temperature, bean type, as well as blend. Caffeine content varies as a function of hot brewing method and the coffee-to-water ratio (Blumberg et al., 2010; Fuller and Rao, 2017). The solubility of caffeine in water increases with increasing temperature. At 25 °C, 80 °C and 100 °C, it is 22, 180, and 670 mg/mL, respectively. Hence, caffeine is a potential by-product that can be extracted from the coffee waste.

With the objective of minimizing the toxic components in the waste (like SCG), the natural dissolvable, nontoxic chemicals and sustainable material can be extracted to be used for subsequent green synthesis of components such as NPs. This ensures detoxification of the extracted SCG making it suitable for applications such as soil conditioning. Coffee grounds are reported to be used in the garden along with manure accounting for plant development due to enhanced C–N (carbon, nitrogen) ratio, maintained soil temperature and moisture content, and enhanced soil structure with reduced weed development (Yamane et al., 2014; Chalker-Scott, L., 2016). The addition of SCG in lettuce and rice showed increased yield and nutrient content (Cervera-Mata et al., 2018; Morikawa and Saigusa, 2011), while decreasing the load of landfill waste. This study is an outcome of the research work done in the Microbial Technology Group for exploring the possibility of economic by-product generation from SCG leading to environmental protection.

21.2   VALUE-ADDED PRODUCT DEVELOPMENT

Natural products are important sources for bioactive compounds. There is potential for developing novel therapeutic agent from them. Extraction from dried SCG was done using solid–liquid extraction method (Paikara et al., 2015) with some modification. SCG was dried in hot air oven maintained at 60 °C for overnight. 15 g of dried SCG added to 300 mL of distilled water was boiled for 15 min at 100 °C with occasional stirring (Brewing). After boiling, it was cooled and filtered through Whatman filter paper (Grade 1, 110 mm, cat. No. 9701296). The filtrate was then centrifuged at 14,000 rpm (Eppendorf 5418R), and the supernatant was collected for further analysis. Qualitative analysis of SCG extract revealed the presence of alkaloids, phenols, tannins, proteins, carbohydrates, and terpenoids as per standard procedure (Paikara et al., 2015). The extract contained 5.57 μg of phenol and 23.7 μg of protein per gram of SCG. Polyphenols and proteins are important components for NP generation. The elemental content analysis through energy-dispersive X-ray fluorescence (EDXRF) analysis of SCG extract before and after extraction revealed the presence of elements such as Mg, Ca, Al, Na, Fe, Mn, Ba, S, P, K, Cu, Zn, Sr, Pb, Cr, Rb, Ni, Br, As, and V, as shown in Figure 21.1a–c. There was signifcant change in their composition pre- and postextraction in case of Mg (p value = 0.026), Mn (p value = 0.001), P (p value = 0.03), Cr (p value = 0.03) but insignificant change in case of Ca (p value = 0.408), Al (p value = 0.789), Na (p value = 0.100), Fe (p value = 1.325), Ba (p value = 0.609), S (p value = 0.199), K (p value = 0.280), Cu (p value = 0.577), Zn (p value = 0.060), Sr (p value = 0.906), Pb (p value = 0.508), Rb (p value = 0.07), Ni (p value = 0.62), Br (p value = 0.233), As (p value = 0.430), and V (p value = 0.42). Presence of elements such as S, P, Mg, and Ca indicate toward the ability of the extracted SCG to support plant growth. The extract was analyzed using micro X-ray fluorescence to further reconfirm the presence of elements such as S, P, Si, Ca, Mg, K, Al, Fe, and Cu (see Figure 21.1d). These elements increase the availability of essential plant nutrients such as nitrogen, phosphorus, iron, and zinc (Kitou and Okuno 1999; Morikawa and Saigusa 2011).

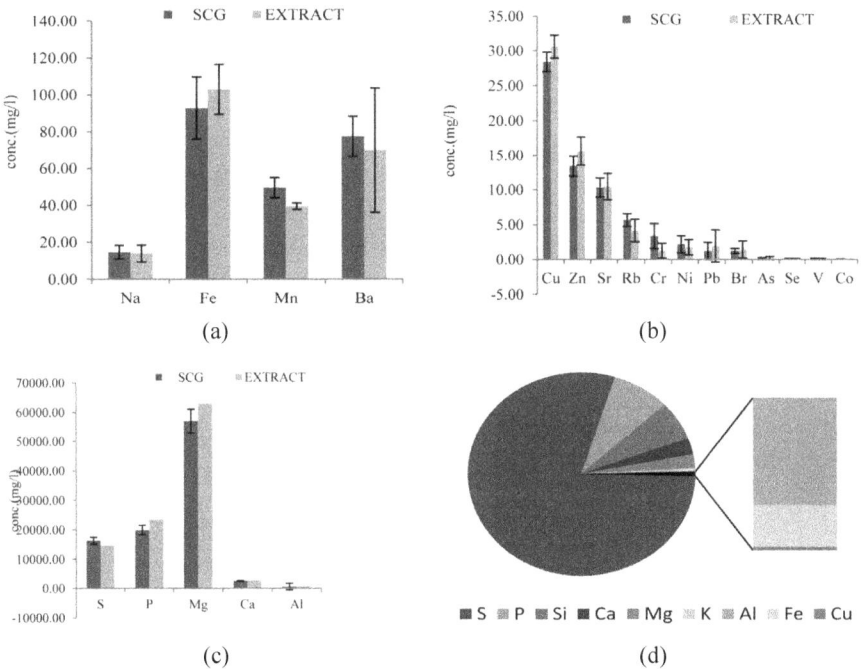

FIGURE 21.1 (a–c) Bar diagram showing the elemental content of SCG before and after extraction with water for different elements analyzed using energy dispersive X-ray fluorescence analysis. (d) Pie chart showing elemental content in SCG extract using micro-XRF analysis.

21.2.1   SCG AS SOIL CONDITIONER FOR PLANT GROWTH

Coffee grounds used as mulches or additives have positive effects on soil (Yamane et al. 2014). It moderates soil temperature and increases soil water retention ability (Ballesteros et al. 2014). Porosity added to the soil due to mixing of SGC aids in movement of air and water through the soil, which is important because it affects the supply of air, moisture, and nutrients to root zone, hence making it available for plant uptake. Field emission scanning electron microscopic [Supra 55 (Serial number 4132), Carl Zeiss; software SmartSEMver 5.05] imaging revealed coarse rough-textured surface for soil particle (see Figure 21.2a), relatively smoother surface for extracted SCG (see Figure 21.2b), and porous structure of the mixture of soil and extracted SCG in 1:2 ratio (see Figure 21.2c) proving the enhanced porosity in case of the latter.

Mung bean (*Vigna radiata* var Meha) were grown in germination tray with four different treatments, in which 44 Mung bean seeds were sown in each of the treatment. The first set had the slots filled with only soil, the second set with extracted SCG and soil mixture in 1:2 ratio, and the third set had unextracted SCG followed by the fourth set that had extracted SGC. The seeds were grown for 20 days. The shoot lengths and root lengths were measured along with the extent of survival. The result showed significant (p value = 0.05) variation in shoot and root length (see Figure 21.3a) along with higher survival (see Figure 21.3b) when sown in extracted SCG and soil mixture as compared to soil alone, unextracted SCG, or extracted SCG alone. The enhanced porosity of the mixture with the rich elemental content (measured through EDXRF analysis) of the extracted SGC was an important plant growth promoting factor. This approach could be used for production of soil conditioner from the SCG to support horticulture with recovery of essential resources, which otherwise would have been wasted in landfills.

21.2.2   CAFFEINE EXTRACTION FROM SCG

Caffeine is a widely consumed psychoactive substance that acts as a mild central nervous system stimulant. It stimulates the heart, respiratory system, and central nervous system while functioning as a diuretic. Caffeine can be extracted from SCG. Caffeine extraction was done according to standard procedure (a laboratory manual, University of Pittsburgh). Brewing at high

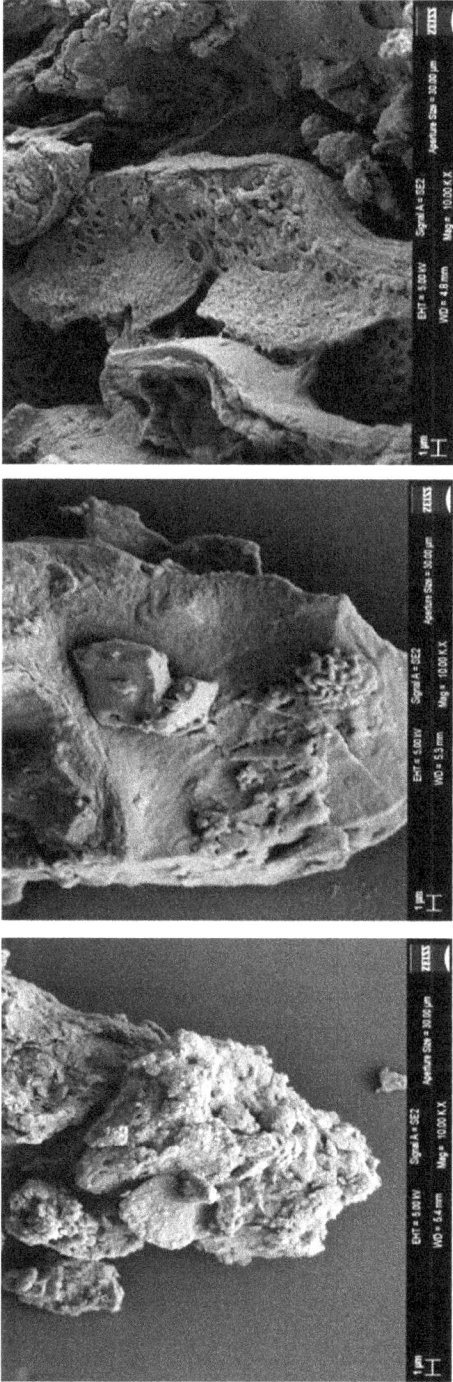

FIGURE 21.2 Soil structure under scanning electron microscope: (a) extracted SGC under scanning electron microscope; (b) scanning electron microscopic image of extracted SGC and soil (1:2) (c).

temperature ensures that caffeine gets dissolved in water along with tannins. Solubility of caffeine in water is lower than in chloroform, whereas tannins are insoluble in chloroform. Hence, the water extract upon extraction with chloroform selectively enriches caffeine in the organic layer leaving behind the tannin salts in the aqueous layer. Evaporation of the chloroform leaves behind the solid crude caffeine. 37.3 mg of caffeine could be extracted per gram of SCG (see Figure 21.4). The chloroform extracted solid (1.5 mg/mL) during wavelength (200–500 nm) scanning revealed a major peak at 272 nm confirming it to be caffeine (Bhawani et al., 2015), which was further reconfirmed through liquid chromatography mass spectrometry showing a m/z value of 195 eluted at 1.20 min. A solid:liquid ratio of 3:10 during extraction gave optimum yield of caffeine. The caffeine so extracted could find application in cosmetics as well as pharmaceutical industry.

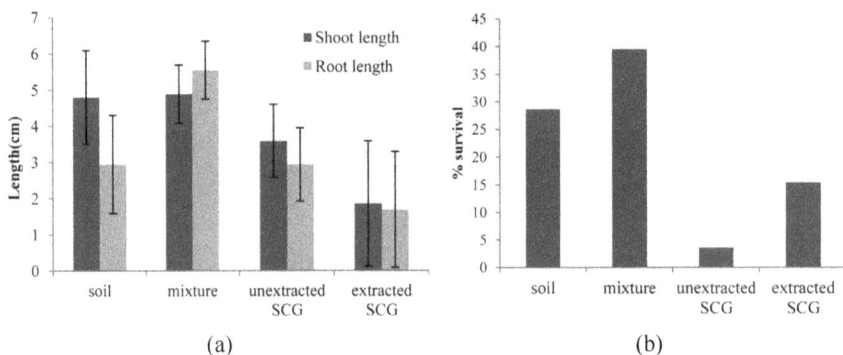

(a)   (b)

FIGURE 21.3   (a) Bar diagram showing variation in shoot length and root length during the four treatments. (b) Bar diagram showing percentage survival during the four treatments.

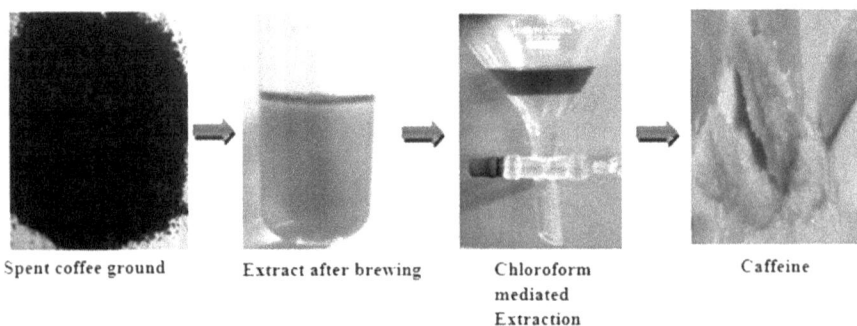

Spent coffee ground   Extract after brewing   Chloroform   Caffeine
                                                        mediated
                                                        Extraction

FIGURE 21.4   Pictorial representation of the steps of extraction of caffeine from SCG.

21.2.3 BIOSYNTHESIS OF SILVER NP AND ITS APPLICATION

Metallic NPs are promising owing to their antibacterial properties due to their high surface-area-to-volume ratio. One of the frequently used materials for synthesizing NPs is silver owing to its high electrical conductivity, optical properties, and antibacterial effect (Khan et al., 2017). Stabilization of NPs is a must to avoid aggregation, which otherwise results in loss of chemical and physical properties of NPs. This is achieved by coating the particles with a capping agent. Polyphenols are a type of antioxidant, found in coffee grounds that allow the coffee extract to reduce metal ions to form metal NPs. Various stabilizers such as alkaloids, phenolic acids, and protein as well as reducing agents such as terpenoids, polyphenols, and proteins play significant role in the reduction of Ag ions to Ag-NPs (Makarov et al., 2014). The presence of such components in the SCG extract hinted toward their functioning as the bioresource for NP generation.

Response surface methodology using a Doehlert design confirms the optimum condition for phenol extraction from SGC (see Figure 21.5a–f). The optimum extraction was attained at pH $(x1)$ of 7.85, temperature $(x2)$ of incubation of 63.18 °C, time $(x3)$ of extraction of 54.57 min, and solid:water $(x4)$ ratio of 0.5415. A mathematical model was suggested to predict the percentage extraction of phenol, which can be expressed in the following equation:

$$R = 41.26 + 13.39*x1 + 8.77*x2 + 2.68*x3 + 2.94*x4 - 25.54*x1^2 - 10.15*x2^2 - 22.21*x3^2 - 10.42*x4^2 + 0.45*x1*x2 - 2.36*x1*x3 + 4.15*x1*x4 + 3.32*x2*x3 - 13.29*x2*x4 - 0.96*x3*x4.$$

The maximum extraction of phenol at the above condition was 45.28 µg per gram of SCG. Literature reveals optimum phenol extraction from grape solid residue to be at higher temperature (60 °C) (Spigno et al. 2007) while that from fresh coffee or leaf to be at lower temperature (40–45 °C) (Akowuah et al. 2009). In this case, the major portion of the phenols had been extracted during brewing of the roasted coffee ground. What is left is needs higher extraction temperature and moderate incubation time to get into the solution. Hence, it is in agreement with the earlier literature (Spigno et al. 2007). The pH had minimum influence on the extraction, as has also been reported by other groups. However, maximum extraction occurs in the neutral range.

The phenol extracted as mentioned above was used for production of Ag-NP at 37 °C with stirring using silver nitrate solution. There was a

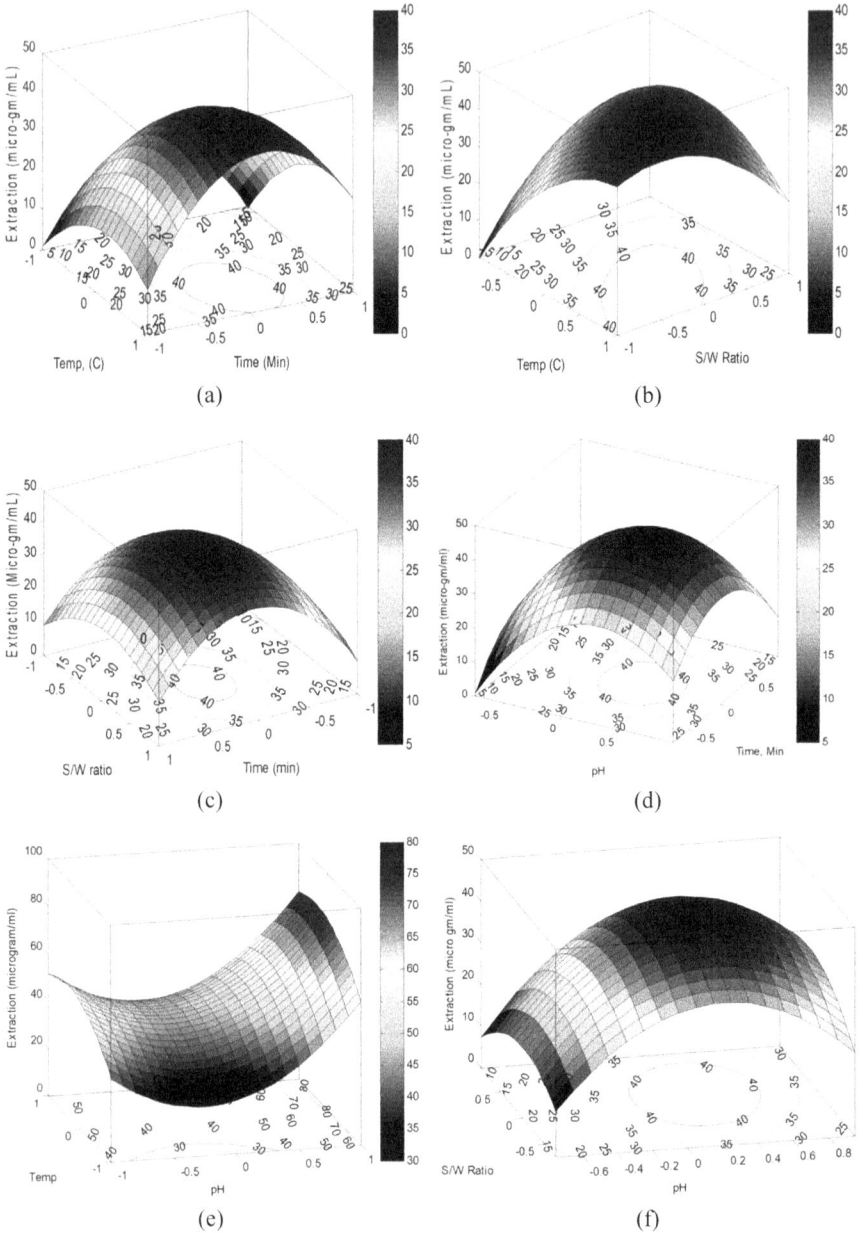

FIGURE 21.5 Effect of (a) temperature and time; (b) temperature and solid/water (s/w) ratio; (c) s/w ratio and contact time; (d) pH and contact time; (e) pH and temperature as well as (f) s/w ratio to pH on the phenol extraction from SCG studied through RSM to optimize phenol extraction.

change in color of the solution of silver nitrate from light yellow to brown indicating NP synthesis. Ag-NP synthesis was done with water as well as hydroalcoholic SCG extract. In case of hydroalcoholic extract, UV–Vis spectra showed surface plasmon resonance band of silver at 320–340 nm instead of at 440–500 nm. Scanning electron microscopy (ZEISS EVO-MA 10) revealed near-spherical and spherical well-dispersed Ag-NPs in case of synthesis from SCG hydroalcoholic and water extract, respectively, following 30 min of incubation. To visualize the effect of production of Ag-NPs with increase in reaction time, the UV–Vis spectra (SHIMADZU UV-1800) of the absorbance (300–600 nm) of silver nitrate solution was recorded after 1, 3, 6, 12, and 24 h (see Figure 21.6) of incubation with stirring showing a sharp peak between 405 and 440 nm indicating formation of NPs in higher quantity. The Ag-NPs produced after 24 h of incubation were larger in size and oval in shape, while those produced at the third hour of incubation were small and spherical (see Figure 21.7a–c). Optimization of the process revealed highest production using water extract after 24 h of incubation at 50 °C and pH 11.

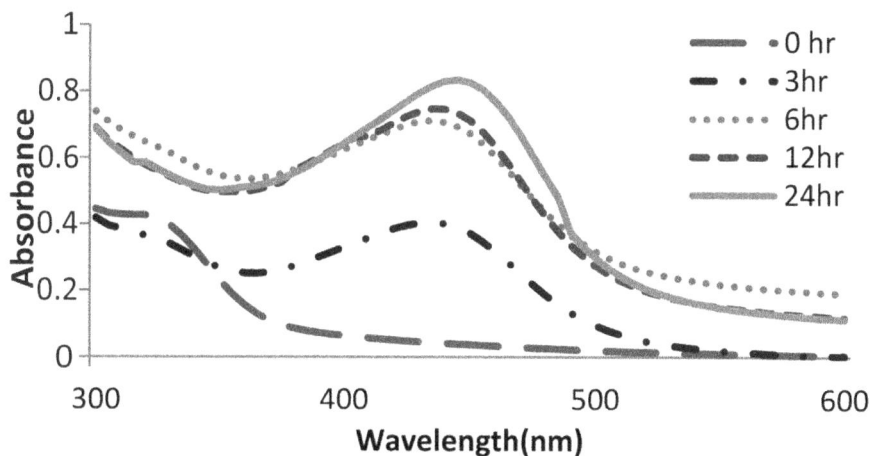

FIGURE 21.6 Absorbance of water extract with silver nitrate solution at 300–600 nm indicating silver nanoparticle synthesis up to 24 h of incubation. The increasing peak height around 440 nm indicates relatively larger quantity of nanoparticle production with increasing incubation time.

21.2.3.1 Ag-NP APPLICATION ON BACTERIAL BIOFILMS

Ag-NP is one of the well-known antimicrobial agent. Nanotechnology provides the solution for biofilm reduction through "nanofunctionalization"

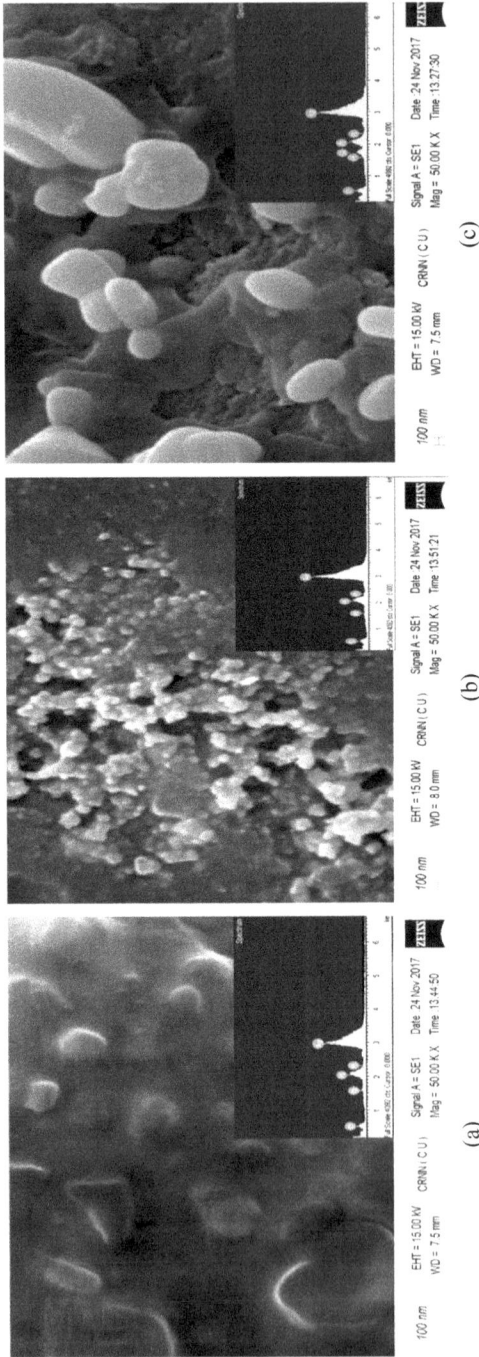

FIGURE 21.7 Scanning electron microscopy images of silver nanoparticles from SCG hydro-alcoholic extract and water extract. (a) Image showing Ag-NP synthesised after 30 min incubation using SCG hydro-alcoholic extract. (b) Image showing Ag-NP synthesis after 30 min incubation using SCG water extract. (c) Image showing Ag-NP synthesis after 24 h incubation from SCG water extract.

surface techniques that prevent the biofilm formation. Ag-NP can prevent the formation of life-threatening biofilms on medical devices. It has recently been established that Ag-NP hydrogel hybrid with different sizes of Ag-NPs can be effectively deployed for their antibacterial effects (Mohan et al., 2007). In this study, the effect of the synthesized Ag-NPs were studied on a well-characterized *Bacillus* sp. ARTU001 (GenBank accession no. MK318642). The isolate was allowed to form biofilm under sterile condition with incubation at 37 °C for 6 and 12 h simultaneously (see Figure 21.8a and b). A third set was incubated with the isolate and Ag-NPs for 6 h to see the effect of Ag-NPs on biofilm formation (see Figure 21.8c), while for the fourth set, Ag-NPs were added aseptically after 6 h of biofilm formation and incubated further for 6 h to check the impact of NPs on preformed biofilms (see Figure 21.8d). Biofilms from these four sets were observed under a scanning electron microscope (ZEISS EVO-MA 10), which revealed thick mat of biofilm produced after 6 h (see Figure 21.8a) and 12 h (see Fig. 21.8b) of incubation at 37 °C. No biofilm formation could be seen when the bacterial isolate was incubated with the NP (see Figure 21.8c) revealing strong antimicrobial effect against the planktonic cells. There were surface stuck Ag-NPs observed in the micrograph. Inhibition of biofilm and biofilm breakdown was evident from Figure 21.8d, which had much less density of biofilm when compared to 6- and 12-h grown biofilm without Ag-NPs. This indicated that the NPs not only prevent biofilm formation, but also disrupt existing biofilms. However, complete eradication was not observed, which might be a dose-dependent response. Thus, the silver NP could act both as antiadhesive against biofilm formation and aid in breakdown of preformed biofilm, which is in accordance with the existing literature (Namasivayam et al., 2013; Markowska et al., 2013).

21.3 CONCLUSION

Food waste has attracted an increasing attention, since it needs to be managed due to its environmental, social, and economic impacts. Global production of coffee is increasing because of higher consumption, resulting in generation of millions of tons of residues worldwide. Considering the high amounts of waste generated, alternatives have been sought for its utilization. In this study, the SCG has been used for the extraction of caffeine, which has application in the pharmaceutical as well as cosmetic industry. The SCG could be used for green synthesis of Ag-NPs, hence having an application in the field of nanotechnology as well as pharmaceutical industry. The green synthesis minimizes the use of chemicals as well as energy for the purpose while utilizing phenols, carbohydrate, alkaloids, tannins, and terpenoids from the extract. The extracted SCG acts as a useful bioresource for supplying essential elements to the soil, enhancing porosity, water retention capacity, and providing temperature buffering action to the soil. All this together facilitates plant growth and, hence, promotes horticultural practices. These findings lead to the proposal of SCG being reused for numerous applications making coffee brewing a zero waste discharge process with economically important by-product development.

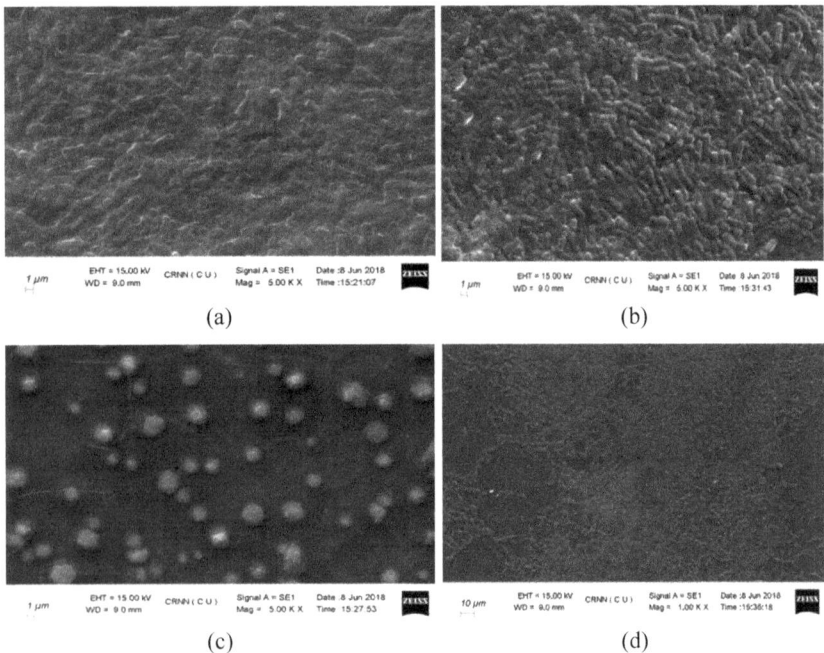

FIGURE 21.8 Scanning electron micrographic images showing biofilm formation by *Bacillus* sp after 6 h (a) and 12 h, (b) of incubation in the absence of Ag-NPs. (c) Represents effect of the addition of Ag-NPs at 0 h along with bacterial inoculums. It indicates complete eradication of planktonic cells by Ag-NPs. (d) Degradation of preformed biofilm within 6 h of application of Ag-NPs.

ACKNOWLEDGEMENT

The authors would like to thank the Indian Institute of Science Education and Research, Kolkata, India, for FESEM facility; the Center for Research in Nanoscience and Nanotechnology, Kolkata, for SEM analysis; Tripura University for providing the laboratory and computation facility; the Centre of Excellence in Environmental Technology and Management, Maulana Abul Kalam Azad University of Technology, West Bengal, for providing the laboratory facility; the Ministry of Human Resource Development, Government of India, under the Frontier Area of Science and Technology for the student fellowships; and the UGC-DAE, Kolkata Centre for the EDXRF analysis, funding for carrying out research and local hospitality for student during experimentation at Kolkata. The authors are thankful to Baba's beans for providing raw materials for the work and Prof. A. R. Thakur for associating them with Baba's beans.

KEYWORDS

- **coffee waste**
- **caffeine**
- **polyphenol**
- **silver nanoparticle biosynthesis**
- **spent coffee ground**

REFERENCES

A laboratory manual, University of Pittsburgh. Caffeine Extraction from Tea. https, //www.coursehero.com/file/9443994/chem023-caffeine-extraction-from-tea1/ (accessed July 07, 2019).

Akowuah, G.A.; Mariam, A.; Chin, J.H. The effect of extraction temperature on total phenols and antioxidant activity of *Gynuraprocumbens* leaf. *Phcog. Mag.* **2009**, *5*, 81–85.

Baek, M.H.; Ijagbemi, C.O.; O, S.J.; Kim, D.S. Removal of Malachite Green from aqueous solution using degreased coffee bean. *J. Hazard. Mater.* **2010**, *176*, 820–828.

Ballesteros, L.F.; Teixeira, J. A.; Mussatto, S.I. Chemical, functional, and structural properties of spent coffee grounds and coffee silverskin. *Food Bioprocess Technol.* **2014**, *7*, 3493–3503.

Baucells, M.; Ferrer, N.; Gomez, P.; Lacort, G.; Roura, M. Quantitative analysis of caffeine applied to pharmaceutical industry. *J. Mol. Struct.* **1993**, *294*, 219–222.

Bhawani, S.A.; Fong, S.S.; Ibrahim M.N.M. Spectrophotometric analysis of caffeine. *Int. J. Anal. Chem.* **2015**, *2015*, 1–7.

Blinova, L.; Bartosova, A.; Sirotiak, M. Biodiesel production from spent coffee grounds. *Research Papers* **2017**, *25*, 113–121.

Blumberg, S.; Frank, O.; Hofmann, T. Quantitative studies on the influence of the bean roasting parameters and hot water percolation on the concentrations of bitter compounds in coffee brew. *J. Agric. Food Chem.* **2010**, *58*, 3720–3728.

Caetano, N.S.; Silva, V.F.M.; Melo, A.C.; Martins, A.A.; Mata, T.M. Spent coffee grounds for biodiesel production and other applications. *Clean Techn. Environ. Policy* **2014**, *16*, 1423–1430.

Cervera-Mata, A.; Pastoriz, S.; Rufian-Henares, J.A.; Parragaa, J.; Martin-Garciaa, J.M.; Delgado, G. Impact of spent coffee grounds as organic amendment on soil fertility and lettuce growth in two Mediterranean agricultural soils. *Arch. Agron. Soil Sci.* **2018**, *64*, 790–804.

Chalker-Scott, L. Using coffee grounds in gardens and landscapes. Home Garden Series 2016, 1–6. http://cru.cahe.wsu.edu/CEPublications/FS207E/FS207E.pdf (accessed July 07, 2019).

Chinmai, K.; Hamsa, B.C.; D'souza, K.D.; Mahesh Chandra B. R.; Shilpa B.S. Feasibility Studies on Spent Coffee Grounds Biochar as an Adsorbent for Color Removal. *Int. J. Appl. Innov. Eng. Manag.* **2014**, *3*, 9–13.

Coffee Market Report. International Coffee Organization 2018. http://www.ico.org/documents/cy2017–18/cmr-0518-e.pdf (accessed on July 07, 2019).

Cruz, R.; Cardoso, M.M.; Fernandes, L.; Oliveira, M.; Mendes, E.; Baptista, P.; Morais, S.; Casal, S. Espresso coffee residues, a valuable source of unextracted compounds. *J. Agric. Food Chem.* **2012**, *60*, 7777–7784.

Davila-Guzman, N.E.; Cerino-Cordova, F.J.; Loredo-Cancino, M.; Rangel-Mendez, J.R.; Gomez-Gonzalez, R.; Soto-Regalado, E. Studies of adsorption of heavy metals onto spent coffee ground, equilibrium, regeneration, and dynamic performance in a fixed-bed column. *Int. J. Chem. Eng.* **2016**, *2016*, 1–11.

Farah, A.; Donangelo, C.M. Phenolic compounds in coffee. *Braz. J. Plant Physiol.* 2006, 18, 23–36.

Farah, A.; Paulis, T.D.; Moreira, D.P.; Trugo, L.C.; Martin, P.R. Chlorogenic acids and lactones in regular and water-decaffeinated arabica coffees. *J. Agric. Food Chem.* **2006**, *54*, 374–381.

Fonseca, B.S.D.; Vilao, A.; Galhano, C.; Simao, J. A. R. Reusing coffee waste in manufacture of ceramics for construction. *Adv. Appl. Ceram.* **2014**, *113*, 159–166.

Fuller, M.; Rao, N.Z. The effect of time, Roasting Temperature, and grind size on caffeine and chlorogenic acid concentrations in cold brew coffee. *Sci. Rep.* **2017**, *7*, 1–9.

Iravani, S.; Zolfaghari, B. Green synthesis of silver nanoparticles using *Pinuseldarica* bark extract. *BioMed Res. Int.* **2013**, *2013*, 1–5.

Kemp,K.C.; Baek, S.B.; Lee, W.G.; Meyyappan,M.; Kim, K.S. Activated carbon derived from waste coffee grounds for stable methane storage. *Nanotechnology* **2015**, *26*, 1–11.

Khan, I.; Saeed, K.; Khan, I. Nanoparticles, properties, applications and toxicities. *Arab. J. Chem.* **2017**. https://doi.org/10.1016/j.arabjc.2017.05.011

Kimura, T. Natural products and biological activity of the pharmacologically active cauliflower mushroom *Sparassis crispa. BioMed Res. Int.* **2013**, *2013*, 1–9.

Kitou, M.; Okuno, S. Decomposition of coffee residue in soil. *Soil Sci. Plant Nutr.* **1999**, 45, 981–985.

Kwon, E.E.; Yi, H.; Jeon, Y.J. Sequential co-production of biodiesel and bioethanol with spent coffee grounds. *Bioresour. Technol.* **2013**, *136*, 475–480.

Makarov, V.V.; Love, A.J.; Sinitsyna, O.V.; Makarova, S.S.; Yaminsky, I.V.; Taliansky, M.E.; N. O. Kalinina. Green nanotechnologies, synthesis of metal nanoparticles using plants. *Acta Naturae* **2014**, *6*, 35–44.

Markowska, K.; Grudniak, A.M.; Wolska, K. I. Silver nanoparticles as an alternative strategy against bacterial biofilms. *Acta Biochim. Pol.* **2013**, *60*, 523–530.

Moeenfard, M.; Rocha, L.; Alves, A. Quantification of caffeoylquinic acids in coffee brews by HPLC-DAD. *J. Anal. Meth. Chem.* **2014,** *2014,* 1–10.

Mohan, Y.M.; Lee, K.; Premkumar, T.; Geckeler, K.E. Hydrogel networks as nonreactors: A novel approach to silver nanoparticles for antibacterial applications. *Polymer* **2007,** *48,* 158–164.

Morikawa, C.K.; Saigusa, M. Recycling coffee grounds and tea leaf wastes to improve the yield and mineral content of grains of paddy rice. *J. Sci. Food Agric.* **2011,** *91,* 2108–2111.

Morones-Ramirez, J.R.; Winkler, J.A.; Spina, C.S.; Collins, J.J. Silver enhances antibiotic activity against gram-negative bacteria. *Sci. Transl. Med.* 2013, *5,* 1–21.

Mussatto, S.I.; Ballesteros, L.F.; Martins, S.; Teixeira, J.A. Extraction of antioxidant phenolic compounds from spent coffee grounds. *Sep. Purif. Technol.* **2011,** 83, 173–179.

Namasivayam, S.K.R.; Christo, B.B.; Arasu, S.M.K.; Kumar, K.A.M.; Deepak, K. Anti biofilm effect of biogenic silver nanoparticles coated medical devices against biofilm of clinical isolate of staphylococcus aureus. *Global J. Med. Res.* **2013,** *13.*

Paikara, D.; Singh, S.; Pandey, B. Phytochemical analysis of leave extract of *Nyctanthes arbortristis*. *IOSR J. Environ. Sci. Toxicol. Food Technol.* **2015,** *1,* 39–42.

Rai, M.; Yadav, A.; Gade, A. Silver nanoparticles as a new generation of antimicrobials. *Biotechnol. Adv.* **2009,** *27,* 76–83.

Roh, J.; Umh, J.N.; Yoo, C.M.; Rengaraj, S.; Lee, B.; Kim, Y. Waste coffee-grounds as potential biosorbents for removal of acid dye 44 from aqueous solution. *Korean J. Chem. Eng.* **2012,** *29,* 903–907.

Sanghi, R.; Verma, P. Biomimetic synthesis and characterisation of protein capped silver nanoparticles. *Bioresour. Technol.* **2009,** *100,* 501–504.

Santos, C.; Fonseca, J.; Aires, A.; Coutinho, J.; Trindade, H. Effect of different rates of spent coffee grounds (SCG) on composting process, gaseous emissions and quality of end-product. *Waste Manag.* **2017,** *59,* 37–47.

Spigno, G.; Tramelli, L.; Faveri, D.M.D. Effects of extraction time, temperature and solvent on concentration and antioxidant activity of grape marc phenolics. *J. Food Eng.* **2007,** *81,* 200–208.

Yamane, K.; Kono, M.; Fukunaga, T.; Iwai, K.; Sekine, R.; Watanabe, Y.; Iijima, M. Field evaluation of coffee grounds application for crop growth enhancement, weed control, and soil improvement. *Plant Prod. Sci.* **2014,** *17,* 93–102.

Zuorro, A.; Lavecchia, R. Spent coffee grounds as a valuable source of phenolic compounds and bioenergy. *J. Clean. Prod.* **2012,** *34,* 49–56.

CHAPTER 22

Bioremediation: An Ecofriendly Technology for Sustainable Aquaculture

S. K. NAYAK[1*] and P. K. NANDA[2]

[1]*Department of Biotechnology, Maharaja Sriram Chandra Bhanja Deo University, Takatpur, Mayurbhanj, Odisha 757003, India*

[2]*Indian Veterinary Research Institute, Eastern Regional Station, Belgachia Road, Kolkata 700037, India*

Corresponding author. E-mail: sukantanayak@rediffmail.com

ABSTRACT

In recent years, there has been a significant increase in pollution/contamination level in aquaculture cultured ponds. Like any other ecosystem, the pond environment in an aquaculture system shares various physical, chemical, and ecological characteristics with that of other aquatic ecosystems. Furthermore, various nutrient cycles such as nitrogen, sulfur, carbon, and phosphorus cycles are operating in an aquatic ecosystem for the movement and exchange of organic and inorganic matters. However, in a closed pond environment, wastes from various sources often accumulated, which, in turn, disturb these cycles and overall health status of cultured organisms, and, therefore, poses serious challenges to the system. Recently, biological methods involving application of "bioremediation" technology is gaining popularity day by day. Bioremediation is based on the principle of biodegradation, either partial or total transformation or detoxification, of contaminants of soil and water. In this process, living organisms such as microorganisms (bacteria, actinomycetes, cyanobacteria, fungi, and yeast), plants, and/or even enzymes are used to detoxify or clean up contaminated areas. Bioremediating techniques, using microbes either alone and/or in combination, have now been adopted in aquaculture practices. Several types of bioremediation techniques such as bioaugmentation, biostimulation, biofloc system, microbial mat, and integrated multitrophic aquaculture, etc., are currently used in aquaculture practices. Although all these concepts differ from each other, yet the overall outcome of each degradation process depends on a number of intrinsic and extrinsic factors. Important among them are microbes (biomass and its diversity and concentration, and enzyme activities), substrate (physico-chemical characteristics, composition/structure, and concentration), and environmental factors, including pH, temperature, moisture content, availability of carbon, and energy sources, etc. Common predominant microbes are nitrifiers, sulfur bacteria, *Bacillus, Aeromonas,* and *Pseudomonas* species. In this chapter, the details of each process have been discussed.

22.1   INTRODUCTION

Bioremediation is a biological mechanism of recycling wastes to another form(s) that can be used and reused by other organisms. The concept of bioremediation involves biodegradation, either through partial or total transformation or detoxification of contaminants by microorganisms and plants for elimination, attenuation, or transformation of polluting or contaminating substances. In this process, living organisms such as microorganisms (bacteria, actinomycetes, cyanobacteria, fungi, and yeast), plants, and/or even enzymes are used to detoxify or clean up contaminated soil and water. As per the United States Environmental Protection Agency, bioremediation agents are microbiological cultures, enzyme, or nutrient additives that significantly increase the rate of biodegradation to mitigate the effects of the discharge (Nichols, 2001). The Office of Technological Assessment (1991) defined this process as "the act of adding materials to contaminated environments to accelerate the natural biodegradation processes". The whole purpose of bioremediation is to reduce pollutant levels to undetectable, nontoxic, or within levels of acceptable or permissible limits as set by the regulatory agencies of different countries (Pointing, 2001). During the process, bioremediation enhances the rate of the natural microbial degradation of contaminants not only by supplementing the indigenous microorganisms (bacteria or fungi) with nutrients, carbon sources, or electron donors (biostimulation, biorestoration), but also by adding an enriched culture of microorganisms that have characteristics to degrade the desired contaminant at a much quicker rate.

Over the years, there has been a significant increase in pollution/ contamination level in different ecosystems, and aquatic systems are no exception to this. The water bodies are often contaminated with various types of toxic/waste materials due to several reasons such as urbanization, population explosion, and rapid industrialization. Because of this, almost all aquatic ecosystems irrespective of type, that is, fresh, brackish, and marine, are contaminated with several types of anthropogenic waste materials and increasingly becoming unsuitable for various uses, including aquaculture. The pond environment in an aquaculture system shares various physical (gas balance, sedimentation resuspension, and water circulation), chemical (pH and organic matter decomposition), and ecological (food web structure and prey–predator relationship) characteristics with other aquatic ecosystems. Furthermore, to maintain a balance in the aquatic ecosystem, various nutrient cycles such as nitrogen, sulfur, carbon, and phosphorus cycles are operating in an aquatic ecosystem for the movement and exchange of organic and inorganic matters. Accumulation of wastes from various sources often disturbs these cycles, and the overall health status of cultured organisms poses serious challenges to this closed environment practiced system. The intensive type of aquaculture practices, although contributes significantly for the nation development and eliminating poverty in many parts of world, often creates adverse impact on the aquatic environment due to high stocking density and excessive application of inputs such as feed, fertilizers, chemicals/disinfectants, etc. Therefore, like any other terrestrial counterpart, wastes (waste products, unutilized feed, excreta, *etc.*) are major obstacle in development and sustainability of aquaculture. Therefore, it is of paramount importance to maintain a proper nutrient cycling through natural bacteria for removal or degradation of different compounds, contaminants, and pollutants in order to improve the quality of water, facilitating better growth of cultured species.

Recently, technological advancements have been made in treating different types of discharged/accumulated wastes to make the water bodies fit for culture of aquatic organisms. As the scope of bioremediation concept is very broad and applicable to land and aquatic systems, its application through various microbial bioremediation processes to mitigate the impact of waste on aquatic environment for sustainable aquaculture is discussed here.

22.2   SCOPE OF BIOREMEDIATION IN AQUACULTURE

Aquaculture the sunrise sector has been supporting human demands for aquatic products and is an important industry throughout the world. To cater to the growing demand, the aquaculture industry, especially that of freshwater, brackish, and mariculture, has been growing tremendously. The current aquaculture practice, mostly of fresh and brackish water, is done with high inputs (stocking density and feed) and minimal or zero water exchange. Such types of culture practices usually affect the physic-chemical as well as nutrient cycle of the pond environment, which, in turn, can affect the sustainability of this industry (Cao et al., 2007). Furthermore, accumulation of residues of various chemicals, disinfectants, pesticides, fertilizers, antibiotics, dead organisms, fecal matters, etc., also adds to the deterioration process. The major changes in water quality include increased chemical oxygen demand, biochemical oxygen demand, enhanced ammonia, nitrite, nitrate, phosphate, and hydrogen sulfide (H_2S) level at pond bottom. Since aquaculture practices generate considerable amount of wastes, which differs not only in quality but also in quantity of components depending on the species farmed and the farming practices adopted, a proper mechanism is needed to manage the wastes.

22.3   WASTE PRODUCTION IN AQUACULTURE

Aquaculture, as compared to agriculture or livestock-based farming, is much more diverse and complex. Basically, three different types/forms of cultures, that is, extensive, semi-intensive, and intensive are being practiced in aquaculture. Out of these, the traditional or extensive method, because of optimum utilization of farm resources, is often termed as eco-friendly, but production is less compared to other methods. In contrast, recent developments in aquaculture through intensification of farming have no doubt increased the production but altered the quality of natural habitats through increased effluent discharges, which contains high quantities of materials of both organic and inorganic forms, leading to severe problems in cultured species, including production loss due to diseases. Supplementation of formulated commercialized feed along with several other feed additives is indispensable for enhancing the production and maintaining sustainability in intensive and semi-intensive aquaculture. The indiscriminate use of commercial feeds is the major waste contributor in aquatic farms (Iwama, 1991). Besides, leftover feed and other feed additives, including fertilizers, prophylactic, and therapeutic drugs, fecal matter, metabolic by-products, residues of biocides, wastes generated during molting (in shrimp/crustacean culture), collapsed algal blooms, etc., also lead to the generation of wastes. All these outputs generate considerable amounts of waste leading to the deterioration of water quality. The potential impact of effluent discharged from aquaculture in terms of organic matter, nitrogen, sulfur, phosphorous, etc., into the environment is very high.

Although it depends on the feed conversion ratio and production of fish per unit area, it is gauged that the utilization percentage of nitrogen and phosphorous in the diets by aquatic animals are only 13.9 and 25.4, respectively (He and Wu, 2003), leaving the rest deposited in sediment. It is the nutrients that are stated to be responsible for undesirable changes in aquatic system such as eutrophication or algal bloom, which, in turn, depletes oxygen resources and lowers productivity (Jang et al., 2004; Cao et al., 2007). It is, therefore, imperative to prevent build-up of organic detritus and slime as sludge as a result of decomposition of unutilized feed, fecal matter, and dead algae present in the water column and bottom sediments. It is imperative to maintain a healthy ecosystem for sustainable aquafarming through improving the quality of culture pond water for better production without affecting the environment. This can only be achieved by adopting eco-friendly techniques such as bioremediation, which not only improves the aquatic environment but also helps in eliminating infectious agents/pathogens, thereby preventing/ controlling diseases besides improving the overall health status of the cultured organisms.

22.4   BIOREMEDIATION OF AQUACULTURE WASTES

Basically, two major processes, namely chemical and biological, are adopted for treatment of wastewater in aquaculture. The chemical treatment method is more effective for the removal of nonbiodegradable materials, but its disadvantages are massive, as expensive operational costs and initial investments limit their widespread application. In contrast, biological methods involving application of "bioremediation" technology is gaining popularity day by day. This technology involves microorganisms in treating wastewater so as to improve the quality of water suitable from the aquaculture point of view. Bioremediation is a process whereby organic wastes are degraded to harmless products through biological means under controlled conditions or to levels below permissible limits as established by regulatory authorities. The process involves intriguing microorganisms in ponds to escalate the mineralization of organic matter and get rid of undesirable waste compounds, thereby lowering the accumulation of slime or organic matter in the pond bottom. This helps in better penetration of oxygen into the sediment offering a conducive environment for the farmed stock. Microbes used as bioremediating agents are mixed with sand or clay and broadcasted in sufficient quantities enabling them to be deposited in the bottom of the culture system. In many instances, *Lactobacillus* species are also used along with *Bacillus* species to break down the organic detritus quickly, which, in turn, not only helps in the removal of large organic compounds, but also reduces the water turbidity to a greater extent.

A successful bioremediation process involves the optimizing nitrification rates to lower ammonia concentration and denitrification rates to completely remove excess nitrogen from ponds in the form of nitrogen gas, maximize sulfide oxidation to lower build-up of hydrogen sulfide (H_2S), and minimize sludge accumulation through mineralization of carbon to carbon dioxide. This increases primary productivity that stimulates the production of secondary crops and maintains stable pond community, where unwanted species do not become dominant (Bratwold et al., 1997). This, as a process, is effective when conducive environmental conditions permit microbial growth and often involve the manipulation of environmental parameters that enable the microbial growth and degradation to progress at a faster rate and, in turn, helps in improving water quality and maintaining the health and stability of aquaculture systems.

22.4.1 BIOREMEDIATION OF ORGANIC MATTERS

The solid organic matters (dissolved and/or suspended), containing mainly carbon chains and abundantly available in aquaculture systems, are mainly algae (phytoplankton and zooplankton) and microbiota. To effectively clear the carbonaceous wastes from water, the bioremediation agent selected must contain microbes that multiply rapidly. Furthermore, these bacteria must produce a variety of enzymes and have good enzymatic capability that break down carbonaceous organic matter to small molecules only to be taken up as energy sources by other organisms. Gram positive spore forming bacteria, due to their potency for stable bacterial preparations, are mostly preferred. Important among them are the members of genus *Bacillus* (e.g., *B. subtilis, B. cereus, B. licheniformes, and B. coagulans*) and *Phenibacillus* (e.g., *P. polymyxa*).

22.4.2 BIOREMEDIATION OF NITROGENOUS COMPOUNDS

In water, ammonia occurs in two forms, that is, ionized ammonia (NH_4^+-N) and unionized ammonia (NH_3-N), together called the total ammonia nitrogen (TAN), is the principal nitrogenous metabolic waste. Ammonia is produced by fish due to catabolism of amino acids in the liver and excreted as principal excretory product. The excreted ammonia in dissolved form is frequently accumulated as waste in the water body, which, in turn, deteriorates the water quality causing toxicity to the species cultured. It is toxic to cultured organisms since this metabolite increases blood pH and lowers oxygen availability in the blood, thereby affecting gills. Ammonia is reported to cause stress and reduced feed intake. These factors enhance disease susceptibility and may cause mortality of fish in extreme concentrations. Besides, the microbial decomposition of unconsumed feed, trash fish, chicken offal, and kitchen wastes, shell molts of crustaceans, dead algae, zooplankton, decomposition of animal and plant tissues, etc., applied/generated during intensive aquaculture practices not only lead to the production of ammonia, but also get accumulated in pond water (see Figure 22.1). As TAN is a key limiting factor to judge water quality parameter after oxygen and affect fish reared in intensive systems, there is a need for effective and economical management method for treating ammonia. The concentration of ammonia is dependent upon various parameters, including pH, temperature, and a lesser extent to the salinity of water. At the pH value of higher than 7.5, levels of unionized ammonia present in water increase. The ammonia in this form is toxic in nature and a critical water quality parameter for the organisms, including fish and invertebrates living in it and, therefore, should be maintained below 0.3 ppm. In acute and chronic cases, it causes physiological imbalances, impairs growth, and damages organs. All these changes cause stress, decrease resistance resulting in susceptibility to infectious agents, and may cause mortality of fish in extreme concentrations.

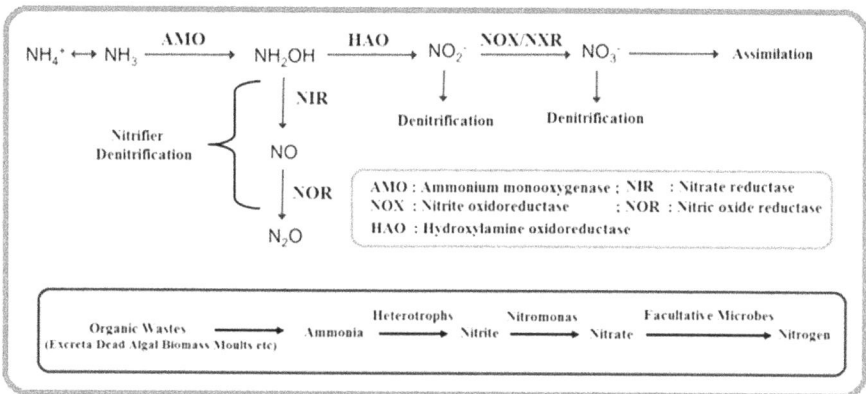

FIGURE 22.1 Generation of nitrogen by various processes by microorganisms in an aquatic ecosystem.

The toxic effects of nitrogenous compounds are often nullified to some extent by physical methods such as filtration and aeration, frequent water exchange, etc. Although, several chemicals and oxidizing and chelating agents are also used, but these methods have both beneficial and negative impacts. To overcome this, nowadays, the nitrification process using microbes is widely used for removal or control of the ammonia content in pond culture systems. Nitrification is an obligate aerobic, oxidizing process, where ammonia is converted into nitrite by autotrophic bacteria. This occurs in an environment, which contains oxygen and bacteria get the carbon for cell growth from carbon dioxide. The most common ammonia oxidizers are *Nitrosomonas, Nitrosovibrio, Nitrosococcus, Nitrolobus,* and *Nitrospira.* These bacteria are important from aquaculture perspective, as it is difficult to maintain healthy environmental conditions in ponds in their absence. The nitrifying bacteria utilize ammonia–nitrogen as the source of energy producing nitrite-nitrogen (NO_2–N).

Nitrite, although not that much toxic compared to unionized ammonia nitrogen, is a critical water quality parameter and can be harmful to fish causing anoxia, a common mode of toxicity and, therefore, needs to be maintained below 0.1 ppm level. Nitrite poisoning lead to the inhibition of oxygen uptake ability of red blood cells, a condition known as "methemoglobinemia," where the hemoglobin is converted into methemoglobin, thereby causing brown blood disease. Several nitrifying bacteria such as Nitrobacter species can utilize nitrite as energy source and produce nitrate-nitrogen (NO_3–N) and the process is known as nitrification. In this process, the pH of the cultured water alters toward acidic range which in turn facilitates the availability of soluble materials. The primary producers (algae and aquatic plants) and phytoplankton utilize nitrate.

The concentration of NO_3–N is not considered as a critical water quality factor, as most of the aquatic species can tolerate concentrations as high as 200 ppm and a level beyond 200 ppm is hardly encountered during aquaculture practice, as nitrate is removed from the system during water exchanges, through passive denitrification in anaerobic pockets or filtration systems (van Rijn, 1996, Tal et al., 2006). The denitrification process develops an anaerobic area where only anaerobic bacteria can grow, which reduce nitrate to nitrogen gas. The difference between nitrification and denitrification is the type of bacteria, which perform the processes. The nitrifying bacteria (*Nitrosomonas* and *Nitrobacter* sp.) belongs to the family *Nitrobacteriacaea.* Although efficient to convert ammonia at a much faster rate compared to heterotrophs, they take up long hours to double. In spite of their obligate aerobic nature, they fail to multiply or convert ammonia due to lack of oxygen.

In contrast, denitrification is an anaerobic and reducing process. It occurs in environments without oxygen and during the process transforms nitrate to dinitrogen, a gas that is harmless and bubbles out of the system. This involves a multistep process producing three intermediary products like nitrite (NO_2^-), nitric oxide (NO), and nitrous oxide (N_2O), before the end product (dinitrogen) is generated to complete the process. In most cases, these intermediate products are toxic such as ammonia and nitrite. Therefore, the ammonium released by the heterotrophic bacteria and also by cultured species must be taken up by the algae or rapidly oxidized again by nitrifying bacteria to prevent toxicity to the cultured animals in the culture ponds. If the concentration of nitrate becomes too high, it will create problem with rapid depletion of oxygen due to limited diffusion in the grow-out pond bottom detritus layer. Furthermore, oxygen can also be consumed by bacteria, other microbes, and aquatic or benthic organisms living therein. In such a scenario, deficit of oxygen occurs during the grow-out stage. In the absence of oxygen, nitrate present is used for respiration producing nitrite, ammonia, and nitrogen gas. Because of this, the cultured species are often exposed to sublethal concentrations of toxic compounds such as sulphate, nitrite, and ammonia. Furthermore, nitrogen compounds such as ammonium and nitrite if present at sufficiently high concentration are toxic to cultured organisms, and nitrate may even cause "blue baby syndrome," fostering a potentially life threatening public health issue (Nora'aini et al., 2005).

22.4.3 BIOREMEDIATION OF PHOSPHOROUS COMPOUNDS

Phosphorus is a key nutrient required for every organism, and fish is no exception. In fish, its role in growth and other metabolic activities is well demonstrated. A phosphate level of 0.06 ppm is desirable for fish culture (Stone and Thomeforde, 2004). However, fish exhibit an increased level of stress when phosphorous level goes beyond 3 ppm; a level higher than 0.7 ppm is harmful. In the pond culture system, the accumulation of phosphorus in the culture system is mainly through the excretion, excreta either dissolved or particulate forms, industrial and municipal sewage contamination, unused feed, etc. The unabsorbed phosphorus in the form of uneaten fish feed or excreted in solid form as feces gets accumulated, as a consequence of which eutrophication of ponds occur. Therefore, concentration of total phosphorus (total-P) is an indicator of organic load in the pond system (see Figure 22.2).

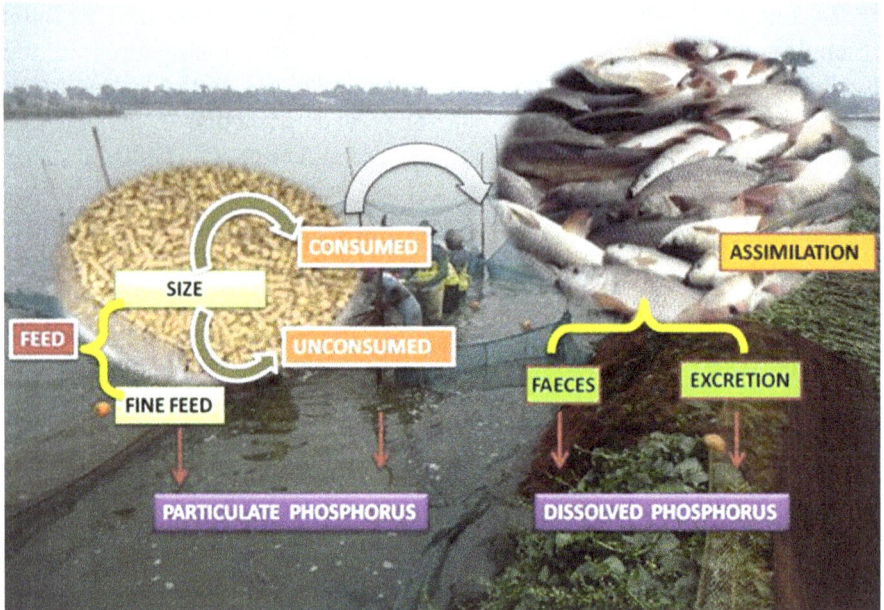

FIGURE 22.2   Generation of phosphorous in an aquatic ecosystem through feed.

Furthermore, phosphorous is generated from organic compound as PO_4 by certain phosphotases and phytases producing bacteria. Any deviation from the normal NO_3/PO_4 ratio can influence the rate of nitrification process or bacterial regeneration of phosphorous. Fishmeal is one of the major contributors of phosphorous, and its overuse in feed often leads to excretion of high dissolved P along with N; the key lies in limiting its availability in feed to reduce the content of phosphorous. Native bacteria are capable of liberating PO_4 from these compounds through the production of organic and mineral acids. Different bacterial species belonging to *Bacilus*, *Pseudomonas*, *Rhadobacter*, and even yeast such as *Saccharomyces cerevisiae* are found to be useful in reducing phosphate content in water (Usharani et al., 2009; Patricia et al., 2012).

22.4.4   *BIOREMEDIATION OF SULFUR COMPOUNDS*

Sulfur is normally present in the form of sulfate ion in most of the water bodies, including aquaculture systems. Accumulation of organic matter in aquatic ecosystem from various sources, including anthropogenic and aquaculture activities, tend to elevate the H_2S level. The level of H_2S should be restricted within 2 μg/L, beyond which it can cause adverse consequences in freshwater fish when exposed for longer periods. It causes anoxic conditions of sediments and reduces the diversity of benthic fauna. In aerobic conditions, organic sulfur decomposes to sulfide, which, in turn, gets oxidized to sulphate. While unionized H_2S is extremely toxic to fish and may occur in natural waters and aquaculture farms as well, HS^- is regarded as nontoxic. H_2S being soluble in water can cause gill damage and other ailments in fish/shrimp. Any detectable concentration of H_2S is, therefore, detrimental to fish production. To maintain a favorable environment at pond bottom, the photosynthetic benthic bacteria that break H_2S and belonging to *Amoebobacter, Chlorobium, Chromatium, Clathrochloris, Lamprocystis, Prosthecochloris, Pelodictyon, Rhodospirillum, Rhodopseudomonas, Thiocystis, Thiospirillum, Thiocapsa, Thiodictyon, Thiopedia*, etc., are the common bioremediating agents that are applied throughout the world for aquaculture practices.

Besides, the purple and green sulfur bacteria, which contain bacterio-chlorophyll and perform photosynthesis under anaerobic conditions in the sediment–water interface, are also used. The photosynthetic purple nonsulfur bacteria can decompose organic matter, H_2S, NO_2, and harmful wastes of ponds. In contrast, photosynthetic sulfur bacteria belonging to families of *Chromatiaceae* and *Chlorobiaceae* favor anaerobic conditions for growth by utilizing solar energy and sulfide and, therefore, most preferred bioremediating agents for combating the H_2S toxicity in aquaculture. Such autotrophic and photosynthetic bacteria are usually adsorbed onto the sand grains and applied to the pond bottom to ameliorate H_2S toxicity.

22.5 COMMERCIAL MICROBES AS BIOREMEDIATING AGENTS

Bioremediating techniques, using microbes either alone and/or in combination, have now been adopted in aquaculture practices in various parts of world (see Table 22.1). Predominant microbes among them are nitrifiers, sulfur bacteria, *Bacillus sp.*, and *Pseudomonas sp.* Amongst, *Bacillus* is the most commonly used organism followed by *Aeromonas* and *Pseudomonas sp.* Common bioremediators and their use in aquaculture are summarized in a tabular form. Recently, various types of products involving live bacterial inoculum, enzyme preparations, plant, and yeast extracts are commercially available for improving water quality in aquaculture (see Table 22.2). For example, commercial formulations such as "EM" flora (involving *Bacillus*, photosynthetic bacteria, and Actinomycetes) have been developed for the treatment and maintenance of neutral pH of pond water.

TABLE 22.1 Common Bioremediating Agents used in Aquaculture

Name of the Agent	Culture Type	Culturable Species	Method of Application
Bacillus sp.	Fish	*Centropomus undecimalis*	Added to water
Bacillus sp.	Shrimp	Penaeids	Spread in pond water
Aeromonas sp.	Fish	*Crassostrea gigas*	Spread in pond water
Aeromonas sp.	Fish	*Crassostrea gigas*	Spread in pond water
Nitrosomonas sp.	Fish	*Cyprinus carpio*	Simulated condition
Nitrobacter sp.	Fish	*Cyprinus carpio*	Simulated condition
Roseobacter sp.	Fish	*Oncorhynchus mykiss*	Spread in pond water

TABLE 22.2 Different Types of Commercial Bioremediating Agents used in Aquaculture

Sl. No	Commercial Product	Principal Microbial Agent
1	BRF 4	*Nitrobacter* sp.
2	BRF 13A	*Nitromonas* sp.
3	Nitroclear	
4	BZT aquaculture	Nitrifiers
5	ABIL nitrifying package	
6	Bactaclean	
7	Remus	Nitrifying bacteria
8	BRF-4	
9	Alken Clear Flo1100	
10	Ammonix	
11	Biostart	*Bacillus* sp.
12	Pronto	
13	Detrodigest	
14	Alken Clear flo1002	
15	ProbacBC	
16	Biogreen	*Bacillus subtilis*
17	Ps-1	*Pseudomonas* sp.
18	Super PS	Sulfur bacteria

22.6 PROBIOTICS AS BIOREMEDIATING AGENTS IN AQUACULTURE

Probiotics, a term coined by Parkar (1974), is considered as method of biological control or "Biocontrol" and increasingly being used now days with the demand for more environmental friendly aquaculture practices. Probiotics, as bioremediating agents, are used to regulate the microbiota of aquaculture water, enhance decomposition of the undesirable organic substances, and improve ecological environment by reducing the toxic gases such as ammonia, nitrite, hydrogen sulfide, methane, etc. Furthermore, probiotics improve the nutrition level of cultured species by increasing the population of food organisms. Research findings indicate that disease problems in aquaculture systems can be eliminated or minimized through the biocontrol process so that better yield can be achieved through higher survival and growth rate. Probiotic formulations such as Biostart, Liqualife, Pond pro VC, Nitro clear, Eutro clear etc., are marketed nowadays with elucidation that bacteria, which improve water quality, may be good for animal health.

22.7 TYPES OF BIOREMEDIATION PROCESSES

Several types of bioremediation techniques are currently in use in aquaculture practices to mitigate the adverse effects of the generated wastes. However, the overall outcome of each degradation process depends on microbes (biomass concentration, population diversity, and enzyme activities), substrate (physicochemical characteristics, molecular structure, and concentration), environmental factors (pH, temperature, and moisture content), availability of carbon and energy sources, etc. Some of the major bioremediation techniques that are used as in aquaculture are as follows.

22.7.1 BIOFLOC SYSTEM

Bioflocs refer to the aggregates (flocs) of algae, bacteria, protozoans, and several types of particulate organic matters such as unutilized/unconsumed feed materials. This technology plays a crucial role in maintaining the stability of aquatic ecosystem and has immense potential to reduce harmful toxic metabolites in hatcheries and grow-out systems. This technology, which is practiced under intense aeration and mixing, facilitates degradation, assimilation, and nitrification of organic wastes by a wide microbial biofloc forming community. Further in this technology, an additional organic carbon source such as molasses in a regulated carbon-to-nitrogen ratio is used to facilitate the growth and multiplication of various microbes, which, in turn, act as supplementary proteins, vitamins, and minerals source for cultured host species. The most advantage of this technology is that it helps in preventing deterioration of water quality even with high-nutrient supplementation due to high mechanical aeration, which, in turn, favors the degradation of toxic nitrogen and other compounds by the microbial community. Therefore, it is very suitable for intensive culture systems, which maintain high stocking density and use inputs such as feed with minimal or no water exchange. Furthermore, it helps in reducing the organic load as well as increases the heterotrophic bacteria, which utilize toxic nitrogenous matters as a substrate for their growth. Besides maintaining water quality at optimal levels, it is effective against pathogens and enhances the survivability and growth of aquatic animals also, thereby improving production. However, application of this technology is restricted to cultures with limited or no water exchange systems. Furthermore, selecting microbes for application in such systems is challenging, since multiple studies point out conflicting reports on the efficacy of various microbes for the same variables.

22.7.2 MICROBIAL MATS

Microbial mats (often called "algal mats" and "bacterial mats") are usually a consortium of cohesive biofilm-type microbial communities and occur in nature as stratified communities of cyanobacteria and bacteria in a multi-layered sheet, either on the surfaces of sediment or as floating masses in waters. The mats are usually embedded and bound to their substrates by slimy extracellular polymeric matrix of gel substances secreted by them. The microbial communities mostly include anoxygenic photoautotrophs and sulfur-reducing bacteria. Among photoautotrophs, cyanobacteria such as *Chroococcus, Lyngbya, Oscillatoria, Rhodopseudomonas* sp., etc., and proteobacteria such as eubacteria are predominant. Such types of mats mostly contain both nitrogen fixing and photosynthetic bacteria. Besides, they are self-sufficient, and self-sustaining, and/or have limited growth requirements.

The mats play an important role in transformation of nutrient-enriched effluents/wastewater from aquaculture. Furthermore, these mats can degrade and completely mineralize organic contaminants and can precipitate metals by surface absorption. These microbial mats are usually cultured on large scale and also manipulated for a variety of functions. For example, protein produced via nitrogen fixation can be used for supplying nutrition to fish. Despite these, application of biofilms and microbial mats for bioremediation of aquaculture wastes is not yet popular. As a combination of factors such as selection of suitable fish species, optimum substrate density, and economic viability of the potential substrate materials plays an important role determining the success rate in terms of fish production, more site-specific research is needed to refine this technology under field conditions and thereafter apply in large areas. Furthermore, benthic algal mats rarely grow on bottoms in highly eutrophic environments due to limitation in light penetration and, therefore, need some hard substrate in euphotic layer of the rearing system to grow.

22.7.3 BIOSTIMULATION

Biostimulation is a natural remediation process that involves the addition of limiting nutrients that are otherwise available in quantities low enough to sustain microbial activities. This is accomplished by addition of phosphorous, nitrogen, or carbon sources to stimulate the growth of native microorganisms. Besides, oxygen or other electron donors and acceptors are added to stimulate the microbial population. Therefore, apart from fulfilling the primary objective of providing dissolved oxygen to cultured animals, aeration also improves conditions for organic matter decomposition and nitrification. Only by adding these supplemental nutrients in right concentrations and controlling other parameters such as temperature and pH, the degrading microbes are able to achieve maximum growth rate; as a result, the maximum rate of pollutant degradation takes place. However, the rate of success depends upon the type, concentration, and even spread or distribution of nutrients/additives when covering a vast area, competition between native microbes and nondegraders (heterotrophic bacteria) for nutrients, etc.

22.7.4 INTEGRATED MULTITROPHIC AQUACULTURE (IMTA)

The monoculture-based farming system mostly involves culture of species that may be the fed-aquaculture species or organic/inorganic extractive aquaculture species such as fish, shrimp, bivalves, aquatic plants, etc., and are independently cultured. Earlier in the age-old aquatic practices, co-culture of different fish species from the same trophic level was also practiced. On contrary to this, multitrophic form of culture, that is, combination of species from different trophic or nutritional levels in the same culture system, is now being adopted. The concepts such as IMTA, which is based on the recycling of the by-products from one species to become inputs, viz., fertilizers, food, and energy for another cocultured species. In this concept, the fed species such as finfish/shrimps are cocultured with extractive species that utilize the organic extractive aquaculture species suspension feeders/deposit feeders/herbivorous fish, etc., as well as inorganic extractive aquaculture species such as seaweeds for their growth. Comprehensive field and laboratory feasibility studies conducted to investigate the potential of IMTA with fish species indicate that the load of organic wastes can be significantly reduced by coculturing bivalves such as mussels, oysters, etc., with fish species. The most advantage of this concept is that this process not only reduces the production cost, but also promotes diversification of species.

The IMTA concept is very flexible and can either be land or open-water (marine or freshwater) based system. It involves several combinations such as fish/seaweed/shellfish, fish/shrimp and seaweed/shrimp, etc. Freshwater IMTA (otherwise known as aquaponics), integrated agriculture–aquaculture systems, integrated periurban aquaculture systems, and integrated fisheries–aquaculture systems are some of the forms of the IMTA concept. Despite this, large-scale application of this technology is greatly limited due to intricacy and involvement of high cost.

22.7.5 BIOAUGMENTATION

The bioaugmentation technique is based on the principle of supplementation or addition of microorganisms and/or its metabolites in the polluted environment. Although most commonly used to remediate soils contaminated with hazardous organic pollutants such as crude oil, petroleum products, or pesticides, it is also used as one of the recent bioremediation strategies to improve water quality, where indigenous microorganisms are either not identified in the soil or they do not have the ability to carry out metabolic activity during the remediation process. The process rate is increased by adding allochthonous source of microorganisms to the system. This process can detoxify, degrade, or remove pollutants, especially organic matter and unutilized nutrients and thereby accelerating the removal of contaminants. It helps in reducing the accumulation of organic wastes to environmentally safe levels in a closed or zero water exchange system. In this system, both the parameters (physicochemical and biological) of water remain stable throughout the culture period causing minimal stress to the cultured animals.

Unlike the bioflocs technique, bioaugmentation does not need organic matter to control ammonia concentration, which perhaps accomplishes the process of organic matter oxidation more efficiently. Rather, several digestive enzymes producing microorganisms such as *Bacillus* sp., *Nitrobacter* sp., *Aspergillus niger,* and *Trichoderma* sp. when used as bioaugmenting agents are reported to increase the growth and survival percentage of the cultured species. Besides improving survival, this technique is also helpful in reducing pathogenic bacteria in aquaculture. For example, *Bacillus* species as bioagumenting agents are found to reduce the pathogen load from the water as well as lower the concentrations of nutrients in the water, thus acting as both bioaugmentation and biocontrol agents. The usefulness of bioaugmentating agents in aquaculture is promising, but needs further in-depth research is necessary in selecting the effective indigenous microbial strains or genetically engineered microbes.

22.8 ADVANTAGES OF BIOREMEDIATION

Bioremediation is a very effective and successful process, wherein microbes are used for degrading, detoxifying, or transforming the aquatic wastes to harmless residues/products including water, carbon dioxide, and cell biomass. One important aspect of this technique is that it can simultaneously treat both soil and water of the culture pond. Furthermore, it is environment friendly and less expensive as compared to other types of treatments and can be very effective against a broad range of compounds and wastes, both organic and inorganic nature. Besides, other advantage of this technique is that the entire process is easy to implement and maintain, requires very less effort, and can be carried out in both on-site and off-site without hampering normal activities. This also eliminates the need and risk of transporting harmful wastes and/or products off-site, thus averting potential threats not only to human health but also the environment that may arise during transportation and land filling.

22.9 LIMITATIONS OF BIOREMEDIATION

The process of bioremediation is influenced and controlled by various factors. Bioremediation, as whole, is a time consuming as compared to other treatments or application of additives, nutrients, surfactants, oxygen, etc., and mostly depends on climatic conditions. Being a biological process, the rate of success largely depends upon many predisposing factors such as used microbial strain(s) and its metabolic potential, availability of nutrients, physicochemical, and other environmental factors, nature and concentration of contaminants, etc. However, many factors, including depletion of preferential substrates, lack of nutrients, level of toxicity, and solubility of contaminants, could well hamper the bioremediation process.

The bioremediation process is effective but limited to only biodegradable compounds and at times fails to degrade the substances completely leaving harmful effects on the food chain. There are also some concerns that the undegraded and/or partially degraded products could be more persistent or toxic than the original compounds. Nonetheless, bioremediation can only be effective when environmental conditions help in achieving the optimum microbial growth and activity. However, only a few species of bacteria and fungi have shown their ability as potent pollutant degrader, and at times, the manipulation of environmental parameters is required to facilitate the microbial growth and degradation to proceed at an accelerated rate. Furthermore, the products selected should be inexpensive, and the strains selected should have the unique desired properties to degrade a wide range of pollutants of aquaculture origin. Above all, evaluating the performance of bioremediation is difficult in the absence of clear guidelines and definitions.

22.10 FUTURE PERSPECTIVES AND CONCLUDING REMARKS

Nowadays, the intensive and superintensive aquaculture practices are facing an uphill task over the years in maintaining environmental stability and sustainability due to accumulation of farm waste. Many intrinsic and extrinsic factors, either individually or in combination, are stated to be responsible for this. As far as waste treatment and water quality improvement are concerned, a number of *in situ* remediation strategies are currently being implemented. Bioremediation technology is ecofriendly, and advances in research and development in various scientific fields such as microbiology, chemistry, molecular biology, and chemical and environmental engineering, among others, have actively contributed to the development of bioremediation progress in recent years. Furthermore, the role of bioremediating agents in waste management and improving water quality vis-à-vis controlling pathogenic organisms has become crucial, especially due to the increasing emergence of antibiotic resistant pattern among the aquatic bacteria. Several types of bioremediation techniques such as bioaugmentation, biostimulation, biofloc system, microbial mat, IMTA, etc., are currently used in aquaculture practices. However, widespread application of these concepts and their success depends upon several factors, including area/region, strain used, etc. To overcome the shortcomings, region- or area-specific programs may be developed to get desired results so as to reap maximum benefits through application of bioremediation in aquaculture.

KEYWORDS

- **aquaculture**
- **bioagumentation**
- **bioflocs**
- **bioremediation**
- **probiotics**

REFERENCES

Abatenh, E.; Gizaw, B.; Tsegaye, Z; Wassie, M. Application of microorganisms in bioremediation-review. *J. Environ. Microbiol.* **2017**, *1(1)*,2–9.

Akinsemolu, A. A. The role of microorganisms in achieving the sustainable development goals. *J. Clean. Prod.* **2018**, *182*, 139–155.

Alavandi, S.V. Mitigating nitrogenous wastes in aquaculture, ENVIS, **2010**, 8(3 & 4), 5–9.

Antony, S.P.; Philip, R. Bioremediation in shrimp culture systems. *NAGA, Worldfish Center Quat.*, **2006**, *29(3 & 4)*, 62–66.

Barik, P.; Ram, R.; Haldar, C.; Vardia, H. K. Study on nitrifying bacteria as bioremediator of ammonia in simulated aquaculture system. *J. Entomol. Zool. Stud.*, **2018**, *6(3)*, 1200–1206.

Barik, P.; Vardia, H. K.; Gupta, S. B. Bioremediation of ammonia and nitrite in polluted water. *Int. J. Fish. Aquac.*, **2011**, *3(7)*, 136–142.

Bender, J.;, P. Microbial mats for multiple applications in aquaculture and bioremediation. *Bioresour. Technol.*, **2004**, *94(3)*, 229–238.

Bentzon-Tilia, M.; Sonnenschein, E.C.; Gram, L. Monitoring and managing microbes in aquaculture – towards a sustainable industry. *Microb. Biotechnol.*, **2016**, *9(5)*, 576–584.

Boyd, C. E. Chemistry and efficacy of amendments used to treat water and soil quality imbalances in shrimp ponds. In *Swimming Through Troubled Water: Proceedings of the Special Session on Shrimp Farming*; Browdy C. L. & Hopkins, J.S. Eds.), Aquaculture 95, World Aquaculture Society, Baton Rouge, LA, USA, **1995**; pp. 183–199.

Bratwold, D.; Browdy, C.L.; Hopkins J.S. Microbial ecology for shrimp ponds towards zero discharge. Abstract of the 1997 Annual meeting of the World Aquaculture Society, Seattle, WA, USA, **1997**; p. 54.

Brito, L. O.; Junior, L.; de Oliveira C.; Lavander, H. D.; Abreu, J.; Lima de W. S.; Gálvez, A. O. Bioremediation of shrimp biofloc wastewater using clam, seaweed and fish. *Chem. Ecol.*, **2018**, *34(10)*, 901–913.

Cao, L.; Wang, W.; Yang, Y.; Yang, C.; Yuan, Z.; Xiong, S.; Diana, J. Environmental impact of aquaculture and counter measures to aquaculture pollution in China. *Environ. Sci.. Pol. Res.*, **2007**, *14(7)*, 452–462.

Chavez-Crooker, P.; Obreque, J. Bioremediation of aquaculture wastes. *Curr. Opin. Biotechnol.*, **2010**, *21*, 313–317.

Divya, M.; Aanand, S.; Srinivasan, A.; Ahilan, B. Bioremediation—An eco-friendly tool for effluent treatment: A review. *Int. J. Appl. Res.*, **2015**, *1(12),* 530–537.

Ghosh, S.; Rao, M. V. H.; Ranjan, R.; Xavier, B.; Edward, L. L.; Menon, M.; Behera, P. R.; Naik, N. R. Bioremediation, A novel tool for sustainable shrimp culture. In *Aquaculture and Fisheries Environment.* Gupta, S. K., Bharti, P. K., Eds., Discovery Publishing House Pvt. Ltd., New Delhi, India, **2014**; pp. 1–207.

Hanh, D. N.; Rajbhandari, B.K.; Annachhatre, A. P. Bioremediation of sediments from intensive aquaculture shrimp farms by using calcium peroxide as slow oxygen release agent. *Environ. Technol.*, **2005**, *26(5),* 581–590.

He, F.; Wu, Z. B. Application of aquatic plants in sewage treatment and water quality improvement. *Chinese Bull. Bot.*, **2003**, *6(20),* 641–647.

Iwama, G. K. Interactions between aquaculture and the environment. *Crit. Rev. Environ. Con.*, **1991**, *21(2),* 177–216.

Jang. J. D.; Barford, J. P.; Renneberg, R. Application of biochemical oxygen demand (BOD) biosensor for optimization of biological carbon and nitrogen removal from synthetic wastewater in a sequencing batch reactor system. *Biosens. Bioelectron.*, **2004**, *19*,805–812.

Manan, H.; Moh, J. H. Z.; Kasan, N. A.; Suratman, S.; Ikhwanuddin, Mhd. Identification of biofloc microscopic composition as the natural bioremediation in zero water exchange of Pacific white shrimp, *Penaeus vannamei,* culture in closed hatchery system. *Appl. Water Sci.* **2017**, *7*; 2437–2446.

Martinez-Porchas, M.; Luis Rafael Martinez-Cordova, L. R.; Lopez-Elias, J. A.; Porchas-Cornejo, M. A. Bioremediation of aquaculture effluents. In *Microbial Biodegradation and Bioremediation*; Das, S., Ed.; 1st ed., Elsevier Inc., Waltham, MA, USA, **2014**; pp. 539–553.

Musyoka, S. Concept of microbial bioremediation in aquaculture wastes; Review. *Int. J. Adv. Sci. Tech. Res.* **2016**, *6 (5)*,1–10.

Muthukrishnan, S.; Vikineswary, V.; Tan, G. A.; Chong, V. C. Identification of indigenous bacteria isolated from shrimp aquaculture wastewater with bioremediation application, total ammoniacal nitrogen (TAN) and nitrite removal. *Sains Malays.* **2015**, *44(8)*,1103–1110.

Nemutanzhela, M. E.; Roets, Y.; Gardiner, N.; Lalloo, R. The use and benefits of Bacillus based biological agents in aquaculture, 2014. https, //www.intechopen.com/books/sustainable-aqua-culture-techniques/the-use-and-benefits-of-bacillus-based-biological-agents-in-aquaculture.

Nichols, W.J. The U.S. Environmental Protect Agency, National Oil and Hazardous Substances Pollution Contingency Plan, Subpart J Product Schedule (40 CFR 300.900). *Proceedings of 2001 International Oil Spill Conference. American Petroleum Institute*, Washington, DC, USA, 2001; pp. 1479–1483.

Nogueira, S. M. S.; Junior, J. S.; Maia, H. D.; Saboya, J. P. S.; Farias, W. R. L. Use of *Spirulina platensis* in treatment of fish farming wastewater. *Rev. Ciênc. Agron.*, **2017**, *49 (4),* 599–606.

Nora'aini, A.; Wahab Mohammad, A.; Jusoh, A.; Hasan, M.R.; Ghazali, N.; Kamaruzaman, K. Treatment of aquaculture wastewater using ultra-low pressure asymmetric polyethersulfone (PES) membrane. *Desalination*, **2005**, *185(1–3),* 317–326.

Office of Technology Assessment. Bioremediation of marine oil spills: An analysis of oil spill response technologies, OTA-BP-O-70, **1991**, Office of Technology Assessment, Washington, DC, USA.

Office of Technology Assessment. Coping with an oiled sea: An analysis of oil spill response technologies, OTA-BP-O-63, **1990**, Office of Technology Assessment, Washington, DC, USA.

Panigrahi, A.; Azad, I. S. Microbial intervention for better fish health in aquaculture: The Indian scenario. *Fish Physiol. Biochem.*, **2007**, *33,* 429–440.

Parker, R. B. Probiotics. The other half of the antibiotics story. *Anim. Nutr. Health*, **1974**; *29*,4–8.

Pointing, S.B. Feasibility of bioremediation by white-rot fungi. *Appl. Microbiol. Biotechnol.* **2001**, *57*, 20–33.

Queiroz, J.F.; Boyd, C.E. **1988**. Effects of a bacterial inoculum in channel catfish ponds. *J. World Aquac. Soc.*, **1988**, *29 (1)*, 67–73.

Rijn, J. V.; Tal, Y.; Schreier, H. J. Denitrification in recirculating systems: Theory and applications. *Aquacult. Eng.* **2006**, *34*, 364–376.

Robinson, G.; Caldwell, G.; Wade, M. J.; Free, A.; Jones, C. L. W.; Stead, S. M. Profiling bacterial communities associated with sediment-based aquaculture bioremediation systems under contrasting redox regimes. *Sci. Rep.* **2016**, *6*, 38850.

Sharma, S. Bioremediation, features, strategies and applications. *Asian J. Pharm. Life Sci.* **2012**, *2(2)*, 202–213.

Shukla, D. P.; Mishra, A. Y.; Vaghela, K. B.; Jain, N. K. Eco-friendly approach for environment pollution: A review on bioremediation. *Int. J. Curr. Adv. Res.* **2017**, *6 (10)*, 6956–6961.

Sirakov, I. N.; Velichkova, K. N. Bioremediation of wastewater originate from aquaculture and biomass production from microalgae species-*Nannochloropsis oculata* and *Tetraselmis chuii*. *Bulg. J. Agric. Sci.* **2014**, *20 (1)*, 66–72.

Subasinghe, R. P.; Curry, D.; McGladdery, S. E.; Bartley, D. Recent technological innovations in aquaculture. In *Review of the State of World Aquaculture*. FAO Fisheries Circular No. 886, (Revision.2), Rome, Italy, **2003**; pp. 59–74.

Tal, Y.; Watts, J.E.M.; Schreier, H. J. Anaerobic ammonium-oxidizing (annamox) bacteria in associated activity in fixed-film biofilters of a marine recirculating aquaculture system. *Appl. Environ. Microbiol.* **2006**. *72(4)*, 2896–2904.

Md Yusoff, F.; Banerjee, S.; Khatoon, H.; Mohamed, S. Biological approaches in management of nitrogenous compounds in aquaculture systems. *Dyn. Biochem. Pro. Biotechnol. Mol. Biol.* **2011**, *5 (1)*, 21–31.

## CHAPTER 23

Biochar: An Advanced Remedy for Environmental Management and Water Treatment

CHINMAYEE ACHARYA[1,2], ANEEYA K. SAMANTARA[3],
ABHISEK SASMAL[4], CHITTARANJAN PANDA[2*], and
HRUDAYANATH THATOI[1]

[1]*Department of Biotechnology, North Odisha University, Takatpur, Baripada, Odisha 757003, India*

[2]*Environment and Sustainability Department, CSIR-Institute of Minerals and Materials Technology, Bhubaneswar, Odisha 751013, India*

[3]*National Institute of Science Education and Research, Jatni, Khordha 752050, India*

[4]*College of basic Science and Humanities, OUAT, Bhubaneswar-751003*

Corresponding author. E-mail: drpanda-cr@yahoo.com

ABSTRACT

Environmental pollution from hazardous metals can arise from natural as well as anthropogenic sources. Natural sources are seepage from rocks into water, volcanic activity, forest fires, etc. They are mainly introduced into the environment by rapid industrialization, consumerist life style, and anthropogenic sources (discharges from mining, metal plating, battery, and paper industries). Among others, lead, copper, chromium, cadmium, arsenic, and nickel are the most toxic and carcinogenic heavy metals that could cause serious environmental and health problems. In contrast, managing the agriculture waste poses a significant environmental burden leading to pollution of ground and surface waters demanding development of an urgent remedy. Although various water treatment technologies have been developed, the adsorption process is proven to be an efficient one due to its cost effectiveness, ease of operation, and efficiency. For this, many adsorbents are developed, but the biochars act as an efficient adsorbent among others. These biochars are mainly derived from the agricultural residues by following different heating techniques. Furthermore, the efficiency of these biochars for the adsorption has observed to be increased by conjugating with different metal/metal oxide nanoparticles. In this chapter, a broad discussion on biochar synthesis, their characterization, and composite preparation will be aimed to discuss. In addition, a detailed discussion on the adsorption processes and their working principles will be presented.

23.1   INTRODUCTION

Agricultural waste is usually handled as a liability, which is responsible for environmental issues that ultimately cause ground and surface water pollution due to nonavailability of adequate and efficient technologies to convert the agricultural biomasses into an asset. Crop residues such as wheat, millet, sorghum, oilseeds, (castor, mustard), and corn cobs can be the source of reasonable crop management issues as they increase. Additionally, other agro-industrial residues may include groundnut shell, coconut shell and coir, rice husk, tamarind shell, cassava peels, coffee husk, etc. By far and large, some of the most common agricultural by-products available in bulk include tea waste, silk cotton shell, oil palm fiber and shells, cashew nut shells, rice husk, bagasse, coconut shells, coir pith, etc. (Sugumaran and Sheshadri, 2009). Therefore, transformation of organic waste to develop biochar is one of the most sustainable options that can improve natural rates of carbon sequestration in the soil and lowers the farm waste and recover the soil quality (Srinivasarao et al., 2013).

The past few decades have witnessed a surge in terms of magnitude with the sources of waste becoming increasingly diverse. To consume the resources efficiently, it is essential to develop various types of recycling technologies. It is important to discover novel techniques to maintain the conventional technologies. Pyrolysis or carbonizations are the most common methods for energy production from waste plants materials. The off-gases produced throughout the carbonization process in limited- or low-oxygen medium are used for their energy potential. The residue that is derived is generally rich in elemental carbon with a possibility of being used for a variety of applications, viz., for water filtration purposes. Biochar is a stable carbon dominant material, which is achieved by thermal or hydrothermal processes called pyrolysis, that is, when biomass is heated at elevated temperature with limited oxygen environment (Ahmed et al., 2015; Klinar, 2016). Biochar is not a pure carbon, but rather mix of carbon (C), hydrogen (H), oxygen (O), nitrogen (N), sulfur (S), and ash in different proportions (Masek, 2009). The main quality that makes char/biochar as a potential candidate for soil and water amendment purposes is their highly porous structures, which are greatly responsible for their increased water retention abilities and increased surface area. Recently, considerable research efforts have also been conducted on biochar for its potential applications in a different of areas where biochar is being used in environmental management include (1) soil improvement, (2) waste management, (3) climate change mitigation, and (4) energy production (Lehmann and Joseph, 2009). Different varieties of biomass such as agricultural crop residues, forestry residues, wood waste, organic portion of municipal solid waste, and animal manures have been recommended as feedstock for biochar production. The convenience of different biomass as feedstock is dependent on the properties, chemical composition, and environmental, along with economic and logistical, causes (Verheijen et al., 2010). Recently, considerable research attempts have been aimed on biochar-based adsorbents for removal of contaminants, which can be used for beneficial effects for both carbon sequestration and water pollution (Ahmad et al., 2014; Tan et al., 2015). Biochar has a great potential to adsorb contaminants from water due to abundance of feedstock materials, cost benefit and favorable physical/chemical surface properties (Tan et al., 2015). Various factors including the source and type of the biomass as well as the pyrolysis conditions for syntheses do have a strong influence on the quality of the biochar there by also affecting its potential applications (Chen et al., 2011).. Biochar is a carbonaceous (65%–90%) solid by-product of biomass pyrolysis, which contains several pores, oxygen functional groups, and aromatic surfaces. The porous structure of char increases the water-holding capacities and nutrient retention of soil, and also microbial accumulation. Considering the positive aspect, biochar is used as a soil conditioner that is very favorable for agricultural applications. The organic carbon pool of the soil is degrading on a day-by-day basis owing to exploitative agricultural practices. Thus, charging the soil with a resistant biochar fraction might enhance the total carbon pool of the soil, consequently increasing the soil fertility (Qambrani et al., 2017). Biochar is emerging as an economical substitute that has wider environmental applications due to its individual component, for example, maximum adsorption capacity, large specific surface area, microporosity, and ion exchange capacity (Ahmad et al., 2014; Mohanty et al., 2013).

This chapter critically reviews the current updates on several environmental applications of biochar, including a brief discussion of its production and properties and specific mechanisms involved in the removal of certain types of organic and inorganic pollutants from aqueous phases.

23.2 BIOCHAR MANAGEMENT AND RESEARCH

23.2.1 *SYNTHESIS OF BIOCHAR*

Biochar is a carbonaceous solid residue obtained from the thermochemical conversion of carbon-rich biomass under O_2-limited and low temperatures (<700 °C), through the process called low-temperature pyrolysis (Lehmann and Joseph, 2009). The thermochemical technologies for converting biomass into renewable energy products and biochar can be classified into the general categories of pyrolysis, gasification, hydrothermal carbonization, and chemical treatments. It has invoked noteworthy interest as a prospective candidate for improving soil fertility, crop productivity, and nutrient retentive properties as well as for serving to replenish the carbon stock (Chan et al., 2008; Lehmann, 2007; Van Zwieten et al., 2010; Woolf et al., 2010). Biochar is produced specifically for application toward water treatment and soil for environmental management. Energy-efficient, environmentally sustainable, and scalable production of biochars can be effective for heavy metals adsorption in wastewater and contaminated soils.

When low-cost biomass, mostly agricultural by-products, is used for biochar production, the cost of biochar production is mainly related to the machinery and heating, which is only about $4 per gigajoule (Lehmann, 2007). The most common technique to synthesis biochar is pyrolysis. Pyrolysis can be classified into slow pyrolysis and fast pyrolysis depending on the heating rate and residence time. Slow pyrolysis is characteristically a thermal conversion process marked by long residence time and low heating rates that generates roughly equal compositions of solid, liquid, and gaseous products. This method has been used to synthesize charcoal for centuries. Basically, any form of waste biomass can be considered for fast pyrolysis, which is a thermal conversion process that is characterized by short residence times, fast heating rates, and moderate temperatures (500°C–1000 °C), and provides high yields of bio oil from biomass, together with noncondensable gasses and solid biochars (Sanchez et al., 2009). The major difference between the two pyrolysis conditions is that the fast pyrolysis method favors a high yield of bio-oil, whereas slow pyrolysis favors high yield of biochar. Biochar is not a carbon component, but a mixture of carbon (C), hydrogen (H), oxygen (O), nitrogen (N), sulfur (S), and ash in different amounts (Masek, 2009). Biochar can be derived from many different sources comprising wood components, agricultural residues, forest residues, and wastes from food, sugar, or juice processing (McKay, 2002; (Novak, 2009). The choice of feedstock as well as the pyrolysis temperature is the major cause that affects the physical and chemical properties of the resultant biochar (Zhao et al., 2013).

23.3 FEEDSTOCK MATERIALS

23.3.1 *AGRICULTURAL RESIDUES*

The increasing use of agricultural and forest residues has been proposed as feedstock for biochar production because of their availability and economically cost effective. Furthermore, conversion of waste biomass to value-added biochar compound can decrease processing cost related to discarding the abundant agricultural waste and forest residues.

However, availability of large-scale agricultural by-products, such as cereal straw, rice husk, groundnut shells, coir pith, and wheat straw, may be suitable for high biochar yield due to the maximum levels of K and Zn (Raveendran et al., 1995). The hemicellulose, cellulose, and lignin content can impact the ratio of volatile carbon in oil and gas and the amount of carbon fixed in biochar. The most suitable biomass with high lignin concentration produces the highest biochar when pyrolyzed (Demirbas, 2006; Fushimi et al., 2003).

Pyrolysis temperature plays a significant role in varying biochar characteristics (Uchimiya et al., 2011a). The carbon content of biochars depends on the raw material and production conditions, that is, at high pyrolysis temperature; woody and herbaceous biomass normally gives more carbon-rich biochar in comparison to other biomass such as sewage sludge and animal manures. Generally, there exists organic phase of biochars of carbonized organic matter or more aromatic groups and more stable and noncarbonized organic matter that is relatively more aliphatic and less stable (Novak, 2009; Bruun et al., 2011).

23.3.2 *INDUSTRIAL BY-PRODUCTS*

Biochar production from industrial wastes and by-products is receiving growing attention recently. The feasibility of employing waste materials collected from digested residues and sewage sludge for biochar production have been reported (Phuengprasop et al., 2011; Yao et al., 2011a; 2011b). Anaerobic digestion comprises biodegradation of waste residues by different types of microorganisms. The carbonaceous substance degrades and results in concentrating cationic or metallic elements in the residues, which can be converted into biochars, may have high ion exchange capacity for the adsorption of heavy metals from aqueous solutions (Wong et al., 2004; Hanay et al., 2008).

23.3.3   *NONCONVENTIONAL MATERIALS*

The nonconventional materials such as waste tires (Karakoyun et al., 2011), invasive plants (Liao et al., 2013), municipal solid waste (Hwang et al., 2008), bioenergy residues (Yao et al., 2015), and food wastes (Ahmed and Gupta, 2010; Rhee and Park, 2010) have also been reported for the production of activated carbon for sorption. These wastes can be used as feedstock for biochar production, and pyrolysis of wastes reduces the operation cost and waste disposal.

23.4   PHYSICOCHEMICAL PROPERTIES OF BIOCHARS

23.4.1   *PHYSICAL PROPERTIES*

The pore volume and size, specific surface area, and particle size of biochar are the major parameters used in defining the physical properties of biochar.

23.4.1.1   *SURFACE AREA, PORE VOLUME, YIELD, AND ASH CONTENT OF BCS*

The surface area, pore volume, and yield of biochar increased with increasing pyrolysis temperatures, due to the emission of volatile matters including cellulose, hemicelluloses, and lignin from biomass, which forms structures during the pyrolysis (Ahmad et al., 2012; Chan et al., 2012; Kim et al., 2013). These structures facilitate to improve the specific surface area and pore structure of biochars (Lin et al., 2018). The decreased pore size forms the internal pore structure and increase porosity as the discharge of volatile during carbonization takes place (Ahmad et al. 2012). The biochar structure can be affected by numerous factors; commonly, biochar has abundant surface functional groups (hydroxyl, carboxyl, carbonyl, and methyl), the increased pore structure, the high surface area, and the stable molecular structure (Beesley et al., 2011). Particularly, for organic molecules, surface area may have some influence on the adsorption of heavy metals. The biochars have highly porous structures, especially microporous structures, with a high a surface area. Generally, the structures of biochars are not homogeneous, and irregular pores with different shapes and sizes were observed (Shinogi and Kanri, 2003). Surface area plays the major role in absorption of organic molecules. The carbon residues are highly porous and have large porosity. Pyrolysis temperature has an important role in altering biochar properties; therefore, pyrolysis temperature increases the surface area, and the degree of porosity gradually increases, while the biochar (BC) yield decreases with an increase in the pyrolysis temperature (Yoder et al., 2011). Biochar porosity, which determines the surface area, shows pore size distribution that is highly variable and encompasses nano- (<0.9 nm), micro- (<2 nm), to macropores (>50 nm) (Downie et al., 2009). The surface area and porosity of biochar may have profound effects on its nutrient retentive properties (Chan and Xu, 2012). The porosity volume, size of the pores, specific surface area, and the particle size of the biochar are by far the most important parameters that may define the physical properties of biochar. High volatile matter contents in the feedstock could motivate the enlargement of porous structures and the reactivity of biochar (Pacioni et al., 2016). The specific surface area of biochar is defined as the ratio of the total pore surface area to the total biochar particle mass, and it is well correlated with its porosity. The physical properties may affect the chemical properties of gasification biochar. For example, larger surface area and micropore volume found to be correlated with maximum total polycyclic aromatic hydrocarbons on biochar surface (Rollinson, 2016). Alkali and alkaline earth metallic species such as K, Na, Ca, Fe, and Mg are commonly found in biomass. These metallic species have essential role in the gasification process for determining the difference of gasification products and effectiveness of gasification, which is likely due to the potential catalytic effects associated by the common alkaline earth metallic species (Yip et al., 2009).

23.4.2 CHEMICAL PROPERTIES

The chemical properties of biochar, which are likely related to applications, include carbon and ash contents, functional groups, aromaticity, and pH. The thermochemical production conditions are related to composition and reactivity of biochar (e.g., temperature, gasifying agent, and equivalence ratio) and the types of biomass (Naisse et al., 2013). The pH values of gasification biochar generally belong to the alkaline range ($7 < $ pH $ < 12$) (Hansen et al., 2016; Shackley et al., 2012; Wiedner et al., 2013) and directly related to their metal salt and ash content and high degree of carbonization (Griffith et al., 2013; Shen et al., 2016).

23.4.2.1   ELEMENTAL COMPOSITION

The elemental compositions of BCs are carbon, oxygen, hydrogen, and H/C and O/C molar ratios for different pyrolysis temperatures. The carbon content had the tendency to be enhanced by increasing the pyrolysis temperature from 100 to 950 °C. Likewise, the oxygen and hydrogen contents exhibited a decreasing trend with an increase in the pyrolysis temperature, which could have been due to differences in the source of the agricultural biomasses as well as pyrolysis conditions drawing strength from the observations of Fuertes et al. (2010) who found higher C content in hardwood as compared to corn stover; in separate studies, Chun et al. (2004) and Chen et al. (2008) indicated that the C content increased with increasing pyrolysis temperature. The degree of carbonization can be defined by the molar H/C ratio, because H is primarily correlated with plant organic matter (Kuhlbusch, 1995). The higher H/C ratio value describes that biochars may hold a definite amount of original plant organic residues like cellulose. The molar-oxygen-to-carbon (O/C) ratio of biochar may be used as a substitute for the surface hydrophilicity as it contains indicative of polar group, basically derived from carbohydrates (Chun et al., 2004).

23.5   BIOCHAR PRODUCTION

There are many types of feedstock resources such as agricultural crop residues, agricultural by-products, forest remains, wood waste, organic part of municipal solid waste, and manures. Biomass obtained from agricultural residues categorized into two groups: (i) primarily formed biomass as a resource of bioenergy and biochar and (ii) by products as waste biomass. Waste biomass has been used widely for biochar synthesis because of the low cost types of biomass compared to other (Brick, 2010).

The significant abundance of lignocellulosic biomasses generally makes it a potential feedstock for biochar production as they are mainly composed of cellulose, hemicellulose, lignin, and trace amounts of extractives, with an average elemental composition of $CH_{1.4}O_{0.6}$ (Agblevor, 2007). Cellulose and hemicellulose are carbohydrate polymers and collectively known as holocellulose, while lignin fractions are nonsugar-type macromolecules. Cellulose is the largest fraction and constitutes about 38–50% by weight of lignocellulosic biomass and generally exists as microfibrils, that is, thread-like fiber structure embedded on a matrix, which is composed of hemicelluloses and lignin, a linear polysaccharide made of anhydro-D-glucose monomeric units with a β (1 → 4) glycosidic linkage. Hemicellulose is a branched polysaccharide consisting of different sugar monomers such as glucose, galactose, xylose, mannose, arabinose, and uronic acids. Hemicellulose is a form of cross-linking glucans, as it forms linkage between cellulose and lignin through hydrogen bonds. Lignin is considered as a group of amorphous, high-molecular-weight, chemically related compounds, which is an aromatic polymer composed of phenolic macromolecule having high cross-linking degree between the phenyl propane units. The high cross-linking degree in lignin is thermally more stable in comparison to hemicellulose. Lignin acts as the cementing material, which provides structural and mechanical strength as well as elasticity to the lignocellulosic biomass (Kataki et al., 2015). The composition and amount of these elements vary with the type of biomass. The particular biomass that is convenient as a potential feedstock for biochar production depends on different characteristics such as moisture content, calorific value, fixed carbon, oxygen, hydrogen, nitrogen, volatiles, ash content, and cellulose/lignin ratio (McHenry, 2009; Sanchez, 2009; Mohan et al., 2014). The physicochemical properties of biochar vary with the types of biomass, the particle size of the feedstock, the temperature (increase in rate of temperature), the residence time for pyrolysis, and the modification conditions (Tan et al., 2015, Sun et al., 2012). Biochar is a solid product can be synthesized by thermochemical decomposition of biomass at temperatures above 300 °C in the presence of limited or no oxygen, which is commonly known as pyrolysis (Demirbas and Arin, 2002). Biochar is not a pure carbon, but it consists of carbon (C), hydrogen (H), oxygen (O), nitrogen (N), sulfur (S), and ash. Pyrolysis is normally divided into fast, intermediate, and slow depending on the residence time and temperature.

Fast pyrolysis, at a short residence time (<2 s), is often used to produce bio-oil from biomass yielding about 75% bio-oil (Mohan et al., 2006).

Slow and intermediate pyrolysis of biomass with a residence time ranging from a few minutes to several days normally favors biochar production (25%–35%) (Laird et al., 2009). However, gasification is different from the general pyrolysis process. In this process, the biomass is converted into gases having high concentration of carbon monoxide and hydrogen by pyrolysis of biomass at high temperature (>700 °C) in a controlled oxygen environment. The resulting gas mixture is known as synthetic gas or syngas (Mohan et al., 2006).

23.5.1 MODIFICATION AND FICTIONALIZATION OF BIOCHAR

During the past years, so many attempts have been made to construct high-efficiency, environment friendly, and cost-effective sorbents and black carbon to remove inorganic pollutants from wastewater, but the efficiency of sorption can be enhanced by activation or modification. This chapter evaluates different techniques to increase the sorption efficiency of biochar. It is important to develop a novel adsorbent with cost-effective and maximum removal performance for pollutants. The adsorption capacity of biochar is generally determined by the physicochemical properties such as special functional groups, specific surface area, pore properties, and surface charges (Mohan et al., 2014). The properties of biochar depend on the type of feedstock and the pyrolysis conditions such as residence time, temperature, heating rate, and type of pyrolysis furnace.

The development of methods for the modification of biochars is required to enhance the sorption capacity of inorganic pollutants that can be removed from aqueous solution. Physical and chemical activation of biochar can increase the surface area and the oxygenated functional groups to increase cation adsorption. The surface functionalization of biochar can alter the properties of the biochar by forming biochar-based composites, which enables the adsorption of different pollutants. There are various modification methods for biochar functionalization through which biochar can be chemically, physically, or biologically modified by treating the feedstock (Rajapaksha et al., 2016). Biochar functionalization generally leads to the variations of biochar surface properties, including specific surface area, surface charge, functional groups, pore volume, and pore size distribution. These methodologies include the chemical modification, such as acid modification (hydrochloric acid, nitric acid, etc.), chemical oxidation (hydrogen peroxide, potassium permanganate, etc.) and amination (ammonia, NH_4OH, $(NH_4)_2S_2O_8$), which have shown that a biochar increases the adsorption capacity of organic and inorganic contaminants (Rajapaksha et al., 2016).

Advanced methods have been developed for the modification of biochar to enhance the adsorption of inorganic pollutants from water. Modifications can be done either pre- or postpyrolysis of the biochar. The prepyrolysis biochar modifications comprise the treatment of the feedstock, and a postpyrolysis modification involves the treatment of the biochar after production through pyrolysis. Figure 23.1 provides a classification system of the most accepted modification techniques reported in the literature.

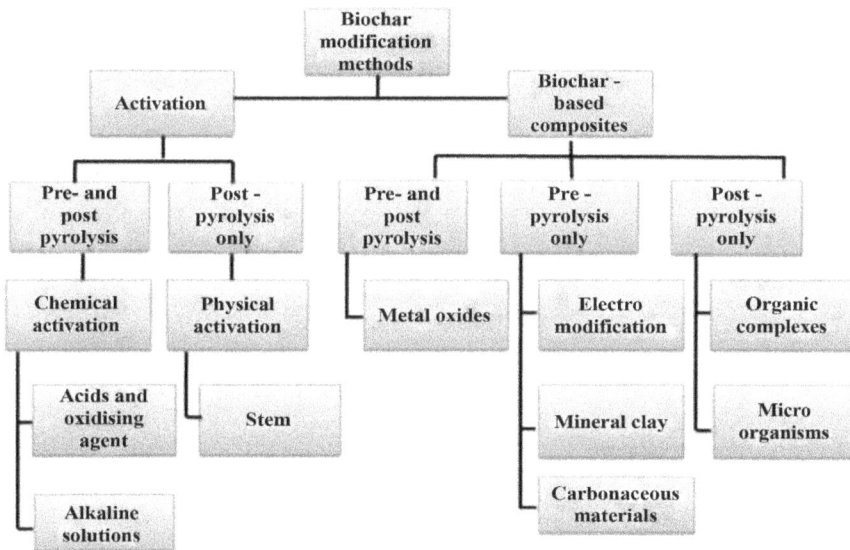

FIGURE 23.1 Classification systems of biochar modification methods to physically and chemically activate biochars and produce biochar-based composites (adapted from Sizmur et al., 2017).

Further improvements of technology for all modification methods is to enhance the sorption efficacy by which the biochar adsorbs pollutants from wastewater, normally by changing its physical or chemical properties, such as surface area and surface functionality because most of the biochars are produced directly from the precursors without any modifications. The biochar properties can be improved by physical and chemical modification. However, for chemical activation, acid and alkali treatments are the most common methods and physical activation, and steam is the primary treatment method.

Biochar-based composites are synthesized by impregnation or coating the surface of the biochar with metal oxides (Michalekova-Richveisova et al., 2017); carbon-based nanocomposites, such as graphene (Shang et al., 2016) and carbon nanotubes (Inyang et al., 2015), can be found in functionalized or nonfunctionalized forms. For example, graphene and carbon nanotubes can be functionalized with pendant groups such as $-NH_2$, $-OH$, and $-COOH$ groups, etc., through chemical treatments producing functionalized materials, which are highly dispersible compared to the pristine (nonfunctionalized) materials. Water treatment applications include Fe/Fe_3O_4, Al/Al_2O_3, TiO_2, and Ag (Guo et al., 2012; Shaari et al., 2012; Pyrzynska et al., 2010; Kim et al., 2013; Gupta et al., 2011; He et al., 2010; Lee et al., 2012), complex organic compounds, such as chitosan (Zhou et al., 2013) or amino acids (Yang and Jiang, 2014), or inoculation with microorganisms (Frankel et al., 2016) (see Figure 23.2).

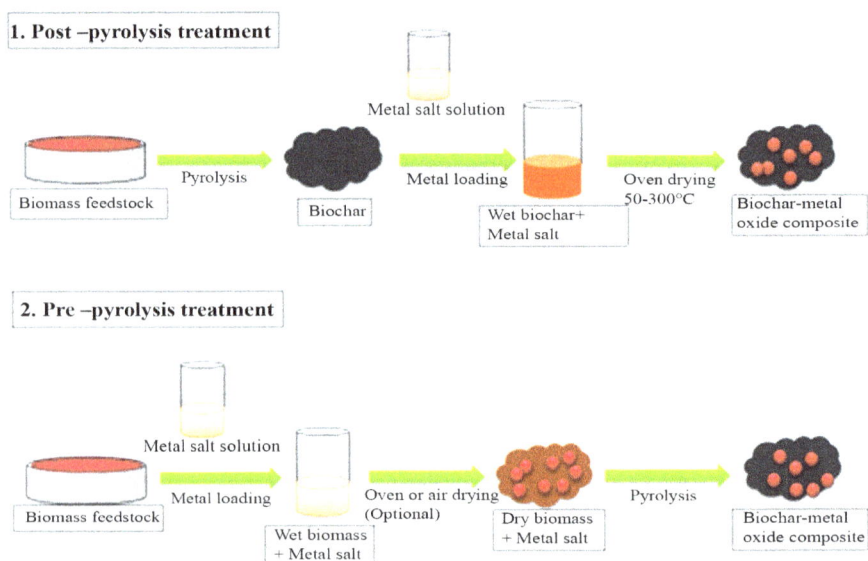

FIGURE 23.2   Diagram outlining the pre-pyrolysis and post-pyrolysis techniques used to modify biochars with metal salts to produce metal oxide biochar-based composites (adapted from Sizmur et al., 2017).

23.5.2   *PHYSICAL AND CHEMICAL ACTIVATION*

In the physical activation system, biomass is exposed to high temperature, which results in the increase in the structural porosity of the biochar and surface area. In chemical activation methods, the biochars are treated with acidic or alkaline solutions, which form more oxygenated functional groups to biochar surfaces and increase the surface properties of biochar to bind positively charged pollutants through adsorption (Hadjittofi et al., 2014).

23.5.3 DOPING AND MODIFICATION

In surface oxidation, oxygenated functional groups such as C=O, OH, and COOH are vital for enhancing biochar properties for different applications. For heavy metal removal biochar surface, OH and COOH groups can significantly increase the adsorption capacity. These groups interact with the metals via hydrogen bonding and complexation (Zhou et al., 2013). In addition, the introduction of oxygenated functional groups on the surface can greatly influence the sorption of heavy metals.

23.5.4 IMPREGNATION METHODS

Impregnation can be done by collaborating metal salts and metal oxides with biochar to ease the physical or chemical bond of metal ions on the biochar surface. The impregnation can be carried out by suspending the biomass in different concentration of salts, followed by pyrolysis, and another method is pyrolysis of biochars followed by the impregnation of metal salts or oxides. BC modifications have already been done by applying ammonium chloride (Shen et al., 2015b), nitrates (Shen et al., 2015a), and carbonates (Agrafioti et al., 2014). The impregnation process of biochar increases the physical and chemical properties by developing the structure of composites, which increases the yield and sorption points of the biochar composite.

23.5.5 ACID MODIFICATION

Acidic modification can improve the physicochemical properties of biochars; this can be activated by various oxidants, which may increase the acidic property of sorbents by eliminating minerals and increasing the hydrophilic nature of biochar (Shen et al., 2008). Acidic modification is generally carried out by suspending biochar in acid solutions in a ratio of 1:10 (biochar and acid) at room temperature to 120 °C (Zhou et al., 2013; Jin et al., 2014; Zhang et al., 2015). BC oxidation forms a number of acidic functional groups on its surface; for example, increasing amount of oxygen leads to increase the O/C and H/C molar ratios with sulfuric acid treatment. The increased O/C and H/C ratios show reduced hydrophobicity (Vithanage et al., 2015). Nitric acid treatment degrades the pore wall and converts the micropores into meso- or macropores; therefore, the resulting acid-modified biochar tends to occupy more acidic functional groups such as –OH, –COOH, C=O, and other oxygen containing groups (Xue et al., 2012; Hadjittofi et al., 2014; Li et al., 2014). The acid-modified biochars may have lower surface area than other type of modifications, which could be possibly due to the breakage of pore structures.

23.5.6   ALKALI MODIFICATION

Alkaline modification improves the physical and chemical properties of BCs. Group I metal hydroxides (KOH or NaOH) are commonly used for the alkali treatment of BCs. Activation of biochars through metal hydroxides increases adsorption capacity by increasing the surface area, porosity, and by creating more number of oxygenated functional groups on the surface of the biochar. These oxygenated functional groups have proton donating exchange sites, through which metal cations such as Pb^{2+} can be chemically adsorbed (Petrovic et al., 2016 and Goswami et al. 2016).

23.6   APPLICATIONS OF BIOCHAR

Environmental (soil and water) amendment is one of the priority applications of biochar. Biochar quality plays a vital role in removal of contaminants, which is often commanded by pyrolysis temperature and the source of feedstock. For example, biochars produced at higher pyrolysis temperatures (>500 °C) have an increased penchant for organic pollutants owing to larger surface area, high degree of porosity, organophilicity (Ahmad et al., 2014; Keiluweit et al., 2010), high carbon-to-nitrogen (C/N) ratio, high pH (Keiluweit et al., 2010; Ronsse et al., 2013), and low dissolved organic carbon (Uchimiya et al., 2013), whereas biochar partially carbonized at lower pyrolysis temperatures (<500 °C) contains a high content of dissolved organic carbon and O-bearing pendant groups, comparatively low porosity and C/N ratio, and hence more competent for removal of inorganic contaminants (Ahmad et al., 2014; Keiluweit et al., 2010; Kong et al., 2011). With respect to source of feedstock, biochar derived from crop residues and woody biomass exhibited a higher surface area as compared to biochar derived from solid wastes and animal manure, all of which were processed at a high pyrolysis temperatures. Other factors such as pH, residence time, application rate, and contaminant type also affects the contaminant removal efficiency of biochar (Ahmad et al., 2014).

23.6.1 REMOVAL OF ORGANIC POLLUTANTS

Over the decade, application of biochar for elimination of various organic contaminants from water and soil has received a significant impetus (Zhang et al., 2013). Some of the targeted organic pollutant species include agrochemicals (Mandal et al., 2017) antibiotics/drugs (Ahmed et al., 2015), and industrial chemicals, including aromatic hydrocarbons (PAHs) (Li et al., 2014), volatile organic compounds (Inyang and Dickenson, 2015), and cationic aromatic dyes (Lin it al., 2018).

The mechanisms for removal of pollutants are mostly controlled by the interactions of pollutants with various characteristics of biochar. In case of organic pollutants, the removal is primarily achieved via chemisorption and physisorption (viz., pore diffusion, hydrophobic interactions, electrostatic interactions via π–π electron donor–acceptor, and H-bonding) through– COOH, –OH, and R–OH functional groups (Ahmad et al., 2014; Teixido et al., 2011; Xie et al., 2015). Other mechanisms including partitioning in noncarbonized phase (due to reduction of substrate polarity), chemical transformations (via reductive reactions or electrical conductivity), and most of the bound contaminants are ultimately mineralized (by microbial action on the surface and within the matrix of the biochar) (Beesley et al., 2010; Kong et al., 2011; Xu et al., 2013a,b; Ahmad et al., 2014). A schematic of possible chemical interactions involved in the biochar-mediated removal of organic pollutants has illustrated in Figure 23.3.

Biochar interactions with organic pollutants are affected by various factors, viz., pH, pyrolysis temperature, source of feedstock, and proportion of pollutant-to-biochar. As already discussed, higher pyrolysis temperature leads to higher surface area and porosity in biochar. While these features are desirable for the removal of nonpolar organic pollutants, biochar produced at lower temperature is deficit of such properties (Kong et al., 2011; Ahmad et al., 2014). High pyrolysis temperature (> 500 °C) results in higher aromaticity, low polarity, and low acidity on biochar due to the loss of –O- and –H-based pendant groups (Keiluweit et al., 2010). Decreased –O-bearing functional groups normally facilitate hydrophobic interactions, whereas biochar produced at temperatures <500 °C predominantly contains –O- and –H-based functional groups with high organophilicity toward polar compounds (Ahmad et al., 2012).

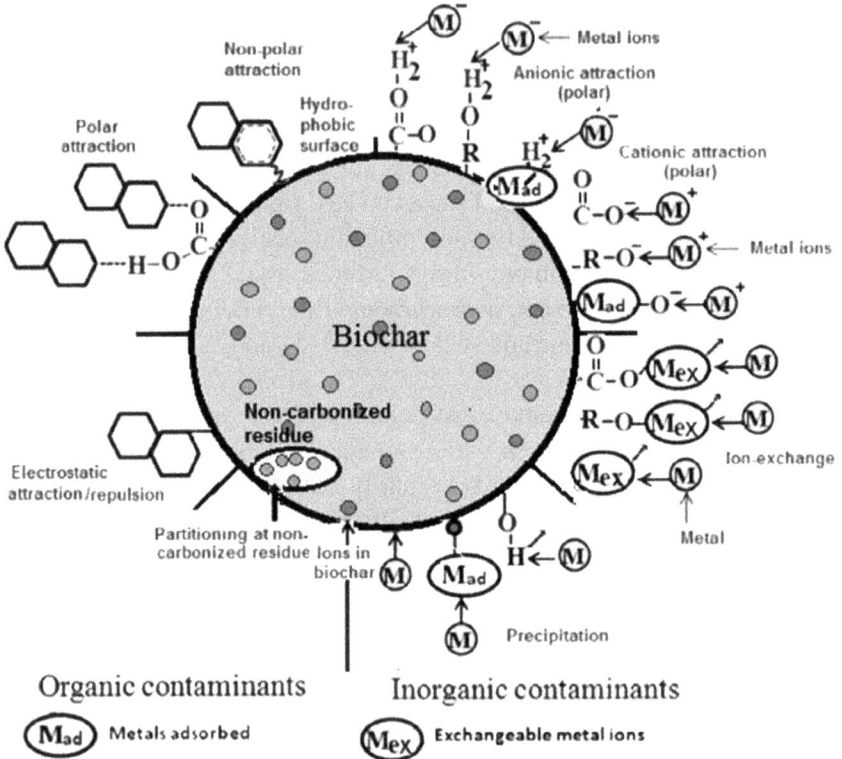

FIGURE 23.3 Different mechanisms of biochar interactions with organic and inorganic pollutants (adapted from Oliveira et al., 2017).

23.6.2 *BIOCHAR FOR REMOVAL OF HEAVY METAL SPECIES*

The effect of feedstock source for removal of inorganic pollutants has also been an active area of focus. Heavy metals, viz., Pb^{2+}, Cu^{2+}, As^{3+}, Zn^{2+}, Ni^{2+}, Cr^{2+}, Hg^{2+}, and Cd^{2+}, have been evaluated for their effective removal from soil through application of biochar derived from various agricultural biomasses (Hartley et al., 2009; Beesley et al., 2010; Lu et al., 2012). The biochars were chiefly derived from green waste, sewage sludge, hardwood, wood chips, coconut shell, and cottonseed hull at pyrolysis temperatures of ≤550 °C (Beesley et al., 2010; Uchimiya et al., 2011a,b; Zhang et al., 2013; Nelissen et al., 2014). Hardwood-derived biochar was identified as the most efficient for removal of Cu^{2+}, Cd^{2+}, and As^{3+} via metal complexation. Removal of Cu^{2+}, Cd^{2+}, and As^{3+} in soil was attributed to high content of dissolved organic carbon of biochar and a high soil pH > 7.0 corresponding to high biochar dosage (Beesley et al., 2010). Biochar also facilitated the reduction of As^{5+} to As^{3+}, thereby reducing the solubility and enhancing As^{3+} immobilization in soil (Zhang et al., 2013). The removal of Cu^{2+}, Zn^{2+}, Na^+, Ca^{2+}, Mg^{2+}, K^+, Sr^{2+}, and B^+ from soil was high and low NO_3 phytoavailability with amendment of willow biochar, whereas Fe^{2+} and Ti^{3+} removal and NO_3 phytoavailability were considerably higher with pine-derived biochar application (Nelissen et al., 2014).

Lead produced from automobiles and other petro-based exhausts is a gravely poisonous heavy metal pollutant frequently registered in urban soil profiles. Alkaline biochar has been very promisingly used for removing lead from contaminated sites. The mechanisms of Pb^{2+} removal from contaminated soils are chiefly attributed to surface co-precipitation and inner-complexation reactions, which occur on the structures of organic and mineral oxides of sewage sludge- and farm-manure-derived biochars (Cao et al., 2009; Ho et al., 2017; Wang et al., 2017). Several groups have reported the removal efficiency of Pb^{2+} around 64–92% in soil by using wood chip and coconut shell derived biochars (Reddy et al., 2014; Paranavithana et al., 2016). High removal efficiency of biochar was due to a rise in soil pH and its precipitation in various forms, viz., oxides, hydroxides, carbonates, and phosphates. Other removal mechanisms included formation of –COOH– and –OH-based complexes (Park et al., 2016), surface complexation via K^+/Na^+ electrostatic exchange, and precipitation as lead-phosphate silicates (Lu et al., 2012).

Cadmium (Cd) in the recent past has been identified as another pollutant species owing to its toxicity at remarkably low concentrations, persistence and widespread occurrence. Beyond permissible limits, it causes anaemia and renal damage and is dubious for its inter-relation with the itai–itai disease (Bodek et al., 1998; Kalderis et al., 2008). Goswami et al. (2016) demonstrated use of a low-cost Ipomoea derived biochar for removal of Cd from aqueous solution with low contact time period and adsorbent dosage. The alkaline biochars that produced enhanced porous structures resulting in increased adsorption sites available to Cd were found to be highly efficient in removal of Cd from water samples. The mechanism for high adsorption efficiency is attributed to surface complexation of Cd^{2+} ions with biochar particles.

23.6.3 *BIOCHAR-MEDIATED SOIL CARBON SEQUESTRATION AND REMEDIATION*

The basic principle behind using biochar for carbon sequestration is associated with the role of soils in the carbon cycle. As per the NASA report, 2008, the worldwide flux of CO_2 from soils to the atmosphere is in the range of 60 Gt of carbon per year. The CO_2 evolved is mainly due to the result of microbial respiration within the soil system by microbial decomposition of soil organic matter (SOM). Biochar components are believed to be considerably persistent than SOM and as such are decomposed very slowly over a time frame, which can be assessed over centuries or millennia. This means that charging of soils with biochar greatly enhances the carbon input compared to the carbon output through soil microbial respiration, and it precisely is the basis behind biochar's possible carbon negativity and hence a potential climate change mitigation option. In this regard, scientific groups have focused their research worldwide on biochar-mediated remediation strategies including nutrient retention, water retention, liming effects, and yield improvement.

The phytoavailability of soil nutrients is a key factor regulating crop productivity. Biochar can effectively condition soils, consequently enhancing plant growth by not only supplying but at the same time retaining nutrients and by providing other services such as improving physical and biological characteristics of the soil (Glaser et al. 2002; Lehmann et al. 2003; Lehmann and Rondon 2005). While there is minimal nutrient value in biochar for example, many biochars contain little or no N, yet biochar can still improve N levels in soil (Jeffery et al., 2015). Biochar has a porous structure that expands by several thousand-fold during pyrolysis (Ippolito et al., 2015). To the extent that N occurs in a soil (or is added), the porous structure of biochar can retain N and prevent the N-leaching that is typical in many soils. This N-retention not only improves N phytoavailability, but also augments the benefits of future N applications (Jeffery et al., 2015). Biochar's porosity also allows for a variety of nutrients to be stored.

Cation exchange capacity (CEC) is biochar's ability to electrostatically attract cations and is developed when the feedstock is exposed to water and oxygen, resulting in an oxygenated surface (Ippolito et al., 2015). Biochar tends to increase soil CEC, which allows a soil to hold more phytonutrients, especially calcium, magnesium, and potassium (Cornell Cooperative Extension, 2007). This increased CEC also reduces the concentration of iron and aluminum, which makes phosphorous more available to plants (Jeffery et al., 2015).

Another interesting property of biochar important for crop production is retaining soil moisture for plant uptake, thus promoting crop growth or reducing the need for irrigation. Biochar water adsorption is the adhesion of a thin layer of water molecules to the surface of biochar. Along with nutrient retention, the porous structure of biochar also promotes water adsorption (Kizito et al., 2015). An indicator of plant water stress is high leaf proline content, which has been shown to be reduced in plants grown in biochar-amended soils (Kammann and Graber, 2015). Some studies have demonstrated that biochar application mitigates water stress for multiple growing seasons after a single application, with increased effects for subsequent applications. However, in other cases, biochar application rates were not correlated with water-stress amelioration effects (Baronti et al., 2014; de Melo Carvalho et al., 2014).

In areas with acidic soil, such as Massachusetts, lime is often used to raise soil pH to a level suitable for crops. Biochar can serve as a liming agent or as a substitute for incorporating limestone in soils (Berek, et al., 2011). For the purpose of this economic analysis, raising soil pH also has a clear market value replacing the cost of applied limestone.

23.6.4 BIOCHAR APPLICATION FOR REMOVAL OF MISCELLANEOUS POLLUTANTS

Biochar has also been considered ideal for removal of toxic species from industrial wastewaters, for instance, from wastewater containing furans (Li et al., 2014) and phenolics (Brebu and Vasile, 2010), generated during thermochemical pretreatment of lignocellulosic biomass, wastewaters from pharmaceutical, agrochemical industries, oil and mines, electroplating, metal processing, and electronic and landfill leachate, among others. Comparative assessments using hardwood- and softwood-derived biochars for removal of toxic species from wastewater from oil refineries and mining belts, and drugs and chemical industries have been very recently conducted (Gunatilake, 2015; Frankel et al., 2016). Biochar derived from bamboo, cottonseed hull, ground-flax, cotton, etc., at temperature of 200–800 °C has been used for the removal of furfural and 5-hydoxy methyl furfural (HMF). However, activation of biochar with steam, $KMnO_4$, HNO_3, and NaOH resulted in the removal efficiency of over 99% for furfural and hydroxyl methyl furfural (Klasson et al., 2011; Li et al., 2014). Biochar has also been useful for ameliorating the stability and fulfillment of biological processes by removing the potential toxic compounds from the aqueous phases. Application of 1%–5% activated commercial biochar (Darco G60) and rice-husk-derived biochar (derived from anaerobic digestion of lignocellulosic biomass) greatly improved the microbial growth (of mainly cellulose degrading bacteria), further fostering the CH_4 and H_2 yield by 17–35%, volatile solids degradation by 30%–34%, and chemical oxygen demand removal by 55%–69% (Inthapanya et al., 2012; Kumar et al., 1987).

Petroleum hydrocarbon (PHC) degradation is a serious concern for effective remediation of the soil contaminated due to petroleum spillage. This project is even more challenging in cold condition. PHC degradation slows down during the winter, which considerably increases the soil remediation time in Arctic land farms. Meat-and-bone-meal-derived biochar (3% w/w) amendment found to increase aromatic PHC degradation in petroleum-contaminated soil at severe cold condition. The positive effect of biochar attributed to increase in the liquid water content in frozen soil and proliferation of PHC degrading microbes (Karppinen et al., 2017).

23.6.5   *BIOCHAR AND SOIL BIOTA*

The effects of biochar addition to soils on soil biota have been extensively studied by Lehmann et al. (2006), who concluded that the knowledge on the shifts in microbial consortia is limited and that their knowledge of biochar effects on soil biota is limited too. This is even more so when confining the discussion to N. Since the review by Lehmann et al. (2009), the study of Jones et al. (2011) has measured higher growth rates of bacteria and fungi after incorporating biochar, but this effect has not been observed after storage of soil in the laboratory, leading the authors to speculate that the effect was the result of an indirect rhizosphere effect. Dempster et al. (2012) found that the addition of a eucalyptus-based biochar at 25 t/ha altered the ammonia oxidizer community structure when it was present with inorganic-N, with lower nitrification rates ensuing. The latter was assumed due to a negative priming effect on the SOM resulting in lower NH_4^+ concentrations, since the potential for NO_3^- adsorption to remove NO_3^- was at minimum values when biochar was added to the soil. Anderson et al. (2011) examined biochar-induced soil microbial community changes in pasture soils where biochar had been incorporated and found that the bacterial families, viz., *Brady rhizobiaceae* and *Hypho microbiaceae,* were abundant in the soil biota as compared to the control soils where no biochar had been added. During anaerobic phases, members of these families can utilize NO_3^-, N_2, and NH_3, and they are capable of N_2 fixation and denitrification. This result may explain the enhanced N_2 fixation previously observed in bean crops with biochar present. Regardless of a meta-analysis showing root nodulation increases with biochar addition, which was believed to be due to soil pH and phosphate availability and became more suitable for efficient N fixation, inadequate information still prevails on both the long- and short-term effects of biochar on N2 fixation. Anderson et al. (2011) concluded that adding biochar to the soil potentially increased microbial N cycling, especially the abundance of those organisms that may decrease N_2O fluxes and NH_4^+ concentrations. Contrary reports by Yoo and Kang (2012) suggested that the higher N_2O fluxes observed in the presence of swine manure-derived biochar in paddy soils were partially a consequence of higher denitrifier abundance. Noguera et al. (2012) hypothesized earthworms and biochar to have a synergistic effect on nutrient availability and plant growth. However, while differences in mineral N were observed to be dependent on soil types and but no interactions could be established between earthworms and biochar with respect to mineral N content owing to the short-term nature of the mesocosm study.. Augustenborg et al. (2012) reported biochar-reduced earthworm-induced N_2O fluxes, although the possible mechanisms for the same have not been cleared. The present scenario warrants a systematic and rigorous experimentation to evaluate biochar-induced effects on soil biota with regard to soil N cycling. Furthermore, biochar N studies should also aim to elucidate the effects and potential risks, if any, that biochar may have in the future by investigating long-term analogs such as charcoal-rich soils or aged versus fresh biochars.

23.7 CONCLUDING REMARKS

Biochar is a novel renewable resource, which has a potential to address several environmental issues that we have been encountering in recent years, including remediation of pollutants in soil, water, and gaseous media. This could synergistically improve soil, water, and air quality, carbon sequestration, etc. Since the biochar quality and performance varies significantly depending on feedstock types and pyrolysis conditions, future progress in biochar development is expected to center around "tuning" the properties for tailored applications. Biochar activation is another important area to tailor the application of biochar for removal of specific contaminants. For example, tannery waste activation of pine-wood-derived biochar attributed to enhance adsorption of ammonium-N and other organic and inorganic pollutants from wastewater; chemical activation of biochar showed increased adsorption of pollutants with low desorption. Further research is needed to identify various activation methods and adsorption and desorption mechanisms of diverse contaminants. Microbial communities and their distribution in biochar-amended soil have not been well examined, especially with respect to biochar properties (e.g., pH, particle size, microporosity, nutrient content, and ion exchange capacity). With growing concern of soil pollution and infertility, biochar application could open up new avenues not only for their remediation, but also as a source of macro- and micronutrients (captured from waste material) in nutrient-deficient soils and slow release fertilizers. Further study is needed to critically examine the role of microbes in soil remediation and mineralization processes. Another important research area could be toward agronomic welfare, biochar surface chemistry, and surface interaction with various soil constituents especially "micronutrients" binding on biochar and their exchange mechanisms. There is also a growing interest in applications of biochar in wastewater treatment, especially removal of toxic compounds from various industries. Thus, biochar treatment could serve as a pretreatment for the removal of toxic compounds for subsequent biological treatment. However, even these areas require further research since the use of biochar as a mitigation tool demands a deeper mechanistic understanding and at the same time an increase in our ability to predict net effects over time. As a new exciting field of study, both research potential and uncertainties still exist. Future directions demand relevant and rigorous investigations in order to explore the potentials and close the knowledge gaps as well.

KEYWORDS

- **biochar**
- **feedstock**
- **bioremediation**
- **physicochemical properties**
- **production**

REFERENCES

Agrafioti, E.; Kalderis, D.; Diamadopoulos, E. Arsenic and chromium removal from water using biochars derived from rice husk, organic solid wastes and sewage sludge. *J. Environ. Manage.* **2014**, 133, 309–314.

Ahmad, M.; Lee, S. S.; Dou, X.; Mohan, D.; Sung, J. K.; Yang, J. E.; Ok, Y. S. Effects of pyrolysis temperature on soybean stover-and peanut shell-derived biochar properties and TCE adsorption in water. *Bioresour. Technol.* **2012**, 118, 536–544.

Ahmad, M.; Rajapaksha, A. U.; Lim, J. E.; Zhang, M.; Bolan, N.; Mohan, D.; Vithanage, M.; Lee, S. S.; Ok, Y. S. Biochar as a sorbent for contaminant management in soil and water, a review. *Chemosphere.* **2014**, 99, 19–33.

Ahmed, I. I.; Gupta, A. K. Pyrolysis and gasification of food waste, Syngas characteristics and char gasification kinetics. *Appl. Energy.* **2010**, 87(1), 101–108.

Ahmed, M. B.; Zhou, J. L.; Ngo, H. H.; Guo, W. Adsorptive removal of antibiotics from water and wastewater, progress and challenges. *Sci. Total Environ.* **2015**, 532, 112–126.

Anderson, C. R.; Condron, L. M.; Clough, T. J.; Fiers, M.; Stewart, A.; Hill, R. A.; Sherlock, R. R. Biochar induced soil microbial community change, implications for biogeochemical cycling of carbon, nitrogen and phosphorus. *Pedobiologia.* **2011**, 54(5–6), 309–320.

Augustenborg, C. A.; Hepp, S.; Kammann, C.; Hagan, D.; Schmidt, O.; Muller, C. Biochar and earthworm effects on soil nitrous oxide and carbon dioxide emissions. *J. Environ. Qual.* **2012**, 41(4), 1203–1209.

Baronti, S.; Vaccari, F. P.; Miglietta, F.; Calzolari, C.; Lugato, E.; Orlandini, S.; Pini, R.; Zulian, C.; Genesio, L. Impact of biochar application on plant water relations in *Vitis vinifera* (L.). *Eur. J. Agron.* **2014**, 53, 38–44.

Beesley, L.; Moreno-Jiménez, E.; Gomez-Eyles, J. L. Effects of biochar and greenwaste compost amendments on mobility, bioavailability and toxicity of inorganic and organic contaminants in a multi-element polluted soil. *Environ. Pollut.* **2010**, 158(6), 2282–2287.

Beesley, L.; Moreno-Jiménez, E.; Gomez-Eyles, J. L.; Harris, E.; Robinson, B.; Sizmur, T. A review of biochars' potential role in the remediation, revegetation and restoration of contaminated soils. *Environ. Pollut.* **2011**, 159(12), 3269–3282.

Berek, A. K., Hue, N.; Ahmad, A. Beneficial use of biochar to correct soil acidity. *Hanai Ai/ The Food Provider*, 2011.

Bodek, Itamar. *Environmental Inorganic Chemistry, Properties, Processes, and Estimation Methods..* Amsterdam, The Netherlands: Elsevier, 1988.

Brebu, M.; Vasile, C. Thermal degradation of lignin—A review. *Cellul. Chem. Technol.* **2010**, 44(9), 353.

Brick, S.; Lyutse, S. Biochar, Assessing the promise and risks to guide US policy. Natural Resources Defense Council, NRDC Issue Paper. 2010.

Bruun, E. W.; Muller-Stover, D.; Ambus, P.; Hauggaard-Nielsen, H. Application of biochar to soil and N2O emissions, potential effects of blending fast-pyrolysis biochar with anaerobically digested slurry. *Eur. J. Sol. Sci.* **2011**, 62(4), 581–589.

Cao, X.; Ma, L.; Gao, B.; Harris, W. Dairy-manure derived biochar effectively sorbs lead and atrazine. *Environ. Sci. Technol.* **2009**, 43, 3285–3291.

Chan, K.Y.; Van Zwieten, L.; Meszaros, I.; Downie, A.; Joseph, S. Agronomic values of green waste biochar as a soil amendment. *Soil Res.* **2008**. 45(8), 629–634.

Chan, K.Y.; Xu, Z. Biochar, nutrient properties and their enhancement. In *Biochar for Environmental Management.* 99–116, Abingdon, UK: Routledge, 2012.

Chen, B.; Zhou, D.; Zhu, L. Transitional adsorption and partition of nonpolar and polar aromatic contaminants by biochars of pine needles with different pyrolytic temperatures. *Environ. Sci. Technol.* **2008**, 42(14), 5137–5143.

Chen, X.; Chen, G.; Chen, L.; Chen, Y.; Lehmann, J.; McBride, M. B.; Hay, A. G. Adsorption of copper and zinc by biochars produced from pyrolysis of hardwood and corn straw in aqueous solution. *Bioresour. Technol.* **2011**, 102(19), 8877–8884.

Chun, Y.; Sheng, G.; Chiou, C. T.; Xing, B. Compositions and sorptive properties of crop residue-derived chars. *Environ. Sci. Technol.* **2004**, 38(17), 4649–4655.

de MeloCarvalho, M. T.; de HolandaNunes Maia, A.; Madari, B. E.; Bastiaans, L.; Van Oort, P. A. J.; Heinemann, A. B.; Soler da Silva, M. A.; Petter, F. A.; Marimon Jr., B. H.; Meinke, H. Biochar increases plant-available water in a sandy loam soil under an aerobic rice crop system. *Solid Earth.* **2014**, 5(2), 939–952.

Demirbas, A. Production and characterization of bio-chars from biomass via pyrolysis. *Energy Sources.* Part A, **2006**, 28(5), 413–422.

Demirbas, A.; Arin, G. An overview of biomass pyrolysis. *Energy Sources,* **2002**, 24(5), 471–482.

Dempster, D. N.; Gleeson, D. B.; Solaiman, Z. I.; Jones, D. L.; Murphy, D. V. Decreased soil microbial biomass and nitrogen mineralisation with Eucalyptus biochar addition to a coarse textured soil. *Plant Soil.* **2012**, 354(1–2), 311–324.

Frankel, M. L.; Bhuiyan, T. I.; Veksha, A.; Demeter, M. A.; Layzell, D. B.;Helleur, R. J.; Hill, J. M.; Turner, R. J. Removal and biodegradation of naphthenic acids by biochar and attached environmental biofilms in the presence of co-contaminating metals. *Bioresour. Technol.* **2016**, 216, 352–361.

Fuertes, A. B.; Arbestain, M. C.; Sevilla, M.; Macia-Agullo, J. A.; Fiol, S.; Lopez, R.; Smernik, R. J.; Aitkenhead, W. P.; Arce, F.; Macias, F. Chemical and structural properties of carbonaceous products obtained by pyrolysis and hydrothermal carbonisation of corn stover. *Soil Res.* **2010**, 48(7), 618–626.

Fushimi, C.; Araki, K.; Yamaguchi, Y.; Tsutsumi, A. Effect of heating rate on steam gasification of biomass. 2. Thermogravimetric-mass spectrometric (TG-MS) analysis of gas evolution. *Ind. Eng. Chem. Res.* **2003**, 42(17), 3929–3936.

Glaser, B.; Lehmann, J.; Zech, W. Ameliorating physical and chemical properties of highly weathered soils in the tropics with charcoal—A review. *Biol. Fert. Soils.* **2002**, 35(4), 219–230.

Goswami, R.; Shim, J.; Deka, S.; Kumari, D.; Kataki, R.; Kumar, M. Characterization of cadmium removal from aqueous solution by biochar produced from Ipomoea fistulosa at different pyrolytic temperatures. *Ecol. Eng.* **2016**, 97, 444–451.

Griffith, S. M.; Banowetz, G. M.; Gady, D. Chemical characterization of chars developed from thermochemical treatment of Kentucky bluegrass seed screenings. *Chemosphere.* **2013**, 92(10), 1275–1279.

Gunatilake, S. K. Adsorptive behavior of arsenic(III) ions from aqueous solution onto forestry and agricultural waste biochar pyrolyzed at 400°C. *Sabaragamuwa Univ. J.* **2015**, 14(2), 108.

Guo, J.; Wang, R.; Tjiu, W. W.; Pan, J.; Liu, T. Synthesis of Fe nanoparticles@ graphene composites for environmental applications. *J. Hazard. Mater.* **2012**, 225, 63–73.

Gupta, V. K.; Agarwal, S.; Saleh, T. A. Chromium removal by combining the magnetic properties of iron oxide with adsorption properties of carbon nanotubes. *Water Res.* **2011**, 45(6), 2207–2212.

Hadjittofi, L.; Prodromou, M.; Pashalidis, I. Activated biochar derived from cactus fibres—preparation, characterization and application on Cu(II) removal from aqueous solutions. *Bioresour. Technol.* **2014**, 159, 460–464.

Hanay, O.; Hasar, H.; Kocer, N. N. and Aslan, S. Evaluation for agricultural usage with speciation of heavy metals in a municipal sewage sludge. *Bull. Environ. Contamination Toxicol.* **2008**, 81(1), 42–46.

Hansen, V.; Muller-Stover, D.; Munkholm, L. J.; Peltre, C.; Hauggaard-Nielsen, H.; Jensen, L. S. The effect of straw and wood gasification biochar on carbon sequestration, selected soil fertility indicators and functional groups in soil: An incubation study. *Geoderma.* **2016**, 269, 99–107.

Hartley, W.; Dickinson, N. M.; Riby, P.; Lepp, N. W. Arsenic mobility in brownfield soils amended with green waste compost or biochar and planted with Miscanthus. *Environ. Pollut.* **2009**, 157(10), 2654–2662.

He, F.; Fan, J.; Ma, D.; Zhang, L.; Leung, C.; Chan, H. L. The attachment of Fe_3O_4 nanoparticles to graphene oxide by covalent bonding. *Carbon*, **2010**, 48(11), 3139–3144.

Ho, S. H.; Yang, Z. K.; Nagarajan, D.; Chang, J. S.; Ren, N. Q. High-efficiency removal of lead from wastewater by biochar derived from anaerobic digestion sludge. *Bioresour. Technol.* **2017**, 246, 142–149.

Inthapanya, S.; Preston, T. R.; Leng, R. A. Biochar increases biogas production in a batch digester charged with cattle manure. *Livest. Res. Rural Dev.* **2012**, 24(12), 212.

Inyang, M.; Dickenson, E. The potential role of biochar in the removal of organic and microbial contaminants from potable and reuse water: A review. *Chemosphere.* **2015**, 134, 232–240.

Inyang, M.; Gao, B.; Zimmerman, A.; Zhou, Y.; Cao, X. Sorption and cosorption of lead and sulfapyridine on carbon nanotube-modified biochars. *Environ. Sci. Pollut. Res. Int.* **2015**, 22(3), 1868–1876.

Ippolito, J. A.; Spokas, K. A.; Novak, J. M.; Lentz, R. D.; Cantrell, K. B. Biochar elemental composition and factors influencing nutrient retention. In *Biochar for Envrionmental Management: Science*. 171–196. Abingdon, UK: Routledge, **2015**.

Jeffery, S.; Abalos, D.; Spokas, K. A.; Verheijen, F. G. Biochar Effects on Crop Yield Biochar for Environmental Management, Science and Technology. Earthscan Books Ltd, London. **2015**, 301–325.

Jin, H.; Capareda, S.; Chang, Z.; Gao, J.; Xu, Y.; Zhang, J. Biochar pyrolytically produced from municipal solid wastes for aqueous As(V) removal, adsorption property and its improvement with KOH activation. *Bioresour. Technol.* **2014**, 169, 622–629.

Jones, D. L.; Jones, G. E.; Murphy, D. V. Biochar mediated alterations in herbicide breakdown and leaching in soil. *Soil Biol. Biochem.* **2011**, 43, 804–813.

Kalderis, D.; Koutoulakis, D.; Paraskeva, P.; Diamadopoulos, E.; Otal, E.; del Valle, J. O.; Fernandez-Pereira, C. Adsorption of polluting substances on activated carbons prepared from rice husk and sugarcane bagasse. *Chem. Eng. J.* **2008**, 144(1), 42–50.

Kammann, C.; Graber, E. R. *Biochar for Environmental Management, Science, Technology and Implementation.* 391–420, Abingdon, UK: Routledge, 2015.

Karakoyun, N.; Kubilay, S.; Aktas, N.; Turhan, O.; Kasimoglu, M.; Yilmaz, S.; Sahiner, N. Hydrogel–biochar composites for effective organic contaminant removal from aqueous media. *Desalination.* **2011**, 280(1–3), 319–325.

Karppinen, E. M.; Stewart, K. J.; Farrell, R. E.; Siciliano, S. D. Petroleum hydrocarbon remediation in frozen soil using a meat and bonemeal biochar plus fertilizer. *Chemosphere.* **2017**, 173, 330–339.

Kataki, R.; Chutia, R. S.; Mishra, M.; Bordoloi, N.; Saikia, R.; Bhaskar, T. Feedstock suitability for thermochemical processes. In *Recent Advances in Thermo-Chemical Conversion of Biomass.* 31–74, Amsterdam, The Netherlands: Elsevier, **2015**.

Keiluweit, M.; Nico, P. S.; Johnson, M. G.; Kleber, M. Dynamic molecular structure of plant biomass-derived black carbon (biochar). *Environ. Sci. Technol.* **2010**, 44(4), 1247–1253.

Kim, J. D.; Yun, H.; Kim, G. C.; Lee, C. W.; Choi, H. C. Antibacterial activity and reusability of CNT-Ag and GO-Ag nanocomposites. *Appl. Surf. Sci.* **2013**, 283, 227–233.

Kizito, S.; Wu, S.; Kirui, W. K.; Lei, M.; Lu, Q.; Bah, H.; Dong, R. Evaluation of slow pyrolyzed wood and rice husks biochar for adsorption of ammonium nitrogen from piggery manure anaerobic digestate slurry. *Sci. Total Environ.* **2015**, 505, 102–112.

Klasson, K. T.; Uchimiya, M.; Lima, I; Boihem Jr., L. Feasibility of removing furfurals from sugar solutions using activated biochars made from agricultural residues. *Bioresources.* **2011,** 6(3), 3242–3251.

Klinar, D. Universal model of slow pyrolysis technology producing biochar and heat from standard biomass needed for the techno-economic assessment. *Bioresour. Technol.* **2016,** 206, 112–120.

Kong, H.; He, J.; Gao, Y.; Wu, H.; Zhu, X. Cosorption of phenanthrene and mercury (II) from aqueous solution by soybean stalk-based biochar. *J. Agric. Food Chem.* **2011,** 59(22), 12116–12123.

Kuhlbusch, T. A. J. Method for determining black carbon in residues of vegetation fires. *Environ. Sci. Technol.* **1995,** 29(10), 2695–2702.

Kumar, S.; Jain, M. C.; Chhonkar, P. K. A note on stimulation of biogas production from cattle dung by addition of charcoal. *Biological Wastes.* **1987,** 20(3), 209–215.

Laird, D. A.; Brown, R. C.; Amonette, J. E.; Lehmann, J. Review of the pyrolysis platform for coproducing bio-oil and biochar. *Biofuel Bioprod. Biorefin.* **2009,** 3(5), 547–562.

Lee, Y. C.; Yang, J. W. Self-assembled flower-like TiO$_2$ on exfoliated graphite oxide for heavy metal removal. *J. Ind. Eng. Chem.* **2012,** 18(3), 1178–1185.

Lehmann J., Joseph S. *Biochar for Environmental Management. Science and Technology.* London, UK: Earthscan, Ltd., 2009.

Lehmann, J.; Czimczik, C.; Laird, D.; Sohi, S. Stability of biochar in soil. Biochar for environmental management. *Sci. Technol.* **2009,** 183–206.

Lehmann, J.; da Silva, J. P.; Steiner, C.; Nehls, T.; Zech, W.; Glaser, B. Nutrient availability and leaching in an archaeological Anthrosol and a Ferralsol of the Central Amazon basin, fertilizer, manure and charcoal amendments. *Plant Soil.* **2003,** 249(2), 343–357.

Lehmann, J.; Rondon, M. Bio-char soil management on highly weathered soils in the humid tropics. In *Biological Approaches to Sustainable Soil Systems*. Boca Raton, FL, USA: CRC Press, **2006.**

Lehmann, J. A. handful of carbon. *Nature.* **2007,** 443, 143–144.

Li, Y.; Shao, J.; Wang, X.; Deng, Y.; Yang, H.; Chen, H. Characterization of modified biochars derived from bamboo pyrolysis and their utilization for target component (furfural) adsorption. *Energy Fuels.* **2014,** 28(8), 5119–5127.

Liao, R.; Gao, B.; Fang, J. Invasive plants as feedstock for biochar and bioenergy production. *Bioresour. Technol.* **2013,** 140, 439–442.

Lin, Y. C.; Ho, S. H.; Zhou, Y.; Ren, N. Q. Highly efficient adsorption of dyes by biochar derived from pigments-extracted macroalgaepyrolyzed at different temperature. *Bioresour. Technol.* **2018,** 259, 104–110.

Lu, H.; Zhang, W.; Yang, Y.; Huang, X.; Wang, S.; Qiu, R. Relative distribution of Pb2$^+$ sorption mechanisms by sludge-derived biochar. *Water Res.* **2012,** 46(3), 854–862.

Mandal, S.; Sarkar, B.; Bolan, N.; Ok, Y. S.; Naidu, R. Enhancement of chromate reduction in soils by surface modified biochar. *J Environ. Manage.* **2017,** 186, 277–284.

Masek, O. Biochar production technologies. **2009.** [Online]. Available: http://www.geos.ed.ac.uk/sccs/ biochar/ documents/BiocharLaunch-OMasek.pdf

McHenry, M. P. Agricultural bio-char production, renewable energy generation and farm carbon sequestration in Western Australia, Certainty, uncertainty and risk. *Agriculture, Ecosystems & Environment.* **2009,** 129(1–3), 1–7.

McKay, G. Dioxin characterisation, formation and minimisation during municipal solid waste (MSW) incineration. *Chem. Eng. J.* **2002,** 86(3), 343–368.

Michalekova-Richveisova, B.; Fritak, V.; Pipiska, M.; Duriska, L.; Moreno-Jimenez, E. Soja, G. Iron-impregnated biochars as effective phosphate sorption materials. *Environ. Sci. Pollut. Res.* **2017,** 24(1), 463–475.

Mohan, D.; Kumar, H.; Sarswat, A.; Alexandre-Franco, M. Pittman Jr., C. U. Cadmium and lead remediation using magnetic oak wood and oak bark fast pyrolysis bio-chars. *Chem. Eng. J.* **2014,** 236, 513–528.

Mohan, D.; Pittman Jr., C. U.; Steele, P. H. Pyrolysis of wood/biomass for bio-oil: A critical review. *Energy Fuels.* **2006,** 20(3), 848–889.

Mohanty, P.; Nanda, S.; Pant, K. K.; Naik, S.; Kozinski, J. A.; Dalai, A. K. Evaluation of the physiochemical development of biochars obtained from pyrolysis of wheat straw, timothy grass and pinewood, effects of heating rate. *J. Anal. Appl. Pyrolysis.* **2013,** 104, 485–493.

Naisse, C.; Alexis, M.; Plante, A.; Wiedner, K.; Glaser, B.; Pozzi, A.; Carcaillet, C.; Criscuoli, I.; Rumpel, C. Can biochar and hydrochar stability be assessed with chemical methods? *Org. Geochem.* **2013,** 60, 40–44.

Nelissen, V.; Ruysschaert, G.; Muller-Stover, D.; Bode, S.; Cook, J.; Ronsse, F.; Shackley, S.; Boeckx, P.; Hauggaard-Nielsen, H. Short-term effect of feedstock and pyrolysis temperature on biochar characteristics, soil and crop response in temperate soils. *Agronomy.* **2014,** 4(1), 52–73.

Noguera, D.; Barot, S.; Laossi, K. R.; Cardoso, J.; Lavelle, P.; de Carvalho, M. C. Biochar but not earthworms enhances rice growth through increased protein turnover. *Soil Biol. Biochem.* **2012,** *52*, 13–20.

Novak, J. M.; Busscher, W. J.; Laird, D. L.; Ahmedna, M.; Watts, D. W.; Niandou, M. A. Impact of biochar amendment on fertility of a southeastern coastal plain soil. *Soil Sci.* **2009,** 174(2), 105–112.

Oliveira, F. R.; Patel, A. K.; Jaisi, D. P.; Adhikari, S.; Lu, H.; Khanal, S. K. Environmental application of biochar. Current status and perspectives. *Bioresour. Technol.* **2017,** 246, 110–122.

Pacioni, T. R.; Soares, D.; Di Domenico, M.; Rosa, M. F.; Moreira, R. D. F. P. M.; Jose, H. J. Bio-syngas production from agro-industrial biomass residues by steam gasification. *Waste Manage.* **2016,** 58, 221–229.

Paranavithana, G. N.; Kawamoto, K.; Inoue, Y.; Saito, T.; Vithanage, M.; Kalpage, C. S.; Herath, G. B. B. Adsorption of Cd^{2+} and Pb^{2+} onto coconut shell biochar and biochar-mixed soil. *Environ. Earth. Sci.* **2016,** 75(6), 484.

Park, J. H.; Ok, Y. S.; Kim, S. H.; Cho, J. S.; Heo, J. S.; Delaune, R. D.; Seo, D. C. Competitive adsorption of heavy metals onto sesame straw biochar in aqueous solutions. *Chemosphere.* **2016,** 142, 77–83.

Petrovic, J. T.; Stojanovic. M. D.; Milojkovic, J. V.; Petrovic, M. S.; Sostaric, T. D.; Lausevic, M. D.; Mihajlovic, M. L. Alkali modified hydrochar of grape pomace as a perspective adsorbent of Pb^{2+} from aqueous solution. *J. Environ. Manage.* 2016, 182, 292–300.

Phuengprasop, T.; Sittiwong, J.; Unob, F. Removal of heavy metal ions by iron oxide coated sewage sludge. *J. Hazard. Mater.* **2011,** 186, 502–507.

Pyrzynska, K.; Bystrzejewski, M. Comparative study of heavy metal ions sorption onto activated carbon, carbon nanotubes, and carbon-encapsulated magnetic nanoparticles. *Colloids Surf. A.* **2010,** 362(1–3), 102–109.

Qambrani, N. A.; Rahman, M. M.; Won, S.; Shim, S.; Ra, C. Biochar properties and eco-friendly applications for climate change mitigation, waste management, and wastewater treatment: A review. *Renew. Sust. Energ. Rev.* **2017,** 79, 255–273.

Rajapaksha, A. U.; Chen, S. S.; Tsang, D. C. W.; Zhang, M.; Vithanage, M.; Mandal, S.; Gao, B.; Bolan, N. S.; Ok, Y. S. Engineered/designer biochar for contaminant removal/ immobilization from soil and water, Potential and implication of biochar modification. *Chemosphere.* **2016**, 148, 276–291.

Raveendran, K.; Ganesh, A.; Khilar, K. C. Influence of mineral matter on biomass pyrolysis characteristics. *Fuel.* **1995**, 74, 1812–1822.

Reddy, K. R.; Xie, T.; Dastgheibi, S. Evaluation of biochar as a potential filter media for the removal of mixed contaminants from urban storm water runoff. *J. Environ. Eng.* **2014**, 140 (12), 04014043.

Rhee, S. W.; Park, H. S. Effect of mixing ratio of woody waste and food waste on the characteristics of carbonization residue. *J. Mater. Cycles Waste Manage.* **2010**, 12, 220–226.

Rollinson, A. N. Gasification reactor engineering approach to understanding the formation of biochar properties. *Proc. R. Soc. A. Math., Phys. Eng. Sci.* **2016**, 472, 20150841.

Ronsse, F.; Van Hecke, S.; Dickinson, D.; Prins, W. Production and characterization of slow pyrolysis biochar, influence of feedstock type and pyrolysis conditions. *GCB Bioenergy.* **2013**, 5(2), 104–115.

Sanchez C. Lignocellulosic residues, biodegradation and bioconversion by fungi. *Biotechnol Adv.* **2009**, 27, 185–194.

Sanchez, M. E.; Lindao, E.; Margaleff, D.; Martinez, O.; Moran, A. Pyrolysis of agricultural residues from rape and sunflowers. Production and characterization of bio-fuels and biochar soil management. *J. Anal. Appl. Pyrolysis.* **2009**, 85, 142–144.

Shaari N.; Tan S. H.; Mohamed A. R. Synthesis and characterization of CNT/Ce-TiO$_2$ nanocomposite for phenol degradation. *J. Rare Earth.* **2012**, 30, 651–658.

Shackley, S.; Carter, S.; Knowles, T.; Middelink, E.; Haefele, S.; Sohi, S.; Cross, A.; Haszeldine, S. Sustainable gasification-biochar systems? A case-study of rice-husk gasification in Cambodia, Part I, context, chemical properties, environmental and health and safety issues. *Energy Policy.* **2012**, 42, 49–58.

Shen, B.; Chen, J.; Yue, S.; Li, G. A comparative study of modified cotton biochar and activated carbon based catalysts in low temperature SCR. *Fuel.* **2015a**, 156, 47–53.

Shen, B.; Li, G.; Wang, F.; Wang, Y.; He, C.; Zhang, M.; Singh, S. Elemental mercury removal by the modified bio-char from medicinal residues. *Chem. Eng. J.* **2015b**, 272, 28–37.

Shen, W.; Li, Z.; Liu, Y. Surface chemical functional groups modification of porous carbon. *Recent Patents Chem. Eng.* **2008**, 1 (1), 27–40.

Shen, Y.; Linville, J. L.; Ignacio-de Leon, P. A. A.; Schoene, R. P.; Urgun-Demirtas, M. Towards a sustainable paradigm of waste-to-energy process, enhanced anaerobic digestion of sludge with woody biochar. *J. Clean. Prod.* **2016**, 135, 1054–1064.

Shinogi, Y.; Kanri, Y. Pyrolysis of plant, animal and human waste, physical and chemical characterization of the pyrolytic products. *Bioresour. Technol.* **2003**, 90(3), 241–247.

Sizmur, T.; Fresno, T.; Akgul, G.; Frost, H.; Moreno-Jimenez, E. Biochar modification to enhance sorption of inorganics from water. *Bioresour. Technol.* **2017**, 246, 34–47.

Sohi, S. P.; Krull, E.; Lopez-Capel, E.; Bol, R. A review of biochar and its use and function in soil. *Adv. Agronomy.* **2010**, 105, 47–82.

Srinivasarao, Ch.; Venkateswarlu, B.; Lal, R.; Singh, A. K.; Kundu S. Sustainable management of soils of dryland ecosystems for enhancing agronomic productivity and sequestering carbon. *Adv. Agronomy.* **2013**, 121, 253–329.

Sugumaran, P.; Sheshadri, S. Evaluation of selected biomass for charcoal production. *J. Sci. Ind. Res.* **2009**, 68, 719–723.

Sun, H.; Hockaday, W. C.; Masiello, C. A.; Zygourakis, K. Multiple controls on the chemical and physical structure of biochars. *Ind. Eng. Chem. Res.* **2012,** 51(9), 3587–3597.

Tan, X. F.; Liu, Y. G.; Zeng, G.; Wang, X.; Hu, X.; Gu, Y.; Yang, Z. Application of biochar for the removal of pollutants from aqueous solutions. *Chemosphere.* **2015,** 125, 70–85.

Teixido, M.; Pignatello, J. J.; Beltran, J. L.; Granados, M.; Peccia, J. Speciation of the ionizable antibiotic sulfamethazine on black carbon (biochar). *Environ. Sci. Technol.* **2011,** 45, 10020–10027.

Uchimiya, M.; Chang, S.; Klasson, K. T. Screening biochars for heavy metal retention in soil, Role of oxygen functional groups. *J. Hazard. Mater.* **2011a,** 190, 432–441.

Uchimiya, M.; Klasson, K. T.; Wartelle, L. H.; Lima, I. M. Influence of soil properties on heavy metal sequestration by biochar amendment: 1. Copper sorption isotherms and the release of cations. *Chemosphere.* **2011b,** 82, 1431–1437.

Uchimiya, M.; Ohno, T.; He, Z. Pyrolysis temperature-dependent release of dissolved organic carbon from plant, manure, and biorefinery wastes. *J. Anal. Appl. Pyrol.* **2013,** 104, 84–94.

Van Zwieten, L.; Kimber, S.; Morris, S.; Chan, K. Y.; Downie, A.; Rust, J.; Joseph, S; Cowie, A. Effects of biochar from slow pyrolysis of papermill waste on agronomic performance and soil fertility. *Plant Soil.* **2010,** 327(1–2), 235–246.

Verheijen, F.; Jeffery, S.; Bastos, A. C.; van der Velde, M.; Diafas I. Biochar application to soils—A critical scientific review of effects on soil properties, processes and functions. European Commission, Brussels, Belgium, **2010.** [Online]. Available: http://eusoils.jrc. ec.europa.eu/esdbarchive/eusoilsdocs/other/EUR24099.pdf

Vithanage, M.; Rajapaksha, A. U.; Zhang, M.; Thiele-Bruhn, S.; Lee, S. S.; Ok, Y. S. Acid-activated biochar increased sulfamethazine retention in soils. *Environ. Sci. Pollut. Res.* **2015,** 22(3), 2175–2186.

Wang, S.; Guo, W.; Gao, F.; Yang, R. Characterization and Pb (II) removal potential of corn straw-and municipal sludge-derived biochars. *Roy. Soc. Open. Sci.* **2017,** 4(9), 170402.

Wiedner, K.; Rumpel, C.; Steiner, C.; Pozzi, A.; Maas, R.; Glaser, B. Chemical evaluation of chars produced by thermochemical conversion (gasification, pyrolysis and hydrothermal carbonization) of agro-industrial biomass on a commercial scale. *Biomass Bioenergy.* **2013,** 59, 264–278.

Wong, J. W. C.; Xiang, L.; Gu, X. Y.; Zhou, L. X. Bioleaching of heavy metals from anaerobically digested sewage sludge using FeS_2 as an energy source. *Chemosphere.* **2004,** 55(1), 101–107.

Woolf, D.; Amonette, J. E.; Street-Perrott, F. A.; Lehmann, J.; Joseph, S., Sustainable biochar to mitigate global climate change. *Nat. Commun.* **2010,** 1, 56.

Xie, T.; Reddy, K. R.; Wang, C.; Yargicoglu, E.; Spokas, K. Characteristics and applications of biochar for environmental remediation: A review. *Crit. Rev. Environ. Sci. Technol.* **2015,** 45(9), 939–969.

Xu, W. Q.; Pignatello, J. J.; Mitch, W. A. The role of black carbon electrical conductivity in mediating hexahydro-1,3,5-trinitro-1,3,5-triazine (RDX) transformation on carbon surfaces by sulfides. *Environ. Sci. Technol.* **2013a,** 47, 7129–7136.

Xu, X.; Cao, X.; Zhao, L. Comparison of rice husk-and dairy manure-derived biochars for simultaneously removing heavy metals from aqueous solutions, role of mineral components in biochars. *Chemosphere.* **2013b,** 92(8), 955–961.

Xue, Y.; Gao, B.; Yao, Y.; Inyang, M.; Zhang, M.; Zimmerman, A. R.; Ro, K. S. Hydrogen peroxide modification enhances the ability of biochar (hydrochar) produced from hydrothermal carbonization of peanut hull to remove aqueous heavy metals, batch and column tests. *Chem. Eng. J.* **2012,** 200, 673–680.

Yang, G. X.; Jiang, H. Amino modification of biochar for enhanced adsorption of copper ions from synthetic wastewater. *Water Res*. **2014**, 48, 396–405.

Yao, Y.; Gao, B.; Inyang, M.; Zimmerman, A. R.; Cao, X. D.; Pullammanappallil, P.; Yang, L. Biochar derived from anaerobically digested sugar beet tailings, Characterization and phosphate removal potential. *Bioresour. Technol*. **2011b**, 102, 6273–6278.

Yao, Y.; Gao, B.; Inyang, M.; Zimmerman, A. R.; Cao, X. D.; Pullammanappallil, P.; Yang, L. Y. Removal of phosphate from aqueous solution by biochar derived from anaerobically digested sugar beet tailings. *J. Hazard. Mater.* **2011a**, 190, 501–507.

Yao, Y.; Gao, B.; Wu, F.; Zhang, C. Z.; Yang, L. Y. Engineered Biochar from Biofuel Residue, Characterization and its silver removal potential. *ACS Appl. Mater. Int.* **2015**, 7, 10634–10640.

Yip, K.; Tian, F.; Hayashi, J. I.; Wu, H. Effect of alkali and alkaline earth metallic species on biochar reactivity and syngas compositions during steam gasification. *Energy Fuels.* **2009**, 24(1), 173–181.

Yoder, J.; Galinato, S.; Granatstein, D.; Garcia-Perez, M. Economic tradeoff between biochar and bio-oil production via pyrolysis. *Biomass. Bioenerg*. **2011**, 35(5), 1851–1862.

Yoo, G.; Kang, H. Effects of biochar addition on greenhouse gas emissions and microbial responses in a short-term laboratory experiment. *J. Environ. Qual*. **2012**, 41(4), 1193–1202.

Zhang, K.; Cheng, X.; Dang, H.; Ye, C.; Zhang, Y.; Zhang, Q. Linking litter production, quality and decomposition to vegetation succession following agricultural abandonment. *Soil Biol. Biochem*. **2013**, 57, 803–813.

Zhang, M. M.; Liu, Y. G.; Li, T. T.; Xu, W. H.; Zheng, B. H.; Tan, X. F.; Wang, H.; Guo, Y. M.; Guo, F. Y.; Wang, S. F. Chitosan modification of magnetic biochar produced from Eichhorniacrassipes for enhanced sorption of Cr(VI) from aqueous solution. *RSC Adv.* **2015**, 5(58), 46955–46964.

Zhao, L.; Cao, X.; Masek, O.; Zimmerman, A. Heterogeneity of biochar properties as a function of feedstock sources and production temperatures. *J. Hazard. Mater.* **2013**, 256, 1–9.

Zhou, Y.; Gao, B.; Zimmerman, A. R.; Fang, J.; Sun, Y.; Cao, X. Sorption of heavy metals on chitosan-modified biochars and its biological effects. *Chem. Eng. J.* **2013**, 231, 512–518.

Index

For Product Safety Concerns and Information please contact our EU
representative GPSR@taylorandfrancis.com
Taylor & Francis Verlag GmbH, Kaufingerstraße 24, 80331 München, Germany

www.ingramcontent.com/pod-product-compliance
Lightning Source LLC
Chambersburg PA
CBHW060419220326
41598CB00021BA/2226

9 781774 638132